STUDENT SOLUTIONS MANUAL, VOLUME TWO, FOR STEWART'S

CALCULUS

THIRD EDITION

JAMES STEWART
McMaster University

BARBARA FRANK
St. Andrews Presbyterian College

with the assistance of
Andy Bulman-Fleming

Brooks/Cole Publishing Company

I(T)P™ An International Thomson Publishing Company

Pacific Grove • Albany • Bonn • Boston • Cincinnati • Detroit • London • Madrid • Melbourne
Mexico City • New York • Paris • San Francisco • Singapore • Tokyo • Toronto • Washington

Sponsoring Editor: *Elizabeth Rammel*
Editorial Associate: *Carol Ann Benedict*
Production Coordinator: *Dorothy Bell*
Cover Design: *Katherine Minerva, Vernon T. Boes*
Cover Sculpture: *Christian Haase*
Cover Photo: *Ed Young*
Typesetting: *Andy Bulman-Fleming*
Printing and Binding: *Malloy Lithographing, Inc.*

For more information, contact:

BROOKS/COLE PUBLISHING COMPANY
511 Forest Lodge Rd.
Pacific Grove, CA 93950
USA

International Thomson Editores
Campos Eliseos 385, Piso 7
Col. Polanco
11560 México D. F. México

International Thomson Publishing Europe
Berkshire House 168-173
High Holborn
London WC1V 7AA
England

International Thomson Publishing GmbH
Königwinterer Strasse 418
53227 Bonn
Germany

Thomas Nelson Australia
102 Dodds Street
South Melbourne, 3205
Victoria, Australia

International Thomson Publishing Asia
221 Henderson Road
#05–10 Henderson Building
Singapore 0315

Nelson Canada
1120 Birchmount Road
Scarborough, Ontario
Canada M1K 5G4

International Thomson Publishing Japan
Hirakawacho Kyowa Building, 3F
2-2-1 Hirakawacho
Chiyoda-ku, Tokyo 102
Japan

Printed in the United States of America

10 9 8 7 6 5 4 3 2 1

ISBN 0-534-21803-2

Preface

I have edited this solutions manual by comparing the solutions provided by Barbara Frank with my own solutions and those of McGill University students Andy Bulman-Fleming and Alex Taler. Andy also produced this book using EXP Version 3.0 for Windows. I thank him and the staff of TECHarts for producing the diagrams.

JAMES STEWART

CONTENTS

CONTENTS

CONTENTS

CHAPTER TEN

EXERCISES 10.1

1. $a_n = \dfrac{n}{2n+1}$, so the sequence is $\left\{ \dfrac{1}{3}, \dfrac{2}{5}, \dfrac{3}{7}, \dfrac{4}{9}, \dfrac{5}{11}, \dots \right\}$.

3. $a_n = \dfrac{1 \cdot 3 \cdot 5 \cdot \dots \cdot (2n-1)}{n!}$, so the sequence is $\left\{ 1, \dfrac{3}{2}, \dfrac{5}{2}, \dfrac{35}{8}, \dfrac{63}{8}, \dots \right\}$.

5. $a_n = \sin \dfrac{n\pi}{2}$, so the sequence is $\{1, 0, -1, 0, 1, \dots\}$.

7. $a_n = \dfrac{1}{2^n}$ 9. $a_n = 3n - 2$ 11. $a_n = (-1)^{n+1} \left(\dfrac{3}{2}\right)^n$

13. $\lim\limits_{n\to\infty} \dfrac{1}{4n^2} = \dfrac{1}{4} \lim\limits_{n\to\infty} \dfrac{1}{n^2} = \dfrac{1}{4} \cdot 0 = 0$. Convergent

15. $\lim\limits_{n\to\infty} \dfrac{n^2-1}{n^2+1} = \lim\limits_{n\to\infty} \dfrac{1-1/n^2}{1+1/n^2} = 1$. Convergent

17. $\{a_n\}$ diverges since $\dfrac{n^2}{n+1} = \dfrac{n}{1+1/n} \to \infty$ as $n \to \infty$.

19. $\lim\limits_{n\to\infty} |a_n| = \lim\limits_{n\to\infty} \dfrac{n^2}{1+n^3} = \lim\limits_{n\to\infty} \dfrac{1/n}{(1/n^3)+1} = 0$, so by Theorem 5, $\lim\limits_{n\to\infty} (-1)^n \left(\dfrac{n^2}{1+n^3}\right) = 0$.

21. $\{a_n\} = \{0, -1, 0, 1, 0, -1, 0, 1, \dots\}$. This sequence oscillates among $0, -1,$ and 1 and so diverges.

23. $a_n = \left(\dfrac{\pi}{3}\right)^n$ so $\{a_n\}$ diverges by Equation 7 with $r = \dfrac{\pi}{3} > 1$.

25. $0 < \dfrac{3+(-1)^n}{n^2} \le \dfrac{4}{n^2}$ and $\lim\limits_{n\to\infty} \dfrac{4}{n^2} = 0$, so $\left\{ \dfrac{3+(-1)^n}{n^2} \right\}$ converges to 0 by the Squeeze Theorem.

27. $\lim\limits_{x\to\infty} \dfrac{\ln(x^2)}{x} = \lim\limits_{x\to\infty} \dfrac{2\ln x}{x} \overset{\text{H}}{=} \lim\limits_{x\to\infty} \dfrac{2/x}{1} = 0$, so by Theorem 2, $\left\{ \dfrac{\ln(n^2)}{n} \right\}$ converges to 0.

29. $\sqrt{n+2} - \sqrt{n} = \left(\sqrt{n+2} - \sqrt{n}\right) \dfrac{\sqrt{n+2}+\sqrt{n}}{\sqrt{n+2}+\sqrt{n}} = \dfrac{2}{\sqrt{n+2}+\sqrt{n}} < \dfrac{2}{2\sqrt{n}} = \dfrac{1}{\sqrt{n}} \to 0$ as $n \to \infty$. So

by the Squeeze Theorem $\left\{\sqrt{n+2} - \sqrt{n}\right\}$ converges to 0.

31. $\lim\limits_{x\to\infty} \dfrac{x}{2^x} \overset{\text{H}}{=} \lim\limits_{x\to\infty} \dfrac{1}{(\ln 2)2^x} = 0$, so by Theorem 2 $\{a_n\}$ converges to 0.

33. Let $y = x^{-1/x}$. Then $\ln y = -(\ln x)/x$ and $\lim\limits_{x\to\infty} (\ln y) \overset{\text{H}}{=} \lim\limits_{x\to\infty} -(1/x)/1 = 0$, so $\lim\limits_{x\to\infty} y = e^0 = 1$, and so $\{a_n\}$

converges to 1.

35. $0 \le \dfrac{\cos^2 n}{2^n} \le \dfrac{1}{2^n}$ $\left[\text{since } 0 \le \cos^2 n \le 1\right]$, so since $\lim\limits_{n\to\infty} \dfrac{1}{2^n} = 0$, $\{a_n\}$ converges to 0 by the Squeeze Theorem.

37. The series converges, since $a_n = \dfrac{1+2+3+\dots+n}{n^2} = \dfrac{n(n+1)/2}{n^2}$ [Theorem 3]

$= \dfrac{n+1}{2n} = \dfrac{1+1/n}{2} \to \dfrac{1}{2}$ as $n \to \infty$.

1

39. $a_n = \dfrac{1}{2} \cdot \dfrac{2}{2} \cdot \dfrac{3}{2} \cdot \cdots \cdot \dfrac{(n-1)}{2} \cdot \dfrac{n}{2} \geq \dfrac{1}{2} \cdot \dfrac{n}{2} = \dfrac{n}{4} \to \infty$ as $n \to \infty$, so $\{a_n\}$ diverges.

41.

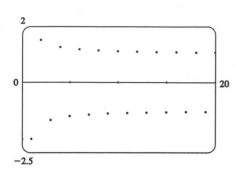

From the graph, we see that the sequence is divergent, since it oscillates between 1 and -1 (approximately).

43.

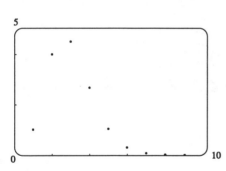

From the graph, it appears that the sequence converges to about 0.78.

$$\lim_{n\to\infty} \frac{2n}{2n+1} = \lim_{n\to\infty} \frac{2}{2+1/n} = 1, \text{ so}$$

$$\lim_{n\to\infty} \arctan\left(\frac{2n}{2n+1}\right) = \arctan 1 = \frac{\pi}{4}.$$

45.

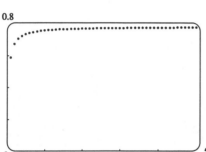

From the graph, it appears that the sequence converges to 0. $\quad 0 < a_n = \dfrac{n^3}{n!}$

$$= \frac{n}{n} \cdot \frac{n}{(n-1)} \cdot \frac{n}{(n-2)} \cdot \frac{1}{(n-3)} \cdot \cdots \cdot \frac{1}{3} \cdot \frac{1}{2} \cdot \frac{1}{1}$$

$$\leq \frac{n^2}{(n-1)(n-2)(n-3)} \quad \text{(for } n \geq 4\text{)}$$

$$= \frac{1/n}{(1-1/n)(1-2/n)(1-3/n)} \to 0 \text{ as } n \to \infty,$$

so by the Squeeze Theorem, $\{a_n\}$ converges to 0.

47.

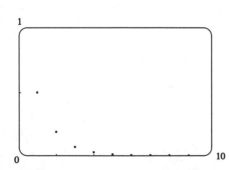

From the graph, it appears that the sequence approaches 0.

$$0 < a_n = \frac{1 \cdot 3 \cdot 5 \cdot \cdots \cdot (2n-1)}{(2n)^n}$$

$$= \frac{1}{2n} \cdot \frac{3}{2n} \cdot \frac{5}{2n} \cdot \cdots \cdot \frac{2n-1}{2n}$$

$$\leq \frac{1}{2n} \cdot (1) \cdot (1) \cdot \cdots \cdot (1) = \frac{1}{2n} \to 0 \text{ as}$$

$n \to \infty$, so by the Squeeze Theorem $\{a_n\}$ converges to 0.

49. If $|r| \geq 1$, then $\{r^n\}$ diverges by (7), so $\{nr^n\}$ diverges also since $|nr^n| = n|r^n| \geq |r^n|$. If $|r| < 1$ then

$$\lim_{x \to \infty} xr^x = \lim_{x \to \infty} \frac{x}{r^{-x}} \overset{H}{=} \lim_{x \to \infty} \frac{1}{(-\ln r)r^{-x}} = \lim_{x \to \infty} \frac{r^x}{-\ln r} = 0, \text{ so } \lim_{n \to \infty} nr^n = 0, \text{ and hence } \{nr^n\} \text{ converges}$$

whenever $|r| < 1$.

51. $3(n+1) + 5 > 3n + 5$ so $\dfrac{1}{3(n+1)+5} < \dfrac{1}{3n+5}$ $\Leftrightarrow$ $a_{n+1} < a_n$ so $\{a_n\}$ is decreasing.

53. $\left\{\dfrac{n-2}{n+2}\right\}$ is increasing since $a_n < a_{n+1}$ $\Leftrightarrow$ $\dfrac{n-2}{n+2} < \dfrac{(n+1)-2}{(n+1)+2}$ $\Leftrightarrow$ $(n-2)(n+3) < (n+2)(n-1)$

$\Leftrightarrow$ $n^2 + n - 6 < n^2 + n - 2$ $\Leftrightarrow$ $-6 < -2$, which is of course true.

55. $a_1 = 0 > a_2 = -1 < a_3 = 0$, so the sequence is not monotonic.

57. $\left\{\dfrac{n}{n^2+n-1}\right\}$ is decreasing since $a_{n+1} < a_n$ $\Leftrightarrow$ $\dfrac{n+1}{(n+1)^2+(n+1)-1} < \dfrac{n}{n^2+n-1}$ $\Leftrightarrow$

$(n+1)(n^2+n-1) < n(n^2+3n+1)$ $\Leftrightarrow$ $n^3 + 2n^2 - 1 < n^3 + 3n^2 + n$ $\Leftrightarrow$

$0 < n^2 + n + 1 = \left(n + \frac{1}{2}\right)^2 + \frac{3}{4}$, which is obviously true.

59. $a_1 = 2^{1/2}, a_2 = 2^{3/4}, a_3 = 2^{7/8}, \ldots, a_n = 2^{(2^n-1)/2^n} = 2^{(1-1/2^n)}$. $\displaystyle\lim_{n \to \infty} a_n = \lim_{n \to \infty} 2^{(1-1/2^n)} = 2^1 = 2$.

Alternate Solution: Let $L = \displaystyle\lim_{n \to \infty} a_n$ (We could show the limit exists by showing that $\{a_n\}$ is bounded and

increasing.) So L must satisfy $L = \sqrt{2 \cdot L} \Rightarrow L^2 = 2L \Rightarrow L(L-2) = 0$ ($L \neq 0$ since the sequence

increases) so $L = 2$.

61. We show by induction that $\{a_n\}$ is increasing and bounded above by 3. Let $P(n)$ be the proposition that

$a_{n+1} > a_n$ and $0 < a_n < 3$. Clearly $P(1)$ is true. Assume $P(n)$ is true. Then $a_{n+1} > a_n$ $\Rightarrow$ $\dfrac{1}{a_{n+1}} < \dfrac{1}{a_n}$

$\Rightarrow$ $-\dfrac{1}{a_{n+1}} > -\dfrac{1}{a_n}$ $\Rightarrow$ $a_{n+2} = 3 - \dfrac{1}{a_{n+1}} > 3 - \dfrac{1}{a_n} = a_{n+1}$ $\Leftrightarrow$ $P(n+1)$. This proves that $\{a_n\}$ is

increasing and bounded above by 3, so $1 = a_1 < a_n < 3$, that is, $\{a_n\}$ is bounded, and hence convergent by

Theorem 10.

If $L = \displaystyle\lim_{n \to \infty} a_n$, then $\displaystyle\lim_{n \to \infty} a_{n+1} = L$ also, so L must satisfy $L = 3 - \dfrac{1}{L}$, so $L^2 - 3L + 1 = 0$ and the quadratic

formula gives $L = \frac{3 \pm \sqrt{5}}{2}$. But $L > 1$, so $L = \frac{3 + \sqrt{5}}{2}$.

63. **(a)** Let a_n be the number of rabbit pairs in the nth month. Clearly $a_1 = 1 = a_2$. In the nth month, each pair

that is 2 or more months old (that is, a_{n-2} pairs) will have a pair of children to add to the a_{n-1} pairs

already present. Thus $a_n = a_{n-1} + a_{n-2}$, so that $\{a_n\} = \{f_n\}$, the Fibonacci sequence.

(b) $a_{n-1} = \dfrac{f_n}{f_{n-1}} = \dfrac{f_{n-1} + f_{n-2}}{f_{n-1}} = 1 + \dfrac{f_{n-2}}{f_{n-1}} = 1 + \dfrac{1}{a_{n-2}}$. If $L = \displaystyle\lim_{n \to \infty} a_n$, then L must satisfy $L = 1 + \dfrac{1}{L}$

or $L^2 - L - 1 = 0$, so $L = \frac{1+\sqrt{5}}{2}$ (since L must be positive.)

65. **(a)**

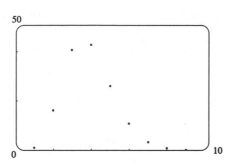

From the graph, it appears that the sequence $\left\{\dfrac{n^5}{n!}\right\}$ converges to 0, that is, $\displaystyle\lim_{n\to\infty}\dfrac{n^5}{n!}=0.$

(b)

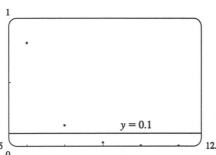

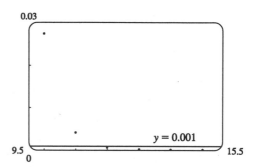

From the first graph, it seems that the smallest possible value of N corresponding to $\epsilon = 0.1$ is 9, since $n^5/n! < 0.1$ whenever $n \ge 10$, but $9^5/9! > 0.1$. From the second graph, it seems that for $\epsilon = 0.001$, the smallest possible value for N is 11.

67. If $\displaystyle\lim_{n\to\infty}|a_n| = 0$ then $\displaystyle\lim_{n\to\infty}-|a_n| = 0$, and since $-|a_n| \le a_n \le |a_n|$, we have that $\displaystyle\lim_{n\to\infty}a_n = 0$ by the Squeeze Theorem.

69. **(a)** First we show that $a > a_1 > b_1 > b$.

$a_1 - b_1 = \dfrac{a+b}{2} - \sqrt{ab} = \tfrac{1}{2}\left(a - 2\sqrt{ab} + b\right) = \tfrac{1}{2}\left(\sqrt{a} - \sqrt{b}\right)^2 > 0$ (since $a > b$) $\quad\Rightarrow\quad a_1 > b_1$.

Also $a - a_1 = a - \tfrac{1}{2}(a+b) = \tfrac{1}{2}(a-b) > 0$ and $b - b_1 = b - \sqrt{ab} = \sqrt{b}\left(\sqrt{b} - \sqrt{a}\right) < 0$, so

$a > a_1 > b_1 > b$. In the same way we can show that $a_1 > a_2 > b_2 > b_1$ and so the given assertion is true

for $n = 1$. Suppose it is true for $n = k$, that is, $a_k > a_{k+1} > b_{k+1} > b_k$. Then

$a_{k+2} - b_{k+2} = \tfrac{1}{2}(a_{k+1} + b_{k+1}) - \sqrt{a_{k+1}b_{k+1}} = \tfrac{1}{2}\left(a_{k+1} - 2\sqrt{a_{k+1}b_{k+1}} + b_{k+1}\right)$

$= \tfrac{1}{2}\left(\sqrt{a_{k+1}} - \sqrt{b_{k+1}}\right)^2 > 0$ and $a_{k+1} - a_{k+2} = a_{k+1} - \tfrac{1}{2}(a_{k+1} + b_{k+1}) = \tfrac{1}{2}(a_{k+1} - b_{k+1}) > 0,$

$b_{k+1} - b_{k+2} = b_{k+1} - \sqrt{a_{k+1}b_{k+1}} = \sqrt{b_{k+1}}\left(\sqrt{b_{k+1}} - \sqrt{a_{k+1}}\right) < 0 \quad\Rightarrow\quad a_{k+1} > a_{k+2} > b_{k+2} > b_{k+1},$

so the assertion is true for $n = k + 1$. Thus it is true for all n by mathematical induction.

(b) From part (a) we have $a > a_n > a_{n+1} > b_{n+1} > b_n > b$, which shows that both sequences are monotonic and bounded. So they are both convergent by Theorem 10.

(c) Let $\displaystyle\lim_{n\to\infty}a_n = \alpha$ and $\displaystyle\lim_{n\to\infty}b_n = \beta$. Then $\displaystyle\lim_{n\to\infty}a_{n+1} = \lim_{n\to\infty}\dfrac{a_n + b_n}{2} \quad\Rightarrow\quad \alpha = \dfrac{\alpha + \beta}{2} \quad\Rightarrow\quad$

$2\alpha = \alpha + \beta \quad\Rightarrow\quad \alpha = \beta.$

71. (a) $2\cos\theta - 1 = \dfrac{1 + 2\cos 2\theta}{1 + 2\cos\theta}$ $\;\leftrightarrow\;$ $(2\cos\theta + 1)(2\cos\theta - 1) = 1 + 2\cos 2\theta$ (provided $\cos\theta \neq -1$), and

this is certainly true since the LHS $= 4\cos^2\theta - 1$ and the RHS $= 1 + 2(2\cos^2\theta - 1) = 4\cos^2\theta - 1$.

(b) By part (a), we can write each a_k as $2\cos(\theta/2^k) - 1 = \dfrac{1 + 2\cos(\theta/2^{k-1})}{1 + 2\cos(\theta/2^k)}$, so we get

$$b_n = \frac{1 + 2\cos\theta}{1 + 2\cos(\theta/2)} \cdot \frac{1 + 2\cos(\theta/2)}{1 + 2\cos(\theta/4)} \cdot \cdots \cdot \frac{1 + 2\cos(\theta/2^{n-2})}{1 + 2\cos(\theta/2^{n-1})} \cdot \frac{1 + 2\cos(\theta/2^{n-1})}{1 + 2\cos(\theta/2^n)} = \frac{1 + 2\cos\theta}{1 + 2\cos(\theta/2^n)}$$

(telescoping product). So $\displaystyle\lim_{n\to\infty} b_n = \lim_{n\to\infty} \frac{1 + 2\cos\theta}{1 + 2\cos(\theta/2^n)} = \frac{1 + 2\cos\theta}{1 + 2\cos 0} = \tfrac{1}{3}(1 + 2\cos\theta)$.

EXERCISES 10.2

1.

n	s_n
1	3.33333
2	4.44444
3	4.81481
4	4.93827
5	4.97942
6	4.99314
7	4.99771
8	4.99924
9	4.99975
10	4.99992
11	4.99997
12	4.99999

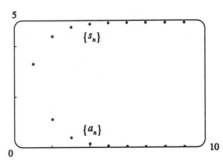

From the graph, it seems that the series converges. In fact, it is a geometric series with $a = \frac{10}{3}$ and $r = \frac{1}{3}$, so its sum is

$$\sum_{n=1}^{\infty} \frac{10}{3^n} = \frac{10/3}{1 - 1/3} = 5.$$

3.

n	s_n
1	0.50000
2	1.16667
3	1.91667
4	2.71667
5	3.55000
6	4.40714
7	5.28214
8	6.17103
9	7.07103
10	7.98012

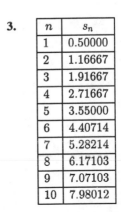

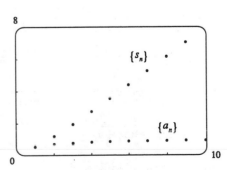

The series diverges, since its terms do not approach 0.

5.

n	s_n
1	0.6464
2	0.8075
3	0.8750
4	0.9106
5	0.9320
6	0.9460
7	0.9558
8	0.9630
9	0.9684
10	0.9726

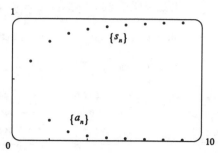

From the graph, it seems that the series converges to 1. To find the sum, we write

$$s_n = \sum_{i=1}^{n}\left(\frac{1}{i^{1.5}} - \frac{1}{(i+1)^{1.5}}\right) = \left(1 - \frac{1}{2^{1.5}}\right) + \left(\frac{1}{2^{1.5}} - \frac{1}{3^{1.5}}\right) + \left(\frac{1}{3^{1.5}} - \frac{1}{4^{1.5}}\right) + \cdots + \left(\frac{1}{n^{1.5}} - \frac{1}{(n+1)^{1.5}}\right)$$

$= 1 - 1/(n+1)^{1.5}$. So the sum is $\lim_{n\to\infty} s_n = 1$.

7. $\displaystyle\sum_{n=1}^{\infty} 4\left(\frac{2}{5}\right)^{n-1}$ is a geometric series with $a = 4, r = \frac{2}{5}$, so it converges to $\dfrac{4}{1-2/5} = \dfrac{20}{3}$.

9. $\displaystyle\sum_{n=1}^{\infty} \frac{2}{3}\left(-\frac{1}{3}\right)^{n-1}$ is geometric with $a = \frac{2}{3}, r = -\frac{1}{3}$, so it converges to $\dfrac{2/3}{1-(-1/3)} = \dfrac{1}{2}$.

11. $a = 2, r = \dfrac{3}{4} < 1$, so the series converges to $\dfrac{2}{1-3/4} = 8$.

13. $a = \dfrac{5e}{3}, r = \dfrac{e}{3} < 1$, so the series converges to $\dfrac{5e/3}{1-e/3} = \dfrac{5e}{3-e}$.

15. $a = 1, r = \dfrac{5}{8} < 1$, so the series converges to $\dfrac{1}{1-5/8} = \dfrac{8}{3}$.

17. $a = \frac{64}{3}, r = \frac{8}{3} > 1$, so the series diverges.

19. This series diverges, since if it converged, so would $2 \cdot \displaystyle\sum_{n=1}^{\infty} \frac{1}{2n} = \sum_{n=1}^{\infty} \frac{1}{n}$ [by Theorem 8(a)], which we know

diverges (Example 7).

21. Converges. $s_n = \displaystyle\sum_{i=1}^{n} \frac{1}{(3i-2)(3i+1)} = \sum_{i=1}^{n}\left[\frac{1/3}{3i-2} - \frac{1/3}{3i+1}\right]$ (partial fractions)

$$= \left[\frac{1}{3}\cdot 1 - \frac{1}{3}\cdot\frac{1}{4}\right] + \left[\frac{1}{3}\cdot\frac{1}{4} - \frac{1}{3}\cdot\frac{1}{7}\right] + \left[\frac{1}{3}\cdot\frac{1}{7} - \frac{1}{3}\cdot\frac{1}{10}\right] + \cdots$$

$$+ \left[\frac{1}{3}\cdot\frac{1}{3n-2} - \frac{1}{3}\cdot\frac{1}{3n+1}\right] = \frac{1}{3} - \frac{1}{3(3n+1)} \quad \text{(telescoping series)}$$

$$\Rightarrow \quad \lim_{n\to\infty} s_n = \frac{1}{3} \quad \Rightarrow \quad \sum_{n=1}^{\infty} \frac{1}{(3n-2)(3n+1)} = \frac{1}{3}$$

23. Converges by Theorem 8.

$$\sum_{n=1}^{\infty}[2(0.1)^n + (0.2)^n] = 2\sum_{n=1}^{\infty}(0.1)^n + \sum_{n=1}^{\infty}(0.2)^n = 2\left(\frac{0.1}{1-0.1}\right) + \frac{0.2}{1-0.2} = \frac{2}{9} + \frac{1}{4} = \frac{17}{36}$$

25. Diverges by the Test for Divergence. $\displaystyle\lim_{n\to\infty}\frac{n}{\sqrt{1+n^2}}=\lim_{n\to\infty}\frac{1}{\sqrt{1+1/n^2}}=1\neq 0.$

27. Converges. $\displaystyle s_n=\sum_{i=1}^{n}\frac{1}{i(i+2)}=\sum_{i=1}^{n}\left[\frac{1/2}{i}-\frac{1/2}{i+2}\right]$ (partial fractions)

$$=\left[\frac{1}{2}-\frac{1}{6}\right]+\left[\frac{1}{4}-\frac{1}{8}\right]+\left[\frac{1}{6}-\frac{1}{10}\right]+\cdots+\left[\frac{1}{2n-2}-\frac{1}{2n+2}\right]+\left[\frac{1}{2n}-\frac{1}{2n+4}\right]$$

$$=\frac{1}{2}+\frac{1}{4}-\frac{1}{2n+2}-\frac{1}{2n+4}\quad\text{(telescoping series)}.$$

Thus $\displaystyle\sum_{n=1}^{\infty}\frac{1}{n(n+2)}=\lim_{n\to\infty}\left[\frac{1}{2}+\frac{1}{4}-\frac{1}{2n+2}-\frac{1}{2n+4}\right]=\frac{3}{4}.$

29. Converges. $\displaystyle\sum_{n=1}^{\infty}\frac{3^n+2^n}{6^n}=\sum_{n=1}^{\infty}\left[\left(\frac{1}{2}\right)^n+\left(\frac{1}{3}\right)^n\right]=\frac{1/2}{1-1/2}+\frac{1/3}{1-1/3}=\frac{3}{2}$

31. Converges. $\displaystyle s_n=\left(\sin 1-\sin\frac{1}{2}\right)+\left(\sin\frac{1}{2}-\sin\frac{1}{3}\right)+\cdots+\left[\sin\frac{1}{n}-\sin\frac{1}{n+1}\right]=\sin 1-\sin\frac{1}{n+1}$, so

$$\sum_{n=1}^{\infty}\left(\sin\frac{1}{n}-\sin\frac{1}{n+1}\right)=\lim_{n\to\infty}s_n=\sin 1-\sin 0=\sin 1$$

33. Diverges since $\displaystyle\lim_{n\to\infty}\arctan n=\frac{\pi}{2}\neq 0.$

35. $s_n=(\ln 1-\ln 2)+(\ln 2-\ln 3)+(\ln 3-\ln 4)+\cdots+[\ln n-\ln(n+1)]=\ln 1-\ln(n+1)=-\ln(n+1)$

(telescoping series). Thus $\displaystyle\lim_{n\to\infty}s_n=-\infty$, so the series is divergent.

37. $0.\overline{5}=0.5+0.05+0.005+\cdots=\dfrac{0.5}{1-0.1}=\dfrac{5}{9}$

39. $0.\overline{307}=0.307+0.000307+0.000000307+\cdots=\dfrac{0.307}{1-0.001}=\dfrac{307}{999}$

41. $0.12\overline{3456}=\dfrac{123}{1000}+\dfrac{0.000456}{1-0.001}=\dfrac{123}{1000}+\dfrac{456}{999,000}=\dfrac{123,333}{999,000}=\dfrac{41,111}{333,000}$

43. $\displaystyle\sum_{n=0}^{\infty}(x-3)^n$ is a geometric series with $r=x-3$, so it converges whenever $|x-3|<1\quad\Rightarrow$

$-1<x-3<1\quad\Leftrightarrow\quad 2<x<4.$ The sum is $\dfrac{1}{1-(x-3)}=\dfrac{1}{4-x}.$

45. $\displaystyle\sum_{n=2}^{\infty}\left(\frac{x}{5}\right)^n$ is a geometric series with $r=\dfrac{x}{5}$, so converges whenever $\left|\dfrac{x}{5}\right|<1\quad\Leftrightarrow\quad -5<x<5.$ The sum is

$\dfrac{(x/5)^2}{1-x/5}=\dfrac{x^2}{25-5x}.$

47. $\displaystyle\sum_{n=0}^{\infty}(2\sin x)^n$ is geometric so converges whenever $|2\sin x|<1\quad\Leftrightarrow\quad -\frac{1}{2}<\sin x<\frac{1}{2}\quad\Leftrightarrow$

$n\pi-\frac{\pi}{6}<x<n\pi+\frac{\pi}{6}$, where the sum is $\dfrac{1}{1-2\sin x}.$

7

49. After defining f, We use `convert(f,parfrac);` in Maple or `Apart` in Mathematica to find that the general

term is $\dfrac{1}{(4n+1)(4n-3)} = -\dfrac{1/4}{4n+1} + \dfrac{1/4}{4n-3}$. So the nth partial sum is

$$s_n = \sum_{k=1}^{n}\left(-\frac{1/4}{4k+1} + \frac{1/4}{4k-3}\right)$$

$$= \frac{1}{4}\left[\left(-\frac{1}{5}+1\right) + \left(-\frac{1}{9}+\frac{1}{5}\right) + \left(-\frac{1}{13}+\frac{1}{9}\right) + \cdots + \left(-\frac{1}{4n+1}+\frac{1}{4n-3}\right)\right] = \frac{1}{4}\left(1 - \frac{1}{4n+1}\right).$$

The series converges to $\lim\limits_{n\to\infty} s_n = \frac{1}{4}$. This can be confirmed by directly computing the sum using

`sum(f,1..infinity);` (in Maple) or `Sum[f, {n,0,Infinity}]` (in Mathematica).

51. Plainly $a_1 = 0$ since $s_1 = 0$. For $n \neq 1$, $a_n = s_n - s_{n-1} = \dfrac{n-1}{n+1} - \dfrac{(n-1)-1}{(n-1)+1}$

$= \dfrac{(n-1)n - (n+1)(n-2)}{(n+1)n} = \dfrac{2}{n(n+1)}$. Also $\sum\limits_{n=1}^{\infty} a_n = \lim\limits_{n\to\infty} s_n = \lim\limits_{n\to\infty} \dfrac{1-1/n}{1+1/n} = 1$.

53. **(a)** The first step in the chain occurs when the local government spends D dollars. The people who receive it

spend a fraction c of those D dollars, that is, Dc dollars. Those who receive the Dc dollars spend a

fraction c of it, that is, Dc^2 dollars. Continuing in this way, we see that the total spending after n

transactions is $S_n = D + Dc + Dc^2 + \cdots + Dc^{n-1} = \dfrac{D(1-c^n)}{1-c}$ by (3).

(b) $\lim\limits_{n\to\infty} S_n = \lim\limits_{n\to\infty} \dfrac{D(1-c^n)}{1-c} = \dfrac{D}{1-c}\lim\limits_{n\to\infty}(1-c^n) = \dfrac{D}{1-c} = \dfrac{D}{s} = kD$, since $0 < c < 1 \Rightarrow$

$\lim\limits_{n\to\infty} c^n = 0$. If $c = 0.8$, then $s = 1 - c = 0.2$ and the multiplier is $k = 1/s = 5$.

55. $\sum_{n=2}^{\infty}(1+c)^{-n}$ is a geometric series with $a = (1+c)^{-2}$ and $r = (1+c)^{-1}$, so the series converges when

$\left|(1+c)^{-1}\right| < 1 \Rightarrow |1+c| > 1 \Rightarrow 1+c > 1 \text{ or } 1+c < -1 \Rightarrow c > 0 \text{ or } c < -2$. We calculate

the sum of the series and set it equal to 2: $\dfrac{(1+c)^{-2}}{1-(1+c)^{-1}} = 2 \Leftrightarrow \left(\dfrac{1}{1+c}\right)^2 = 2 - 2\left(\dfrac{1}{1+c}\right) \Leftrightarrow$

$1 - 2(1+c)^2 + 2(1+c) = 0 \Leftrightarrow 2c^2 + 2c - 1 = 0 \Leftrightarrow c = \dfrac{-2\pm\sqrt{12}}{4} = \dfrac{\pm\sqrt{3}-1}{2}$. However, the negative

root is inadmissible because $-2 < \dfrac{-\sqrt{3}-1}{2} < 0$. So $c = \dfrac{\sqrt{3}-1}{2}$.

57. Let d_n be the diameter of C_n. We draw lines from
the centers of the C_i to the center of D (or C), and
using the Pythagorean Theorem, we can write

$1^2 + \left(1 - \frac{1}{2}d_1\right)^2 = \left(1 + \frac{1}{2}d_1\right)^2 \Leftrightarrow$

$1 = \left(1 + \frac{1}{2}d_1\right)^2 - \left(1 - \frac{1}{2}d_1\right)^2 = 2d_1$ (difference of squares)

$\Rightarrow d_1 = \frac{1}{2}$. Similarly,

$1 = \left(1 + \frac{1}{2}d_2\right)^2 - \left(1 - d_1 - \frac{1}{2}d_2\right)^2 = (2 - d_1)(d_1 + d_2)$

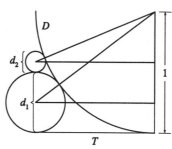

$\Leftrightarrow d_2 = \dfrac{(1-d_1)^2}{2-d_1}, 1 = \left(1 + \frac{1}{2}d_3\right)^2 - \left(1 - d_1 - d_2 - \frac{1}{2}d_3\right)^2 \Leftrightarrow d_3 = \dfrac{[1-(d_1+d_2)]^2}{2-(d_1+d_2)}$, and in general,

$d_{n+1} = \dfrac{\left(1 - \sum_{i=1}^{n} d_i\right)^2}{2 - \sum_{i=1}^{n} d_i}$.

If we actually calculate d_2 and d_3 from the formulas above, we find that they are $\frac{1}{6} = \frac{1}{2 \cdot 3}$ and $\frac{1}{12} = \frac{1}{3 \cdot 4}$ respectively, so we suspect that in general, $d_n = \frac{1}{n(n+1)}$. To prove this, we use induction: assume that for all $k \le n$, $d_k = \frac{1}{k(k+1)} = \frac{1}{k} - \frac{1}{k+1}$. Then $\sum_{i=1}^{n} d_i = 1 - \frac{1}{n+1} = \frac{n}{n+1}$ (telescoping sum). Substituting this into our formula for d_{n+1}, we get $d_{n+1} = \dfrac{\left[1 - \dfrac{n}{n+1}\right]^2}{2 - \left(\dfrac{n}{n+1}\right)} = \dfrac{\dfrac{1}{(n+1)^2}}{\dfrac{n+2}{n+1}} = \dfrac{1}{(n+1)(n+2)}$, and the induction is complete.

Now, we observe that the partial sums $\sum_{i=1}^{n} d_i$ of the diameters of the circles approach 1 as $n \to \infty$; that is, $\sum_{n=1}^{\infty} d_n = \sum_{n=1}^{\infty} \frac{1}{n(n+1)} = 1$, which is what we wanted to prove.

59. The series $1 - 1 + 1 - 1 + 1 - 1 + \cdots$ diverges (geometric series with $r = -1$) so we cannot say $0 = 1 - 1 + 1 - 1 + 1 - 1 + \cdots$.

61. $\sum_{n=1}^{\infty} ca_n = \lim_{n \to \infty} \sum_{i=1}^{n} ca_i = \lim_{n \to \infty} c \sum_{i=1}^{n} a_i = c \lim_{n \to \infty} \sum_{i=1}^{n} a_i = c \sum_{n=1}^{\infty} a_n$, which exists by hypothesis.

63. Suppose on the contrary that $\sum(a_n + b_n)$ converges. Then by Theorem 8(c), so would $\sum[(a_n + b_n) - a_n] = \sum b_n$, a contradiction.

65. The partial sums $\{s_n\}$ form an increasing sequence, since $s_n - s_{n-1} = a_n > 0$ for all n. Also, the sequence $\{s_n\}$ is bounded since $s_n \le 1000$ for all n. So by Theorem 10.1.10 , the sequence of partial sums converges, that is, the series $\sum a_n$ is convergent.

67. (a) At the first step, only the interval $\left(\frac{1}{3}, \frac{2}{3}\right)$ (length $\frac{1}{3}$) is removed. At the second step, we remove the intervals $\left(\frac{1}{9}, \frac{2}{9}\right)$ and $\left(\frac{7}{9}, \frac{8}{9}\right)$, which have a total length of $2 \cdot \left(\frac{1}{3}\right)^2$. At the third step, we remove 2^2 intervals, each of length $\left(\frac{1}{3}\right)^3$. In general, at the nth step we remove 2^{n-1} intervals, each of length $\left(\frac{1}{3}\right)^n$, for a length of $2^{n-1} \cdot \left(\frac{1}{3}\right)^n = \frac{1}{3}\left(\frac{2}{3}\right)^{n-1}$. Thus, the total length of all removed intervals is $\sum_{n=1}^{\infty} \frac{1}{3}\left(\frac{2}{3}\right)^{n-1} = \frac{1/3}{1 - 2/3} = 1$ (geometric series with $a = \frac{1}{3}$ and $r = \frac{2}{3}$).

Notice that at the nth step, the leftmost interval that is removed is $\left(\left(\frac{1}{3}\right)^n, \left(\frac{2}{3}\right)^n\right)$, so we never remove 0, so 0 is in the Cantor set. Also, the rightmost interval removed is $\left(1 - \left(\frac{2}{3}\right)^n, 1 - \left(\frac{1}{3}\right)^n\right)$, so 1 is never removed. Some other numbers in the Cantor set are $\frac{1}{3}, \frac{2}{3}, \frac{1}{9}, \frac{2}{9}, \frac{7}{9}$, and $\frac{8}{9}$.

(b) The area removed at the first step is $\frac{1}{9}$; at the second step, $2^3 \cdot \left(\frac{1}{9}\right)^2$; at the third step, $\left(2^3\right)^2 \cdot \left(\frac{1}{9}\right)^3$. In general, the area removed at the nth step is $\left(2^3\right)^{n-1}\left(\frac{1}{9}\right)^n = \frac{1}{9}\left(\frac{8}{9}\right)^{n-1}$, so the total area of all removed squares is $\sum_{n=1}^{\infty} \frac{1}{8}\left(\frac{8}{9}\right)^{n-1} = \frac{1/9}{1 - 8/9} = 1$.

69. (a) $s_1 = \dfrac{1}{1\cdot 2} = \dfrac{1}{2}$, $s_2 = \dfrac{1}{2} + \dfrac{1}{1\cdot 2\cdot 3} = \dfrac{5}{6}$, $s_3 = \dfrac{5}{6} + \dfrac{3}{1\cdot 2\cdot 3\cdot 4} = \dfrac{23}{24}$, $s_4 = \dfrac{23}{24} + \dfrac{4}{1\cdot 2\cdot 3\cdot 4\cdot 5} = \dfrac{119}{120}$.

The denominators are $(n+1)!$ so a guess would be $s_n = \dfrac{(n+1)! - 1}{(n+1)!}$.

(b) For $n = 1$, $s_1 = \dfrac{1}{2} = \dfrac{2! - 1}{2!}$, so the formula holds for $n = 1$. Assume $s_k = \dfrac{(k+1)! - 1}{(k+1)!}$. Then

$$s_{k+1} = \frac{(k+1)! - 1}{(k+1)!} + \frac{k+1}{(k+2)!} = \frac{(k+1)! - 1}{(k+1)!} + \frac{k+1}{(k+1)!(k+2)} = \frac{(k+2)! - (k+2) + k + 1}{(k+2)!}$$

$$= \frac{(k+2)! - 1}{(k+2)!}$$

Thus the formula is true for $n = k + 1$. So by induction, the guess is correct.

(c) $\displaystyle\lim_{n\to\infty} s_n = \lim_{n\to\infty} \frac{(n+1)! - 1}{(n+1)!} = \lim_{n\to\infty}\left[1 - \frac{1}{(n+1)!}\right] = 1$ and so $\displaystyle\sum_{n=0}^{\infty} \frac{n}{(n+1)!} = 1$.

EXERCISES 10.3

1. $\displaystyle\sum_{n=1}^{\infty} \frac{2}{\sqrt[3]{n}} = 2\sum_{n=1}^{\infty} \frac{1}{n^{1/3}}$, which is a p-series, $p = \frac{1}{3} < 1$, so it diverges.

3. $\displaystyle\sum_{n=5}^{\infty} \frac{1}{n^{1.0001}}$ is a p-series, $p = 1.0001 > 1$, so it converges.

5. $\displaystyle\sum_{n=5}^{\infty} \frac{1}{(n-4)^2} = \sum_{n=1}^{\infty} \frac{1}{n^2}$ is a p-series, $p = 2 > 1$, so it converges.

7. Since $\dfrac{1}{\sqrt{x}+1}$ is continuous, positive, and decreasing on $[0, \infty)$ we can apply the Integral Test.

$$\int_1^\infty \frac{1}{\sqrt{x}+1}\, dx = \lim_{t\to\infty}\left[2\sqrt{x} - 2\ln(\sqrt{x}+1)\right]_1^t \quad \begin{bmatrix} \text{using the substitution } u = \sqrt{x}+1,\ \text{so} \\ x = (u-1)^2 \text{ and } dx = 2(u-1)du \end{bmatrix}$$

$$= \lim_{t\to\infty}\left(\left[2\sqrt{t} - 2\ln\left(\sqrt{t}+1\right)\right] - (2 - 2\ln 2)\right).$$

Now $2\sqrt{t} - 2\ln\left(\sqrt{t}+1\right) = 2\ln\left(\dfrac{e^{\sqrt{t}}}{\sqrt{t}+1}\right)$ and so $\displaystyle\lim_{t\to\infty}\left[2\sqrt{t} - 2\ln\left(\sqrt{t}+1\right)\right] = \infty$ (using l'Hospital's

Rule) so both the integral and the original series diverge.

9. $f(x) = xe^{-x^2}$ is continuous and positive on $[1, \infty)$, and since $f'(x) = e^{-x^2}(1 - 2x^2) < 0$ for $x > 1$, f is

decreasing as well. We can use the Integral Test. $\displaystyle\int_1^\infty xe^{-x^2}\, dx = \lim_{t\to\infty}\left[-\frac{1}{2}e^{-x^2}\right]_1^t = 0 - \left(-\frac{e^{-1}}{2}\right) = \frac{1}{2e}$ so the

series converges.

11. $f(x) = \dfrac{x}{x^2+1}$ is continuous and positive on $[1, \infty)$, and since $f'(x) = \dfrac{1 - x^2}{(x^2+1)^2} < 0$ for $x > 1$, f is also

decreasing. Using the Integral Test, $\displaystyle\int_1^\infty \frac{x}{x^2+1}\, dx = \lim_{t\to\infty}\left[\frac{\ln(x^2+1)}{2}\right]_1^t = \infty$, so the series diverges.

13. $f(x) = \dfrac{1}{x \ln x}$ is continuous and positive on $[2, \infty)$, and also decreasing since $f'(x) = -\dfrac{1 + \ln x}{x^2 (\ln x)^2} < 0$ for

$x > 2$, so we can use the Integral Test. $\displaystyle\int_2^\infty \dfrac{1}{x \ln x}\, dx = \lim_{t\to\infty} \left[\ln (\ln x)\right]_2^t = \lim_{t\to\infty}\left[\ln (\ln t) - \ln (\ln 2)\right] = \infty$, so the

series diverges.

15. $f(x) = \dfrac{\arctan x}{1 + x^2}$ is continuous and positive on $[1, \infty)$. $f'(x) = \dfrac{1 - 2x \arctan x}{(1 + x^2)^2} < 0$ for $x > 1$, since

$2x \arctan x \geq \frac{\pi}{2} > 1$ for $x \geq 1$. So f is decreasing and we can use the Integral Test.

$\displaystyle\int_1^\infty \dfrac{\arctan x}{1 + x^2}\, dx = \lim_{t\to\infty}\left[\tfrac{1}{2}\arctan^2 x\right]_1^t = \dfrac{(\pi/2)^2}{2} - \dfrac{(\pi/4)^2}{2} = \dfrac{3\pi^2}{32}$, so the series converges.

17. $f(x) = \dfrac{1}{x^2 + 2x + 2}$ is continuous and positive on $[1, \infty)$, and $f'(x) = -\dfrac{2x + 2}{(x^2 + 2x + 2)^2} < 0$ for $x \geq 1$, so f

is decreasing and we can use the Integral Test. $\displaystyle\int_1^\infty \dfrac{1}{x^2 + 2x + 2}\, dx = \int_1^\infty \dfrac{1}{(x+1)^2 + 1}\, dx$

$= \lim_{t\to\infty}\left[\arctan(x + 1)\right]_1^t = \frac{\pi}{2} - \arctan 2$, so the series converges also.

19. We have already shown that when $p = 1$ the series diverges (in Exercise 13 above), so assume $p \neq 1$.

$f(x) = \dfrac{1}{x(\ln x)^p}$ is continuous and positive on $[2, \infty)$, and $f'(x) = -\dfrac{p + \ln x}{x^2(\ln x)^{p+1}} < 0$ if $x > e^{-p}$, so that f is

eventually decreasing and we can use the Integral Test.

$\displaystyle\int_2^\infty \dfrac{1}{x(\ln x)^p}\, dx = \lim_{t\to\infty}\left[\dfrac{(\ln x)^{1-p}}{1 - p}\right]_2^t \quad \text{(for } p \neq 1) \quad = \lim_{t\to\infty}\left[\dfrac{(\ln t)^{1-p}}{1 - p}\right] - \dfrac{(\ln 2)^{1-p}}{1 - p}.$

This limit exists whenever $1 - p < 0 \quad \Leftrightarrow \quad p > 1$, so the series converges for $p > 1$.

21. Clearly the series cannot converge if $p \geq -\frac{1}{2}$, because then $\lim_{n\to\infty} n(1 + n^2)^p \neq 0$. Also, if $p = -1$ the series

diverges (see Exercise 11 above.) So assume $p < -\frac{1}{2}$, $p \neq -1$. Then $f(x) = x(1 + x^2)^p$ is continuous,

positive, and eventually decreasing on $[1, \infty)$, and we can use the Integral Test.

$\displaystyle\int_1^\infty x(1 + x^2)^p\, dx = \lim_{t\to\infty}\left[\dfrac{1}{2} \cdot \dfrac{(1 + x^2)^{p+1}}{p + 1}\right]_1^t = \lim_{t\to\infty}\dfrac{1}{2} \cdot \dfrac{(1 + t^2)^{p+1}}{p + 1} - \dfrac{2^p}{p + 1}$. This limit exists and is finite

$\Leftrightarrow \quad p + 1 < 0 \quad \Leftrightarrow \quad p < -1$, so the series converges whenever $p < -1$.

23. Since this is a p-series with $p = x$, $\zeta(x)$ is defined when $x > 1$.

25. **(a)** $f(x) = \dfrac{1}{x^2}$ is positive and continuous and $f'(x) = \dfrac{-2}{x^3}$ is negative for $x > 1$, and so the Integral Test

applies. $\displaystyle\sum_{n=1}^\infty \dfrac{1}{n^2} \approx s_{10} = \dfrac{1}{1^2} + \dfrac{1}{2^2} + \dfrac{1}{3^2} + \cdots + \dfrac{1}{10^2} \approx 1.54977.$

$R_{10} \leq \displaystyle\int_{10}^\infty \dfrac{1}{x^2}\, dx = \lim_{t\to\infty}\left[\dfrac{-1}{x}\right]_{10}^t = \lim_{t\to\infty}\left(-\dfrac{1}{t} + \dfrac{1}{10}\right) = \dfrac{1}{10}$, so the error is at most 0.1.

(b) $s_{10} + \displaystyle\int_{11}^\infty \dfrac{1}{x^2}\, dx \leq s \leq s_{10} + \int_{10}^\infty \dfrac{1}{x^2}\, dx \quad \Rightarrow \quad s_{10} + \dfrac{1}{11} \leq s \leq s_{10} + \dfrac{1}{10} \quad \Rightarrow$

$1.549768 + 0.090909 = 1.640677 \leq s \leq 1.549768 + 0.1 = 1.649768$, so we get $s \approx 1.64522$ with

error ≤ 0.005.

(c) $R_n \leq \int_n^\infty (1/x^2)\, dx = 1/n$. So $R_n < 0.001$ for $n > 1000$.

27. $f(x) = x^{-3/2}$ is positive and continuous and $f'(x) = -\frac{3}{2}x^{-5/2}$ is negative for $x > 1$, so the Integral Test applies. Using (5), we need n such that

$$0.01 > \frac{1}{2}\left(\int_n^\infty x^{-3/2}\,dx - \int_{n+1}^\infty x^{-3/2}\,dx\right) = \frac{1}{2}\left(\lim_{t\to\infty}\left[\frac{-2}{\sqrt{x}}\right]_n^t - \lim_{t\to\infty}\left[\frac{-2}{\sqrt{x}}\right]_{n+1}^t\right) = \frac{1}{\sqrt{n}} - \frac{1}{\sqrt{n+1}} \quad\Leftrightarrow$$

$n > 13$. Then, again from (5),

$s \approx s_{14} + \frac{1}{2}\left(\int_{14}^\infty x^{-3/2}\,dx + \int_{15}^\infty x^{-3/2}\,dx\right) = 2.0872 + \frac{1}{\sqrt{14}} + \frac{1}{\sqrt{15}} \approx 2.6127$. Any larger value of n will also work. For instance, $s \approx s_{30} + \frac{1}{\sqrt{30}} + \frac{1}{\sqrt{31}} \approx 2.6124$.

29. **(a)** From (1), with $f(x) = \frac{1}{x}$, $\frac{1}{2} + \frac{1}{3} + \frac{1}{4} + \cdots + \frac{1}{n} \leq \int_1^n \frac{1}{x}\,dx = \ln n$, so

$$s_n = 1 + \frac{1}{2} + \frac{1}{3} + \frac{1}{4} + \cdots + \frac{1}{n} \leq 1 + \ln n.$$

   (b) By part (a), $s_{10^6} \leq 1 + \ln 10^6 \approx 14.82 < 15$ and $s_{10^9} \leq 1 + \ln 10^9 \approx 21.72 < 22$.

31. $b^{\ln n} = e^{\ln b \ln n} = n^{\ln b} = \dfrac{1}{n^{-\ln b}}$. This is a p-series, which converges for all b such that $-\ln b > 1$ $\quad\Leftrightarrow$ $\ln b < -1$, so for $b < 1/e$.

EXERCISES 10.4

1. $\dfrac{1}{n^3 + n^2} < \dfrac{1}{n^3}$ since $n^3 + n^2 > n^3$ for all n, and since $\displaystyle\sum_{n=1}^\infty \frac{1}{n^3}$ is a convergent p-series ($p = 3 > 1$), $\displaystyle\sum_{n=1}^\infty \frac{1}{n^3 + n^2}$ converges also by the Comparison Test [part (a).]

3. $\dfrac{3}{n2^n} \leq \dfrac{3}{2^n}$. $\displaystyle\sum_{n=1}^\infty \frac{3}{2^n}$ is a geometric series with $|r| = \frac{1}{2} < 1$, and hence converges, so $\displaystyle\sum_{n=1}^\infty \frac{3}{n2^n}$ converges also, by the Comparison Test.

5. $\dfrac{1 + 5^n}{4^n} > \dfrac{5^n}{4^n} = \left(\dfrac{5}{4}\right)^n$. $\displaystyle\sum_{n=0}^\infty \left(\frac{5}{4}\right)^n$ is a divergent geometric series ($|r| = \frac{5}{4} > 1$) so $\displaystyle\sum_{n=0}^\infty \frac{1 + 5^n}{4^n}$ diverges by the Comparison Test.

7. $\dfrac{3}{n(n+3)} < \dfrac{3}{n^2}$. $\displaystyle\sum_{n=1}^\infty \frac{3}{n^2} = 3\sum_{n=1}^\infty \frac{1}{n^2}$ is a convergent p-series ($p = 2 > 1$) so $\displaystyle\sum_{n=1}^\infty \frac{3}{n(n+3)}$ converges by the Comparison Test.

9. $\dfrac{\sqrt{n}}{n-1} > \dfrac{\sqrt{n}}{n} = \dfrac{1}{n^{1/2}}$. $\displaystyle\sum_{n=2}^\infty \frac{1}{n^{1/2}}$ is a divergent p-series ($p = \frac{1}{2} < 1$) so $\displaystyle\sum_{n=2}^\infty \frac{\sqrt{n}}{n-1}$ diverges by the Comparison Test.

11. $n^3 + 1 > n^3 \;\Rightarrow\; \dfrac{1}{n^3+1} < \dfrac{1}{n^3} \;\Rightarrow\; \dfrac{n}{n^3+1} < \dfrac{n}{n^3} \;\Rightarrow\; \dfrac{n-1}{n^3+1} < \dfrac{n}{n^3} = \dfrac{1}{n^2}$. Now $\displaystyle\sum_{n=1}^{\infty} \dfrac{1}{n^2}$ is a

convergent p-series ($p = 2 > 1$) so $\displaystyle\sum_{n=1}^{\infty} \dfrac{n-1}{n^3+1}$ converges by the Comparison Test.

13. $\dfrac{3 + \cos n}{3^n} \le \dfrac{4}{3^n}$ since $\cos n \le 1$. $\displaystyle\sum_{n=1}^{\infty} \dfrac{4}{3^n}$ is a geometric series with $|r| = \frac{1}{3} < 1$ so it converges, and so

$\displaystyle\sum_{n=1}^{\infty} \dfrac{3 + \cos n}{3^n}$ converges by the Comparison Test.

15. $\dfrac{n}{\sqrt{n^5 + 4}} < \dfrac{n}{\sqrt{n^5}} = \dfrac{1}{n^{3/2}}$. $\displaystyle\sum_{n=1}^{\infty} \dfrac{1}{n^{3/2}}$ is a convergent p-series ($p = \frac{3}{2} > 1$) so $\displaystyle\sum_{n=1}^{\infty} \dfrac{n}{\sqrt{n^5 + 4}}$ converges by the

Comparison Test.

17. $\dfrac{2^n}{1 + 3^n} < \dfrac{2^n}{3^n} = \left(\dfrac{2}{3}\right)^n$. $\displaystyle\sum_{n=1}^{\infty} \left(\dfrac{2}{3}\right)^n$ is a convergent geometric series ($|r| = \frac{2}{3} < 1$), so $\displaystyle\sum_{n=1}^{\infty} \dfrac{2^n}{1 + 3^n}$ converges by

the Comparison Test.

19. Let $a_n = \dfrac{1}{1 + \sqrt{n}}$ and $b_n = \dfrac{1}{\sqrt{n}}$. Then $\displaystyle\lim_{n\to\infty} \dfrac{a_n}{b_n} = \lim_{n\to\infty} \dfrac{\sqrt{n}}{1 + \sqrt{n}} = 1 > 0$. Since $\displaystyle\sum_{n=1}^{\infty} \dfrac{1}{\sqrt{n}}$ is a divergent

p-series ($p = \frac{1}{2} < 1$), $\displaystyle\sum_{n=1}^{\infty} \dfrac{1}{1 + \sqrt{n}}$ also diverges by the Limit Comparison Test.

21. Let $a_n = \dfrac{n^2 + 1}{n^4 + 1}$ and $b_n = \dfrac{1}{n^2}$. Then $\displaystyle\lim_{n\to\infty} \dfrac{a_n}{b_n} = \lim_{n\to\infty} \dfrac{n^4 + n^2}{n^4 + 1} = 1$. Since $\displaystyle\sum_{n=1}^{\infty} \dfrac{1}{n^2}$ is a convergent p-series

($p = 2 > 1$), so is $\displaystyle\sum_{n=1}^{\infty} \dfrac{n^2 + 1}{n^4 + 1}$ by the Limit Comparison Test.

23. Let $a_n = \dfrac{n^2 - n + 2}{\sqrt[4]{n^{10} + n^5 + 3}}$ and $b_n = \dfrac{1}{\sqrt{n}}$. Then

$\displaystyle\lim_{n\to\infty} \dfrac{a_n}{b_n} = \lim_{n\to\infty} \dfrac{n^{5/2} - n^{3/2} + 2n^{1/2}}{\sqrt[4]{n^{10} + n^5 + 3}} = \lim_{n\to\infty} \dfrac{1 - n^{-1} + 2n^{-2}}{\sqrt[4]{1 + n^{-5} + 3n^{-10}}} = 1$. Since $\displaystyle\sum_{n=1}^{\infty} \dfrac{1}{\sqrt{n}}$ is a divergent p-series

($p = \frac{1}{2} < 1$), $\displaystyle\sum_{n=1}^{\infty} \dfrac{n^2 - n + 2}{\sqrt[4]{n^{10} + n^5 + 3}}$ diverges by the Limit Comparison Test.

25. Let $a_n = \dfrac{n + 1}{n2^n}$ and $b_n = \dfrac{1}{2^n}$. Then $\displaystyle\lim_{n\to\infty} \dfrac{a_n}{b_n} = \lim_{n\to\infty} \dfrac{n + 1}{n} = 1$. Since $\displaystyle\sum_{n=1}^{\infty} \dfrac{1}{2^n}$ is a convergent geometric series

($|r| = \frac{1}{2} < 1$), $\displaystyle\sum_{n=1}^{\infty} \dfrac{n + 1}{n2^n}$ converges by the Limit Comparison Test.

27. Let $a_n = \dfrac{\ln n}{n^3}$ and $b_n = \dfrac{1}{n^2}$. Then $\displaystyle\lim_{n\to\infty} \dfrac{a_n}{b_n} = \lim_{n\to\infty} \dfrac{\ln n}{n} = \lim_{n\to\infty} \dfrac{1/n}{1} = 0$. So since $\displaystyle\sum_{n=1}^{\infty} \dfrac{1}{n^2}$ converges (p-series,

$p = 2 > 1$), so does $\displaystyle\sum_{n=1}^{\infty} \dfrac{\ln n}{n^3}$ by part (b) of the Limit Comparison Test.

29. Clearly $n! = n(n-1)(n-2)\cdots(3)(2) \geq 2 \cdot 2 \cdot 2 \cdot \cdots \cdot 2 \cdot 2 = 2^{n-1}$, so $\dfrac{1}{n!} \leq \dfrac{1}{2^{n-1}}$. $\displaystyle\sum_{n=1}^{\infty} \dfrac{1}{2^{n-1}}$ is a convergent

geometric series ($|r| = \tfrac{1}{2} < 1$) so $\displaystyle\sum_{n=1}^{\infty} \dfrac{1}{n!}$ converges by the Comparison Test.

31. Let $a_n = \sin\left(\dfrac{1}{n}\right)$ and $b_n = \dfrac{1}{n}$. Then $\displaystyle\lim_{n\to\infty} \dfrac{a_n}{b_n} = \lim_{n\to\infty} \dfrac{\sin(1/n)}{1/n} = \lim_{\theta\to 0} \dfrac{\sin\theta}{\theta} = 1$, so since $\displaystyle\sum_{n=1}^{\infty} b_n$ is the harmonic

series (which diverges), $\displaystyle\sum_{n=1}^{\infty} \sin\left(\dfrac{1}{n}\right)$ diverges as well by the Limit Comparison Test.

33. $\displaystyle\sum_{n=1}^{10} \dfrac{1}{n^4 + n^2} = \dfrac{1}{2} + \dfrac{1}{20} + \dfrac{1}{90} + \cdots + \dfrac{1}{10{,}100} \approx 0.567975$. Now $\dfrac{1}{n^4 + n^2} < \dfrac{1}{n^4}$, so using the reasoning and

notation of Example 7, the error is $R_{10} \leq T_{10} = \displaystyle\sum_{n=11}^{\infty} \dfrac{1}{n^4} \leq \int_{10}^{\infty} \dfrac{dx}{x^4} = \lim_{t\to\infty}\left[-\dfrac{x^{-3}}{3}\right]_{10}^{t} = \dfrac{1}{3000} = 0.000\overline{3}$.

35. $\displaystyle\sum_{n=1}^{10} \dfrac{1}{1 + 2^n} = \dfrac{1}{3} + \dfrac{1}{5} + \dfrac{1}{9} + \cdots + \dfrac{1}{1025} \approx 0.76352$. Now $\dfrac{1}{1 + 2^n} < \dfrac{1}{2^n}$, so the error is

$R_{10} \leq T_{10} = \displaystyle\sum_{n=11}^{\infty} \dfrac{1}{2^n} = \dfrac{1/2^{11}}{1 - 1/2}$ (geometric series) ≈ 0.00098.

37. Since $\dfrac{d_n}{10^n} \leq \dfrac{9}{10^n}$ for each n, and since $\displaystyle\sum_{n=1}^{\infty} \dfrac{9}{10^n}$ is a convergent geometric series ($|r| = \tfrac{1}{10} < 1$),

$0.d_1 d_2 d_3\ldots = \displaystyle\sum_{n=1}^{\infty} \dfrac{d_n}{10^n}$ will always converge by the Comparison Test.

39. Since $\sum a_n$ converges, $\displaystyle\lim_{n\to\infty} a_n = 0$, so there exists N such that $|a_n - 0| < 1$ for all $n > N \quad \Rightarrow \quad 0 \leq a_n < 1$

for all $n > N \quad \Rightarrow \quad 0 \leq a_n^2 \leq a_n$. Since $\sum a_n$ converges, so does $\sum a_n^2$ by the Comparison Test.

41. We wish to prove that if $\displaystyle\lim_{n\to\infty} \dfrac{a_n}{b_n} = \infty$ and $\sum b_n$ diverges, then so does $\sum a_n$. So suppose on the contrary that

$\sum a_n$ converges. Since $\displaystyle\lim_{n\to\infty} \dfrac{a_n}{b_n} = \infty$, we have that $\displaystyle\lim_{n\to\infty} \dfrac{b_n}{a_n} = 0$, so by part (b) of the Limit Comparison Test

(proved in Exercise 40), if $\sum a_n$ converges, so must $\sum b_n$. But this contradicts our hypothesis, so $\sum a_n$ must

diverge.

43. $\displaystyle\lim_{n\to\infty} n a_n = \lim_{n\to\infty} \dfrac{a_n}{1/n}$, so we apply the Limit Comparison Test with $b_n = \dfrac{1}{n}$. Since $\displaystyle\lim_{n\to\infty} n a_n > 0$ we know that

either both series converge or both series diverge, and we also know that $\displaystyle\sum_{n=0}^{\infty} \dfrac{1}{n}$ diverges (p-series with $p = 1$).

Therefore $\sum a_n$ must be divergent.

45. Yes. Since $\sum a_n$ converges, its terms approach 0 as $n \to \infty$, so $\displaystyle\lim_{n\to\infty} \dfrac{\sin a_n}{a_n} = 1$ by Theorem 2.4.4. Thus

$\sum \sin a_n$ converges by the Limit Comparison Test.

EXERCISES 10.5

1. $\sum_{n=1}^{\infty}(-1)^{n-1}\dfrac{3}{n+4}$. $b_n = \dfrac{3}{n+4} > 0$ and $b_{n+1} < b_n$ for all n; $\lim_{n\to\infty} b_n = 0$ so the series converges by the

   Alternating Series Test.

3. $\sum_{n=1}^{\infty}(-1)^{n}\dfrac{n}{n+1}$. $\lim_{n\to\infty}\dfrac{n}{n+1} = 1$ so $\lim_{n\to\infty}(-1)^n\dfrac{n}{n+1}$ does not exist and the series diverges by the

   Test for Divergence.

5. $\sum_{n=1}^{\infty}(-1)^{n-1}\dfrac{1}{n^2}$. $b_n = \dfrac{1}{n^2} > 0$ and $b_{n+1} < b_n$ for all n, and $\lim_{n\to\infty}\dfrac{1}{n^2} = 0$, so the series converges by the

   Alternating Series Test.

7. $\sum_{n=1}^{\infty}(-1)^{n+1}\dfrac{n}{5n+1}$. $\lim_{n\to\infty}\dfrac{n}{5n+1} = \dfrac{1}{5}$ so $\lim_{n\to\infty}(-1)^{n+1}\dfrac{n}{5n+1}$ does not exist and the series diverges by the

   Test for Divergence.

9. $\sum_{n=1}^{\infty}(-1)^{n}\dfrac{n}{n^2+1}$. $b_n = \dfrac{n}{n^2+1} > 0$ for all n. $b_{n+1} < b_n \iff \dfrac{n+1}{(n+1)^2+1} < \dfrac{n}{n^2+1} \iff$

   $(n+1)(n^2+1) < \left[(n+1)^2+1\right]n \iff n^3+n^2+n+1 < n^3+2n^2+2n \iff 0 < n^2+n-1$, which

   is true for all $n \geq 1$. Also $\lim_{n\to\infty}\dfrac{n}{n^2+1} = \lim_{n\to\infty}\dfrac{1/n}{1+1/n^2} = 0$. Therefore the series converges by the

   Alternating Series Test.

11. $\sum_{n=1}^{\infty}(-1)^{n-1}\dfrac{\sqrt{n}}{n+4}$. $b_n = \dfrac{\sqrt{n}}{n+4} > 0$ for all n. Let $f(x) = \dfrac{\sqrt{x}}{x+4}$. Then $f'(x) = \dfrac{4-x}{2\sqrt{x}(x+4)^2} < 0$ if $x > 4$,

   so $\{b_n\}$ is decreasing after $n = 4$. $\lim_{n\to\infty}\dfrac{\sqrt{n}}{n+4} = \lim_{n\to\infty}\dfrac{1}{\sqrt{n}+4/\sqrt{n}} = 0$. So the series converges by the

   Alternating Series Test.

13. $\sum_{n=2}^{\infty}(-1)^{n}\dfrac{n}{\ln n}$. $\lim_{n\to\infty}\dfrac{n}{\ln n} = \lim_{n\to\infty}\dfrac{1}{1/n} = \infty$ so the series diverges by the Test for Divergence.

15. $\sum_{n=1}^{\infty}\dfrac{\cos n\pi}{n^{3/4}} = \sum_{n=1}^{\infty}\dfrac{(-1)^n}{n^{3/4}}$. $b_n = \dfrac{1}{n^{3/4}}$ is decreasing and positive, and $\lim_{n\to\infty}\dfrac{1}{n^{3/4}} = 0$ so the series converges by

   the Alternating Series Test.

17. $\sum_{n=1}^{\infty}(-1)^{n}\sin\left(\dfrac{\pi}{n}\right)$. $b_n = \sin\left(\dfrac{\pi}{n}\right) > 0$ for $n \geq 2$ and $\sin\left(\dfrac{\pi}{n}\right) \geq \sin\left(\dfrac{\pi}{n+1}\right)$, and $\lim_{n\to\infty}\sin\left(\dfrac{\pi}{n}\right) = \sin 0 = 0$, so

   the series converges by the Alternating Series Test.

19. $\dfrac{n^n}{n!} = \dfrac{n\cdot n\cdots\cdots n}{1\cdot 2\cdots\cdots n} \geq n \implies \lim_{n\to\infty}\dfrac{n^n}{n!} = \infty \implies \lim_{n\to\infty}\dfrac{(-1)^n n^n}{n!}$ does not exist. So the series diverges by

   the Test for Divergence.

21. Let $\sum b_n$ be the series for which $b_n = 0$ if n is odd and $b_n = 1/n^2$ if n is even. Then $\sum b_n = \sum 1/(2n)^2$ clearly converges (by comparison with the p-series for $p = 2$). So suppose that $\sum (-1)^{n-1} b_n$ converges. Then by Theorem 10.2.8(b), so does $\sum [(-1)^{n-1} b_n + b_n] = 1 + \frac{1}{3} + \frac{1}{5} + \cdots = \sum \frac{1}{2n-1}$. But this diverges by comparison with the harmonic series, a contradiction. Therefore $\sum (-1)^{n-1} b_n$ must diverge. The Alternating Series Test does not apply since $\{b_n\}$ is not decreasing.

23. Clearly $b_n = \dfrac{1}{n+p}$ is decreasing and eventually positive and $\lim\limits_{n\to\infty} b_n = 0$ for any p. So the series will converge (by the Alternating Series Test) for any p for which every b_n is defined, that is, $n + p \neq 0$ for $n \geq 1$, or p is not a negative integer.

25. If $b_n = \dfrac{1}{n^2}$, then $b_{11} = \dfrac{1}{121} < 0.01$, so by Theorem 1, $\displaystyle\sum_{n=1}^{\infty} \frac{1}{n^2} \approx \sum_{n=1}^{10} \frac{1}{n^2} \approx 0.82$.

27. $\displaystyle\sum_{n=0}^{\infty} (-1)^n \frac{2^n}{n!}$. Since $\dfrac{2}{n} < \dfrac{2}{3}$ for $n \geq 4$, $0 < \dfrac{2^n}{n!} < \dfrac{2}{1} \cdot \dfrac{2}{2} \cdot \dfrac{2}{3} \cdot \left(\dfrac{2}{3}\right)^{n-3} \to 0$ as $n \to \infty$, so by the Squeeze Theorem, $\lim\limits_{n\to\infty} \dfrac{2^n}{n!} = 0$, and hence $\displaystyle\sum_{n=0}^{\infty} (-1)^n \frac{2^n}{n!}$ is a convergent alternating series. $\dfrac{2^8}{8!} = \dfrac{256}{40,320} < 0.01$, so
$$\sum_{n=0}^{\infty} (-1)^n \frac{2^n}{n!} \approx \sum_{n=0}^{7} (-1)^n \frac{2^n}{n!} \approx 0.13.$$

29. $\displaystyle\sum_{n=1}^{\infty} \frac{(-1)^{n-1}}{(2n-1)!}$. $b_5 = \dfrac{1}{(2 \cdot 5 - 1)!} = \dfrac{1}{362,880} < 0.00001$, so $\displaystyle\sum_{n=1}^{\infty} \frac{(-1)^{n-1}}{(2n-1)!} \approx \sum_{n=1}^{4} \frac{(-1)^{n-1}}{(2n-1)!} \approx 0.8415$.

31. $\displaystyle\sum_{n=0}^{\infty} \frac{(-1)^n}{2^n n!}$. $b_6 = \dfrac{1}{2^6 6!} = \dfrac{1}{46,080} < 0.000022$, so $\displaystyle\sum_{n=0}^{\infty} \frac{(-1)^n}{2^n n!} \approx \sum_{n=0}^{5} \frac{(-1)^n}{2^n n!} \approx 0.6065$.

33. $\displaystyle\sum_{n=1}^{\infty} \frac{(-1)^{n-1}}{n} = 1 - \frac{1}{2} + \frac{1}{3} - \frac{1}{4} + \cdots + \frac{1}{49} - \frac{1}{50} + \frac{1}{51} - \frac{1}{52} + \cdots$. The 50th partial sum of this series is an underestimate, since $\displaystyle\sum_{n=1}^{\infty} \frac{(-1)^{n-1}}{n} = s_{50} + \left(\frac{1}{51} - \frac{1}{52}\right) + \left(\frac{1}{53} - \frac{1}{54}\right) + \cdots$, and the terms in parentheses are all positive. The result can be seen geometrically in Figure 1.

35. (a) We will prove this by induction. Let $P(n)$ be the proposition that $s_{2n} = h_{2n} - h_n$. $P(1)$ is true by an easy calculation. So suppose that $P(n)$ is true. We will show that $P(n+1)$ must be true as a consequence.
$$h_{2n+2} - h_{n+1} = \left(h_{2n} + \frac{1}{2n+1} + \frac{1}{2n+2}\right) - \left(h_n + \frac{1}{n+1}\right)$$
$$= (h_{2n} - h_n) + \frac{1}{2n+1} - \frac{1}{2n+2} = s_{2n} + \frac{1}{2n+1} - \frac{1}{2n+2} = s_{2n+2},$$
which is $P(n+1)$, and proves that $s_{2n} = h_{2n} - h_n$ for all n.

(b) We know that $h_{2n} - \ln 2n \to \gamma$ and $h_n - \ln n \to \gamma$ as $n \to \infty$. So
$$s_{2n} = h_{2n} - h_n = (h_{2n} - \ln 2n) - (h_n - \ln n) + (\ln 2n - \ln n), \text{ and}$$
$$\lim_{n\to\infty} s_{2n} = \gamma - \gamma + \lim_{n\to\infty} [\ln 2n - \ln n] = \lim_{n\to\infty} (\ln 2 + \ln n - \ln n) = \ln 2.$$

EXERCISES 10.6

1. $\displaystyle\sum_{n=1}^{\infty} \frac{1}{n\sqrt{n}} = \sum_{n=1}^{\infty} \frac{1}{n^{3/2}}$ is a convergent p-series ($p = \frac{3}{2} > 1$), so the given series is absolutely convergent.

3. $\displaystyle\lim_{n\to\infty} \left|\frac{a_{n+1}}{a_n}\right| = \lim_{n\to\infty}\left|\frac{(-3)^{n+1}/(n+1)^3}{(-3)^n/n^3}\right| = 3\lim_{n\to\infty}\left(\frac{n}{n+1}\right)^3 = 3 > 1$, so the series diverges by the Ratio Test.

5. $\displaystyle\sum_{n=1}^{\infty} \frac{1}{2n+1}$ diverges (use the Integral Test or the Limit Comparison Test with $b_n = 1/n$), but since

$\displaystyle\lim_{n\to\infty} \frac{1}{2n+1} = 0, \sum_{n=1}^{\infty} \frac{(-1)^{n+1}}{2n+1}$ converges by the Alternating Series Test, and so is conditionally convergent.

7. $\displaystyle\lim_{n\to\infty}\left|\frac{a_{n+1}}{a_n}\right| = \lim_{n\to\infty} \frac{1/(2n+1)!}{1/(2n-1)!} = \lim_{n\to\infty}\frac{1}{(2n+1)2n} = 0$, so by the Ratio Test the series is absolutely

convergent.

9. $\displaystyle\sum_{n=1}^{\infty} \frac{n}{n^2+4}$ diverges (use the Limit Comparison Test with $b_n = 1/n$). But since $0 \le \frac{n+1}{(n+1)^2+4} < \frac{n}{n^2+4}$

$\Leftrightarrow\quad n^3+n^2+4n+4 < n^3+2n^2+5n \quad\Leftrightarrow\quad 0 < n^2+n-4$ (which is true for $n \ge 2$), and since

$\displaystyle\lim_{n\to\infty} \frac{n}{n^2+4} = 0, \sum_{n=1}^{\infty}(-1)^n\frac{n}{n^2+4}$ converges (conditionally) by the Alternating Series Test.

11. $\displaystyle\lim_{n\to\infty} \frac{2n}{3n-4} = \frac{2}{3}$, so $\displaystyle\sum_{n=1}^{\infty}(-1)^n\frac{2n}{3n-4}$ diverges by the Test for Divergence.

13. $\displaystyle\left|\frac{\sin 2n}{n^2}\right| \le \frac{1}{n^2}$ and $\displaystyle\sum_{n=1}^{\infty}\frac{1}{n^2}$ converges (p-series, $p = 2 > 1$), so $\displaystyle\sum_{n=1}^{\infty}\frac{\sin 2n}{n^2}$ converges absolutely by the

Comparison Test.

15. $\displaystyle\lim_{n\to\infty}\left|\frac{a_{n+1}}{a_n}\right| = \lim_{n\to\infty}\left|\frac{2^{n+1}/[(n+1)3^{n+2}]}{2^n/(n3^{n+1})}\right| = \frac{2}{3}\lim_{n\to\infty}\frac{n}{n+1} = \frac{2}{3} < 1$ so the series converges absolutely by the

Ratio Test.

17. $\displaystyle\lim_{n\to\infty}\left|\frac{a_{n+1}}{a_n}\right| = \lim_{n\to\infty}\frac{(n+2)5^{n+1}/[(n+1)3^{2(n+1)}]}{(n+1)5^n/(n3^{2n})} = \lim_{n\to\infty}\frac{5n(n+2)}{9(n+1)^2} = \frac{5}{9} < 1$ so the series converges absolutely

by the Ratio Test.

19. $\displaystyle\lim_{n\to\infty}\left|\frac{a_{n+1}}{a_n}\right| = \lim_{n\to\infty}\frac{(n+1)!/10^{n+1}}{n!/10^n} = \lim_{n\to\infty}\frac{n+1}{10} = \infty$, so the series diverges by the Ratio Test.

21. $\displaystyle\frac{|\cos(n\pi/3)|}{n!} < \frac{1}{n!}$ and $\displaystyle\sum_{n=1}^{\infty}\frac{1}{n!}$ converges (Exercise 10.4.29), so the given series converges absolutely by the

Comparison Test.

23. $\displaystyle\lim_{n\to\infty}\left|\frac{a_{n+1}}{a_n}\right| = \lim_{n\to\infty}\frac{(n+1)^{n+1}/5^{2n+5}}{n^n/5^{2n+3}} = \lim_{n\to\infty}\frac{1}{25}\left(\frac{n+1}{n}\right)^n(n+1) = \infty$, so the series diverges by the

Ratio Test.

25. $\lim\limits_{n\to\infty} \sqrt[n]{|a_n|} = \lim\limits_{n\to\infty} \left| \dfrac{1-3n}{3+4n} \right| = \frac{3}{4} < 1$, so the series converges absolutely by the Root Test.

27. $\lim\limits_{n\to\infty} \left| \dfrac{a_{n+1}}{a_n} \right| = \lim\limits_{n\to\infty} \dfrac{(n+1)!/[1\cdot 3\cdot 5\cdot \,\cdots\, \cdot (2n+1)]}{n!/[1\cdot 3\cdot 5\cdot \,\cdots\, \cdot (2n-1)]} = \lim\limits_{n\to\infty} \dfrac{n+1}{2n+1} = \frac{1}{2} < 1$, so the series converges

absolutely by the Ratio Test.

29. $\sum\limits_{n=1}^{\infty} \dfrac{2\cdot 4\cdot 6\cdot \,\cdots\, \cdot (2n)}{n!} = \sum\limits_{n=1}^{\infty} \dfrac{2^n n!}{n!} = \sum\limits_{n=1}^{\infty} 2^n$ which diverges by the Test for Divergence since $\lim\limits_{n\to\infty} 2^n = \infty$.

31. $\lim\limits_{n\to\infty} \left| \dfrac{a_{n+1}}{a_n} \right| = \lim\limits_{n\to\infty} \dfrac{(n+3)!/[(n+1)!\,10^{n+1}]}{(n+2)!/(n!\,10^n)} = \frac{1}{10} \lim\limits_{n\to\infty} \dfrac{n+3}{n+1} = \frac{1}{10} < 1$ so the series converges absolutely by

the Ratio Test.

33. By the recursive definition, $\lim\limits_{n\to\infty} \left| \dfrac{a_{n+1}}{a_n} \right| = \lim\limits_{n\to\infty} \left| \dfrac{5n+1}{4n+3} \right| = \frac{5}{4} > 1$, so the series diverges by the Ratio Test.

35. **(a)** $\lim\limits_{n\to\infty} \left| \dfrac{1/(n+1)^3}{1/n^3} \right| = \lim\limits_{n\to\infty} \dfrac{n^3}{(n+1)^3} = \lim\limits_{n\to\infty} \dfrac{1}{(1+1/n)^3} = 1$. So inconclusive.

(b) $\lim\limits_{n\to\infty} \left| \dfrac{(n+1)}{2^{n+1}} \cdot \dfrac{2^n}{n} \right| = \lim\limits_{n\to\infty} \dfrac{n+1}{2n} = \lim\limits_{n\to\infty} \left(\dfrac{1}{2} + \dfrac{1}{2n} \right) = \dfrac{1}{2}$. So conclusive (convergent).

(c) $\lim\limits_{n\to\infty} \left| \dfrac{(-3)^n}{\sqrt{n+1}} \cdot \dfrac{\sqrt{n}}{(-3)^{n-1}} \right| = 3 \lim\limits_{n\to\infty} \sqrt{\dfrac{n}{n+1}} = 3 \lim\limits_{n\to\infty} \sqrt{\dfrac{1}{1+1/n}} = 3$. So conclusive (divergent).

(d) $\lim\limits_{n\to\infty} \left| \dfrac{\sqrt{n+1}}{1+(n+1)^2} \cdot \dfrac{1+n^2}{\sqrt{n}} \right| = \lim\limits_{n\to\infty} \left[\sqrt{1+\dfrac{1}{n}} \cdot \dfrac{1/n^2+1}{1/n^2+(1+1/n)^2} \right] = 1$. So inconclusive.

37. **(a)** $\lim\limits_{n\to\infty} \left| \dfrac{a_{n+1}}{a_n} \right| = \lim\limits_{n\to\infty} \dfrac{|x|^{n+1}/(n+1)!}{|x|^n/n!} = |x| \lim\limits_{n\to\infty} \dfrac{1}{n+1} = 0$, so by the Ratio Test the series converges for

all x.

(b) Since the series of part (a) always converges, we must have $\lim\limits_{n\to\infty} \dfrac{x^n}{n!} = 0$ by Theorem 10.2.6.

39. **(a)** $s_5 = \sum\limits_{n=1}^{5} \dfrac{1}{n2^n} = \dfrac{1}{2} + \dfrac{1}{8} + \dfrac{1}{24} + \dfrac{1}{64} + \dfrac{1}{160} = \dfrac{661}{960} \approx 0.68854$. Now the ratios

$r_n = \dfrac{a_{n+1}}{a_n} = \dfrac{n2^n}{(n+1)2^{n+1}} = \dfrac{n}{2(n+1)}$ form an increasing sequence, since

$r_{n+1} - r_n = \dfrac{n+1}{2(n+2)} - \dfrac{n}{2(n+1)} = \dfrac{(n+1)^2 - n(n+2)}{2(n+1)(n+2)} = \dfrac{1}{2(n+1)(n+2)} > 0$. So by Exercise

38(b), the error is less than $\dfrac{a_6}{1 - \lim\limits_{n\to\infty} r_n} = \dfrac{1/(6\cdot 2^6)}{1 - 1/2} = \dfrac{1}{192} \approx 0.00521$.

(b) The error in using s_n as an approximation to the sum is $R_n = \dfrac{a_{n+1}}{1/2} = \dfrac{2}{(n+1)2^{n+1}}$. We want

$R_n < 0.00005 \quad \Leftrightarrow \quad \dfrac{1}{(n+1)2^n} < 0.00005 \quad \Leftrightarrow \quad (n+1)2^n > 20{,}000$. To find such an n we can use

trial and error or a graph. We calculate $(11+1)2^{11} = 24{,}576$, so $s_{11} = \sum\limits_{n=1}^{11} \dfrac{1}{n2^n} \approx 0.693109$ is within

0.00005 of the actual sum.

41. By the Triangle Inequality (see Exercise 4.1.42, or 5.1.42 in the Early Transcendentals version) we have

$$\left|\sum_{i=1}^{n} a_i\right| \le \sum_{i=1}^{n} |a_i| \quad\Rightarrow\quad -\sum_{i=1}^{n} |a_i| \le \sum_{i=1}^{n} a_i \le \sum_{i=1}^{n} |a_i| \quad\Rightarrow\quad -\lim_{n\to\infty}\sum_{i=1}^{n} |a_i| \le \lim_{n\to\infty}\sum_{i=1}^{n} a_i \le \lim_{n\to\infty}\sum_{i=1}^{n} |a_i| \quad\Rightarrow$$

$$-\sum_{n=1}^{\infty} |a_n| \le \sum_{n=1}^{\infty} a_n \le \sum_{n=1}^{\infty} |a_n| \quad\Rightarrow\quad \left|\sum_{n=1}^{\infty} a_n\right| \le \sum_{n=1}^{\infty} |a_n|.$$

43. **(a)** Since $\sum a_n$ is absolutely convergent, and since $|a_n^+| \le |a_n|$ and $|a_n^-| \le |a_n|$ (because a_n^+ and a_n^- each equal either a_n or 0), we conclude by the Comparison Test that both $\sum a_n^+$ and $\sum a_n^-$ must be absolutely convergent. (Or use Theorem 10.2.8.)

(b) We will show by contradiction that both $\sum a_n^+$ and $\sum a_n^-$ must diverge. For suppose that $\sum a_n^+$ converged. Then so would $\sum \left(a_n^+ - \tfrac{1}{2}a_n\right)$ by Theorem 10.2.8. But

$$\sum \left(a_n^+ - \tfrac{1}{2}a_n\right) = \sum \left[\tfrac{1}{2}(a_n + |a_n|) - \tfrac{1}{2}a_n\right] = \tfrac{1}{2}\sum |a_n|,$$ which diverges because $\sum a_n$ is only conditionally convergent. Hence $\sum a_n^+$ can't converge. Similarly, neither can $\sum a_n^-$.

EXERCISES 10.7

1. Use the Comparison Test, with $a_n = \dfrac{\sqrt{n}}{n^2 + 1}$ and $b_n = \dfrac{1}{n^{3/2}}$: $\dfrac{\sqrt{n}}{n^2 + 1} < \dfrac{\sqrt{n}}{n^2} = \dfrac{1}{n^{3/2}}$, and $\displaystyle\sum_{n=1}^{\infty} \dfrac{1}{n^{3/2}}$ is a

convergent p-series $(p = \tfrac{3}{2} > 1)$, so $\displaystyle\sum_{n=1}^{\infty} a_n = \sum_{n=1}^{\infty} \dfrac{\sqrt{n}}{n^2 + 1}$ converges as well.

3. $\displaystyle\sum_{n=1}^{\infty} \dfrac{4^n}{3^{2n-1}} = 3\sum_{n=1}^{\infty} \left(\dfrac{4}{9}\right)^n$ which is a convergent geometric series $(|r| = \tfrac{4}{9} < 1.)$

5. The series converges by the Alternating Series Test, since $a_n = \dfrac{1}{(\ln n)^2}$ is decreasing ($\ln x$ is an increasing function) and $\displaystyle\lim_{n\to\infty} a_n = 0$.

7. $\displaystyle\sum_{k=1}^{\infty} \dfrac{1}{k^{1.7}}$ is a convergent p-series $(p = 1.7 > 1)$.

9. $\displaystyle\lim_{n\to\infty}\left|\dfrac{a_{n+1}}{a_n}\right| = \lim_{n\to\infty}\dfrac{(n+1)/e^{n+1}}{n/e^n} = \dfrac{1}{e}\lim_{n\to\infty}\dfrac{n+1}{n} = \dfrac{1}{e} < 1$, so the series converges by the Ratio Test.

11. Use the Limit Comparison Test with $a_n = \dfrac{n^3 + 1}{n^4 - 1}$ and $b_n = \dfrac{1}{n}$. $\displaystyle\lim_{n\to\infty}\dfrac{a_n}{b_n} = \lim_{n\to\infty}\dfrac{n^4 + n}{n^4 - 1} = \lim_{n\to\infty}\dfrac{1 + 1/n^3}{1 - 1/n^4} = 1$,

and since $\displaystyle\sum_{n=2}^{\infty} b_n$ diverges (harmonic series), so does $\displaystyle\sum_{n=2}^{\infty} \dfrac{n^3 + 1}{n^4 - 1}$.

13. Let $f(x) = \dfrac{2}{x(\ln x)^3}$. $f(x)$ is clearly positive and decreasing for $x \ge 2$, so we apply the Integral Test.

$$\int_2^{\infty} \dfrac{2}{x(\ln x)^3}\,dx = \lim_{t\to\infty}\left[\dfrac{-1}{(\ln x)^2}\right]_2^t = 0 - \dfrac{-1}{(\ln 2)^2}, \text{ which is finite, so } \sum_{n=2}^{\infty} \dfrac{2}{n(\ln n)^3} \text{ converges.}$$

15. $\lim\limits_{n\to\infty}\left|\dfrac{a_{n+1}}{a_n}\right| = \lim\limits_{n\to\infty}\dfrac{3^{n+1}(n+1)^2/(n+1)!}{3^n n^2/n!} = 3\lim\limits_{n\to\infty}\dfrac{n+1}{n^2} = 0$, so the series converges by the Ratio Test.

17. $\dfrac{3^n}{5^n+n} \le \dfrac{3^n}{5^n} = \left(\dfrac{3}{5}\right)^n$. Since $\sum\limits_{n=1}^{\infty}\left(\dfrac{3}{5}\right)^n$ is a convergent geometric series ($|r| = \frac{3}{5} < 1$), $\sum\limits_{n=1}^{\infty}\dfrac{3^n}{5^n+n}$ converges

by the Comparison Test.

19. $\lim\limits_{n\to\infty}\left|\dfrac{a_{n+1}}{a_n}\right| = \lim\limits_{n\to\infty}\dfrac{(n+1)!/[2\cdot5\cdot8\cdots(3n+5)]}{n!/[2\cdot5\cdot8\cdots(3n+2)]} = \lim\limits_{n\to\infty}\dfrac{n+1}{3n+5} = \dfrac{1}{3} < 1$, so the series converges by the

Ratio Test.

21. Use the Limit Comparison Test with $a_i = \dfrac{1}{\sqrt{i(i+1)}}$ and $b_i = \dfrac{1}{i}$. $\lim\limits_{i\to\infty}\dfrac{a_i}{b_i} = \lim\limits_{i\to\infty}\dfrac{i}{\sqrt{i(i+1)}}$

$= \lim\limits_{i\to\infty}\dfrac{1}{\sqrt{1+1/i}} = 1$. Since $\sum\limits_{i=1}^{\infty}b_i$ diverges (harmonic series) so does $\sum\limits_{i=1}^{\infty}\dfrac{1}{\sqrt{i(i+1)}}$.

23. $\lim\limits_{n\to\infty}2^{1/n} = 2^0 = 1$, so $\lim\limits_{n\to\infty}(-1)^n 2^{1/n}$ does not exist and the series diverges by the Test for Divergence.

25. Let $f(x) = \dfrac{\ln x}{\sqrt{x}}$. Then $f'(x) = \dfrac{2-\ln x}{2x^{3/2}} < 0$ when $\ln x > 2$ or $x > e^2$, so $\dfrac{\ln n}{\sqrt{n}}$ is decreasing for $n > e^2$.

By l'Hospital's Rule, $\lim\limits_{n\to\infty}\dfrac{\ln n}{\sqrt{n}} = \lim\limits_{n\to\infty}\dfrac{1/n}{1/(2\sqrt{n})} = \lim\limits_{n\to\infty}\dfrac{2}{\sqrt{n}} = 0$, so the series converges by the

Alternating Series Test.

27. The series diverges since it is a geometric series with $r = -\pi$ and $|r| = \pi > 1$. (Or use the Test for Divergence.)

29. $\sum\limits_{n=1}^{\infty}\dfrac{(-2)^{2n}}{n^n} = \sum\limits_{n=1}^{\infty}\left(\dfrac{4}{n}\right)^n$. $\lim\limits_{n\to\infty}\sqrt[n]{|a_n|} = \lim\limits_{n\to\infty}\dfrac{4}{n} = 0$, so the series converges by the Root Test.

31. $\int_2^{\infty}\dfrac{\ln x}{x^2}dx = \lim\limits_{t\to\infty}\left[-\dfrac{\ln x}{x} - \dfrac{1}{x}\right]_1^t$ (using integration by parts) $= 1$ (by L'Hospital's Rule). So $\sum\limits_{n=1}^{\infty}\dfrac{\ln n}{n^2}$

converges by the Integral Test, and since $\dfrac{k\ln k}{(k+1)^3} < \dfrac{k\ln k}{k^3} = \dfrac{\ln k}{k^2}$, the given series converges by the

Comparison Test.

33. $\lim\limits_{n\to\infty}\left|\dfrac{a_{n+1}}{a_n}\right| = \lim\limits_{n\to\infty}\dfrac{2^{n+1}/(2n+3)!}{2^n/(2n+1)!} = 2\lim\limits_{n\to\infty}\dfrac{1}{(2n+3)(2n+2)} = 0$, so the series converges by the Ratio Test.

35. $0 < \dfrac{\tan^{-1}n}{n^{3/2}} < \dfrac{\pi/2}{n^{3/2}}$. $\sum\limits_{n=1}^{\infty}\dfrac{\pi/2}{n^{3/2}} = \dfrac{\pi}{2}\sum\limits_{n=1}^{\infty}\dfrac{1}{n^{3/2}}$ which is a convergent p-series ($p = \frac{3}{2} > 1$), so

$\sum\limits_{n=1}^{\infty}\dfrac{\tan^{-1}n}{n^{3/2}}$ converges by the Comparison Test.

37. $\lim\limits_{n\to\infty}\sqrt[n]{|a_n|} = \lim\limits_{n\to\infty}\left(\dfrac{n}{n+1}\right)^{n^2/n} = \lim\limits_{n\to\infty}\dfrac{1}{[(n+1)/n]^n} = \dfrac{1}{\lim\limits_{n\to\infty}(1+1/n)^n} = \dfrac{1}{e} < 1$ (see Equation 6.4.9 or

6.4*.8, or 3.4.7 in the Early Transcendentals version), so the series converges by the Root Test.

39. $\lim\limits_{n\to\infty}\sqrt[n]{|a_n|} = \lim\limits_{n\to\infty}(2^{1/n} - 1) = 1 - 1 = 0$, so the series converges by the Root Test.

EXERCISES 10.8

Note: *"R" stands for "radius of convergence" and "I" stands for "interval of convergence" in this section.*

1. **(a)** We are given that the power series $\sum_{n=0}^{\infty} c_n x^n$ is convergent for $x = 4$. So by Theorem 3 it must converge

for at least $-4 < x \le 4$. In particular it converges when $x = -2$, that is, $\sum_{n=0}^{\infty} c_n(-2)^n$ is convergent.

(b) But it does not follow that $\sum_{n=0}^{\infty} c_n(-4)^n$ is necessarily convergent. $\Big[$See the comments after Theorem 3.

An example is $c_n = (-1)^n/(n4^n)$.$\Big]$

3. If $a_n = \dfrac{x^n}{n+2}$, then $\lim_{n\to\infty}\left|\dfrac{a_{n+1}}{a_n}\right| = \lim_{n\to\infty}\left|\dfrac{x^{n+1}}{n+3}\cdot\dfrac{n+2}{x^n}\right| = |x|\lim_{n\to\infty}\dfrac{n+2}{n+3} = |x| < 1$ for convergence (by the

Ratio Test). So $R = 1$. When $x = 1$, the series is $\sum_{n=0}^{\infty}\dfrac{1}{n+2}$ which diverges (Integral Test or Comparison Test),

and when $x = -1$, it is $\sum_{n=0}^{\infty}\dfrac{(-1)^n}{n+2}$ which converges (Alternating Series Test), so $I = [-1, 1)$.

5. If $a_n = nx^n$, then $\lim_{n\to\infty}\left|\dfrac{a_{n+1}}{a_n}\right| = \lim_{n\to\infty}\left|\dfrac{(n+1)x^{n+1}}{nx^n}\right| = |x|\lim_{n\to\infty}\dfrac{n+1}{n} = |x| < 1$ for convergence (by the Ratio

Test). So $R = 1$. When $x = 1$ or -1, $\lim_{n\to\infty} nx^n$ does not exist, so $\sum_{n=0}^{\infty} nx^n$ diverges for $x = \pm 1$. So $I = (-1, 1)$.

7. If $a_n = \dfrac{x^n}{n!}$, then $\lim_{n\to\infty}\left|\dfrac{a_{n+1}}{a_n}\right| = \lim_{n\to\infty}\left|\dfrac{x^{n+1}/(n+1)!}{x^n/n!}\right| = |x|\lim_{n\to\infty}\dfrac{1}{n+1} = 0 < 1$ for all x. So, by the Ratio Test,

$R = \infty$, and $I = (-\infty, \infty)$.

9. If $a_n = \dfrac{(-1)^n x^n}{n2^n}$, then $\lim_{n\to\infty}\left|\dfrac{a_{n+1}}{a_n}\right| = \lim_{n\to\infty}\left|\dfrac{x^{n+1}/[(n+1)2^{n+1}]}{x^n/(n2^n)}\right| = \left|\dfrac{x}{2}\right|\lim_{n\to\infty}\dfrac{n}{n+1} = \left|\dfrac{x}{2}\right| < 1$ for convergence,

so $|x| < 2$ and $R = 2$. When $x = 2$, $\sum_{n=1}^{\infty}\dfrac{(-1)^n x^n}{n2^n} = \sum_{n=1}^{\infty}\dfrac{(-1)^n}{n}$ which converges by the Alternating Series Test.

When $x = -2$, $\sum_{n=1}^{\infty}\dfrac{(-1)^n x^n}{n2^n} = \sum_{n=1}^{\infty}\dfrac{1}{n}$ which diverges (harmonic series), so $I = (-2, 2]$.

11. If $a_n = \dfrac{3^n x^n}{(n+1)^2}$, then $\lim_{n\to\infty}\left|\dfrac{a_{n+1}}{a_n}\right| = \lim_{n\to\infty}\left|\dfrac{3^{n+1} x^{n+1}}{(n+2)^2}\cdot\dfrac{(n+1)^2}{3^n x^n}\right| = 3|x|\lim_{n\to\infty}\left(\dfrac{n+1}{n+2}\right)^2 = 3|x| < 1$ for

convergence, so $|x| < \frac{1}{3}$ and $R = \frac{1}{3}$. When $x = \frac{1}{3}$, $\sum_{n=0}^{\infty}\dfrac{3^n x^n}{(n+1)^2} = \sum_{n=0}^{\infty}\dfrac{1}{(n+1)^2} = \sum_{n=1}^{\infty}\dfrac{1}{n^2}$ which is a

convergent p-series $(p = 2 > 1)$. When $x = -\frac{1}{3}$, $\sum_{n=0}^{\infty}\dfrac{3^n x^n}{(n+1)^2} = \sum_{n=0}^{\infty}\dfrac{(-1)^n}{(n+1)^2}$ which converges by the

Alternating Series Test, so $I = \left[-\frac{1}{3}, \frac{1}{3}\right]$.

13. If $a_n = \dfrac{x^n}{\ln n}$, then $\lim\limits_{n\to\infty}\left|\dfrac{a_{n+1}}{a_n}\right| = \lim\limits_{n\to\infty}\left|\dfrac{x^{n+1}}{\ln(n+1)}\cdot\dfrac{\ln n}{x^n}\right| = |x|\lim\limits_{n\to\infty}\dfrac{\ln n}{\ln(n+1)} = |x|$ (using l'Hospital's Rule), so

$R = 1$. When $x = 1$, $\sum\limits_{n=2}^{\infty}\dfrac{x^n}{\ln n} = \sum\limits_{n=2}^{\infty}\dfrac{1}{\ln n}$ which diverges because $\dfrac{1}{\ln n} > \dfrac{1}{n}$ and $\sum\limits_{n=2}^{\infty}\dfrac{1}{n}$ is the divergent

harmonic series. When $x = -1$, $\sum\limits_{n=2}^{\infty}\dfrac{x^n}{\ln n} = \sum\limits_{n=2}^{\infty}\dfrac{(-1)^n}{\ln n}$ which converges by the Alternating Series Test.

So $I = [-1, 1)$.

15. If $a_n = \dfrac{n}{4^n}(2x-1)^n$, then $\left|\dfrac{a_{n+1}}{a_n}\right| = \left|\dfrac{(n+1)(2x-1)^{n+1}}{4^{n+1}}\cdot\dfrac{4^n}{n(2x-1)^n}\right| = \left|\dfrac{2x-1}{4}\left(1+\dfrac{1}{n}\right)\right| \to \frac{1}{2}|x-\frac{1}{2}|$ as

$n \to \infty$. For convergence, $\frac{1}{2}|x-\frac{1}{2}| < 1 \Rightarrow |x-\frac{1}{2}| < 2 \Rightarrow R = 2$ and $-2 < x - \frac{1}{2} < 2 \Rightarrow -\frac{3}{2} < x < \frac{5}{2}$.

If $x = -\frac{3}{2}$, the series becomes $\sum\limits_{n=0}^{\infty}\dfrac{n}{4^n}(-4)^n = \sum\limits_{n=0}^{\infty}(-1)^n n$ which is divergent by the Test for Divergence.

If $x = \frac{5}{2}$, the series is $\sum\limits_{n=0}^{\infty}\dfrac{n}{4^n}4^n = \sum\limits_{n=0}^{\infty}n$, also divergent by the Test for Divergence. So $I = \left(-\frac{3}{2}, \frac{5}{2}\right)$.

17. If $a_n = \dfrac{(-1)^n(x-1)^n}{\sqrt{n}}$, then $\lim\limits_{n\to\infty}\left|\dfrac{a_{n+1}}{a_n}\right| = \lim\limits_{n\to\infty}\left|\dfrac{(x-1)^{n+1}}{\sqrt{n+1}}\cdot\dfrac{\sqrt{n}}{(x-1)^n}\right| = |x-1|\lim\limits_{n\to\infty}\sqrt{\dfrac{n}{n+1}} = |x-1| < 1$

for convergence, or $0 < x < 2$, and $R = 1$. When $x = 0$, $\sum\limits_{n=1}^{\infty}\dfrac{(-1)^n(x-1)^n}{\sqrt{n}} = \sum\limits_{n=1}^{\infty}\dfrac{1}{\sqrt{n}}$ which is a divergent

p-series ($p = \frac{1}{2} < 1$). When $x = 2$, the series is $\sum\limits_{n=1}^{\infty}\dfrac{(-1)^n}{\sqrt{n}}$ which converges by the Alternating Series Test. So

$I = (0, 2]$.

19. If $a_n = \dfrac{(x-2)^n}{n^n}$, then $\lim\limits_{n\to\infty}\sqrt[n]{|a_n|} = \lim\limits_{n\to\infty}\dfrac{x-2}{n} = 0$, so the series converges for all x (by the Root Test).

$R = \infty$ and $I = (-\infty, \infty)$.

21. If $a_n = \dfrac{2^n(x-3)^n}{n+3}$, then $\lim\limits_{n\to\infty}\left|\dfrac{a_{n+1}}{a_n}\right| = \lim\limits_{n\to\infty}\left|\dfrac{2^{n+1}(x-3)^{n+1}}{n+4}\cdot\dfrac{n+3}{2^n(x-3)^n}\right| = 2|x-3|\lim\limits_{n\to\infty}\dfrac{n+3}{n+4}$

$= 2|x-3| < 1$ for convergence, or $|x-3| < \frac{1}{2} \Leftrightarrow \frac{5}{2} < x < \frac{7}{2}$, and $R = \frac{1}{2}$. When $x = \frac{5}{2}$,

$\sum\limits_{n=0}^{\infty}\dfrac{2^n(x-3)^n}{n+3} = \sum\limits_{n=0}^{\infty}\dfrac{(-1)^n}{n+3}$ which converges by the Alternating Series Test. When $x = \frac{7}{2}$,

$\sum\limits_{n=0}^{\infty}\dfrac{2^n(x-3)^n}{n+3} = \sum\limits_{n=0}^{\infty}\dfrac{1}{n+3} = \sum\limits_{n=3}^{\infty}\dfrac{1}{n}$, the harmonic series, which diverges. So $I = \left[\frac{5}{2}, \frac{7}{2}\right)$.

23. If $a_n = \left(\dfrac{n}{2}\right)^n(x+6)^n$, then $\lim\limits_{n\to\infty}\sqrt[n]{|a_n|} = \lim\limits_{n\to\infty}\dfrac{n(x+6)}{2} = \infty$ unless $x = -6$, in which case the limit is 0. So

by the Root Test, the series converges only for $x = -6$. $R = 0$ and $I = \{-6\}$.

25. If $a_n = \dfrac{(2x-1)^n}{n^3}$, then $\lim\limits_{n\to\infty}\left|\dfrac{a_{n+1}}{a_n}\right| = |2x-1|\lim\limits_{n\to\infty}\left(\dfrac{n}{n+1}\right)^3 = |2x-1| < 1$ for convergence, so

$|x-\frac{1}{2}| < \frac{1}{2} \Leftrightarrow 0 < x < 1$, and $R = \frac{1}{2}$. The series $\sum\limits_{n=1}^{\infty}\dfrac{(2x-1)^n}{n^3}$ converges both for $x = 0$ and $x = 1$ (in

the first case because of the Alternating Series Test and in the second case because we get a p-series with

$p = 3 > 1$). So $I = [0, 1]$.

27. If $a_n = \dfrac{x^n}{(\ln n)^n}$ then $\lim\limits_{n\to\infty} \sqrt[n]{|a_n|} = \lim\limits_{n\to\infty} \dfrac{|x|}{\ln n} = 0 < 1$ for all x, so $R = \infty$ and $I = (-\infty, \infty)$ by the Root Test.

29. If $a_n = \dfrac{(n!)^k}{(kn)!}x^n$, then $\lim\limits_{n\to\infty}\left|\dfrac{a_{n+1}}{a_n}\right| = \lim\limits_{n\to\infty}\dfrac{[(n+1)!]^k(kn)!}{(n!)^k[k(n+1)]!}|x|$

$= \lim\limits_{n\to\infty}\dfrac{(n+1)^k}{(kn+k)(kn+k-1)\cdots(kn+2)(kn+1)}|x| = \lim\limits_{n\to\infty}\left[\dfrac{(n+1)}{(kn+1)}\dfrac{(n+1)}{(kn+2)}\cdots\dfrac{(n+1)}{(kn+k)}\right]|x|$

$= \lim\limits_{n\to\infty}\left[\dfrac{n+1}{kn+1}\right]\lim\limits_{n\to\infty}\left[\dfrac{n+1}{kn+2}\right]\cdots\lim\limits_{n\to\infty}\left[\dfrac{n+1}{kn+k}\right]|x| = \left(\dfrac{1}{k}\right)^k|x| < 1 \quad\Leftrightarrow\quad |x| < k^k$ for convergence, and

the radius of convergence is $R = k^k$.

31. (a) If $a_n = \dfrac{(-1)^n x^{2n+1}}{n!(n+1)!\,2^{2n+1}}$, then $\lim\limits_{n\to\infty}\left|\dfrac{a_{n+1}}{a_n}\right| = \left(\dfrac{x}{2}\right)^2\lim\limits_{n\to\infty}\dfrac{1}{(n+1)(n+2)} = 0$ for all x. So $J_1(x)$

converges for all x; the domain is $(-\infty, \infty)$.

(b), (c) The initial terms of $J_1(x)$ up to $n = 5$ are

$$a_0 = \dfrac{x}{2}, a_1 = -\dfrac{x^3}{16}, a_2 = \dfrac{x^5}{384}, a_3 = -\dfrac{x^7}{18,432}, a_4 = \dfrac{x^9}{1,474,560}, a_5 = -\dfrac{x^{11}}{176,947,200}.$$

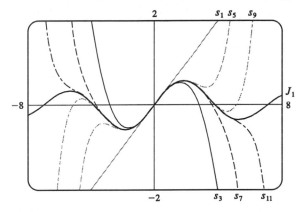

The partial sums seem to approximate $J_1(x)$ well near the origin, but as $|x|$ increases, we need to take a large number of terms to get a good approximation.

33. $s_{2n-1} = 1 + 2x + x^2 + 2x^3 + \cdots + x^{2n-2} + 2x^{2n-1} = (1 + 2x)\left(1 + x^2 + x^4 + \cdots + x^{2n-2}\right)$

$= (1 + 2x)\dfrac{1 - x^{2n}}{1 - x^2} \to \dfrac{1 + 2x}{1 - x^2}$ as $n \to \infty$, when $|x| < 1$.

Also $s_{2n} = s_{2n-1} + x^{2n} \to \dfrac{1 + 2x}{1 - x^2}$ since $x^{2n} \to 0$ for $|x| < 1$. Therefore $s_n \to \dfrac{1 + 2x}{1 - x^2}$ by Exercise 10.1.70(a).

Thus the interval of convergence is $(-1, 1)$ and $f(x) = \dfrac{1 + 2x}{1 - x^2}$.

35. We use the Root Test on the series $\sum c_n x^n$. $\lim\limits_{n\to\infty}\sqrt[n]{|c_n x^n|} = |x|\lim\limits_{n\to\infty}\sqrt[n]{|c_n|} = c|x| < 1$ for convergence, or

$|x| < 1/c$, so $R = 1/c$.

37. $\sum(c_n + d_n)x^n = \sum c_n x^n + \sum d_n x^n$ on the interval $(-2, 2)$, since both series converge there. So the radius of convergence must be at least 2. Now since $\sum c_n x^n$ has $R = 2$, it must diverge either at $x = -2$ or at $x = 2$. So by Exercise 10.2.63, $\sum(c_n + d_n)x^n$ diverges either at $x = -2$ or at $x = 2$, and so its radius of convergence is 2.

EXERCISES 10.9

Note: "R" stands for "radius of convergence" and "I" stands for "interval of convergence" in this section.

1. $f(x) = \dfrac{1}{1+x} = \dfrac{1}{1-(-x)} = \displaystyle\sum_{n=0}^{\infty}(-1)^n x^n$ with $|-x| < 1 \Leftrightarrow |x| < 1$ so $R = 1$ and $I = (-1,1)$.

3. $f(x) = \dfrac{1}{1+4x^2} = \displaystyle\sum_{n=0}^{\infty}(-1)^n\left(4x^2\right)^n$ $\left(\begin{array}{c}\text{substituting } 4x^2 \text{ for } x \text{ in the} \\ \text{series from Exercise 1}\end{array}\right)$ $= \displaystyle\sum_{n=0}^{\infty}(-1)^n 4^n x^{2n}$, with $\left|4x^2\right| < 1$ so

$x^2 < \frac{1}{4} \Leftrightarrow |x| < \frac{1}{2}$, and so $R = \frac{1}{2}$ and $I = \left(-\frac{1}{2}, \frac{1}{2}\right)$.

5. $f(x) = \dfrac{1}{4+x^2} = \dfrac{1}{4}\left(\dfrac{1}{1+x^2/4}\right) = \dfrac{1}{4}\displaystyle\sum_{n=0}^{\infty}(-1)^n\left(\dfrac{x^2}{4}\right)^n$ (using Exercise 1)

$= \displaystyle\sum_{n=0}^{\infty}\dfrac{(-1)^n x^{2n}}{4^{n+1}}$, with $\left|\dfrac{x^2}{4}\right| < 1 \Leftrightarrow x^2 < 4 \Leftrightarrow |x| < 2$, so $R = 2$ and $I = (-2, 2)$.

7. $\dfrac{x}{x-3} = 1 + \dfrac{3}{x-3} = 1 - \dfrac{1}{1-x/3} = 1 - \displaystyle\sum_{n=0}^{\infty}\left(\dfrac{x}{3}\right)^n = -\displaystyle\sum_{n=1}^{\infty}\left(\dfrac{x}{3}\right)^n$. For convergence, $\dfrac{|x|}{3} < 1 \Leftrightarrow$

$|x| < 3$, so $R = 3$ and $I = (-3, 3)$.

Another Method: $\dfrac{x}{x-3} = -\dfrac{x}{3(1-x/3)} = -\dfrac{x}{3}\displaystyle\sum_{n=0}^{\infty}\left(\dfrac{x}{3}\right)^n = -\displaystyle\sum_{n=0}^{\infty}\dfrac{x^{n+1}}{3^{n+1}} = -\displaystyle\sum_{n=1}^{\infty}\dfrac{x^n}{3^n}$

9. $\dfrac{3x-2}{2x^2-3x+1} = \dfrac{3x-2}{(2x-1)(x-1)} = \dfrac{A}{2x-1} + \dfrac{B}{x-1} \Leftrightarrow A + 2B = 3$ and $-A - B = -2 \Leftrightarrow$

$A = B = 1$, so $f(x) = \dfrac{3x-2}{2x^2-3x+1} = \dfrac{1}{2x-1} + \dfrac{1}{x-1} = -\displaystyle\sum_{n=0}^{\infty}(2x)^n - \displaystyle\sum_{n=0}^{\infty}x^n = -\displaystyle\sum_{n=0}^{\infty}(2^n + 1)x^n$, with

$R = \frac{1}{2}$. At $x = \pm\frac{1}{2}$, the series diverges by the Test for Divergence, so $I = \left(-\frac{1}{2}, \frac{1}{2}\right)$.

11. $f(x) = \dfrac{1}{(1+x)^2} = -\dfrac{d}{dx}\left(\dfrac{1}{1+x}\right) = -\dfrac{d}{dx}\left(\displaystyle\sum_{n=0}^{\infty}(-1)^n x^n\right)$ (from Exercise 1)

$= \displaystyle\sum_{n=1}^{\infty}(-1)^{n+1}n x^{n-1} = \displaystyle\sum_{n=0}^{\infty}(-1)^n(n+1)x^n$ with $R = 1$.

13. $f(x) = \dfrac{1}{(1+x)^3} = -\dfrac{1}{2}\dfrac{d}{dx}\left[\dfrac{1}{(1+x)^2}\right] = -\dfrac{1}{2}\dfrac{d}{dx}\left(\displaystyle\sum_{n=0}^{\infty}(-1)^n(n+1)x^n\right)$ (from Exercise 11)

$= -\dfrac{1}{2}\displaystyle\sum_{n=1}^{\infty}(-1)^n(n+1)n x^{n-1} = \dfrac{1}{2}\displaystyle\sum_{n=0}^{\infty}(-1)^n(n+2)(n+1)x^n$ with $R = 1$.

15. $f(x) = \ln(5-x) = -\displaystyle\int\dfrac{dx}{5-x} = -\dfrac{1}{5}\displaystyle\int\dfrac{dx}{1-x/5} = -\dfrac{1}{5}\displaystyle\int\left[\displaystyle\sum_{n=0}^{\infty}\left(\dfrac{x}{5}\right)^n\right]dx$

$= C - \dfrac{1}{5}\displaystyle\sum_{n=0}^{\infty}\dfrac{x^{n+1}}{5^n(n+1)} = C - \displaystyle\sum_{n=1}^{\infty}\dfrac{x^n}{n5^n}$

Putting $x = 0$, we get $C = \ln 5$. The series converges for $|x/5| < 1 \Leftrightarrow |x| < 5$. So $R = 5$.

24

17. $f(x) = \ln(1+x) - \ln(1-x) = \displaystyle\int \frac{dx}{1+x} + \int \frac{dx}{1-x} = \int \left[\sum_{n=0}^{\infty} (-1)^n x^n + \sum_{n=0}^{\infty} x^n \right] dx$

$= \displaystyle\int \sum_{n=0}^{\infty} 2x^{2n}\, dx = \sum_{n=0}^{\infty} \frac{2x^{2n+1}}{2n+1} + C.$

But $f(0) = \ln 1 - \ln 1 = 0$, so $C = 0$ and we have $f(x) = \displaystyle\sum_{n=0}^{\infty} \frac{2x^{2n+1}}{2n+1}$ with $R = 1$.

19. $f(x) = \ln(3+x) = \displaystyle\int \frac{dx}{3+x} = \frac{1}{3}\int \frac{dx}{1+x/3} = \frac{1}{3}\int \sum_{n=0}^{\infty} (-1)^n \left(\frac{x}{3}\right)^n dx$ (from Exercise 1)

$= C + \dfrac{1}{3}\displaystyle\sum_{n=0}^{\infty} \frac{(-1/3)^n}{n+1} x^{n+1} = \ln 3 + \frac{1}{3}\sum_{n=1}^{\infty} \frac{(-1/3)^{n-1}}{n} x^n = \ln 3 + \sum_{n=1}^{\infty} \frac{(-1)^{n-1}}{n3^n} x^n$ $[C = f(0) = \ln 3]$

with $R = 3$. The terms of the series are $a_0 = \ln 3$, $a_1 = \dfrac{x}{3}$, $a_2 = -\dfrac{x^2}{18}$, $a_3 = \dfrac{x^3}{81}$, $a_4 = -\dfrac{x^4}{324}$, $a_5 = \dfrac{x^5}{1215}, \ldots$.

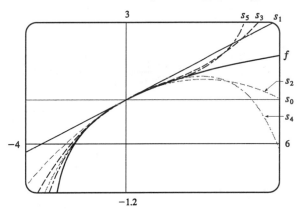

As n increases, $s_n(x)$ approximates f better on the interval of convergence, which is $(-3, 3)$.

21. $\displaystyle\int \frac{dx}{1+x^4} = \int \sum_{n=0}^{\infty} (-1)^n x^{4n}\, dx = C + \sum_{n=0}^{\infty} \frac{(-1)^n x^{4n+1}}{4n+1}$ with $R = 1$.

23. By Example 7, $\arctan x = \displaystyle\sum_{n=0}^{\infty} (-1)^n \frac{x^{2n+1}}{2n+1}$, so

$\displaystyle\int \frac{\arctan x}{x}\, dx = \int \sum_{n=0}^{\infty} (-1)^n \frac{x^{2n}}{2n+1}\, dx = C + \sum_{n=0}^{\infty} (-1)^n \frac{x^{2n+1}}{(2n+1)^2}$ with $R = 1$.

25. We use the representation $\displaystyle\int \frac{dx}{1+x^4} = C + \sum_{n=0}^{\infty} \frac{(-1)^n x^{4n+1}}{4n+1}$ from Exercise 21, with $C = 0$. So

$\displaystyle\int_0^{0.2} \frac{dx}{1+x^4} = \left[x - \frac{x^5}{5} + \frac{x^9}{9} - \frac{x^{13}}{13} + \cdots \right]_0^{0.2} = 0.2 - \frac{0.2^5}{5} + \frac{0.2^9}{9} - \frac{0.2^{13}}{13} + \cdots$. Since the series is

alternating, the error in the nth-order approximation is less than the first neglected term by Theorem 10.5.1. If

we use only the first two terms of the series, then the error is at most $0.2^9/9 \approx 5.7 \times 10^{-8}$. So, to six decimal

places, $\displaystyle\int_0^{0.2} \frac{dx}{1+x^4} \approx 0.2 - \frac{0.2^5}{5} \approx 0.199936$.

27. We substitute x^4 for x in Example 7, and find that $\displaystyle\int x^2 \tan^{-1}(x^4)\,dx = \int x^2 \sum_{n=0}^{\infty}(-1)^n \frac{\left(x^4\right)^{2n+1}}{2n+1}\,dx$

$\displaystyle = \int \sum_{n=0}^{\infty}(-1)^n \frac{x^{8n+6}}{2n+1}\,dx = C + \sum_{n=0}^{\infty}(-1)^n \frac{x^{8n+7}}{(2n+1)(8n+7)}.$ So

$\displaystyle\int_0^{1/3} x^2 \tan^{-1}(x^4)\,dx = \left[\frac{x^7}{7} - \frac{x^{15}}{45} + \cdots\right]_0^{1/3} = \frac{1}{7\cdot 3^7} - \frac{1}{45\cdot 3^{15}} + \cdots.$ The series is alternating, so if we use

only one term, the error is at most $1/(45 \cdot 3^{15}) \approx 1.5 \times 10^{-9}$. So $\int_0^{1/3} x^2 \tan^{-1}(x^4)\,dx \approx 1/(7 \cdot 3^7) \approx 0.000065$

to six decimal places.

29. Using the result of Example 6 with $x = -0.1$, we have

$\displaystyle\ln 1.1 = \ln[1 - (-0.1)] = 0.1 - \frac{0.01}{2} + \frac{0.001}{3} - \frac{0.0001}{4} + \frac{0.00001}{5} - \cdots.$ If we use only the first four terms,

the error is at most $\displaystyle\frac{0.00001}{5} = 0.000002.$ So $\ln 1.1 \approx 0.1 - \frac{0.01}{2} + \frac{0.001}{3} - \frac{0.0001}{4} \approx 0.09531.$

31. **(a)** $\displaystyle J_0(x) = \sum_{n=0}^{\infty} \frac{(-1)^n x^{2n}}{2^{2n}(n!)^2},\ J_0'(x) = \sum_{n=1}^{\infty} \frac{(-1)^n 2n x^{2n-1}}{2^{2n}(n!)^2},$ and $\displaystyle J_0''(x) = \sum_{n=1}^{\infty} \frac{(-1)^n 2n(2n-1) x^{2n-2}}{2^{2n}(n!)^2},$ so

$\displaystyle x^2 J_0''(x) + x J_0'(x) + x^2 J_0(x) = \sum_{n=1}^{\infty} \frac{(-1)^n 2n(2n-1) x^{2n}}{2^{2n}(n!)^2} + \sum_{n=1}^{\infty} \frac{(-1)^n 2n x^{2n}}{2^{2n}(n!)^2} + \sum_{n=0}^{\infty} \frac{(-1)^n x^{2n+2}}{2^{2n}(n!)^2}$

$\displaystyle = \sum_{n=1}^{\infty} \frac{(-1)^n 2n(2n-1) x^{2n}}{2^{2n}(n!)^2} + \sum_{n=1}^{\infty} \frac{(-1)^n 2n x^{2n}}{2^{2n}(n!)^2} + \sum_{n=1}^{\infty} \frac{(-1)^{n-1} x^{2n}}{2^{2n-2}[(n-1)!]^2}$

$\displaystyle = \sum_{n=1}^{\infty}(-1)^n \left[\frac{2n(2n-1) + 2n - 2^2 n^2}{2^{2n}(n!)^2}\right] x^{2n} = \sum_{n=1}^{\infty}(-1)^n \left[\frac{4n^2 - 2n + 2n - 4n^2}{2^{2n}(n!)^2}\right] x^{2n} = 0.$

(b) $\displaystyle\int_0^1 J_0(x)\,dx = \int_0^1 \left[\sum_{n=0}^{\infty} \frac{(-1)^n x^{2n}}{2^{2n}(n!)^2}\right] dx = \int_0^1 dx - \int_0^1 \frac{x^2}{4}\,dx + \int_0^1 \frac{x^4}{64}\,dx - \int_0^1 \frac{x^6}{2304}\,dx + \cdots$

$\displaystyle = \left[x - \frac{x^3}{3\cdot 4} + \frac{x^5}{5\cdot 64} - \frac{x^7}{7\cdot 2304} + \cdots\right]_0^1 = 1 - \frac{1}{12} + \frac{1}{320} - \frac{1}{16,128} + \cdots.$

Since $\frac{1}{16,128} \approx 0.000062$, it follows from Theorem 10.5.1 that, correct to three decimal places,

$\int_0^1 J_0(x)\,dx \approx 1 - \frac{1}{12} + \frac{1}{320} \approx 0.920.$

33. **(a)** We calculate $\displaystyle f'(x) = \sum_{n=1}^{\infty} \frac{n x^{n-1}}{n!} = \sum_{n=1}^{\infty} \frac{x^{n-1}}{(n-1)!} = \sum_{n=0}^{\infty} \frac{x^n}{n!} = f(x).$

(b) By Theorem 6.5.2 (or Theorem 3.5.2 in the Early Transcendentals version), the only solutions to the

differential equation $\displaystyle\frac{df(x)}{dx} = f(x)$ are $f(x) = Ke^x$, but $f(0) = 1$, so $K = 1$ and $f(x) = e^x$.

Or: We could solve the equation $\displaystyle\frac{df(x)}{dx} = f(x)$ as a separable differential equation.

35. If $a_n = \dfrac{x^n}{n^2}$, then $\displaystyle\lim_{n\to\infty}\left|\dfrac{a_{n+1}}{a_n}\right| = |x|\lim_{n\to\infty}\left(\dfrac{n}{n+1}\right)^2 = |x| < 1$ for convergence, so $R = 1$. When $x = \pm 1$,

$\displaystyle\sum_{n=1}^{\infty}\left|\dfrac{x^n}{n^2}\right| = \sum_{n=1}^{\infty}\dfrac{1}{n^2}$ which is a convergent p-series ($p = 2 > 1$), so the interval of convergence for f is $[-1, 1]$. By

Theorem 10.9.2, the radii of convergence of f' and f'' are both 1, so we need only check the endpoints.

$f'(x) = \displaystyle\sum_{n=1}^{\infty}\dfrac{nx^{n-1}}{n^2} = \sum_{n=0}^{\infty}\dfrac{x^n}{n+1}$, and this series diverges for $x = 1$ (harmonic series) and converges for $x = -1$

(Alternating Series Test), so the interval of convergence is $[-1, 1)$. $f''(x) = \displaystyle\sum_{n=1}^{\infty}\dfrac{nx^{n-1}}{n+1}$ diverges at both 1 and

-1 (Test for Divergence) since $\displaystyle\lim_{n\to\infty}\dfrac{n}{n+1} = 1 \neq 0$, so its interval of convergence is $(-1, 1)$.

EXERCISES 10.10

1.

n	$f^{(n)}(x)$	$f^{(n)}(0)$
0	$\cos x$	1
1	$-\sin x$	0
2	$-\cos x$	-1
3	$\sin x$	0
4	$\cos x$	1
...	...	...

$\cos x = f(0) + f'(0)x + \dfrac{f''(0)}{2!}x^2 + \dfrac{f^{(3)}(0)}{3!}x^3 + \dfrac{f^{(4)}(0)}{4!}x^4 + \cdots$

$= 1 - \dfrac{x^2}{2!} + \dfrac{x^4}{4!} - \cdots = \displaystyle\sum_{n=0}^{\infty}\dfrac{(-1)^n x^{2n}}{(2n)!}$

If $a_n = \dfrac{(-1)^n x^{2n}}{(2n)!}$, then

$\displaystyle\lim_{n\to\infty}\left|\dfrac{a_{n+1}}{a_n}\right| = x^2\lim_{n\to\infty}\dfrac{1}{(2n+2)(2n+1)} = 0 < 1$ for all x.

So $R = \infty$.

3.

n	$f^{(n)}(x)$	$f^{(n)}(0)$
0	$(1+x)^{-2}$	1
1	$-2(1+x)^{-3}$	-2
2	$2\cdot 3(1+x)^{-4}$	$2\cdot 3$
3	$-2\cdot 3\cdot 4(1+x)^{-5}$	$-2\cdot 3\cdot 4$
4	$2\cdot 3\cdot 4\cdot 5(1+x)^{-6}$	$2\cdot 3\cdot 4\cdot 5$
...	...	...

So $f^{(n)}(0) = (-1)^n(n+1)!$ and

$\dfrac{1}{(1+x)^2} = \displaystyle\sum_{n=0}^{\infty}\dfrac{(-1)^n(n+1)!}{n!}x^n$

$= \displaystyle\sum_{n=0}^{\infty}(-1)^n(n+1)x^n$.

If $a_n = (-1)^n(n+1)x^n$, then

$\displaystyle\lim_{n\to\infty}\left|\dfrac{a_{n+1}}{a_n}\right| = |x|$, so $R = 1$.

5.

n	$f^{(n)}(x)$	$f^{(n)}(0)$
0	$\sinh x$	0
1	$\cosh x$	1
2	$\sinh x$	0
3	$\cosh x$	1
4	$\sinh x$	0
...	...	...

So $f^{(n)}(0) = \begin{cases} 0 & \text{if } n \text{ is even} \\ 1 & \text{if } n \text{ is odd} \end{cases}$

and $\sinh x = \displaystyle\sum_{n=0}^{\infty}\dfrac{x^{2n+1}}{(2n+1)!}$. If $a_n = \dfrac{x^{2n+1}}{(2n+1)!}$ then

$\displaystyle\lim_{n\to\infty}\left|\dfrac{a_{n+1}}{a_n}\right| = x^2\lim_{n\to\infty}\dfrac{1}{(2n+3)(2n+2)} = 0 < 1$

for all x, so $R = \infty$.

7.

n	$f^{(n)}(x)$	$f^{(n)}(\pi/4)$
0	$\sin x$	$\sqrt{2}/2$
1	$\cos x$	$\sqrt{2}/2$
2	$-\sin x$	$-\sqrt{2}/2$
3	$-\cos x$	$-\sqrt{2}/2$
4	$\sin x$	$\sqrt{2}/2$
...	...	...

$$\sin x = f\left(\tfrac{\pi}{4}\right) + f'\left(\tfrac{\pi}{4}\right)\left(x - \tfrac{\pi}{4}\right) + \frac{f''\left(\tfrac{\pi}{4}\right)}{2!}\left(x - \tfrac{\pi}{4}\right)^2$$
$$+ \frac{f^{(3)}\left(\tfrac{\pi}{4}\right)}{3!}\left(x - \tfrac{\pi}{4}\right)^3 + \frac{f^{(4)}\left(\tfrac{\pi}{4}\right)}{4!}\left(x - \tfrac{\pi}{4}\right)^4 + \cdots$$
$$= \frac{\sqrt{2}}{2}\left[1 + \left(x - \tfrac{\pi}{4}\right) - \tfrac{1}{2!}\left(x - \tfrac{\pi}{4}\right)^2 - \tfrac{1}{3!}\left(x - \tfrac{\pi}{4}\right)^3 + \tfrac{1}{4!}\left(x - \tfrac{\pi}{4}\right)^4 + \cdots\right]$$
$$= \frac{\sqrt{2}}{2}\sum_{n=0}^{\infty}\frac{(-1)^{n(n-1)/2}\left(x - \tfrac{\pi}{4}\right)^n}{n!}$$

If $a_n = \dfrac{(-1)^{n(n-1)/2}\left(x - \tfrac{\pi}{4}\right)^n}{n!}$, then $\lim\limits_{n\to\infty}\left|\dfrac{a_{n+1}}{a_n}\right| = \lim\limits_{n\to\infty}\dfrac{\left|x - \tfrac{\pi}{4}\right|}{n+1} = 0 < 1$

for all x, so $R = \infty$.

9.

n	$f^{(n)}(x)$	$f^{(n)}(1)$
0	x^{-1}	1
1	$-x^{-2}$	-1
2	$2x^{-3}$	2
3	$-3 \cdot 2x^{-4}$	$-3 \cdot 2$
4	$4 \cdot 3 \cdot 2x^{-5}$	$4 \cdot 3 \cdot 2$
...	...	...

So $f^{(n)}(1) = (-1)^n n!$, and
$$\frac{1}{x} = \sum_{n=0}^{\infty}\frac{(-1)^n n!}{n!}(x - 1)^n = \sum_{n=0}^{\infty}(-1)^n(x - 1)^n.$$
If $a_n = (-1)^n(x - 1)^n$ then
$$\lim_{n\to\infty}\left|\frac{a_{n+1}}{a_n}\right| = |x - 1| < 1 \text{ for convergence,}$$
so $0 < x < 2$ and $R = 1$.

11. Clearly $f^{(n)}(x) = e^x$, so $f^{(n)}(3) = e^3$ and $e^x = \sum\limits_{n=0}^{\infty}\dfrac{e^3}{n!}(x - 3)^n$. If $a_n = \dfrac{e^3}{n!}(x - 3)^n$ then

$$\lim_{n\to\infty}\left|\frac{a_{n+1}}{a_n}\right| = \lim_{n\to\infty}\frac{|x - 3|}{n+1} = 0 \text{ for all } x, \text{ so } R = \infty.$$

13. If $f(x) = \cos x$, then by Formula 9, $R_n(x) = \dfrac{f^{(n+1)}(z)}{(n+1)!}x^{n+1}$, where $0 < |z| < |x|$. But $f^{(n+1)}(z) = \pm\sin z$ or

$\pm\cos z$. In each case, $\left|f^{(n+1)}(z)\right| \leq 1$, so $|R_n(x)| \leq \dfrac{1}{(n+1)!}x^{n+1} \to 0$ as $n \to \infty$ by Equation 11. So

$\lim\limits_{n\to\infty} R_n(x) = 0$ and, by Theorem 8, the series in Exercise 1 represents $\cos x$ for all x.

15. If $f(x) = \sinh x$, then $R_n(x) = \dfrac{f^{(n+1)}(z)}{(n+1)!}x^{n+1}$, where $0 < |z| < |x|$. But for all n,

$\left|f^{(n+1)}(z)\right| \leq \cosh z \leq \cosh x$ (since all derivatives are either $\sinh$ or $\cosh$, $|\sinh z| < |\cosh z|$ for all z, and

$|z| < |x| \Rightarrow \cosh z < \cosh x$), so $|R_n(z)| \leq \dfrac{\cosh x}{(n+1)!}x^{n+1} \to 0$ as $n \to \infty$ (by Equation 11). So by

Theorem 9, the series represents $\sinh x$ for all x.

17. $e^{3x} = \sum\limits_{n=0}^{\infty}\dfrac{(3x)^n}{n!} = \sum\limits_{n=0}^{\infty}\dfrac{3^n x^n}{n!}$, with $R = \infty$

19. $x^2\cos x = x^2\sum\limits_{n=0}^{\infty}\dfrac{(-1)^n x^{2n}}{(2n)!} = \sum\limits_{n=0}^{\infty}\dfrac{(-1)^n x^{2n+2}}{(2n)!}$, $R = \infty$

21. $x\sin\left(\dfrac{x}{2}\right) = x\sum\limits_{n=0}^{\infty}\dfrac{(-1)^n (x/2)^{2n+1}}{(2n+1)!} = \sum\limits_{n=0}^{\infty}\dfrac{(-1)^n x^{2n+2}}{(2n+1)!2^{2n+1}}$, with $R = \infty$.

23. $\sin^2 x = \tfrac{1}{2}[1 - \cos 2x] = \dfrac{1}{2}\left[1 - \sum\limits_{n=0}^{\infty}\dfrac{(-1)^n(2x)^{2n}}{(2n)!}\right] = \dfrac{1}{2}\left[1 - 1 - \sum\limits_{n=1}^{\infty}\dfrac{(-1)^n(2x)^{2n}}{(2n)!}\right] = \sum\limits_{n=1}^{\infty}\dfrac{(-1)^{n+1}2^{2n-1}x^{2n}}{(2n)!}$,

with $R = \infty$.

25. $\dfrac{\sin x}{x} = \dfrac{1}{x}\displaystyle\sum_{n=0}^{\infty}\dfrac{(-1)^n x^{2n+1}}{(2n+1)!} = \sum_{n=0}^{\infty}\dfrac{(-1)^n x^{2n}}{(2n+1)!}$ and this series also gives the required value at $x = 0$, so $R - \infty$.

27.

n	$f^{(n)}(x)$	$f^{(n)}(0)$
0	$(1+x)^{1/2}$	1
1	$\frac{1}{2}(1+x)^{-1/2}$	$\frac{1}{2}$
2	$-\frac{1}{4}(1+x)^{-3/2}$	$-\frac{1}{4}$
3	$\frac{3}{8}(1+x)^{-5/2}$	$\frac{3}{8}$
4	$-\frac{15}{16}(1+x)^{-7/2}$	$-\frac{15}{16}$
...	...	...

So $f^{(n)}(0) = \dfrac{(-1)^{n-1}1 \cdot 3 \cdot 5 \cdots (2n-3)}{2^n}$ for $n \geq 2$, and

$$\sqrt{1+x} = 1 + \frac{x}{2} + \sum_{n=2}^{\infty}\frac{(-1)^{n-1}1 \cdot 3 \cdot 5 \cdots (2n-3)}{2^n n!}x^n.$$

If $a_n = \dfrac{(-1)^{n+1}1 \cdot 3 \cdot 5 \cdots (2n-3)}{2^n n!}x^n$, then

$$\lim_{n \to \infty}\left|\frac{a_{n+1}}{a_n}\right| = \frac{|x|}{2}\lim_{n \to \infty}\frac{2n-1}{n+1} = |x| < 1 \text{ for}$$

convergence, so $R = 1$.

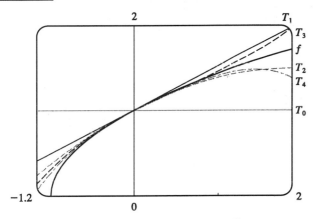

29. $f(x) = (1+x)^{-3} = -\dfrac{1}{2}\dfrac{d}{dx}\left[\dfrac{1}{(1+x)^2}\right] = -\dfrac{1}{2}\dfrac{d}{dx}\left[\displaystyle\sum_{n=0}^{\infty}(-1)^n(n+1)x^n\right]$ (from Exercise 5)

$$= -\frac{1}{2}\sum_{n=1}^{\infty}(-1)^n n(n+1)x^{n-1} = \sum_{n=0}^{\infty}\frac{(-1)^n(n+1)(n+2)x^n}{2},$$

with $R = 1$ since that is the R in Exercise 5.

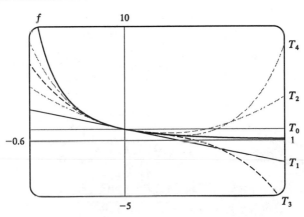

31. $\ln(1+x) = \displaystyle\int \frac{dx}{1+x} = \int \sum_{n=0}^{\infty}(-1)^n x^n \, dx = \sum_{n=1}^{\infty}\frac{(-1)^{n-1}x^n}{n}$ with $R=1$, so $\ln(1.1) = \displaystyle\sum_{n=1}^{\infty}\frac{(-1)^{n-1}(0.1)^n}{n}$.

This is an alternating series with $b_5 = \frac{(0.1)^5}{5} = 0.000002$, so to five decimals,

$\ln(1.1) \approx \displaystyle\sum_{n=1}^{4}\frac{(-1)^{n-1}(0.1)^n}{n} \approx 0.09531$.

33. $\displaystyle\int \sin(x^2)\,dx = \int \sum_{n=0}^{\infty}(-1)^n \frac{(x^2)^{2n+1}}{(2n+1)!}\,dx = \int \sum_{n=0}^{\infty}\frac{(-1)^n x^{4n+2}}{(2n+1)!}\,dx = C + \sum_{n=0}^{\infty}\frac{(-1)^n x^{4n+3}}{(4n+3)(2n+1)!}$

35. Using the series we obtained in Exercise 27, we get

$\sqrt{x^3+1} = 1 + \dfrac{x^3}{2} + \displaystyle\sum_{n=2}^{\infty}\frac{(-1)^{n-1}1\cdot 3\cdot 5\cdots (2n-3)}{2^n\, n!}x^{3n}$, so

$\displaystyle\int \sqrt{x^3+1}\,dx = \int\left(1 + \frac{x^3}{2} + \sum_{n=2}^{\infty}\frac{(-1)^{n-1}1\cdot 3\cdot 5\cdots (2n-3)}{2^n\, n!}x^{3n}\right)dx$

$= C + x + \dfrac{x^4}{8} + \displaystyle\sum_{n=2}^{\infty}\frac{(-1)^{n-1}1\cdot 3\cdot 5\cdots (2n-3)}{2^n n!\,(3n+1)}x^{3n+1}$.

37. Using our series from Exercise 33, we get $\displaystyle\int_0^1 \sin(x^2)\,dx = \sum_{n=0}^{\infty}\left[\frac{(-1)^n x^{4n+3}}{(4n+3)(2n+1)!}\right]_0^1 = \sum_{n=0}^{\infty}\frac{(-1)^n}{(4n+3)(2n+1)!}$

and $|c_3| = \frac{1}{75,600} < 0.000014$, so by Theorem 10.5.1, we have

$\displaystyle\sum_{n=0}^{2}\frac{(-1)^n}{(4n+3)(2n+1)!} \approx \frac{1}{3} - \frac{1}{42} + \frac{1}{1320} \approx 0.310$.

39. We first find a series representation for $f(x) = (1+x)^{-1/2}$, and then substitute.

n	$f^{(n)}(x)$	$f^{(n)}(0)$
0	$(1+x)^{-1/2}$	1
1	$-\frac{1}{2}(1+x)^{-3/2}$	$-\frac{1}{2}$
2	$\frac{3}{4}(1+x)^{-5/2}$	$3/4$
3	$-\frac{15}{8}(1+x)^{-7/2}$	$-15/8$
...	...	...

$\dfrac{1}{\sqrt{1+x}} = 1 - \dfrac{x}{2} + \dfrac{3}{4}\left(\dfrac{x^2}{2!}\right) - \dfrac{15}{8}\left(\dfrac{x^3}{3!}\right) + \cdots \quad\Rightarrow$

$\dfrac{1}{\sqrt{1+x^3}} = 1 - \frac{1}{2}x^3 + \frac{3}{8}x^6 - \frac{5}{16}x^9 + \cdots \quad\Rightarrow$

$\displaystyle\int_0^{0.1}\frac{dx}{\sqrt{1+x^3}} = \left[x - \tfrac{1}{8}x^4 + \tfrac{3}{56}x^7 - \tfrac{1}{32}x^{10} + \cdots\right]_0^{0.1}$

$\approx (0.1) - \tfrac{1}{8}(0.1)^4$ by Theorem 10.28

since $\frac{3}{56}(0.1)^7 \approx 0.0000000054 < 10^{-8}$.

Therefore $\displaystyle\int_0^{0.1}\frac{dx}{\sqrt{1+x^3}} \approx 0.09998750$.

41. As in Example 8(a), we have $e^{-x^2} = 1 - \dfrac{x^2}{1!} + \dfrac{x^4}{2!} + \dfrac{x^6}{3!} + \cdots$ and we know that $\cos x = 1 - \dfrac{x^2}{2!} + \dfrac{x^4}{4!} - \cdots$ from

Equation 16. Therefore $e^{-x^2}\cos x = \left(1 - x^2 + \frac{1}{2}x^4 - \cdots\right)\left(1 - \frac{1}{2}x^2 + \frac{1}{24}x^4 - \cdots\right)$

$= 1 - \frac{1}{2}x^2 + \frac{1}{24}x^4 - x^2 + \frac{1}{2}x^4 + \frac{1}{2}x^4 + \cdots = 1 - \frac{3}{2}x^2 + \frac{25}{24}x^4 + \cdots$

43. From Example 6 in Section 10.9, we have

$$\ln(1 - x) = -x - \tfrac{1}{2}x^2 - \tfrac{1}{3}x^3 - \cdots,$$

$|x| < 1$. Therefore

$$y = \frac{\ln(1 - x)}{e^x} = \frac{-x - \tfrac{1}{2}x^2 - \tfrac{1}{3}x^3 - \cdots}{1 + x + \tfrac{1}{2}x^2 + \tfrac{1}{6}x^3 + \cdots}.$$

So by the long division at right,

$$\frac{\ln(1 - x)}{e^x} = -x + \frac{x^2}{2} - \frac{x^3}{3} + \cdots, \quad |x| < 1.$$

$$
\begin{array}{l}
\phantom{1 + x + \tfrac{1}{2}x^2 + \tfrac{1}{6}x^3 - \cdots)\,}-x + \tfrac{1}{2}x^2 - \tfrac{1}{3}x^3 + \cdots \\
1 + x + \tfrac{1}{2}x^2 + \tfrac{1}{6}x^3 - \cdots \,\big|\, \overline{-x - \tfrac{1}{2}x^2 - \tfrac{1}{3}x^3 - \cdots} \\
\phantom{1 + x + \tfrac{1}{2}x^2 + \tfrac{1}{6}x^3)}\underline{-x -\phantom{\tfrac{1}{2}} x^2 - \tfrac{1}{2}x^3 - \cdots} \\
\phantom{1 + x + \tfrac{1}{2}x^2 + \tfrac{1}{6}x^3)}\tfrac{1}{2}x^2 + \tfrac{1}{6}x^3 - \cdots \\
\phantom{1 + x + \tfrac{1}{2}x^2 + \tfrac{1}{6}x^3)}\underline{\tfrac{1}{2}x^2 + \tfrac{1}{2}x^3 + \cdots} \\
\phantom{1 + x + \tfrac{1}{2}x^2 + \tfrac{1}{6}x^3)0}-\tfrac{1}{3}x^3 + \cdots \\
\phantom{1 + x + \tfrac{1}{2}x^2 + \tfrac{1}{6}x^3)0}\underline{-\tfrac{1}{3}x^3 + \cdots} \\
\phantom{1 + x + \tfrac{1}{2}x^2 + \tfrac{1}{6}x^3)00}\cdots
\end{array}
$$

45. $\displaystyle\sum_{n=0}^{\infty}(-1)^n\frac{x^{4n}}{n!} = \sum_{n=0}^{\infty}\frac{\left(-x^4\right)^n}{n!} = e^{-x^4}$ by (12).

47. $\displaystyle\sum_{n=0}^{\infty}\frac{(-1)^n\pi^{2n+1}}{4^{2n+1}(2n+1)!} = \sum_{n=0}^{\infty}\frac{(-1)^n(\pi/4)^{2n+1}}{(2n+1)!} = \sin\frac{\pi}{4} = \frac{1}{\sqrt{2}}$ by (15).

49. $\displaystyle\sum_{n=0}^{\infty}\frac{x^{n+1}}{(n+1)!} = \frac{x}{1!} + \frac{x^2}{2!} + \frac{x^3}{3!} + \cdots = \left(1 + \frac{x}{1!} + \frac{x^2}{2!} + \frac{x^3}{3!} + \cdots\right) - 1 = e^x - 1$ by (12).

51. By (12), $e^x = 1 + x + \dfrac{x^2}{2!} + \dfrac{x^3}{3!} + \dfrac{x^4}{4!} + \cdots$, but for $x > 0$, all of the terms after the first two on the RHS are

positive, so $e^x > 1 + x$ for $x > 0$.

53. $\displaystyle\lim_{x\to 0}\frac{\sin x - x + \tfrac{1}{6}x^3}{x^5} = \lim_{x\to 0}\frac{\left(x - \tfrac{1}{6}x^3 + \tfrac{1}{5!}x^5 - \tfrac{1}{7!}x^7 + \cdots\right) - x + \tfrac{1}{6}x^3}{x^5}$

$\displaystyle = \lim_{x\to 0}\frac{\tfrac{1}{5!}x^5 - \tfrac{1}{7!}x^7 + \cdots}{x^5} = \lim_{x\to 0}\left(\frac{1}{5!} - \frac{x^2}{7!} + \frac{x^4}{9!} - \cdots\right) = \frac{1}{5!} = \frac{1}{120}$,

since power series are continuous functions.

55. We must show that f equals its Taylor series expansion on I; that is, we must show that $\displaystyle\lim_{n\to\infty}|R_n(x)| = 0$. For

$$x \in I, \quad |R_n(x)| = \left|\frac{f^{(n+1)}(z)}{(n+1)!}(x-a)^{n+1}\right| \le \frac{M \cdot R^{n+1}}{(n+1)!} \to 0 \text{ as } n \to \infty \text{ by (11)}.$$

EXERCISES 10.11

1. $(1+x)^{1/2} = \sum_{n=0}^{\infty} \binom{1/2}{n} x^n = 1 + \left(\frac{1}{2}\right)x + \frac{\left(\frac{1}{2}\right)\left(-\frac{1}{2}\right)}{2!}x^2 + \frac{\left(\frac{1}{2}\right)\left(-\frac{1}{2}\right)\left(-\frac{3}{2}\right)}{3!}x^3 + \cdots$

$= 1 + \frac{x}{2} - \frac{x^2}{2^2 \cdot 2!} + \frac{1 \cdot 3 \cdot x^3}{2^3 \cdot 3!} - \frac{1 \cdot 3 \cdot 5 \cdot x^4}{2^4 \cdot 4!} + \cdots$

$= 1 + \frac{x}{2} + \sum_{n=2}^{\infty} \frac{(-1)^{n-1} 1 \cdot 3 \cdot 5 \cdot \cdots \cdot (2n-3) x^n}{2^n \cdot n!}, \quad R = 1$

3. $[1+(2x)]^{-4} = 1 + (-4)(2x) + \frac{(-4)(-5)}{2!}(2x)^2 + \frac{(-4)(-5)(-6)}{3!}(2x)^3 + \cdots$

$= 1 + \sum_{n=1}^{\infty} \frac{(-1)^n 2^n 4 \cdot 5 \cdot 6 \cdots (n+3)}{n!} x^n = \sum_{n=0}^{\infty} (-1)^n \frac{2^n (n+1)(n+2)(n+3)}{6} x^n,$

and for convergence $|2x| < 1 \iff |x| < \frac{1}{2}$ so $R = \frac{1}{2}$.

5. $[1+(-x)]^{-1/2} = \sum_{n=0}^{\infty} \binom{-1/2}{n}(-x)^n = 1 + \left(-\frac{1}{2}\right)(-x) + \frac{\left(-\frac{1}{2}\right)\left(-\frac{3}{2}\right)}{2!}(-x)^2 + \cdots$

$= 1 + \frac{x}{2} + \frac{1 \cdot 3}{2^2 2!}x^2 + \frac{1 \cdot 3 \cdot 5}{2^3 3!}x^3 + \frac{1 \cdot 3 \cdot 5 \cdot 7}{2^4 4!}x^4 + \cdots = 1 + \sum_{n=1}^{\infty} \frac{1 \cdot 3 \cdot 5 \cdot \cdots \cdot (2n-1)}{2^n n!}x^n,$

so $\frac{x}{\sqrt{1-x}} = x + \sum_{n=1}^{\infty} \frac{1 \cdot 3 \cdot 5 \cdot \cdots \cdot (2n-1)}{2^n n!}x^{n+1}$ with $R = 1$.

7. $(1-x^4)^{1/4} = 1 + \left(\frac{1}{4}\right)(-x^4) + \frac{\left(\frac{1}{4}\right)\left(-\frac{3}{4}\right)}{2!}(-x^4)^2 + \frac{\left(\frac{1}{4}\right)\left(-\frac{3}{4}\right)\left(-\frac{7}{4}\right)}{3!}(-x^4)^3 + \cdots$

$= 1 - \frac{x^4}{4} - \sum_{n=2}^{\infty} \frac{3 \cdot 7 \cdot 11 \cdot \cdots \cdot (4n-5)}{4^n \cdot n!} x^{4n}$ with $R = 1$.

9. $(1-x)^{-5} = 1 + (-5)(-x) + \frac{(-5)(-6)}{2!}(-x)^2 + \frac{(-5)(-6)(-7)}{3!}(-x)^3 + \cdots$

$= 1 + \sum_{n=1}^{\infty} \frac{5 \cdot 6 \cdot 7 \cdots (n+4)}{n!} x^n = \sum_{n=0}^{\infty} \frac{(n+4)!}{4! \cdot n!} x^n \quad \Rightarrow$

$\frac{x^5}{(1-x)^5} = \sum_{n=0}^{\infty} \frac{(n+4)!}{4! \cdot n!} x^{n+5} \left(\text{or} \sum_{n=0}^{\infty} \frac{(n+1)(n+2)(n+3)(n+4)}{24} x^{n+5} \right)$ with $R = 1$.

11. $(8+x)^{-1/3} = \frac{1}{2}\left(1 + \frac{x}{8}\right)^{-1/3}$

$= \frac{1}{2}\left[1 + \left(-\frac{1}{3}\right)\left(\frac{x}{8}\right) + \frac{\left(-\frac{1}{3}\right)\left(-\frac{4}{3}\right)}{2!}\left(\frac{x}{8}\right)^2 + \cdots\right]$

$= \frac{1}{2}\left[1 + \sum_{n=1}^{\infty} \frac{(-1)^n 1 \cdot 4 \cdot 7 \cdot \cdots \cdot (3n-2)}{3^n \cdot n! \, 8^n} x^n\right]$

and $|x/8| < 1 \iff |x| < 8$, so $R = 8$. The first three Taylor polynomials are $T_1(x) = \frac{1}{2} - \frac{1}{48}x$, $T_2(x) = \frac{1}{2} - \frac{1}{48}x + \frac{1}{576}x^2$,

and $T_3(x) = \frac{1}{2} - \frac{1}{48}x + \frac{1}{576}x^2 - \frac{4 \cdot 7}{27 \cdot 6 \cdot 512}x^3$

$= \frac{1}{2} - \frac{1}{48}x + \frac{1}{576}x^2 - \frac{7}{41,472}x^3.$

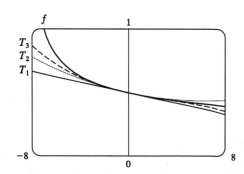

13. **(a)** $(1-x^2)^{-1/2} = 1 + \left(-\frac{1}{2}\right)(-x^2) + \frac{\left(-\frac{1}{2}\right)\left(-\frac{3}{2}\right)}{2!}(-x^2)^2 + \frac{\left(-\frac{1}{2}\right)\left(-\frac{3}{2}\right)\left(-\frac{5}{2}\right)}{3!}(-x^2)^3 + \cdots$

$\qquad = 1 + \sum_{n=1}^{\infty} \frac{1 \cdot 3 \cdot 5 \cdot \cdots \cdot (2n-1)}{2^n \cdot n!} x^{2n}$

(b) $\sin^{-1} x = \int \frac{1}{\sqrt{1-x^2}}\, dx = C + x + \sum_{n=1}^{\infty} \frac{1 \cdot 3 \cdot 5 \cdot \cdots \cdot (2n-1)}{(2n+1)2^n \cdot n!} x^{2n+1}$

$\qquad = x + \sum_{n=1}^{\infty} \frac{1 \cdot 3 \cdot 5 \cdots (2n-1)}{(2n+1)2^n \cdot n!} x^{2n+1}$ since $0 = \sin^{-1} 0 = C$.

15. **(a)** $(1+x)^{-1/2} = 1 + \left(-\frac{1}{2}\right)x + \frac{\left(-\frac{1}{2}\right)\left(-\frac{3}{2}\right)}{2!}x^2 + \frac{\left(-\frac{1}{2}\right)\left(-\frac{3}{2}\right)\left(-\frac{5}{2}\right)}{3!}x^3 + \cdots$

$\qquad = 1 + \sum_{n=1}^{\infty} \frac{(-1)^n 1 \cdot 3 \cdot 5 \cdot \cdots \cdot (2n-1)}{2^n \cdot n!} x^n$

(b) Take $x = 0.1$ in the above series. $\frac{1 \cdot 3 \cdot 5 \cdot 7}{2^4\, 4!}(0.1)^4 < 0.00003$, so

$\qquad \frac{1}{\sqrt{1.1}} \approx 1 - \frac{0.1}{2} + \frac{1 \cdot 3}{2^2 \cdot 2!}(0.1)^2 - \frac{1 \cdot 3 \cdot 5}{2^3 \cdot 3!}(0.1)^3 \approx 0.953$.

17. **(a)** $(1-x)^{-2} = 1 + (-2)(-x) + \frac{(-2)(-3)}{2!}(-x)^2 + \cdots = \sum_{n=0}^{\infty}(n+1)x^n$, so

$\qquad \frac{x}{(1-x)^2} = \sum_{n=0}^{\infty}(n+1)x^{n+1} = \sum_{n=1}^{\infty} nx^n$.

(b) With $x = \frac{1}{2}$ in part (a), we have $\sum_{n=1}^{\infty} \frac{n}{2^n} = \frac{1/2}{(1-1/2)^2} = 2$.

19. **(a)** $(1+x^2)^{1/2} = 1 + \left(\frac{1}{2}\right)x^2 + \frac{\left(\frac{1}{2}\right)\left(-\frac{1}{2}\right)}{2!}(x^2)^2 + \frac{\left(\frac{1}{2}\right)\left(-\frac{1}{2}\right)\left(-\frac{3}{2}\right)}{3!}(x^2)^3 + \cdots$

$\qquad = 1 + \frac{x^2}{2} + \sum_{n=2}^{\infty} \frac{(-1)^{n-1} 1 \cdot 3 \cdot 5 \cdot \cdots \cdot (2n-3)}{2^n \cdot n!} x^{2n}$

(b) The coefficient of x^{10} in the above Maclaurin series is $\frac{f^{(10)}(0)}{10!}$, so $f^{(10)}(0) = 10!\left(\frac{1 \cdot 3 \cdot 5 \cdot 7}{2^5 \cdot 5!}\right) = 99{,}225$.

21. **(a)** $g'(x) = \sum_{n=1}^{\infty} \binom{k}{n} n x^{n-1}$.

$\qquad (1+x)g'(x) = (1+x)\sum_{n=1}^{\infty} \binom{k}{n} n x^{n-1} = \sum_{n=1}^{\infty} \binom{k}{n} n x^{n-1} + \sum_{n=1}^{\infty} \binom{k}{n} n x^n$

$\qquad = \sum_{n=0}^{\infty} \binom{k}{n+1}(n+1)x^n + \sum_{n=0}^{\infty} \binom{k}{n} n x^n$

$\qquad = \sum_{n=0}^{\infty}(n+1)\frac{k(k-1)(k-2)\cdots(k-n)}{(n+1)!}x^n + \sum_{n=0}^{\infty}\left((n)\frac{k(k-1)(k-2)\cdots(k-n+1)}{n!}\right)x^n$

$\qquad = \sum_{n=0}^{\infty} \frac{(n+1)k(k-1)(k-2)\cdots(k-n+1)}{(n+1)!}[(k-n)+n]x^n$

$\qquad = \sum_{n=0}^{\infty} \frac{k^2(k-1)(k-2)\cdots(k-n+1)}{n!}x^n = k\sum_{n=0}^{\infty}\binom{k}{n}x^n = kg(x)$. So $g'(x) = \frac{kg(x)}{1+x}$.

(b) $h'(x) = -k(1+x)^{-k-1}g(x) + (1+x)^{-k}g'(x) = -k(1+x)^{-k-1}g(x) + (1+x)^{-k}\frac{kg(x)}{1+x}$

$\qquad = -k(1+x)^{-k-1}g(x) + k(1+x)^{-k-1}g(x) = 0$

(c) From part (b) we see that $h(x)$ must be constant for $x \in (-1,1)$, so $h(x) = h(0) = 1$ for $x \in (-1,1)$.

$\qquad$ Thus $h(x) = 1 = (1+x)^{-k}g(x) \quad \Leftrightarrow \quad g(x) = (1+x)^k$ for $x \in (-1,1)$.

EXERCISES 10.12

1.

n	$f^{(n)}(x)$	$f^{(n)}\left(\frac{\pi}{6}\right)$
0	$\sin x$	$\frac{1}{2}$
1	$\cos x$	$\frac{\sqrt{3}}{2}$
2	$-\sin x$	$-\frac{1}{2}$
3	$-\cos x$	$-\frac{\sqrt{3}}{2}$

$$T_3(x) = \sum_{n=0}^{3} \frac{f^{(n)}\left(\frac{\pi}{6}\right)}{n!}\left(x - \tfrac{\pi}{6}\right)^n$$

$$= \tfrac{1}{2} + \tfrac{\sqrt{3}}{2}\left(x - \tfrac{\pi}{6}\right) - \tfrac{1}{4}\left(x - \tfrac{\pi}{6}\right)^2 - \tfrac{\sqrt{3}}{12}\left(x - \tfrac{\pi}{6}\right)^3$$

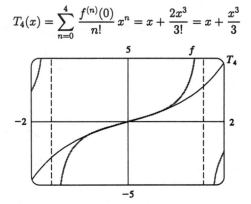

3.

n	$f^{(n)}(x)$	$f^{(n)}(0)$
0	$\tan x$	0
1	$\sec^2 x$	1
2	$2\sec^2 x \tan x$	0
3	$4\sec^2 x \tan^2 x + 2\sec^4 x$	2
4	$8\sec^2 x \tan^3 x + 16\sec^4 x \tan x$	0

$$T_4(x) = \sum_{n=0}^{4} \frac{f^{(n)}(0)}{n!}x^n = x + \frac{2x^3}{3!} = x + \frac{x^3}{3}$$

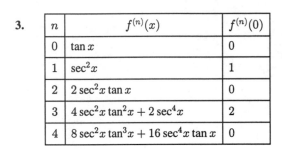

5.

n	$f^{(n)}(x)$	$f^{(n)}(0)$
0	$e^x \sin x$	0
1	$e^x(\sin x + \cos x)$	1
2	$2e^x \cos x$	2
3	$2e^x(\cos x - \sin x)$	2

$$T_3(x) = \sum_{n=0}^{3} \frac{f^{(n)}(0)}{n!}x^n = x + x^2 + \tfrac{1}{3}x^3$$

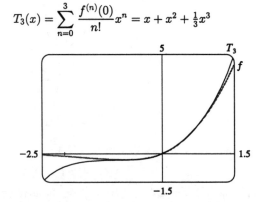

7.

n	$f^{(n)}(x)$	$f^{(n)}(8)$
0	$x^{-1/3}$	$\frac{1}{2}$
1	$-\frac{1}{3}x^{-4/3}$	$-\frac{1}{48}$
2	$\frac{4}{9}x^{-7/3}$	$\frac{1}{288}$
3	$-\frac{28}{27}x^{-10/3}$	$-\frac{7}{6912}$

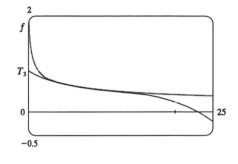

$$T_3(x) = \sum_{n=0}^{3} \frac{f^{(n)}(8)}{n!}(x-8)^n$$
$$= \frac{1}{2} - \frac{1}{48}(x-8) + \frac{1}{576}(x-8)^2 - \frac{7}{41,472}(x-8)^3$$

9.

n	$f^{(n)}(x)$	$f^{(n)}(0)$	$T_n(x)$
0	$\cos x$	1	1
1	$-\sin x$	0	1
2	$-\cos x$	-1	$1 - \frac{1}{2}x^2$
3	$\sin x$	0	$1 - \frac{1}{2}x^2$
4	$\cos x$	1	$1 - \frac{1}{2}x^2 + \frac{1}{24}x^4$

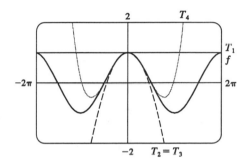

11. In Maple, we can find the Taylor polynomials by the following method: first define `f:=sec(x);` and then set
`T2:=convert(taylor(f,x=0,3),polynom);`, `T4:=convert(taylor(f,x=0,5),polynom);`,
etc. (The third argument in the `taylor` function is one more than the degree of the desired polynomial.)

We must `convert` to the type `polynom` because the output of the `taylor` function contains an error term which we do not want. In Mathematica, we use
`Tn:=Normal[Series[f,{x,0,n}]]`, with n=2, 4, etc. Note that in Mathematica, the "degree" argument is the same as the degree of the desired polynomial. The eighth Taylor polynomial is $T_8(x) = 1 + \frac{1}{2}x^2 + \frac{5}{24}x^4 + \frac{61}{720}x^6 + \frac{277}{8064}x^8$.

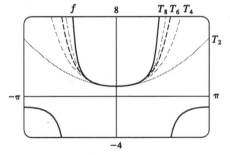

13. $f(x) = (1+x)^{1/2}$ $f(0) = 1$
$f'(x) = \frac{1}{2}(1+x)^{-1/2}$ $f'(0) = \frac{1}{2}$
$f''(x) = -\frac{1}{4}(1+x)^{-3/2}$

(a) $(1+x)^{1/2} \approx T_1(x) = 1 + \frac{1}{2}x$

(b) By Taylor's Formula, the remainder is
$$R_1(x) = \frac{f''(z)}{2!}x^2 = -\frac{1}{8(1+z)^{3/2}}x^2, \text{ where } z \text{ lies between}$$
0 and x. Now $0 \le x \le 0.1 \Rightarrow 0 \le x^2 \le 0.01$
and $0 < z < 0.1 \Rightarrow 1 < 1+z < 1.1$, so
$|R_1(x)| < \frac{0.01}{8.1} = 0.00125$.

(c)

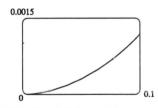

From the graph of
$$|R_1(x)| = |\sqrt{1+x} - (1+\tfrac{1}{2}x)|,$$
it seems that the error is at most 0.0013 on $(0, 0.1)$.

15. $f(x) = \sin x \qquad f\left(\frac{\pi}{4}\right) = \frac{\sqrt{2}}{2}$

$f'(x) = \cos x \qquad f'\left(\frac{\pi}{4}\right) = \frac{\sqrt{2}}{2}$

$f''(x) = -\sin x \qquad f''\left(\frac{\pi}{4}\right) = -\frac{\sqrt{2}}{2}$

$f'''(x) = -\cos x \qquad f'''\left(\frac{\pi}{4}\right) = -\frac{\sqrt{2}}{2}$

$f^{(4)}(x) = \sin x \qquad f^{(4)}\left(\frac{\pi}{4}\right) = \frac{\sqrt{2}}{2}$

$f^{(5)}(x) = \cos x \qquad f^{(5)}\left(\frac{\pi}{4}\right) = \frac{\sqrt{2}}{2}$

$f^{(6)}(x) = -\sin x$

(c)

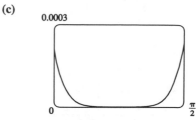

From the graph, it seems that the error is less than 0.00026 on $\left(0, \frac{\pi}{2}\right)$.

(a) $\sin x \approx T_5(x) = \frac{\sqrt{2}}{2} + \frac{\sqrt{2}}{2}\left(x - \frac{\pi}{4}\right) - \frac{\sqrt{2}}{4}\left(x - \frac{\pi}{4}\right)^2 - \frac{\sqrt{2}}{12}\left(x - \frac{\pi}{4}\right)^3 + \frac{\sqrt{2}}{48}\left(x - \frac{\pi}{4}\right)^4 + \frac{\sqrt{2}}{240}\left(x - \frac{\pi}{4}\right)^5$

(b) The remainder is $R_5(x) = \frac{1}{6!}f^{(6)}(z)\left(x - \frac{\pi}{4}\right)^6 = \frac{1}{720}(-\sin z)\left(x - \frac{\pi}{4}\right)^6$, where z lies between $\frac{\pi}{4}$ and x.

Since $0 \le x \le \frac{\pi}{2},\ -\frac{\pi}{4} \le x - \frac{\pi}{4} \le \frac{\pi}{4} \ \Rightarrow \ 0 \le \left(x - \frac{\pi}{4}\right)^6 \le \left(\frac{\pi}{4}\right)^6$, and since $0 < z < \frac{\pi}{2}, 0 < \sin z < 1$,

so $|R_5(x)| < \frac{1}{720}\left(\frac{\pi}{4}\right)^6 \approx 0.00033$.

17. $f(x) = \tan x \qquad\qquad f(0) = 0$

$f'(x) = \sec^2 x \qquad\qquad f'(0) = 1$

$f''(x) = 2\sec^2 x \tan x \qquad\quad f''(0) = 0$

$f'''(x) = 4\sec^2 x \tan^2 x + 2\sec^4 x \qquad f'''(0) = 2$

$f^{(4)}(x) = 8\sec^2 x \tan^3 x + 16\sec^4 x \tan x$

(c)

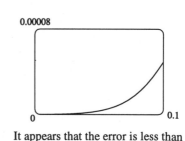

From the graph, it seems that the error is less than 0.006 on $(0, \pi)$

(a) $\tan x \approx T_3(x) = x + \frac{1}{3}x^3$

(b) $R_3(x) = \dfrac{f^{(4)}(z)}{4!}x^4 = \dfrac{8\sec^2 z \tan^3 z + 16\sec^4 z \tan z}{4!}x^4$

$= \dfrac{\sec^2 z \tan^3 z + 2\sec^4 z \tan z}{3}x^4$ where z lies between 0 and x. Now $0 \le x^4 \le \left(\frac{\pi}{6}\right)^4$ and $0 < z < \frac{\pi}{6}$

$\Rightarrow \sec^2 z < \frac{4}{3}$ and $\tan z < \frac{\sqrt{3}}{3}$, so $|R_3(x)| < \dfrac{\frac{4}{3} \cdot \frac{1}{3\sqrt{3}} + 2 \cdot \frac{16}{9} \cdot \frac{1}{\sqrt{3}}}{3}\left(\frac{\pi}{6}\right)^4 = \dfrac{4\sqrt{3}}{9}\left(\frac{\pi}{6}\right)^4 < 0.06$.

19. $f(x) = e^{x^2} \qquad\qquad f(0) = 1$

$f'(x) = e^{x^2}(2x) \qquad\qquad f'(0) = 0$

$f''(x) = e^{x^2}(2 + 4x^2) \qquad\quad f''(0) = 2$

$f'''(x) = e^{x^2}(12x + 8x^3) \qquad f'''(0) = 0$

$f^{(4)}(x) = e^{x^2}(12 + 48x^2 + 16x^4)$

(c)

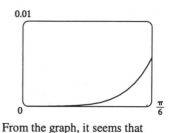

It appears that the error is less than 0.00005 on $(0, 0.1)$.

(a) $e^{x^2} \approx T_3(x) = 1 + x^2$

(b) $R_3(x) = \dfrac{f^{(4)}(z)}{4!}x^4 = \dfrac{e^{z^2}(3 + 12z^2 + 4z^4)}{6}x^4$, where

z lies between 0 and x. $0 \le x \le 0.1 \ \Rightarrow \ |R_3(x)| < \dfrac{e^{0.01}(3 + 0.12 + 0.0004)}{6}(0.0001) < 0.00006$.

21.
$$f(x) = x^{3/4} \qquad f(16) = 8$$
$$f'(x) = \tfrac{3}{4}x^{-1/4} \qquad f'(16) = \tfrac{3}{8}$$
$$f''(x) = -\tfrac{3}{16}x^{-5/4} \qquad f''(16) = -\tfrac{3}{512}$$
$$f'''(x) = \tfrac{15}{64}x^{-9/4} \qquad f'''(16) = \tfrac{15}{32,768}$$
$$f^{(4)}(x) = -\tfrac{135}{256}x^{-13/4}$$

(c)

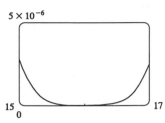

It appears that the error is less
than 3×10^{-6} on $(15, 17)$.

(a) $\quad x^{3/4} \approx T_3(x)$
$$= 8 + \tfrac{3}{8}(x - 16) - \tfrac{3}{1024}(x - 16)^2$$
$$+ \tfrac{5}{65,536}(x - 16)^3$$

(b) $\quad R_3(x) = \dfrac{f^{(4)}(z)}{4!}(x - 16)^4 = -\dfrac{135(x - 16)^4}{256 \cdot 4! \, z^{13/4}}$, where z lies between 16 and x. $|x - 16| \le 1$ and $z > 15$

$\Rightarrow \quad |R_3(x)| < \dfrac{135}{256 \cdot 24 \cdot 15^{13/4}} < 0.0000034.$

23. From Exercise 1, $\sin x = \tfrac{1}{2} + \tfrac{\sqrt{3}}{2}\left(x - \tfrac{\pi}{6}\right) - \tfrac{1}{4}\left(x - \tfrac{\pi}{6}\right)^2 - \tfrac{\sqrt{3}}{12}\left(x - \tfrac{\pi}{6}\right)^3 + R_3(x)$, where $R_3(x) = \dfrac{\sin z}{4!}\left(x - \tfrac{\pi}{6}\right)^4$

and z lies between $\tfrac{\pi}{6}$ and x. Now $35° = \left(\tfrac{\pi}{6} + \tfrac{\pi}{36}\right)$ radians, so the error is $\left|R_3\left(\tfrac{\pi}{36}\right)\right| < \dfrac{(\pi/36)^4}{4!} < 0.000003.$

Therefore, to five decimal places, $\sin 35° \approx \tfrac{1}{2} + \tfrac{\sqrt{3}}{2}\left(\tfrac{\pi}{36}\right) - \tfrac{1}{4}\left(\tfrac{\pi}{36}\right)^2 + \tfrac{\sqrt{3}}{12}\left(\tfrac{\pi}{36}\right)^3 \approx 0.57358.$

25. All derivatives of e^x are e^x, so the remainder term is $R_n(x) = \dfrac{e^z}{(n + 1)!}x^{n+1}$, where $0 < z < 0.1$. So we want

$R_n(0.1) \le \dfrac{e^{0.1}}{(n + 1)!}(0.1)^{n+1} < 0.00001$, and we find that $n = 3$ satisfies this inequality. [In fact

$R_3(0.1) < 0.0000046.$]

27. $\sin x = x - \tfrac{1}{3!}x^3 + \tfrac{1}{5!}x^5 - \cdots$. By the Alternating Series Estimation Theorem, the error in the

approximation $\sin x = x - \tfrac{1}{3!}x^3$ is less than $\left|\tfrac{1}{5!}x^5\right| < 0.01 \quad \Leftrightarrow \quad |x^5| < 1.2 \quad \Leftrightarrow \quad |x| < (1.2)^{1/5} \approx 1.037.$

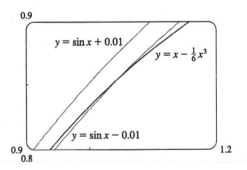

The graph confirms our estimate. Since both the sine function and the given approximation are odd functions,

we only need to check the estimate for $x > 0$.

29. Let $s(t)$ be the position function of the car, and for convenience set $s(0) = 0$. The velocity of the car is

$v(t) = s'(t)$ and the acceleration is $a(t) = s''(t)$, so the second degree Taylor polynomial is

$T_2(t) = s(0) + v(0)t + \dfrac{a(0)}{2}t^2 = 20t + t^2$. We estimate the distance travelled during the next second to be

$s(1) \approx T_2(1) = 20 + 1 = 21$ m. The function $T_2(t)$ would not be accurate over a full minute, since the car could

not possibly maintain an acceleration of $2 \text{ m}/\text{s}^2$ for that long (if it did, its final speed would be

$140 \text{ m}/\text{s} \approx 315 \text{ mi}/\text{h}!$)

(b) To convert to μm, we substitute

$\lambda/10^6$ for λ in both laws.

We can see that the two laws are

very different for short wavelengths

(Planck's Law gives a maximum at

$\lambda \approx 0.5$ μm; the Rayleigh-Jeans Law

gives no extremum.) The two laws

are similar for large λ.

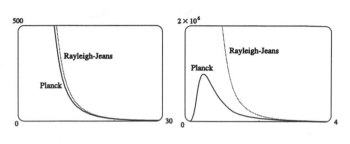

31. $E = \dfrac{q}{D^2} - \dfrac{q}{(D+d)^2} = \dfrac{q}{D^2} - \dfrac{q}{D^2(1+d/D)^2} = \dfrac{q}{D^2}\left[1 - \left(1 + \dfrac{d}{D}\right)^{-2}\right].$

We use the Binomial Series to expand $(1 + d/D)^{-2}$:

$E = \dfrac{q}{D^2}\left[1 - \left(1 - 2\left(\dfrac{d}{D}\right) + \dfrac{2 \cdot 3}{2!}\left(\dfrac{d}{D}\right)^2 - \dfrac{2 \cdot 3 \cdot 4}{3!}\left(\dfrac{d}{D}\right)^3 + \cdots\right)\right]$

$= \dfrac{q}{D^2}\left[2\left(\dfrac{d}{D}\right) - 3\left(\dfrac{d}{D}\right)^2 + 4\left(\dfrac{d}{D}\right)^3 - \cdots\right] \approx 2qd \cdot \dfrac{1}{D^3}$ when D is much larger than d.

33. (a) If the water is deep, then $2\pi d/L$ is large, and we know that $\tanh x \to 1$ as $x \to \infty$. So we can approximate

$\tanh(2\pi d/L) \approx 1$, and so $v^2 \approx gL/(2\pi) \quad \Leftrightarrow \quad v \approx \sqrt{gL/(2\pi)}$.

(b) From the calculations at right, the first term in

the Maclaurin series of $\tanh x$ is x, so if the

water is shallow, we can approximate

$\tanh\dfrac{2\pi d}{L} \approx \dfrac{2\pi d}{L}$, and so

$v^2 \approx \dfrac{gL}{2\pi} \cdot \dfrac{2\pi d}{L} \quad \Leftrightarrow \quad v \approx \sqrt{gd}.$

$f(x) = \tanh x \qquad\qquad f(0) = 0$

$f'(x) = \operatorname{sech}^2 x \qquad\qquad f'(0) = 1$

$f''(x) = -2\operatorname{sech}^2 x \tanh x \qquad f''(0) = 0$

$f'''(x) = 2\operatorname{sech}^2 x(3\tanh^2 x - 1) \quad f'''(0) = -2$

(c) Since $\tanh x$ is an odd function, its Maclaurin series is alternating, so the error in the approximation

$\tanh\dfrac{2\pi d}{L} \approx \dfrac{2\pi d}{L}$ is less than the first neglected term, which is $\dfrac{|f'''(0)|}{3!}\left(\dfrac{2\pi d}{L}\right)^3 = \dfrac{1}{3}\left(\dfrac{2\pi d}{L}\right)^3$. If

$L > 10d$, then $\dfrac{1}{3}\left(\dfrac{2\pi d}{L}\right)^3 < \dfrac{1}{3}\left(2\pi \cdot \dfrac{1}{10}\right)^3 = \dfrac{\pi^3}{375}$, so the error in the approximation $v^2 = gd$ is less than

$\dfrac{gL}{2\pi} \cdot \dfrac{\pi^3}{375} \approx 0.0132gL.$

35. Using Taylor's Formula with $n = 1$, $a = x_n$, $x = r$, we get $f(r) = f(x_n) + f'(x_n)(r - x_n) + R_1(x)$, where $R_1(x) = \frac{1}{2}f''(z)(r - x_n)^2$ and z lies between x_n and r. But r is a root, so $f(r) = 0$ and Taylor's Formula becomes $0 = f(x_n) + f'(x_n)(r - x_n) + \frac{1}{2}f''(z)(r - x_n)^2$. Taking the first two terms to the left side and dividing by $f'(x_n)$, we have $x_n - r - \dfrac{f(x_n)}{f'(x_n)} = \dfrac{1}{2}\dfrac{f''(z)}{f'(x_n)}|x_n - r|^2$. By the formula for Newton's Method, we have $|x_{n+1} - r| = \left| x_n - \dfrac{f(x_n)}{f'(x_n)} - r \right| = \dfrac{1}{2}\dfrac{|f''(z)|}{|f'(x_n)|}|x_n - r|^2 \leq \dfrac{M}{2K}|x_n - r|^2$ since $|f''(z)| \leq M$ and $|f'(x_n)| \geq K$.

REVIEW EXERCISES FOR CHAPTER 10

1. False. See the warning in Note 2 after Theorem 10.2.6.

3. False. For example, take $a_n = (-1)^n/(n6^n)$.

5. False, since $\lim\limits_{n \to \infty} \left| \dfrac{a_{n+1}}{a_n} \right| = \lim\limits_{n \to \infty} \left| \dfrac{n^3}{(n+1)^3} \right| = \lim\limits_{n \to \infty} \dfrac{1}{(1 + 1/n)^3} = 1.$

7. False. See the remarks after Example 3 in Section 10.4.

9. False. A power series has the form $a_0 + a_1 x + a_2 x^2 + a_3 x^3 + \cdots$.

11. True. See Example 8 in Section 10.1.

13. True. By Theorem 10.10.5 the coefficient of x^3 is $\dfrac{f'''(0)}{3!} = \dfrac{1}{3} \Rightarrow f'''(0) = 2.$

Or: Use Theorem 10.9.2 to differentiate f three times.

15. False. For example, let $a_n = b_n = (-1)^n$. Then $\{a_n\}$ and $\{b_n\}$ are divergent, but $a_n b_n = 1$, so $\{a_n b_n\}$ is convergent.

17. True by Theorem 10.6.3. $\left[\sum (-1)^n a_n \text{ is absolutely convergent and hence convergent.} \right]$

19. $\lim\limits_{n \to \infty} \dfrac{n}{2n + 5} = \lim\limits_{n \to \infty} \dfrac{1}{2 + 5/n} = \dfrac{1}{2}$ and the sequence is convergent.

21. $\{2n + 5\}$ is divergent since $2n + 5 \to \infty$ as $n \to \infty$.

23. $\{\sin n\}$ is divergent since $\lim\limits_{n \to \infty} \sin n$ does not exist.

25. $\left\{ \left(1 + \dfrac{3}{n} \right)^{4n} \right\}$ is convergent. Let $y = \left(1 + \dfrac{3}{x} \right)^{4x}$. Then

$$\lim\limits_{x \to \infty} \ln y = \lim\limits_{x \to \infty} 4x \ln(1 + 3/x) = \lim\limits_{x \to \infty} \dfrac{\ln(1 + 3/x)}{1/(4x)} \overset{H}{=} \lim\limits_{x \to \infty} \dfrac{\dfrac{1}{1 + 3/x}\left(-\dfrac{3}{x^2} \right)}{-1/(4x^2)} = \lim\limits_{x \to \infty} \dfrac{12}{1 + 3/x} = 12, \text{ so}$$

$$\lim\limits_{x \to \infty} y = \lim\limits_{n \to \infty} (1 + 3/n)^{4n} = e^{12}.$$

27. We use induction, hypothesizing that $a_{n-1} < a_n < 2$. Note first that $1 < a_2 = \frac{1}{3}(1+5) = \frac{5}{3} < 2$, so the hypothesis holds for $n = 2$. Now assume that $a_{k-1} < a_k < 2$. Then $a_k = \frac{1}{3}(a_{k-1} + 4) < \frac{1}{3}(a_k + 4) < \frac{1}{3}(2+4) = 2$. So $a_k < a_{k+1} < 2$, and the induction is complete. To find the limit of the sequence, we note that $L = \lim\limits_{n\to\infty} a_n = \lim\limits_{n\to\infty} a_{n+1} \quad \Rightarrow \quad L = \frac{1}{3}(L+4) \quad \Rightarrow \quad L = 2$.

29. Use the Limit Comparison Test with $a_n = \dfrac{n^2}{n^3+1}$ and $b_n = \dfrac{1}{n}$. $\lim\limits_{n\to\infty} \dfrac{a_n}{b_n} = \lim\limits_{n\to\infty} \dfrac{n^2/(n^3+1)}{1/n}$

$= \lim\limits_{n\to\infty} \dfrac{1}{1+1/n^3} = 1$. Since $\sum\limits_{n=1}^{\infty} \dfrac{1}{n}$ (the harmonic series) diverges, $\sum\limits_{n=1}^{\infty} \dfrac{n^2}{n^3+1}$ diverges also.

31. An alternating series with $a_n = \dfrac{1}{n^{1/4}}$, $a_n > 0$ for all n, and $a_n > a_{n+1}$. $\lim\limits_{n\to\infty} a_n = \lim\limits_{n\to\infty} \dfrac{1}{n^{1/4}} = 0$, so the series converges by the Alternating Series Test.

33. $\lim\limits_{n\to\infty} \sqrt[n]{|a_n|} = \lim\limits_{n\to\infty} [n/(3n+1)] = \frac{1}{3} < 1$, so series converges by the Root Test.

35. $\dfrac{|\sin n|}{1+n^2} \le \dfrac{1}{1+n^2} < \dfrac{1}{n^2}$ and since $\sum\limits_{n=1}^{\infty} \dfrac{1}{n^2}$ converges (p-series with $p = 2 > 1$), so does $\sum\limits_{n=1}^{\infty} \dfrac{|\sin n|}{1+n^2}$ by the Comparison Test.

37. $\lim\limits_{n\to\infty} \left| \dfrac{a_{n+1}}{a_n} \right| = \lim\limits_{n\to\infty} \dfrac{1 \cdot 3 \cdot 5 \cdots (2n-1)(2n+1)}{5^{n+1}(n+1)!} \cdot \dfrac{5^n\, n!}{1 \cdot 3 \cdot 5 \cdots (2n-1)} = \lim\limits_{n\to\infty} \dfrac{2n+1}{5(n+1)}$

$= \frac{2}{5} < 1$, so the series converges by the Ratio Test.

39. $\lim\limits_{n\to\infty} \left| \dfrac{a_{n+1}}{a_n} \right| = \lim\limits_{n\to\infty} \dfrac{4^{n+1}}{(n+1)3^{n+1}} \cdot \dfrac{n3^n}{4^n} = \frac{4}{3} \lim\limits_{n\to\infty} \dfrac{n}{n+1} = \frac{4}{3} > 1$ so the series diverges by the Ratio Test.

41. Consider the series of absolute values: $\sum_{n=1}^{\infty} n^{-1/3}$ is a p-series with $p = \frac{1}{3} < 1$ and is therefore divergent. But if we apply the Alternating Series Test we see that $a_{n+1} < a_n$ and $\lim\limits_{n\to\infty} n^{-1/3} = 0$. Therefore $\sum_{n=1}^{\infty} (-1)^{n-1} n^{-1/3}$ is conditionally convergent.

43. $\left| \dfrac{a_{n+1}}{a_n} \right| = \left| \dfrac{(-1)^{n+1}(n+2)3^{n+1}}{2^{2n+3}} \cdot \dfrac{2^{2n+1}}{(-1)^n(n+1)3^n} \right| = \dfrac{n+2}{n+1} \cdot \dfrac{3}{4} = \dfrac{1+(2/n)}{1+(1/n)} \cdot \dfrac{3}{4} \to \dfrac{3}{4} < 1$ as $n \to \infty$ so by

the Ratio Test, $\sum\limits_{n=1}^{\infty} \dfrac{(-1)^n(n+1)3^n}{2^{2n+1}}$ is absolutely convergent.

45. Convergent geometric series. $\sum\limits_{n=1}^{\infty} \dfrac{2^{2n+1}}{5^n} = 2\sum\limits_{n=1}^{\infty} \dfrac{4^n}{5^n} = 2\left(\dfrac{4/5}{1-4/5}\right) = 8$.

47. $\sum\limits_{n=1}^{\infty} [\tan^{-1}(n+1) - \tan^{-1}n] = \lim\limits_{n\to\infty} [(\tan^{-1}2 - \tan^{-1}1) + (\tan^{-1}3 - \tan^{-1}2) + \cdots + (\tan^{-1}(n+1) - \tan^{-1}n)]$

$= \lim\limits_{n\to\infty} [\tan^{-1}(n+1) - \tan^{-1}1] = \frac{\pi}{2} - \frac{\pi}{4} = \frac{\pi}{4}$

49. $1.2 + 0.0\overline{345} = \frac{12}{10} + \frac{345/10,000}{1-1/1000} = \frac{12}{10} + \frac{345}{9990} = \frac{4111}{3330}$

51. $\sum\limits_{n=1}^{\infty} \dfrac{(-1)^{n+1}}{n^5} = 1 - \dfrac{1}{32} + \dfrac{1}{243} - \dfrac{1}{1024} + \dfrac{1}{3125} - \dfrac{1}{7776} + \dfrac{1}{16,807} - \dfrac{1}{32,768} + \cdots$.

Since $\dfrac{1}{32,768} < 0.000031$, $\sum\limits_{n=1}^{\infty} \dfrac{(-1)^{n+1}}{n^5} \approx \sum\limits_{n=1}^{7} \dfrac{(-1)^{n+1}}{n^5} \approx 0.9721$.

53. $\sum\limits_{n=1}^{\infty} \dfrac{1}{2+5^n} \approx \sum\limits_{n=1}^{8} \dfrac{1}{2+5^n} \approx 0.18976224.$ To estimate the error, note that $\dfrac{1}{2+5^n} < \dfrac{1}{5^n}$, so the remainder term

is $R_8 = \sum\limits_{n=9}^{\infty} \dfrac{1}{2+5^n} < \sum\limits_{n=9}^{\infty} \dfrac{1}{5^n} = \dfrac{1/5^9}{1-1/5} \approx 6.4 \times 10^{-7}$ (geometric series with $a = 1/5^9$ and $r = \frac{1}{5}$).

55. Use the Limit Comparison Test. $\lim\limits_{n\to\infty} \left| \dfrac{\left(\frac{n+1}{n}\right)a_n}{a_n} \right| = \lim\limits_{n\to\infty} \dfrac{n+1}{n} = \lim\limits_{n\to\infty} \left(1 + \dfrac{1}{n}\right) = 1 > 0.$ Since $\sum |a_n|$ is

convergent, so is $\sum \left| \left(\dfrac{n+1}{n}\right)a_n \right|$ by the Limit Comparison Test.

57. $\lim\limits_{n\to\infty} \left| \dfrac{a_{n+1}}{a_n} \right| = \lim\limits_{n\to\infty} \left| \dfrac{x^{n+1}}{3^{n+1}(n+1)^3} \cdot \dfrac{3^n n^3}{x^n} \right| = \dfrac{|x|}{3} \lim\limits_{n\to\infty} \left(\dfrac{n}{n+1}\right)^3 = \dfrac{|x|}{3} < 1$ for convergence (Ratio Test) $\Rightarrow$

$|x| < 3$ and the radius of convergence is 3. When $x = \pm 3$, $\sum\limits_{n=1}^{\infty} |a_n| = \sum\limits_{n=1}^{\infty} \dfrac{1}{n^3}$ which is a convergent p-series

$(p = 3 > 1)$, so the interval of convergence is $[-3, 3]$.

59. $\lim\limits_{n\to\infty} \left| \dfrac{a_{n+1}}{a_n} \right| = \lim\limits_{n\to\infty} \left| \dfrac{2^{n+1}(x-3)^{n+1}}{\sqrt{n+4}} \cdot \dfrac{\sqrt{n+3}}{2^n(x-3)^n} \right| = 2|x-3| \lim\limits_{n\to\infty} \sqrt{\dfrac{n+3}{n+4}} = 2|x-3| < 1 \quad \Leftrightarrow$

$|x-3| < \frac{1}{2}$ so the radius of convergence is $\frac{1}{2}$. For $x = \frac{7}{2}$ the series becomes $\sum\limits_{n=0}^{\infty} \dfrac{1}{\sqrt{n+3}} = \sum\limits_{n=3}^{\infty} \dfrac{1}{n^{1/2}}$ which

diverges $(p = \frac{1}{2} < 1)$, but for $x = \frac{5}{2}$ we get $\sum\limits_{n=0}^{\infty} \dfrac{(-1)^n}{\sqrt{n+3}}$ which is a convergent alternating series, so the interval

of convergence is $\left[\frac{5}{2}, \frac{7}{2}\right)$.

61. $f(x) = \sin x \qquad f\left(\frac{\pi}{6}\right) = \frac{1}{2} \qquad\qquad f^{(2n)}\left(\frac{\pi}{6}\right) = (-1)^n \cdot \frac{1}{2}$ and

$f'(x) = \cos x \qquad f'\left(\frac{\pi}{6}\right) = \frac{\sqrt{3}}{2} \qquad\qquad f^{(2n+1)}\left(\frac{\pi}{6}\right) = (-1)^n \cdot \frac{\sqrt{3}}{2}.$

$f''(x) = -\sin x \qquad f''\left(\frac{\pi}{6}\right) = -\frac{1}{2}$

$f'''(x) = -\cos x \qquad f'''\left(\frac{\pi}{6}\right) = -\frac{\sqrt{3}}{2} \qquad\qquad \sin x = \sum\limits_{n=0}^{\infty} \dfrac{f^{(n)}\left(\frac{\pi}{6}\right)}{n!}\left(x - \frac{\pi}{6}\right)^n$

$f^{(4)}(x) = \sin x \qquad f^{(4)}\left(\frac{\pi}{6}\right) = \frac{1}{2}$

$\cdots \qquad\qquad \cdots \qquad\qquad = \sum\limits_{n=0}^{\infty} \dfrac{(-1)^n}{2(2n)!}\left(x - \frac{\pi}{6}\right)^{2n} + \sum\limits_{n=0}^{\infty} \dfrac{(-1)^n\sqrt{3}}{2(2n+1)!}\left(x - \frac{\pi}{6}\right)^{2n+1}$

63. $\dfrac{1}{1+x} = \dfrac{1}{1-(-x)} = \sum\limits_{n=0}^{\infty} (-1)^n x^n$ for $|x| < 1 \quad \Rightarrow \quad \dfrac{x^2}{1+x} = \sum\limits_{n=0}^{\infty} (-1)^n x^{n+2}$ with $R = 1$.

65. $\dfrac{1}{1-x} = \sum\limits_{n=0}^{\infty} x^n$ for $|x| < 1 \quad \Rightarrow \quad \ln(1-x) = -\int \dfrac{dx}{1-x} = -\int \sum\limits_{n=0}^{\infty} x^n \, dx = C - \sum\limits_{n=0}^{\infty} \dfrac{x^{n+1}}{n+1}.$

$\ln(1-0) = C - 0 \quad \Rightarrow \quad C = 0 \quad \Rightarrow \quad \ln(1-x) = -\sum\limits_{n=0}^{\infty} \dfrac{x^{n+1}}{n+1} = \sum\limits_{n=1}^{\infty} \dfrac{-x^n}{n}$ with $R = 1$.

67. $\sin x = \sum\limits_{n=0}^{\infty} \dfrac{(-1)^n x^{2n+1}}{(2n+1)!} \Rightarrow \sin(x^4) = \sum\limits_{n=0}^{\infty} \dfrac{(-1)^n (x^4)^{2n+1}}{(2n+1)!} = \sum\limits_{n=0}^{\infty} \dfrac{(-1)^n x^{8n+4}}{(2n+1)!}$ for all x, so the radius of

convergence is ∞.

69. $(16-x)^{-1/4} = \frac{1}{2}\left(1 - \frac{1}{16}x\right)^{-1/4} = \frac{1}{2}\left[1 + \left(-\frac{1}{4}\right)\left(-\frac{x}{16}\right) + \frac{\left(-\frac{1}{4}\right)\left(-\frac{5}{4}\right)}{2!}\left(-\frac{x}{16}\right)^2 + \cdots\right]$

$$= \frac{1}{2} + \sum_{n=1}^{\infty} \frac{1 \cdot 5 \cdot 9 \cdot \cdots \cdot (4n-3)}{2 \cdot 4^n \cdot n! \cdot 16^n} x^n = \frac{1}{2} + \sum_{n=1}^{\infty} \frac{1 \cdot 5 \cdot 9 \cdot \cdots \cdot (4n-3)}{2^{6n+1} \, n!} x^n \text{ for } \left|-\frac{x}{16}\right| < 1 \quad \Rightarrow \quad R = 16.$$

71. $e^x = \sum_{n=0}^{\infty} \frac{x^n}{n!}$ so $\dfrac{e^x}{x} = \dfrac{1}{x} + \sum_{n=1}^{\infty} \dfrac{x^{n-1}}{n!}$ and $\displaystyle\int \dfrac{e^x}{x}\,dx = C + \ln|x| + \sum_{n=1}^{\infty} \dfrac{x^n}{n \cdot n!}$

73. **(a)**

$f(x) = x^{1/2}$ $\qquad\qquad f(1) = 1$

$f'(x) = \frac{1}{2}x^{-1/2}$ $\qquad f'(1) = \frac{1}{2}$

$f''(x) = -\frac{1}{4}x^{-3/2}$ $\qquad f''(1) = -\frac{1}{4}$

$f'''(x) = \frac{3}{8}x^{-5/2}$ $\qquad f'''(1) = \frac{3}{8}$

$f^{(4)}(x) = -\frac{15}{16}x^{-7/2}$

$\sqrt{x} \approx T_3(x) = 1 + \frac{1}{2}(x-1)$

$\qquad\qquad - \frac{1}{8}(x-1)^2 + \frac{1}{16}(x-1)^3$

(b)

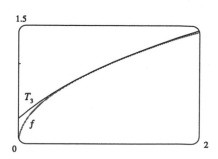

(c) By Taylor's Formula,

$$R_3(x) = \frac{f^{(4)}(z)}{4!}(x-1)^4 = -\frac{5(x-1)^4}{128z^{7/2}},$$

with z between x and 1. If $0.9 \le x \le 1.1$ then

$0 \le |x-1| \le 0.1$ and $z^{7/2} > (0.9)^{7/2}$ so

$$|R_3(x)| < \frac{5(0.1)^4}{128(0.9)^{7/2}} < 0.000006.$$

(d)

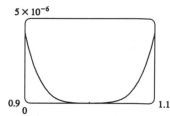

It appears that the error is less than 5×10^{-6} on $(0.9, 1.1)$.

75. $e^x = \sum_{n=0}^{\infty} \dfrac{x^n}{n!} \quad \Rightarrow \quad e^{-1/x^2} = \sum_{n=0}^{\infty} \dfrac{\left(-1/x^2\right)^n}{n!} = 1 - \dfrac{1}{x^2} + \dfrac{1}{2x^4} - \cdots \quad \Rightarrow$

$x^2\left(1 - e^{-1/x^2}\right) = x^2\left(\dfrac{1}{x^2} - \dfrac{1}{2x^4} + \cdots\right) = 1 - \dfrac{1}{2x^2} + \cdots \to 1$ as $x \to \infty$.

77. $f(x) = \sum_{n=0}^{\infty} c_n x^n \quad \Rightarrow \quad f(-x) = \sum_{n=0}^{\infty} c_n(-x)^n = \sum_{n=0}^{\infty}(-1)^n c_n x^n$

(a) If f is an odd function, then $f(-x) = -f(x) \quad \Rightarrow \quad \sum_{n=0}^{\infty}(-1)^n c_n x^n = \sum_{n=0}^{\infty} -c_n x^n$. The coefficients

of any power series are uniquely determined (by Theorem 10.10.5), so $(-1)^n c_n = -c_n$. If n is even, then

$(-1)^n = 1$, so $c_n = -c_n \quad \Rightarrow \quad 2c_n = 0 \quad \Rightarrow \quad c_n = 0$. Thus all even coefficients are 0.

(b) If f is even, then $f(-x) = f(x) \quad \Rightarrow \quad \sum_{n=0}^{\infty}(-1)^n c_n x^n = \sum_{n=0}^{\infty} c_n x^n \quad \Rightarrow \quad (-1)^n c_n = c_n$. If n is

odd, then $(-1)^n = -1$, so $-c_n = c_n \quad \Rightarrow \quad 2c_n = 0 \quad \Rightarrow \quad c_n = 0$. Thus all odd coefficients are 0.

79. Let $f(x) = \sum_{m=0}^{\infty} c_m x^m$ and $g(x) = e^{f(x)} = \sum_{n=0}^{\infty} d_n x^n$. Then $g'(x) = \sum_{n=0}^{\infty} n d_n x^{n-1}$, so $n d_n$ is the coefficient

of x^{n-1}. Also

$g'(x) = e^{f(x)} f'(x) = \left(\sum_{n=0}^{\infty} d_n x^n\right)\left(\sum_{m=1}^{\infty} m c_m x^{m-1}\right)$

$= \left(d_0 + d_1 x + d_2 x^2 + \cdots + d_{n-1}x^{n-1} + \cdots\right)\left(c_1 + 2c_2 x + 3c_3 x^2 + \cdots + n c_n x^{n-1} + \cdots\right)$, so

the coefficient of x^{n-1} is $c_1 d_{n-1} + 2c_2 d_{n-2} + 3c_3 d_{n-3} + \cdots + n c_n d_0 = \sum_{i=1}^{n} i c_i d_{n-i}$. Thus $n d_n = \sum_{i=1}^{n} i c_i d_{n-i}$.

PROBLEMS PLUS (after Chapter 10)

1. It would be far too much work to compute 15 derivatives of f. The key idea is to remember that $f^{(n)}(0)$ occurs in the coefficient of x^n in the Maclaurin series of f. We start with the Maclaurin series for sin:

    $\sin x = x - \dfrac{x^3}{3!} + \dfrac{x^5}{5!} - \cdots$. Then $\sin(x^3) = x^3 - \dfrac{x^9}{3!} + \dfrac{x^{15}}{5!} - \cdots$ and so the coefficient of x^{15} is $\dfrac{f^{(15)}(0)}{15!} = \dfrac{1}{5!}$.

    Therefore, $f^{(15)}(0) = \dfrac{15!}{5!} = 6 \cdot 7 \cdot 8 \cdot 9 \cdot 10 \cdot 11 \cdot 12 \cdot 13 \cdot 14 \cdot 15 = 10{,}897{,}286{,}400$.

3. (a) From Formula 14a in Appendix D, with $x = y = \theta$, we get $\tan 2\theta = \dfrac{2 \tan \theta}{1 - \tan^2\theta}$, so $\cot 2\theta = \dfrac{1 - \tan^2\theta}{2 \tan \theta}$

    $\Rightarrow \quad 2 \cot 2\theta = \dfrac{1 - \tan^2\theta}{\tan \theta} = \cot \theta - \tan \theta$. Replacing θ by $\frac{1}{2}x$, we get $2 \cot x = \cot \frac{1}{2}x - \tan \frac{1}{2}x$, or

    $\tan \frac{1}{2}x = \cot \frac{1}{2}x - 2 \cot x$.

    (b) From part (a) we have $\tan \dfrac{x}{2^n} = \cot \dfrac{x}{2^n} - 2 \cot \dfrac{x}{2^{n-1}}$, so the nth partial sum of the given series is

    $s_n = \dfrac{\tan(x/2)}{2} + \dfrac{\tan(x/4)}{4} + \dfrac{\tan(x/8)}{8} + \cdots + \dfrac{\tan(x/2^n)}{2^n}$

    $= \left[\dfrac{\cot(x/2)}{2} - \cot x \right] + \left[\dfrac{\cot(x/4)}{4} - \dfrac{\cot(x/2)}{2} \right] + \left[\dfrac{\cot(x/8)}{8} - \dfrac{\cot(x/4)}{4} \right]$

    $\quad + \cdots + \left[\dfrac{\cot(x/2^n)}{2^n} - \dfrac{\cot(x/2^{n-1})}{2^{n-1}} \right]$

    $= -\cot x + \dfrac{\cot(x/2^n)}{2^n}$ (telescoping sum).

    Now $\dfrac{\cot(x/2^n)}{2^n} = \dfrac{\cos(x/2^n)}{2^n \sin(x/2^n)} = \dfrac{\cos(x/2^n)}{x} \cdot \dfrac{x/2^n}{\sin(x/2^n)} \to \dfrac{1}{x} \cdot 1 = \dfrac{1}{x}$ as $n \to \infty$ since $\dfrac{x}{2^n} \to 0$ for

    $x \neq 0$. Therefore, if $x \neq 0$ and $x \neq n\pi$, then $\displaystyle\sum_{n=1}^{\infty} \dfrac{1}{2^n} \tan \dfrac{x}{2^n} = \lim_{n \to \infty} \left(-\cot x + \dfrac{1}{2^n} \cot \dfrac{x}{2^n} \right) = -\cot x + \dfrac{1}{x}$.

    If $x = 0$, then all terms in the series are 0, so the sum is 0.

5. (a) At each stage, each side is replaced by four shorter sides, each of length $\frac{1}{3}$ of the side length at the preceding stage. Writing s_0 and ℓ_0 for the number of sides and the length of the side of the initial triangle, we generate the table at right. In general, we have $s_n = 3 \cdot 4^n$ and $\ell_n = \left(\frac{1}{3}\right)^n$, so the length of the perimeter at the nth stage of construction is

    $p_n = s_n \ell_n = 3 \cdot 4^n \cdot \left(\frac{1}{3}\right)^n = 3 \cdot \left(\frac{4}{3}\right)^n$.

$s_0 = 3$	$\ell_0 = 1$
$s_1 = 3 \cdot 4$	$\ell_1 = \dfrac{1}{3}$
$s_2 = 3 \cdot 4^2$	$\ell_2 = \dfrac{1}{3^2}$
$s_3 = 3 \cdot 4^3$	$\ell_3 = \dfrac{1}{3^3}$
$\cdots$	$\cdots$

    (b) $p_n = \dfrac{4^n}{3^{n-1}} = 4 \left(\dfrac{4}{3} \right)^{n-1}$. Since $\frac{4}{3} > 1$, $p_n \to \infty$ as $n \to \infty$.

43

PROBLEMS PLUS

(c) The area of each of the small triangles added at a given stage is one-ninth of the area of the triangle added at the preceding stage. Let a be the area of the original triangle. Then the area a_n of each of the small triangles added at stage n is $a_n = a \cdot \dfrac{1}{9^n} = \dfrac{a}{9^n}$. Since a small triangle is added to each side at every stage, it follows that the total area A_n added to the figure at the nth stage is

$$A_n = s_{n-1} \cdot a_n = 3 \cdot 4^{n-1} \cdot \frac{a}{9^n} = a \cdot \frac{4^{n-1}}{3^{2n-1}}.$$ Then the total area enclosed by the snowflake curve is

$$A = a + A_1 + A_2 + A_3 + \cdots = a + a \cdot \frac{1}{3} + a \cdot \frac{4}{3^3} + a \cdot \frac{4^2}{3^5} + a \cdot \frac{4^3}{3^7} + \cdots. \text{ After the first term, this is a}$$ geometric series with common ratio $\frac{4}{9}$, so $A = a + \dfrac{a/3}{1 - \frac{4}{9}} = a + \dfrac{a}{3} \cdot \dfrac{9}{5} = \dfrac{8a}{5}$. But the area of the original equilateral triangle with side 1 is $a = \frac{1}{2} \cdot 1 \cdot \sin \frac{\pi}{3} = \frac{\sqrt{3}}{4}$.

So the area enclosed by the snowflake curve is $\frac{8}{5} \cdot \frac{\sqrt{3}}{4} = \frac{2\sqrt{3}}{5}$.

7. $x^2 + y^2 \le 4y \iff x^2 + (y-2)^2 \le 4$, so S is part of a circle, as shown in the diagram. The area of S is

$$\int_0^1 \sqrt{4y - y^2}\, dy = \int_{-2}^{-1} \sqrt{4 - v^2}\, dv \quad \text{(put } v = y - 2\text{)}$$
$$= \left[\tfrac{1}{2} v \sqrt{4 - v^2} + \tfrac{1}{2}(4)\sin^{-1}\!\left(\tfrac{1}{2}v\right) \right]_{-2}^{-1} \quad \text{(Formula 30)} \quad = \frac{2\pi}{3} - \frac{\sqrt{3}}{2}$$

Another Method (without calculus): Note that $\theta = \angle ABC = \frac{\pi}{3}$, so the area is

(area of sector AOC) $-$ (area of $\triangle ABC$) $= \frac{1}{2}(2^2)\frac{\pi}{3} - \frac{1}{2}(1)\sqrt{3} = \frac{2\pi}{3} - \frac{\sqrt{3}}{2}$.

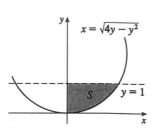

9. $a_{n+1} = \dfrac{a_n + b_n}{2},\ b_{n+1} = \sqrt{b_n a_{n+1}}$. So $a_1 = \cos\theta,\ b_1 = 1 \Rightarrow a_2 = \dfrac{1 + \cos\theta}{2} = \cos^2\dfrac{\theta}{2}$,

$b_2 = \sqrt{b_1 a_2} = \sqrt{\cos^2\dfrac{\theta}{2}} = \cos\dfrac{\theta}{2}$ since $-\dfrac{\pi}{2} \le \theta \le \dfrac{\pi}{2}$. Then

$$a_3 = \frac{1}{2}\left(\cos\frac{\theta}{2} + \cos^2\frac{\theta}{2}\right) = \cos\frac{\theta}{2} \cdot \frac{1}{2}\left(1 + \cos\frac{\theta}{2}\right) = \cos\frac{\theta}{2}\cos^2\frac{\theta}{4} \quad \Rightarrow$$

$$b_3 = \sqrt{b_2 a_3} = \sqrt{\cos\frac{\theta}{2}\cos\frac{\theta}{2}\cos^2\frac{\theta}{4}} = \cos\frac{\theta}{2}\cos\frac{\theta}{4} \quad \Rightarrow$$

$$a_4 = \frac{1}{2}\left(\cos\frac{\theta}{2}\cos^2\frac{\theta}{4} + \cos\frac{\theta}{2}\cos\frac{\theta}{4}\right) = \cos\frac{\theta}{2}\cos\frac{\theta}{4}\cdot\frac{1}{2}\left(1 + \cos\frac{\theta}{4}\right) = \cos\frac{\theta}{2}\cos\frac{\theta}{4}\cos^2\frac{\theta}{8} \quad \Rightarrow$$

$$b_4 = \sqrt{\cos\frac{\theta}{2}\cos\frac{\theta}{4}\cos\frac{\theta}{2}\cos\frac{\theta}{4}\cos^2\frac{\theta}{8}} = \cos\frac{\theta}{2}\cos\frac{\theta}{4}\cos\frac{\theta}{8}.$$ By now we see the pattern:

$b_n = \cos\dfrac{\theta}{2}\cos\dfrac{\theta}{2^2}\cos\dfrac{\theta}{2^3}\cdots\cos\dfrac{\theta}{2^{n-1}}$ and $a_n = b_n\cos\dfrac{\theta}{2^{n-1}}$. (This could be proved by mathematical induction.) By Exercise 8(a), $\sin\theta = 2^{n-1}\sin\dfrac{\theta}{2^{n-1}}\cos\dfrac{\theta}{2}\cos\dfrac{\theta}{4}\cdots\cos\dfrac{\theta}{2^{n-1}}$.

So $b_n = \cos\dfrac{\theta}{2}\cos\dfrac{\theta}{2^2}\cos\dfrac{\theta}{2^3}\cdots\cos\dfrac{\theta}{2^{n-1}} \to \dfrac{\sin\theta}{\theta}$ as $n \to \infty$ by Exercise 8(b), and

$a_n = b_n\cos\dfrac{\theta}{2^{n-1}} \to \dfrac{\sin\theta}{\theta}\cdot 1 = \dfrac{\sin\theta}{\theta}$ as $n \to \infty$. So $\lim\limits_{n\to\infty} a_n = \lim\limits_{n\to\infty} b_n = \dfrac{\sin\theta}{\theta}$.

11. We start with the geometric series $\sum\limits_{n=0}^{\infty} x^n = \dfrac{1}{1-x}$, $|x| < 1$, and differentiate:

$$\sum_{n=1}^{\infty} nx^{n-1} = \frac{d}{dx}\left(\sum_{n=0}^{\infty} x^n\right) = \frac{d}{dx}\left(\frac{1}{1-x}\right) = \frac{1}{(1-x)^2} \text{ for } |x| < 1 \quad \Rightarrow \quad \sum_{n=1}^{\infty} nx^n = x\sum_{n=1}^{\infty} nx^{n-1} = \frac{x}{(1-x)^2}$$

for $|x| < 1$. Differentiate again: $\sum\limits_{n=1}^{\infty} n^2 x^{n-1} = \dfrac{d}{dx}\dfrac{x}{(1-x)^2} = \dfrac{(1-x)^2 - x \cdot 2(1-x)(-1)}{(1-x)^4} = \dfrac{x+1}{(1-x)^3} \quad \Rightarrow$

$$\sum_{n=1}^{\infty} n^2 x^n = \frac{x^2+x}{(1-x)^3} \quad \Rightarrow \quad \sum_{n=1}^{\infty} n^3 x^{n-1} = \frac{d}{dx}\frac{x^2+x}{(1-x)^3}$$

$$= \frac{(1-x)^3(2x+1) - (x^2+x)3(1-x)^2(-1)}{(1-x)^6} = \frac{x^2+4x+1}{(1-x)^4} \quad \Rightarrow \quad \sum_{n=1}^{\infty} n^3 x^n = \frac{x^3+4x^2+x}{(1-x)^4}, \ |x| < 1.$$

The radius of convergence is 1 because that is the radius of convergence for the geometric series we started with. If $x = \pm 1$, the series is $\sum n^3(\pm 1)^n$, which diverges by the Test For Divergence, so the interval of convergence is $(-1, 1)$.

13. (a) Let $a = \arctan x$ and $b = \arctan y$. Then, from the endpapers,

$$\tan(a - b) = \frac{\tan a - \tan b}{1 + \tan a \tan b} = \frac{\tan(\arctan x) - \tan(\arctan y)}{1 + \tan(\arctan x)\tan(\arctan y)} \quad \Rightarrow \quad \tan(a-b) = \frac{x-y}{1+xy} \quad \Rightarrow$$

$$\arctan x - \arctan y = a - b = \arctan\frac{x-y}{1+xy} \text{ since } -\frac{\pi}{2} < \arctan x - \arctan y < \frac{\pi}{2}.$$

(b) From part (a) we have $\arctan\frac{120}{119} - \arctan\frac{1}{239} = \arctan\dfrac{\frac{120}{119} - \frac{1}{239}}{1 + \frac{120}{119} \cdot \frac{1}{239}} = \arctan\dfrac{\frac{28,561}{28,441}}{\frac{28,561}{28,441}} = \arctan 1 = \frac{\pi}{4}$.

(c) Replacing y by $-y$ in the formula of part (a), we get $\arctan x + \arctan y = \arctan\dfrac{x+y}{1-xy}$. So

$$4\arctan\tfrac{1}{5} = 2\left(\arctan\tfrac{1}{5} + \arctan\tfrac{1}{5}\right) = 2\arctan\dfrac{\frac{1}{5} + \frac{1}{5}}{1 - \frac{1}{5} \cdot \frac{1}{5}} = 2\arctan\tfrac{5}{12}$$

$$= \arctan\tfrac{5}{12} + \arctan\tfrac{5}{12} = \arctan\dfrac{\frac{5}{12} + \frac{5}{12}}{1 - \frac{5}{12} \cdot \frac{5}{12}} = \arctan\tfrac{120}{119}.$$

Thus, from part (b), we have $4\arctan\frac{1}{5} - \arctan\frac{1}{239} = \arctan\frac{120}{119} - \arctan\frac{1}{239} = \frac{\pi}{4}$.

(d) From Example 7 in Section 10.9 we have $\arctan x = x - \dfrac{x^3}{3} + \dfrac{x^5}{5} - \dfrac{x^7}{7} + \dfrac{x^9}{9} - \dfrac{x^{11}}{11} + \cdots$, so

$\arctan\dfrac{1}{5} = \dfrac{1}{5} - \dfrac{1}{3 \cdot 5^3} + \dfrac{1}{5 \cdot 5^5} - \dfrac{1}{7 \cdot 5^7} + \dfrac{1}{9 \cdot 5^9} - \dfrac{1}{11 \cdot 5^{11}} + \cdots$. This is an alternating series and the size of the terms decreases to 0, so by Theorem 10.5.1, the sum lies between s_5 and s_6, that is, $0.197395560 < \arctan\frac{1}{5} < 0.197395562$.

(e) From the series in part (d) we get $\arctan\dfrac{1}{239} = \dfrac{1}{239} - \dfrac{1}{3 \cdot 239^3} + \dfrac{1}{5 \cdot 239^5} - \cdots$. The third term is less than 2.6×10^{-13}, so by Theorem 10.5.1 we have, to nine decimal places, $\arctan\frac{1}{239} \approx s_2 \approx 0.004184076$. Thus $0.004184075 < \arctan\frac{1}{239} < 0.004184077$.

(f) From part (c) we have $\pi = 16\arctan\frac{1}{5} - 4\arctan\frac{1}{239}$, so from parts (d) and (e) we have

$16(0.197395560) - 4(0.004184077) < \pi < 16(0.197395562) - 4(0.004184075) \quad \Rightarrow$

$3.141592652 < \pi < 3.141592692$. So, to 7 decimal places, $\pi \approx 3.1415927$.

15. $u = 1 + \dfrac{x^3}{3!} + \dfrac{x^6}{6!} + \dfrac{x^9}{9!} + \cdots, \; v = x + \dfrac{x^4}{4!} + \dfrac{x^7}{7!} + \dfrac{x^{10}}{10!} + \cdots, \; w = \dfrac{x^2}{2!} + \dfrac{x^5}{5!} + \dfrac{x^8}{8!} + \cdots.$

The key idea is to differentiate: $\dfrac{du}{dx} = \dfrac{3x^2}{3!} + \dfrac{6x^5}{6!} + \dfrac{9x^8}{9!} + \cdots = \dfrac{x^2}{2!} + \dfrac{x^5}{5!} + \dfrac{x^8}{8!} + \cdots = w.$

Similarly, $\dfrac{dv}{dx} = 1 + \dfrac{x^3}{3!} + \dfrac{x^6}{6!} + \dfrac{x^9}{9!} + \cdots = u$, and $\dfrac{dw}{dx} = x + \dfrac{x^4}{4!} + \dfrac{x^7}{7!} + \dfrac{x^{10}}{10!} + \cdots = v.$

So $u' = w$, $v' = u$, and $w' = v$. Now differentiate the left hand side of the desired equation:

$$\dfrac{d}{dx}(u^3 + v^3 + w^3 - 3uvw) = 3u^2 u' + 3v^2 v' + 3w^2 w' - 3(u'vw + uv'w + uvw')$$

$$= 3u^2 w + 3v^2 u + 3w^2 v - 3(vw^2 + u^2 w + uv^2) = 0 \quad \Rightarrow \quad u^3 + v^3 + w^3 - 3uvw = C.$$

To find the value of the constant C, we put $x = 1$ in the equation and get $1^3 + 0 + 0 - 3(1 \cdot 0 \cdot 0) = C \quad \Rightarrow$

$C = 1$, so $u^3 + v^3 + w^3 - 3uvw = 1$.

17. $(a^n + b^n + c^n)^{1/n} = \left(c^n \left[\left(\dfrac{a}{c} \right)^n + \left(\dfrac{b}{c} \right)^n + 1 \right] \right)^{1/n} = c \left[\left(\dfrac{a}{c} \right)^n + \left(\dfrac{b}{c} \right)^n + 1 \right]^{1/n}$. Since $0 \le a \le c$, we have

$0 \le a/c \le 1$, so $(a/c)^n \to 0$ or 1 as $n \to \infty$. Similarly, $(b/c)^n \to 0$ or 1 as $n \to \infty$. Thus

$\left(\dfrac{a}{c} \right)^n + \left(\dfrac{b}{c} \right)^n + 1 \to d$, where $d = 1, 2,$ or 3 and so $\left[\left(\dfrac{a}{c} \right)^n + \left(\dfrac{b}{c} \right)^n + 1 \right]^{1/n} \to 1$. Therefore

$\lim\limits_{n \to \infty} (a^n + b^n + c^n)^{1/n} = c.$

19. As in Section 10.9 we have to integrate the function x^x by integrating a series. Writing $x^x = \left(e^{\ln x} \right)^x = e^{x \ln x}$ and

using the Maclaurin series for e^x, we have $x^x = e^{x \ln x} = \sum\limits_{n=0}^{\infty} \dfrac{(x \ln x)^n}{n!} = \sum\limits_{n=0}^{\infty} \dfrac{x^n (\ln x)^n}{n!}$. As with power series,

we can integrate this series term-by-term: $\displaystyle\int_0^1 x^x \, dx = \sum\limits_{n=0}^{\infty} \int_0^1 \dfrac{x^n (\ln x)^n}{n!} \, dx = \sum\limits_{n=0}^{\infty} \dfrac{1}{n!} \int_0^1 x^n (\ln x)^n \, dx$. We

integrate by parts with $u = (\ln x)^n$, $dv = x^n \, dx$, so $du = \dfrac{n(\ln x)^{n-1}}{x} \, dx$ and $v = \dfrac{x^{n+1}}{n+1}$:

$$\int_0^1 x^n (\ln x)^n \, dx = \lim_{t \to 0^+} \int_t^1 x^n (\ln x)^n \, dx = \lim_{t \to 0^+} \left[\dfrac{x^{n+1}}{n+1} (\ln x)^n \right]_t^1 - \lim_{t \to 0^+} \int_t^1 \dfrac{n}{n+1} x^n (\ln x)^{n-1} \, dx$$

$$= 0 - \dfrac{n}{n+1} \int_0^1 x^n (\ln x)^{n-1} \, dx \qquad \text{(where l'Hospital's Rule was used to help evaluate the first limit).}$$

Further integration by parts gives $\displaystyle\int_0^1 x^n (\ln x)^k \, dx = -\dfrac{k}{n+1} \int_0^1 x^n (\ln x)^{k-1} \, dx$ and, combining these steps, we

get $\displaystyle\int_0^1 x^n (\ln x)^n \, dx = \dfrac{(-1)^n \, n!}{(n+1)^n} \int_0^1 x^n \, dx = \dfrac{(-1)^n \, n!}{(n+1)^{n+1}} \quad \Rightarrow$

$$\int_0^1 x^x \, dx = \sum\limits_{n=0}^{\infty} \dfrac{1}{n!} \int_0^1 x^n (\ln x)^n \, dx = \sum\limits_{n=0}^{\infty} \dfrac{1}{n!} \dfrac{(-1)^n \, n!}{(n+1)^{n+1}} = \sum\limits_{n=0}^{\infty} \dfrac{(-1)^n}{(n+1)^{n+1}} = \sum\limits_{n=1}^{\infty} \dfrac{(-1)^{n-1}}{n^n}.$$

21. Call the series S. We group the terms according to the number of digits in their denominators:

$$S = \underbrace{\left(1 + \dfrac{1}{2} + \cdots + \dfrac{1}{8} + \dfrac{1}{9} \right)}_{g_1} + \underbrace{\left(\dfrac{1}{11} + \cdots + \dfrac{1}{99} \right)}_{g_2} + \underbrace{\left(\dfrac{1}{111} + \cdots + \dfrac{1}{999} \right)}_{g_3} + \cdots$$

Now in the group g_n, there are 9^n terms, since we have 9 choices for each of the n digits in the denominator.

Furthermore, each term in g_n is less than $\frac{1}{10^{n-1}}$. So $g_n < 9^n \cdot \frac{1}{10^{n-1}} = 9\left(\frac{9}{10}\right)^{n-1}$.

Now $\sum_{n=1}^{\infty} 9\left(\frac{9}{10}\right)^{n-1}$ is a geometric series with $a = 9$ and $r = \frac{9}{10} < 1$. Therefore, by the Comparison Test,

$$S = \sum_{n=1}^{\infty} g_n < \sum_{n=1}^{\infty} 9\left(\frac{9}{10}\right)^{n-1} = \frac{9}{1 - \frac{9}{10}} = 90.$$

23. If L is the length of a side of the equilateral triangle,
then the area is $A = \frac{1}{2}L \cdot \frac{\sqrt{3}}{2}L = \frac{\sqrt{3}}{4}L^2$ and so
$L^2 = \frac{4}{\sqrt{3}}A$. Let r be the radius of one of the circles
when there are n rows of circles. The figure shows that

$$L = \sqrt{3}r + r + (n-2)(2r) + r + \sqrt{3}r$$
$$= r\left(2n - 2 + 2\sqrt{3}\right), \text{ so } r = \frac{L}{2\left(n + \sqrt{3} - 1\right)}.$$

The number of circles is $1 + 2 + \cdots + n = \frac{n(n+1)}{2}$
and so the total area of the circles is

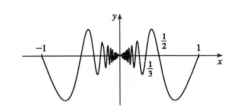

$$A_n = \frac{n(n+1)}{2}\pi r^2 = \frac{n(n+1)}{2}\pi\frac{L^2}{4\left(n + \sqrt{3} - 1\right)^2} = \frac{n(n+1)}{2}\pi\frac{4A/\sqrt{3}}{4\left(n + \sqrt{3} - 1\right)^2} = \frac{n(n+1)}{\left(n + \sqrt{3} - 1\right)^2}\frac{\pi A}{2\sqrt{3}}$$

$$\Rightarrow \frac{A_n}{A} = \frac{n(n+1)}{\left(n + \sqrt{3} - 1\right)^2}\frac{\pi}{2\sqrt{3}} = \frac{1 + 1/n}{\left[1 + (\sqrt{3} - 1)/n\right]^2}\frac{\pi}{2\sqrt{3}} \to \frac{\pi}{2\sqrt{3}} \text{ as } n \to \infty.$$

25. (a) f is continuous when $x \neq 0$ since x, $\sin x$, and π/x are continuous when $x \neq 0$. Also $|\sin(\pi/x)| \leq 1$
$\Rightarrow |x\sin(\pi/x)| \leq |x|$ and $\lim_{x\to 0}|x| = 0$, so by the Squeeze Theorem we have $\lim_{x\to 0}|x\sin(\pi/x)| = 0$, so

$\lim_{x\to 0}f(x) = \lim_{x\to 0}x\sin(\pi/x) = 0 = f(0)$. Therefore f is continuous at 0, and so f is continuous on $(-1, 1)$.

(b) Note that $f(x) = 0$ when $x = 0$ and when $\pi/x = n\pi$ $\Rightarrow$
$x = 1/n$, n an integer. Since $-1 \leq \sin(\pi/x) \leq 1$,
the graph of f lies between the lines $y = x$ and
$y = -x$ and touches these lines when

$$\frac{\pi}{x} = \frac{\pi}{2} + n\pi \Rightarrow x = \frac{1}{n + 1/2}.$$

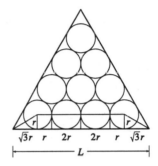

(c) The enlargement of the portion of the graph between $x = \frac{1}{n}$ and $x = \frac{1}{n-1}$
(the case where n is odd is illustrated) shows that the arc length from
$x = \frac{1}{n}$ to $x = \frac{1}{n-1}$ is greater than $|PQ| = \frac{1}{n - 1/2} = \frac{2}{2n - 1}$.

Thus the total length of the graph is greater than $2\sum_{n=1}^{\infty}\frac{2}{2n - 1}$.

This is a divergent series (by comparison with the harmonic series),
so the graph has infinite length.

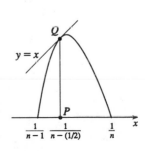

27. We use a method similar to that for Exercise 10.5.35.

Let s_n be the nth partial sum for the given series, and let h_n be the nth partial sum for the harmonic series. From Exercise 10.3.32 we know that $h_n - \ln n \to \gamma$ as $n \to \infty$. So

$$
\begin{aligned}
s_{3n} &= 1 + \frac{1}{2} - \frac{2}{3} + \frac{1}{4} + \frac{1}{5} - \frac{2}{6} + \cdots + \frac{1}{3n-2} + \frac{1}{3n-1} - \frac{2}{3n} \\
&= \left(1 + \frac{1}{2} + \frac{1}{3} + \frac{1}{4} + \cdots + \frac{1}{3n}\right) - \left(\frac{3}{3} + \frac{3}{6} + \frac{3}{9} + \cdots + \frac{3}{3n}\right) = h_{3n} - h_n \\
&= \ln 3 + (h_{3n} - \ln 3n) - (h_n - \ln n) \to \ln 3 + \gamma - \gamma = \ln 3
\end{aligned}
$$

Note: The method suggested in the first printing of the text doesn't quite work. We can differentiate to get

$$
\begin{aligned}
f'(x) &= 1 + x - 2x^2 + x^3 + x^4 - 2x^5 + \cdots \\
&= (1 + x - 2x^2) + x^3(1 + x - 2x^2) + x^6(1 + x - 2x^2) + \cdots \\
&= (1 + x - 2x^2)(1 + x^3 + x^6 + x^9 + \cdots) \\
&= \frac{1 + x - 2x^2}{1 - x^3} \quad (\text{if } |x| < 1) \\
&= \frac{(1 - x)(1 + 2x)}{(1 - x)(1 + x + x^2)} \\
&= \frac{1 + 2x}{1 + x + x^2}.
\end{aligned}
$$

Then $f(x) = \ln(1 + x + x^2) + C$. But $f(0) = 0$, so $C = 0$ and we have shown that

$$
x + \frac{x^2}{2} - \frac{2x^3}{3} + \frac{x^4}{4} + \frac{x^5}{5} - \frac{2x^6}{6} + \cdots = \ln(1 + x + x^2) \text{ for } -1 < x < 1. \text{ As } x \to 1, \text{ the limit of the right side}
$$

is $\ln 3$.

However, it is not so easy to show that the limit of the left side is $1 + \frac{1}{2} - \frac{2}{3} + \frac{1}{4} + \frac{1}{5} - \frac{2}{6} + \cdots$.

29.
$$
\begin{aligned}
&\int_0^1 \left(\frac{-x - 1}{100}\right)\left(\frac{1}{x+1} + \frac{1}{x+2} + \frac{1}{x+3} \cdots + \frac{1}{x+100}\right) dx \\
&= \int_0^1 \frac{(-x-1)(-x-2)(-x-3)\cdots(-x-100)}{100!}\left(\frac{1}{x+1} + \cdots + \frac{1}{x+100}\right) dx \\
&= \frac{1}{100!}\int_0^1 (x+1)(x+2)(x+3)\cdots(x+100)\left(\frac{1}{x+1} + \cdots + \frac{1}{x+100}\right) dx \\
&= \frac{1}{100!}(x+1)(x+2)(x+3)\cdots(x+100)\Big|_0^1 \quad (\text{the Product Rule in reverse}) \\
&= \frac{2 \cdot 3 \cdots\cdots 101}{100!} - \frac{100!}{100!} = 101 - 1 = 100.
\end{aligned}
$$

31. We can write the sum as

$$\sum_{n=1}^{\infty}\frac{1}{n}\left[\sum_{m=1}^{\infty}\frac{1}{m}\left(\frac{1}{m+(n+2)}\right)\right]=\sum_{n=1}^{\infty}\frac{1}{n}\left[\frac{1}{n+2}\sum_{m=1}^{\infty}\left(\frac{1}{m}-\frac{1}{m+(n+2)}\right)\right] \quad \text{(partial fractions)}$$

$$=\frac{1}{2}\sum_{n=1}^{\infty}\left[\left(\frac{1}{n}-\frac{1}{n+2}\right)\sum_{m=1}^{n+2}\left(\frac{1}{m}\right)\right] \quad \left(\begin{array}{l}\text{partial fractions in the outer sum; all terms}\\ \text{beyond the }(n+2)\text{th cancel in the inner sum}\end{array}\right)$$

$$=\frac{1}{2}\left[\sum_{n=1}^{\infty}\left(\frac{1}{n}\sum_{m=1}^{n+2}\frac{1}{m}\right)-\sum_{n=3}^{\infty}\left(\frac{1}{n}\sum_{m=1}^{n}\frac{1}{m}\right)\right] \quad \text{(change the index)}$$

$$=\frac{1}{2}\left[\sum_{n=1}^{2}\left(\frac{1}{n}\sum_{m=1}^{n+2}\frac{1}{m}\right)+\sum_{n=3}^{\infty}\left(\frac{1}{n}\sum_{m=1}^{n+2}\frac{1}{m}-\frac{1}{n}\sum_{m=1}^{n}\frac{1}{m}\right)\right]$$

$$=\frac{1}{2}\left[1\left(1+\frac{1}{2}+\frac{1}{3}\right)+\left(\frac{1}{2}\right)\left(1+\frac{1}{2}+\frac{1}{3}+\frac{1}{4}\right)+\sum_{n=3}^{\infty}\frac{1}{n}\left(\frac{1}{n+1}+\frac{1}{n+2}\right)\right]$$

$$=\frac{1}{2}\left[\frac{11}{6}+\frac{25}{24}+\sum_{n=3}^{\infty}\left(\frac{1}{n}-\frac{1}{n-1}\right)+\sum_{n=3}^{\infty}\left(\frac{1/2}{n}-\frac{1/2}{n-2}\right)\right] \quad \text{(partial fractions)}$$

$$=\frac{1}{2}\left[\frac{11}{6}+\frac{25}{24}+\frac{1}{3}+\frac{1}{2}\left(\frac{1}{3}+\frac{1}{4}\right)\right] \quad \text{(both series telescope)} \quad =\frac{7}{4}.$$

33. (a) We prove by induction that $1 < a_{n+1} < a_n$. For $n = 1$,

$$a_{1+1}=\frac{3\left(\frac{3}{2}\right)^2+4\left(\frac{3}{2}\right)-3}{4\left(\frac{3}{2}\right)^2}=\frac{3\left(\frac{9}{4}\right)+6-3}{9}=\frac{13}{12}, \text{ so } 1<a_2<a_1. \text{ Assume for } k. \text{ Then}$$

$$a_{k+1}=\frac{3a_k^2+4a_k-3}{4a_k^2}>1 \quad \Leftrightarrow \quad 3a_k^2+4a_k-3>4a_k^2 \quad \Leftrightarrow \quad 0>a_k^2-4a_k+3=(a_k-3)(a_k-1)$$

$\Leftrightarrow \quad 1<a_k<3$, which is true by the induction hypothesis. So we have the first inequality. For the

second, $a_{k+1}=\dfrac{3a_k^2+4a_k-3}{4a_k^2}<a_k \quad \Leftrightarrow \quad 3a_k^2+4a_k-3<4a_k^3 \quad \Leftrightarrow$

$0<4a_k^3-3a_k^2-4a_k+3=(4a_k-3)(a_k^2-1)>0$ since $a_k>1$. So both results hold by induction.

(b) $\{a_n\}$ converges by part (a) and Theorem 10.1.10, so let $\lim\limits_{n\to\infty}a_n=L$. Then

$$L=\lim_{n\to\infty}\frac{3a_n^2+4a_n-3}{4a_n^2}=\frac{3\left(\lim\limits_{n\to\infty}a_n\right)^2+4\left(\lim\limits_{n\to\infty}a_n\right)-3}{4\left(\lim\limits_{n\to\infty}a_n\right)^2}=\frac{3L^2+4L-3}{4L^2} \quad \Rightarrow$$

$4L^3-3L^2-4L+3=0 \quad \Rightarrow \quad (4L-3)(L^2-1)=(4L-3)(L-1)(L+1)=0$. Since $a_n>1$ for

all n, $L\geq 1$, but 1 is the only such root of this polynomial, so $L=1$.

(c) Observe that $a_{n+1}=\dfrac{3a_n^2+4a_n-3}{4a_n^2} \quad \Leftrightarrow \quad 4a_n^2\cdot a_{n+1}-3a_n^2=4a_n-3 \quad \Leftrightarrow$

$a_n^2(4a_{n+1}-3)=4a_n-3$. Substituting this for $n=2$ into the same for $n=1$ gives

$a_1^2a_2^2(4a_3-3)=(4a_1-3)$. If we carry on these substitutions we get $a_1^2a_2^2\cdots a_n^2(4a_{n+1}-3)=(4a_1-3)$

$\Leftrightarrow \quad a_1a_2\cdots a_n=\sqrt{\dfrac{4a_1-3}{4a_{n+1}-3}}$. So $\lim\limits_{n\to\infty}a_1a_2\cdots a_n=\lim\limits_{n\to\infty}\sqrt{\dfrac{4a_1-3}{4a_{n+1}-3}}=\sqrt{\dfrac{4\left(\frac{3}{2}\right)-3}{4(1)-3}}=\sqrt{3}.$

CHAPTER ELEVEN

EXERCISES 11.1

1. **(a)**

    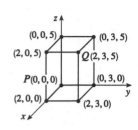

    (b) $|PQ| = \sqrt{(2-0)^2 + (3-0)^2 + (5-0)^2}$

    $\qquad = \sqrt{38}$

3. **(a)**

    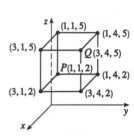

    (b) $|PQ| = \sqrt{(3-1)^2 + (4-1)^2 + (5-2)^2}$

    $\qquad = \sqrt{22}$

5. $|AB| = \sqrt{(3-2)^2 + (3-1)^2 + (4-0)^2} = \sqrt{21}, \quad |BC| = \sqrt{(5-3)^2 + (4-3)^2 + (3-4)^2} = \sqrt{6},$

    $|CA| = \sqrt{(5-2)^2 + (4-1)^2 + (3-0)^2} = \sqrt{27} = 3\sqrt{3}$

    Since no two of the sides are equal in length the triangle isn't isosceles. But $|AB|^2 + |BC|^2 = 27 = |CA|^2$, so the triangle is a right triangle.

7. $|AB| = \sqrt{[5-(-2)]^2 + (4-6)^2 + (-3-1)^2} = \sqrt{49+4+16} = \sqrt{69}$

    $|BC| = \sqrt{(2-5)^2 + (-6+4)^2 + [4-(-3)]^2} = \sqrt{9+100+49} = \sqrt{158}$

    $|CA| = \sqrt{(-2-2)^2 + [6-(-6)]^2 + (1-4)^2} = \sqrt{16+144+9} = \sqrt{169} = 13$

    Since no two sides are of equal length and since $|AB|^2 + |BC|^2 \neq |CA|^2$ the triangle is neither isosceles nor a right triangle.

9. $|PQ| = \sqrt{(0-1)^2 + (3-2)^2 + (7-3)^2} = \sqrt{18} = 3\sqrt{2}$

    $|PR| = \sqrt{(3-1)^2 + (5-2)^2 + (11-3)^2} = \sqrt{4+9+64} = \sqrt{77}$

    $|QR| = \sqrt{(3-0)^2 + (5-3)^2 + (11-7)^2} = \sqrt{9+4+16} = \sqrt{29}$

    Since the sum of the two shortest distances isn't equal to the longest distance, the points aren't collinear. To show that $\sqrt{18} + \sqrt{29} \neq \sqrt{77}$, assume they are equal. Then squaring both sides gives $18 + 29 + 2\sqrt{(18)(29)} = 77$; solving for the radical and squaring again gives $(18)(29) = (15)^2$ or $522 = 225$ which of course is not true, so the values can't be equal.

11. $(x-0)^2 + (y-1)^2 + [z-(-1)]^2 = 4^2$ or $x^2 + (y-1)^2 + (z+1)^2 = 16$

13. $(x+6)^2 + (y+1)^2 + (z-2)^2 = 12$

15. Completing the squares in the equation gives
$$(x^2 + 2x + 1) + (y^2 + 8y + 16) + (z^2 - 4z + 4) = 28 + 1 + 16 + 4 \quad \Rightarrow$$
$$(x+1)^2 + (y+4)^2 + (z-2)^2 = 49 \quad \Rightarrow \quad C(-1, -4, 2), \text{ and } r = 7.$$

17. $\left(x^2 + x + \frac{1}{4}\right) + (y^2 - 2y + 1) + (z^2 + 6z + 9) = 2 + \frac{1}{4} + 1 + 9 \quad \Rightarrow \quad \left(x + \frac{1}{2}\right)^2 + (y-1)^2 + (z+3)^2 = \frac{49}{4}$
$\Rightarrow \quad C\left(-\frac{1}{2}, 1, -3\right)$, and $r = \frac{7}{2}$.

19. $\left(x^2 - x + \frac{1}{4}\right) + y^2 + z^2 = 0 + \frac{1}{4} \Rightarrow \left(x - \frac{1}{2}\right)^2 + (y-0)^2 + (z-0)^2 = \frac{1}{4} \Rightarrow C\left(\frac{1}{2}, 0, 0\right), r = \frac{1}{2}.$

21. Call the given point Q and show that $|P_1Q| = |QP_2| = \frac{1}{2}|P_1P_2|$.
$$|P_1P_2| = \sqrt{(x_2 - x_1)^2 + (y_2 - y_1)^2 + (z_2 - z_1)^2}$$
$$|P_1Q| = \sqrt{\left[\tfrac{1}{2}(x_1 + x_2) - x_1\right]^2 + \left[\tfrac{1}{2}(y_1 + y_2) - y_1\right]^2 + \left[\tfrac{1}{2}(z_1 + z_2) - z_1\right]^2}$$
$$= \tfrac{1}{2}\sqrt{(x_2 - x_1)^2 + (y_2 - y_1)^2 + (z_2 - z_1)^2} = \tfrac{1}{2}|P_1P_2|$$
Similarly $|QP_2| = \frac{1}{2}|P_1P_2|$.

23. From Exercise 21, the midpoints of sides AB, BC and CA are respectively $P_1\left(-\frac{1}{2}, 1, 4\right)$, $P_2\left(1, \frac{1}{2}, 5\right)$ and $P_3\left(\frac{5}{2}, \frac{3}{2}, 4\right)$. Then the lengths of the medians are:
$$|AP_2| = \sqrt{0^2 + \left(\tfrac{1}{2} - 2\right)^2 + (5-3)^2} = \sqrt{\tfrac{9}{4} + 4} = \sqrt{\tfrac{25}{4}} = \tfrac{5}{2},$$
$$|BP_3| = \sqrt{\left(\tfrac{5}{2} + 2\right)^2 + \left(\tfrac{3}{2}\right)^2 + (4-5)^2} = \sqrt{\tfrac{81}{4} + \tfrac{9}{4} + 1} = \sqrt{\tfrac{94}{4}} = \tfrac{1}{2}\sqrt{94}, \text{ and}$$
$$|CP_1| = \sqrt{\left(-\tfrac{1}{2} - 4\right)^2 + (1-1)^2 + (4-5)^2} = \sqrt{\tfrac{81}{4} + 1} = \tfrac{1}{2}\sqrt{85}.$$

25. (a) Since the sphere touches the xy-plane, its radius is the distance from its center, $(2, -3, 6)$, to the xy-plane, namely 6. Therefore $r = 6$ and the equation of the sphere is $(x-2)^2 + (y+3)^2 + (z-6)^2 = 36$.

(b) Here $r =$ distance from center to yz-plane $= 2$. Therefore, the equation is
$$(x-2)^2 + (y+3)^2 + (z-6)^2 = 4.$$

(c) Here $r =$ distance from center to xz-plane $= 3$ Therefore, the equation is
$$(x-2)^2 + (y+3)^2 + (z-6)^2 = 9.$$

27. We need to find a set of points $\{P(x, y, z) \mid |AP| = |BP|\}$.
$$\sqrt{(x+1)^2 + (y-5)^2 + (z-3)^2} = \sqrt{(x-6)^2 + (y-2)^2 + (z+2)^2} \quad \Rightarrow$$
$$(x+1)^2 + (y-5)^2 + (z-3)^2 = (x-6)^2 + (y-2)^2 + (z+2)^2 \quad \Rightarrow$$
$$x^2 + 2x + 1 + y^2 - 10y + 25 + z^2 - 6z + 9 = x^2 - 12x + 36 + y^2 - 4y + 4 + z^2 + 4z + 4 \quad \Rightarrow$$
$14x - 6y - 10z = 9$. Thus the set of points is a plane perpendicular to the line segment joining A and B (since this plane must contain the perpendicular bisector of the line segment AB).

29. A plane parallel to the yz-plane and 9 units in front of it.

31. A half-space containing all points to the right of the plane $y = 2$.

33. A plane perpendicular to the xz-plane and intersecting the xz-plane in the line $x = z$, $y = 0$.

35. A right circular cylinder with radius 1 and axis the z-axis.

37. All points outside the sphere with radius 1 and center $(0, 0, 0)$.

39. Completing the square in z gives $x^2 + y^2 + (z^2 - 2z + 1) < 3 + 1$ or
$x^2 + y^2 + (z - 1)^2 < 4$, all points inside the sphere with radius two and center $(0, 0, 1)$.

41. In the xy-plane the equation $xy = 1$ represents a hyperbola with center at the origin. Since z can assume any value, the region in $\mathbb{R}^3$ is a hyperbolic cylinder.

43. All points on and between the two horizontal planes $z = 2$ and $z = -2$.

45. $y < 0$

47. $r < \sqrt{x^2 + y^2 + z^2} < R$, or $r^2 < x^2 + y^2 + z^2 < R^2$

49. (a)

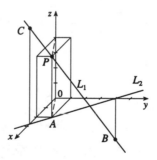

To find the x- and y-coordinates of the point P, we project it onto L_2 and project the resulting point A onto the x- and y-axes. To find the z-coordinate, we project P onto either the xz-plane or the yz-plane (using our knowledge of its x- or y-coordinate) and then project the resulting point onto the z-axis. (Or, we could draw a line parallel to AO from P to the z-axis.) The coordinates of P are $(2, 1, 4)$.

(b) A is the intersection of L_1 and L_2, B is directly below the y-intercept of L_2, and C is directly above the x-intercept of L_2.

EXERCISES 11.2

1. $\mathbf{a} = \langle 4 - 1, 4 - 3 \rangle = \langle 3, 1 \rangle$

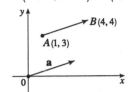

3. $\mathbf{a} = \langle 3 - 3, -3 + 1 \rangle = \langle 0, -2 \rangle$

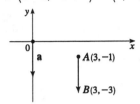

5. $\mathbf{a} = \langle 2, 0, -2 \rangle$

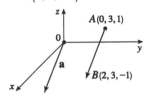

7. $\langle 2, 3 \rangle + \langle 3, -4 \rangle = \langle 5, -1 \rangle$

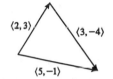

(using position vectors and the parallelogram law)

9. $\langle 1, 0, 1 \rangle + \langle 0, 0, 1 \rangle = \langle 1, 0, 2 \rangle$

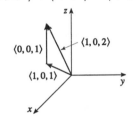

11. $|\mathbf{a}| = \sqrt{5^2 + (-12)^2} = \sqrt{169} = 13$

$\mathbf{a} + \mathbf{b} = \langle 5 - 2, -12 + 8 \rangle = \langle 3, -4 \rangle$

$\mathbf{a} - \mathbf{b} = \langle 5 - (-2), -12 - 8 \rangle = \langle 7, -20 \rangle$

$2\mathbf{a} = \langle 2(5), 2(-12) \rangle = \langle 10, -24 \rangle$

$3\mathbf{a} + 4\mathbf{b} = \langle 15, -36 \rangle + \langle -8, 32 \rangle = \langle 7, -4 \rangle$

13. $|\mathbf{a}| = \sqrt{2^2 + (-3)^2 + 6^2} = \sqrt{49} = 7$ $\mathbf{a} + \mathbf{b} = \langle 3, -2, 10 \rangle$

$\mathbf{a} - \mathbf{b} = \langle 1, -4, 2 \rangle$ $2\mathbf{a} = \langle 4, -6, 12 \rangle$

$3\mathbf{a} + 4\mathbf{b} = \langle 6, -9, 18 \rangle + \langle 4, 4, 16 \rangle = \langle 10, -5, 34 \rangle$

15. $|\mathbf{a}| = \sqrt{1^2 + (-1)^2} = \sqrt{2}$ $\mathbf{a} + \mathbf{b} = (\mathbf{i} - \mathbf{j}) + (\mathbf{i} + \mathbf{j}) = 2\mathbf{i}$

$\mathbf{a} - \mathbf{b} = (\mathbf{i} - \mathbf{j}) - (\mathbf{i} + \mathbf{j}) = -2\mathbf{j}$ $2\mathbf{a} = 2(\mathbf{i} - \mathbf{j}) = 2\mathbf{i} - 2\mathbf{j}$

$3\mathbf{a} + 4\mathbf{b} = 3(\mathbf{i} - \mathbf{j}) + 4(\mathbf{i} + \mathbf{j}) = 3\mathbf{i} - 3\mathbf{j} + 4\mathbf{i} + 4\mathbf{j} = 7\mathbf{i} + \mathbf{j}$

17. $|\mathbf{a}| = \sqrt{1^2 + 1^2 + 1^2} = \sqrt{3}$ $\mathbf{a} + \mathbf{b} = 3\mathbf{i} + 4\mathbf{k}$

$\mathbf{a} - \mathbf{b} = \mathbf{i} + \mathbf{j} + \mathbf{k} - 2\mathbf{i} + \mathbf{j} - 3\mathbf{k} = -\mathbf{i} + 2\mathbf{j} - 2\mathbf{k}$ $2\mathbf{a} = 2\mathbf{i} + 2\mathbf{j} + 2\mathbf{k}$

$3\mathbf{a} + 4\mathbf{b} = (3\mathbf{i} + 3\mathbf{j} + 3\mathbf{k}) + (8\mathbf{i} - 4\mathbf{j} + 12\mathbf{k}) = 11\mathbf{i} - \mathbf{j} + 15\mathbf{k}$

19. $|\langle 1, -2 \rangle| = \sqrt{1^2 + 2^2} = \sqrt{5}$. Thus $\mathbf{u} = \frac{1}{\sqrt{5}} \langle 1, 2 \rangle = \left\langle \frac{1}{\sqrt{5}}, \frac{2}{\sqrt{5}} \right\rangle$.

21. $|\langle -2, 4, 3 \rangle| = \sqrt{(-2)^2 + 4^2 + 3^2} = \sqrt{29}$. Thus $\mathbf{u} = \frac{1}{\sqrt{29}} \langle -2, 4, 3 \rangle = \left\langle -\frac{2}{\sqrt{29}}, \frac{4}{\sqrt{29}}, \frac{3}{\sqrt{29}} \right\rangle$.

23. $|i + j| = \sqrt{1^2 + 1^2} = \sqrt{2}$. Thus $\mathbf{u} = \frac{1}{\sqrt{2}}(i + j) = \frac{1}{\sqrt{2}}i + \frac{1}{\sqrt{2}}j$.

25. $\mathbf{a} = 2i + 3j, \mathbf{b} = i - j \Rightarrow \mathbf{a} + 3\mathbf{b} = 2i + 3j + 3i - 3j = 5i \Rightarrow i = \frac{1}{5}\mathbf{a} + \frac{3}{5}\mathbf{b}$. Substituting this expression for i into $\mathbf{a} = 2i + 3j$ gives $\mathbf{a} = 2(\frac{1}{5}\mathbf{a} + \frac{3}{5}\mathbf{b}) + 3j \Rightarrow \frac{3}{5}\mathbf{a} - \frac{6}{5}\mathbf{b} = 3j \Rightarrow j = \frac{1}{5}\mathbf{a} - \frac{2}{5}\mathbf{b}$.

27. By the triangle law, $\overrightarrow{AB} + \overrightarrow{BC} = \overrightarrow{AC}$, and $\overrightarrow{AC} = -\overrightarrow{CA} \Rightarrow \overrightarrow{AB} + \overrightarrow{BC} + \overrightarrow{CA} = -\overrightarrow{CA} + \overrightarrow{AC} = \mathbf{0}$.

29. **(a), (b)**

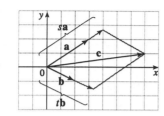

(c) From the sketch, we estimate that $s \approx 1.3$ and $t \approx 1.6$.

(d) $\mathbf{c} = s\mathbf{a} + t\mathbf{b} \Leftrightarrow 7 = 3s + 2t$ and $1 = 2s - t$. Solving these equations gives $s = \frac{9}{7}$ and $t = \frac{11}{7}$.

31. $|\mathbf{F}_1| = 10$ lb and $|\mathbf{F}_2| = 12$ lb.

$\mathbf{F}_1 = -|\mathbf{F}_1|\cos 45°\,i + |\mathbf{F}_1|\sin 45°\,j = -10\cos 45°\,i + 10\sin 45°\,j = -5\sqrt{2}i + 5\sqrt{2}j$

$\mathbf{F}_2 = |\mathbf{F}_2|\cos 30°\,i + |\mathbf{F}_2|\sin 30°\,j = 12\cos 30°\,i + 12\sin 30°\,j = 6\sqrt{3}i + 6j$

$\mathbf{F} = \mathbf{F}_1 + \mathbf{F}_2 = \left(6\sqrt{3} - 5\sqrt{2}\right)i + \left(6 + 5\sqrt{2}\right)j \approx 3.32i + 13.07j$

$|\mathbf{F}| \approx \sqrt{(3.32)^2 + (13.07)^2} \approx 13.5$ lb. $\tan\theta = \dfrac{6 + 5\sqrt{2}}{6\sqrt{3} - 5\sqrt{2}} \Rightarrow \theta = \tan^{-1}\dfrac{6 + 5\sqrt{2}}{6\sqrt{3} - 5\sqrt{2}} \approx 76°$.

33. With respect to the water's surface, the woman's velocity is the vector sum of the velocity of the ship with respect to the water, and her velocity with respect to the ship. If we let north be the positive y-direction, then $\mathbf{v} = \langle 0, 22 \rangle + \langle -3, 0 \rangle = \langle -3, 22 \rangle$. The woman's speed is $|\mathbf{v}| = \sqrt{9 + 484} \approx 22.2$ mi/h. The vector $\mathbf{v}$ makes an angle θ with the east, where $\theta = \tan^{-1}\frac{22}{-3} \approx 98°$. Therefore, the woman's direction is about $\text{N}(98 - 90)°\,\text{W} = \text{N}\,8°\,\text{W}$.

35. $|\mathbf{r} - \mathbf{r}_0|$ is the distance between the points (x, y, z) and (x_0, y_0, z_0), so the set of points is a sphere with radius 1 and center (x_0, y_0, z_0).

Alternate Method: $|\mathbf{r} - \mathbf{r}_0| = 1 \Leftrightarrow \sqrt{(x - x_0)^2 + (y - y_0)^2 + (z - z_0)^2} = 1 \Leftrightarrow$

$(x - x_0)^2 + (y - y_0)^2 + (z - z_0)^2 = 1$, which is the equation of a sphere with radius 1 and center (x_0, y_0, z_0).

37. $\mathbf{a} + (\mathbf{b} + \mathbf{c}) = \langle a_1, a_2 \rangle + (\langle b_1, b_2 \rangle + \langle c_1, c_2 \rangle) = \langle a_1, a_2 \rangle + \langle b_1 + c_1, b_2 + c_2 \rangle$

$= \langle a_1 + b_1 + c_1, a_2 + b_2 + c_2 \rangle = \langle (a_1 + b_1) + c_1, (a_2 + b_2) + c_2 \rangle$

$= \langle a_1 + b_1, a_2 + b_2 \rangle + \langle c_1, c_2 \rangle = (\langle a_1, a_2 \rangle + \langle b_1, b_2 \rangle) + \langle c_1, c_2 \rangle = (\mathbf{a} + \mathbf{b}) + \mathbf{c}$

39. $(c + d)\mathbf{a} = (c + d)\langle a_1, a_2, a_3 \rangle = \langle (c + d)a_1, (c + d)a_2, (c + d)a_3 \rangle$

$= \langle ca_1 + da_1, ca_2 + da_2, ca_3 + da_3 \rangle = \langle ca_1, ca_2, ca_3 \rangle + \langle da_1, da_2, da_3 \rangle = c\mathbf{a} + d\mathbf{a}$.

41. Consider quadrilateral $ABCD$ with sides AB and CD parallel and of equal length; that is, $\overrightarrow{AB} = \overrightarrow{DC}$. Thus $\overrightarrow{AD} = \overrightarrow{AB} + \overrightarrow{BD} = \overrightarrow{DC} + \overrightarrow{BD} \left(\text{since } \overrightarrow{AB} = \overrightarrow{DC}\right) = \overrightarrow{BD} + \overrightarrow{DC} = \overrightarrow{BC}$. This shows that sides AD and BC are parallel and have equal lengths.

EXERCISES 11.3

1. $\mathbf{a} \cdot \mathbf{b} = (2)(-3) + (5)(1) = -1$

3. $\mathbf{a} \cdot \mathbf{b} = (4)(-2) + (7)(1) + (-1)(4) = -5$

5. $\mathbf{a} \cdot \mathbf{b} = (2)(1) + (3)(-3) + (-4)(1) = -11$

7. $\mathbf{a} \cdot \mathbf{b} = (2)(3)\cos\frac{\pi}{3} = 6 \cdot \frac{1}{2} = 3$

9. $\mathbf{a} \cdot \mathbf{i} = \langle a_1, a_2, a_3 \rangle \cdot \langle 1, 0, 0 \rangle = (a_1)(1) + (a_2)(0) + (a_3)(0) = a_1$. Similarly

 $\mathbf{a} \cdot \mathbf{j} = (a_1)(0) + (a_2)(1) + (a_3)(0) = a_2$ and $\mathbf{a} \cdot \mathbf{k} = (a_1)(0) + (a_2)(0) + (a_3)(1) = a_3$.

11. $|\mathbf{a}| = \sqrt{1^2 + 2^2 + 2^2} = 3$, $|\mathbf{b}| = \sqrt{3^2 + 4^2 + 0^2} = 5$, $\mathbf{a} \cdot \mathbf{b} = 3 + 8 + 0 = 11$, $\cos\theta = \frac{11}{3 \cdot 5}$, so

 $\theta = \cos^{-1}\frac{11}{15} \approx 43°$.

13. $|\mathbf{a}| = \sqrt{1^2 + 2^2} = \sqrt{5}$, $|\mathbf{b}| = \sqrt{12^2 + (-5)^2} = \sqrt{13}$, $\mathbf{a} \cdot \mathbf{b} = 12 - 10 = 2$, $\cos\theta = \frac{2}{13\sqrt{5}}$, so

 $\theta = \cos^{-1}\frac{2}{13\sqrt{5}} \approx 86°$.

15. $|\mathbf{a}| = \sqrt{36 + 4 + 9} = 7$, $|\mathbf{b}| = \sqrt{3}$, $\mathbf{a} \cdot \mathbf{b} = 6 - 2 - 3 = 1$, $\cos\theta = \frac{1}{7\sqrt{3}}$, so $\theta = \cos^{-1}\frac{1}{7\sqrt{3}} \approx 85°$.

17. Let a, b and c be the angles at vertices A, B and C respectively. Then a is the angle between vectors $\overrightarrow{AB}$ and $\overrightarrow{AC}$, b is the angle between vectors $\overrightarrow{BA}$ and $\overrightarrow{BC}$, and c is the angle between vectors $\overrightarrow{CA}$ and $\overrightarrow{CB}$. Thus

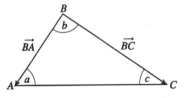

$$\cos a = \frac{\overrightarrow{AB} \cdot \overrightarrow{AC}}{|\overrightarrow{AB}||\overrightarrow{AC}|} = \langle 5, -1, 2 \rangle \cdot \langle -2, -4, -3 \rangle \frac{1}{\sqrt{30 \cdot 29}} = \frac{1}{\sqrt{870}}(-10 + 4 - 6) = -\frac{12}{\sqrt{870}} \text{ and}$$

$a = \cos^{-1}\left(-\frac{12}{\sqrt{870}}\right) \approx 114°$. Similarly

$$\cos b = \frac{\overrightarrow{BA} \cdot \overrightarrow{BC}}{|\overrightarrow{BA}||\overrightarrow{BC}|} = \frac{1}{\sqrt{30 \cdot 83}}\langle -5, 1, -2 \rangle \cdot \langle -7, -3, -5 \rangle = \frac{1}{\sqrt{2490}}(35 - 3 + 10) = \frac{42}{\sqrt{2490}}, \text{ so}$$

$b = \cos^{-1}\frac{42}{\sqrt{2490}} \approx 33°$, and

$$\cos c = \frac{\overrightarrow{CA} \cdot \overrightarrow{CB}}{|\overrightarrow{CA}||\overrightarrow{CB}|} = \frac{1}{\sqrt{29 \cdot 83}}\langle 2, 4, 3 \rangle \cdot \langle 7, 3, 5 \rangle = \frac{1}{\sqrt{2407}}(14 + 12 + 15) = \frac{41}{\sqrt{2407}}, \text{ so}$$

$c = \cos^{-1}\frac{41}{\sqrt{2407}} \approx 33°$.

Alternate Solution: Apply the Law of Cosines three times as follows: $\cos a = \dfrac{\left|\overrightarrow{BC}\right|^2 - \left|\overrightarrow{AB}\right|^2 - \left|\overrightarrow{AC}\right|^2}{2\left|\overrightarrow{AB}\right|\left|\overrightarrow{AC}\right|}$,

$\cos b = \dfrac{\left|\overrightarrow{AC}\right|^2 - \left|\overrightarrow{AB}\right|^2 - \left|\overrightarrow{BC}\right|^2}{2\left|\overrightarrow{AB}\right|\left|\overrightarrow{BC}\right|}$, and $\cos c = \dfrac{\left|\overrightarrow{AB}\right|^2 - \left|\overrightarrow{AC}\right|^2 - \left|\overrightarrow{BC}\right|^2}{2\left|\overrightarrow{AC}\right|\left|\overrightarrow{BC}\right|}$.

19. Since $\mathbf{a} = -2\mathbf{b}$, $\mathbf{a}$ and $\mathbf{b}$ are parallel vectors (and thus not orthogonal). $\mathbf{a} \cdot \mathbf{b} = 8 + (-8) = 0$, so $\mathbf{a}$ and $\mathbf{b}$ are orthogonal.

21. $\mathbf{a} \cdot \mathbf{b} = -2 + 16 + (-15) \neq 0$, so $\mathbf{a}$ and $\mathbf{b}$ aren't orthogonal. Also since $\mathbf{a}$ isn't a scalar multiple of $\mathbf{b}$, $\mathbf{a}$ and $\mathbf{b}$ aren't parallel.

23. $\mathbf{a} \cdot \mathbf{b} = 3 + (-1) + (-2) = 0$, so $\mathbf{a}$ and $\mathbf{b}$ are orthogonal.

25. For the two vectors to be orthogonal we need $(x\mathbf{i} - 2\mathbf{j}) \cdot (x\mathbf{i} + 8\mathbf{j}) = 0 \quad \Rightarrow \quad x^2 - 16 = 0$. So $x = \pm 4$.

27. For the two vectors to be orthogonal, we need $0 = \langle x, 1, 2 \rangle \cdot \langle 3, 4, x \rangle = 3x + 4 + 2x = 5x + 4$ implies $x = -\frac{4}{5}$.

29. Let $\mathbf{a} = a_1\mathbf{i} + a_2\mathbf{j} + a_3\mathbf{k}$ be a vector orthogonal to both $\mathbf{i} + \mathbf{j}$ and $\mathbf{i} + \mathbf{k}$. Then, by definition, $a_1 + a_2 = 0$ and $a_1 + a_3 = 0$, so $a_1 = -a_2 = -a_3$. Furthermore $\mathbf{a}$ is to be a unit vector, so $1 = a_1^2 + a_2^2 + a_3^2 = 3a_1^2$ implies $a_1 = \pm\frac{1}{\sqrt{3}}$. Thus $\mathbf{a} = \frac{1}{\sqrt{3}}\mathbf{i} - \frac{1}{\sqrt{3}}\mathbf{j} - \frac{1}{\sqrt{3}}\mathbf{k}$ and $\mathbf{a} = -\frac{1}{\sqrt{3}}\mathbf{i} + \frac{1}{\sqrt{3}}\mathbf{j} + \frac{1}{\sqrt{3}}\mathbf{k}$ are the two such unit vectors.

31. $|\langle 1, 2, 2 \rangle| = \sqrt{1 + 4 + 4} = 3$, so $\cos\alpha = \frac{1}{3}$, $\cos\beta = \frac{2}{3}$ and $\cos\gamma = \frac{2}{3}$, while $\alpha = \cos^{-1}\frac{1}{3} \approx 71°$ and $\beta = \gamma = \cos^{-1}\frac{2}{3} \approx 48°$.

33. $|-8\mathbf{i} + 3\mathbf{j} + 2\mathbf{k}| = \sqrt{64 + 9 + 4} = \sqrt{77}$, so $\cos\alpha = -\frac{8}{\sqrt{77}}$, $\cos\beta = \frac{3}{\sqrt{77}}$ and $\cos\gamma = \frac{2}{\sqrt{77}}$, while $\alpha = \cos^{-1}\frac{-8}{\sqrt{77}} \approx 156°$, $\beta = \cos^{-1}\frac{3}{\sqrt{77}} \approx 70°$ and $\gamma = \cos^{-1}\frac{2}{\sqrt{77}} \approx 77°$.

35. $|\langle 2, 1.2, 0.8 \rangle| = \sqrt{4 + 1.44 + 0.64} = \frac{5}{5}\sqrt{6.08} = \frac{\sqrt{152}}{5}$, so $\cos\alpha = \frac{10}{\sqrt{152}} = \frac{5}{\sqrt{38}}$, $\cos\beta = \frac{6}{\sqrt{152}} = \frac{3}{\sqrt{38}}$ and $\cos\gamma = \frac{4}{\sqrt{152}} = \frac{2}{\sqrt{38}}$, while $\alpha = \cos^{-1}\frac{5}{\sqrt{38}} \approx 36°$, $\beta = \cos^{-1}\frac{3}{\sqrt{38}} \approx 61°$ and $\gamma = \cos^{-1}\frac{2}{\sqrt{38}} \approx 71°$.

37. $|\mathbf{a}| = \sqrt{4 + 9} = \sqrt{13}$. The scalar projection of $\mathbf{b}$ onto $\mathbf{a}$ is $\text{comp}_{\mathbf{a}}\,\mathbf{b} = \dfrac{\mathbf{a} \cdot \mathbf{b}}{|\mathbf{a}|} = \dfrac{2 \cdot 4 + 3 \cdot 1}{\sqrt{13}} = \dfrac{11}{\sqrt{13}}$.

The vector projection of $\mathbf{b}$ onto $\mathbf{a}$ is $\text{proj}_{\mathbf{a}}\,\mathbf{b} = \dfrac{\mathbf{a} \cdot \mathbf{b}}{|\mathbf{a}|^2}\mathbf{a} = \frac{11}{\sqrt{13}} \cdot \frac{1}{\sqrt{13}}\langle 2, 3 \rangle = \frac{11}{13}\langle 2, 3 \rangle = \left\langle \frac{22}{13}, \frac{33}{13} \right\rangle$.

39. $|\mathbf{a}| = \sqrt{16 + 4 + 0} = 2\sqrt{5}$ so the scalar projection of $\mathbf{b}$ onto $\mathbf{a}$ is $\text{comp}_{\mathbf{a}}\,\mathbf{b} = \dfrac{\mathbf{a} \cdot \mathbf{b}}{|\mathbf{a}|} = \dfrac{1}{2\sqrt{5}}(4 + 2 + 0) = \dfrac{3}{\sqrt{5}}$.

The vector projection of $\mathbf{b}$ onto $\mathbf{a}$ is $\text{proj}_{\mathbf{a}}\,\mathbf{b} = \dfrac{\mathbf{a} \cdot \mathbf{b}}{|\mathbf{a}|^2}\mathbf{a} = \frac{3}{\sqrt{5}} \cdot \frac{1}{2\sqrt{5}}\langle 4, 2, 0 \rangle = \frac{1}{5}\langle 6, 3, 0 \rangle = \left\langle \frac{6}{5}, \frac{3}{5}, 0 \right\rangle$.

41. $|\mathbf{a}| = \sqrt{1 + 0 + 1} = \sqrt{2}$ so the scalar projection of $\mathbf{b}$ onto $\mathbf{a}$ is $\text{comp}_{\mathbf{a}}\,\mathbf{b} = \dfrac{\mathbf{a} \cdot \mathbf{b}}{|\mathbf{a}|} = \frac{1}{\sqrt{2}}(1 + 0 + 0) = \frac{1}{\sqrt{2}}$ while the vector projection is $\text{proj}_{\mathbf{a}}\,\mathbf{b} = \dfrac{\mathbf{a} \cdot \mathbf{b}}{|\mathbf{a}|^2}\mathbf{a} = \frac{1}{\sqrt{2}} \cdot \frac{1}{\sqrt{2}}(\mathbf{i} + \mathbf{k}) = \frac{1}{2}(\mathbf{i} + \mathbf{k})$.

43. $(\text{orth}_{\mathbf{a}}\,\mathbf{b}) \cdot \mathbf{a} = (\mathbf{b} - \text{proj}_{\mathbf{a}}\mathbf{b}) \cdot \mathbf{a} = \mathbf{b} \cdot \mathbf{a} - (\text{proj}_{\mathbf{a}}\,\mathbf{b}) \cdot \mathbf{a} = \mathbf{b} \cdot \mathbf{a} - \dfrac{\mathbf{a} \cdot \mathbf{b}}{|\mathbf{a}|^2}\mathbf{a} \cdot \mathbf{a} = \mathbf{b} \cdot \mathbf{a} - \dfrac{\mathbf{a} \cdot \mathbf{b}}{|\mathbf{a}|^2}|\mathbf{a}|^2 = \mathbf{b} \cdot \mathbf{a} - \mathbf{a} \cdot \mathbf{b} = 0$.

So they are orthogonal by (7).

45. $\text{comp}_{\mathbf{a}}\,\mathbf{b} = \dfrac{\mathbf{a} \cdot \mathbf{b}}{|\mathbf{a}|} = 2 \quad \Leftrightarrow \quad \mathbf{a} \cdot \mathbf{b} = 2|\mathbf{a}| = 2\sqrt{10}$. If $\mathbf{b} = \langle b_1, b_2, b_3 \rangle$ then we need $3b_1 + 0b_2 - 1b_3 = 2\sqrt{10}$.

One possible solution is obtained by taking $b_1 = 0$, $b_2 = 0$, $b_3 = -2\sqrt{10}$.

In general, $\mathbf{b} = \langle s, t, 3s - 2\sqrt{10} \rangle$, $s, t \in \mathbb{R}$.

47. Here $D = (4-2)i + (9-3)j + (15-0)k = 2i + 6j + 15k$ so $W = F \cdot D = 20 + 108 - 90 = 38$ joules.

49. $W = |F||D|\cos\theta = (25)(10)\cos 20°$
≈ 235 ft-lb

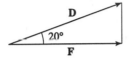

51. **(a)** $a \cdot b$ is a scalar, and the dot product is defined only for vectors, so this expression has no meaning.

(b) This is a scalar multiple of a vector, so it does have meaning.

(c) Both $|a|$ and $b \cdot c$ are scalars, so this is an ordinary product of real numbers, and has meaning.

(d) Both a and $b + c$ are vectors so this dot product has meaning.

(e) $a \cdot b$ is a scalar, but c is a vector, and so they cannot be added and this has no meaning.

(f) $|a|$ is a scalar and the dot product is defined only for vectors, so this has no meaning.

53. First note that $n = \langle a, b \rangle$ is perpendicular to the line, because if $Q_1 = (a_1, b_1)$ and $Q_2 = (a_2, b_2)$ lie on the line, then $n \cdot \overrightarrow{Q_1 Q_2} = aa_2 - aa_1 + bb_2 - bb_1 = 0$, since $aa_2 + bb_2 = -c = aa_1 + bb_1$ from the equation of the line. Let $P_2 = (x_2, y_2)$ lie on the line. Then the distance from P_1 to the line is the absolute value of the scalar projection of $\overrightarrow{P_1 P_2}$ onto n.

$$\text{comp}_n\left(\overrightarrow{P_1 P_2}\right) = \frac{|n \cdot \langle x_2 - x_1, y_2 - y_1 \rangle|}{|n|} = \frac{|ax_2 - ax_1 + by_2 - by_1|}{\sqrt{a^2 + b^2}} = \frac{|ax_1 + by_1 + c|}{\sqrt{a^2 + b^2}} \text{ since}$$

$ax_2 + by_2 = -c$. The required distance is $\dfrac{|3 \cdot -2 + -4 \cdot 3 + 5|}{\sqrt{3^2 + 4^2}} = \dfrac{13}{5}$.

55. For convenience, consider the unit cube positioned so that its back left corner is at the origin, and its edges lie along the coordinate axes. The diagonal of the cube that begins at the origin and ends at $(1, 1, 1)$ has vector representation $\langle 1, 1, 1 \rangle$. The angle θ between this vector and the vector of the edge which also begins at the origin and runs along the x-axis [that is, $\langle 1, 0, 0 \rangle$] is given by $\cos\theta = \dfrac{\langle 1, 1, 1 \rangle \cdot \langle 1, 0, 0 \rangle}{|\langle 1, 1, 1 \rangle||\langle 1, 0, 0 \rangle|} = \dfrac{1}{\sqrt{3}} \Rightarrow$
$\theta = \cos^{-1} \frac{1}{\sqrt{3}} \approx 55°$.

57. Consider the H-C-H combination consisting of the sole carbon atom and the two hydrogen atoms that are at $(1, 0, 0)$ and $(0, 1, 0)$ (or any H-C-H combination, for that matter). Vector representations of the line segments emanating from the carbon atom and extending to these two hydrogen atoms are
$\langle 1 - \frac{1}{2}, 0 - \frac{1}{2}, 0 - \frac{1}{2} \rangle = \langle \frac{1}{2}, -\frac{1}{2}, -\frac{1}{2} \rangle$ and $\langle 0 - \frac{1}{2}, 1 - \frac{1}{2}, 0 - \frac{1}{2} \rangle = \langle -\frac{1}{2}, \frac{1}{2}, -\frac{1}{2} \rangle$. The bond angle, θ, is therefore given by $\cos\theta = \dfrac{\langle \frac{1}{2}, -\frac{1}{2}, -\frac{1}{2} \rangle \cdot \langle -\frac{1}{2}, \frac{1}{2}, -\frac{1}{2} \rangle}{|\langle \frac{1}{2}, -\frac{1}{2}, -\frac{1}{2} \rangle||\langle -\frac{1}{2}, \frac{1}{2}, -\frac{1}{2} \rangle|} = \dfrac{-\frac{1}{4} - \frac{1}{4} + \frac{1}{4}}{\sqrt{\frac{3}{4}}\sqrt{\frac{3}{4}}} = -\dfrac{1}{3} \Rightarrow \theta = \cos^{-1}\left(-\frac{1}{3}\right) \approx 109.5°$.

59. Let $a = \langle a_1, a_2, a_3 \rangle$ and $b = \langle b_1, b_2, b_3 \rangle$.
Property 2: $a \cdot b = \langle a_1, a_2, a_3 \rangle \cdot \langle b_1, b_2, b_3 \rangle = a_1 b_1 + a_2 b_2 + a_3 b_3$
$\qquad = b_1 a_1 + b_2 a_2 + b_3 a_3 = \langle b_1, b_2, b_3 \rangle \cdot \langle a_1, a_2, a_3 \rangle = b \cdot a$
Property 5: $0 \cdot a = \langle 0, 0, 0 \rangle \cdot \langle a_1, a_2, a_3 \rangle = (0)(a_1) + (0)(a_2) + (0)(a_3) = 0$

61. $|\mathbf{a} \cdot \mathbf{b}| = |\|\mathbf{a}\|\mathbf{b}| \cos \theta| = |\mathbf{a}||\mathbf{b}||\cos \theta|$. Since $|\cos \theta| \leq 1$, $|\mathbf{a} \cdot \mathbf{b}| = |\mathbf{a}||\mathbf{b}||\cos \theta| \leq |\mathbf{a}||\mathbf{b}|$.

Note: We have equality in the case of $\cos \theta = \pm 1$, so $\theta = 0$ or $\theta = \pi$, thus equality when **a** and **b** are parallel.

63. (a)

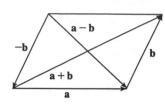

The Parallelogram Law states that the sum of the squares of the lengths of the diagonals of a parallelogram equals the sum of the squares of its (four) sides.

(b) $|\mathbf{a} + \mathbf{b}|^2 = (\mathbf{a} + \mathbf{b}) \cdot (\mathbf{a} + \mathbf{b}) = |\mathbf{a}|^2 + 2(\mathbf{a} \cdot \mathbf{b}) + |\mathbf{b}|^2$ and

$|\mathbf{a} - \mathbf{b}|^2 = (\mathbf{a} - \mathbf{b}) \cdot (\mathbf{a} - \mathbf{b}) = |\mathbf{a}|^2 - 2(\mathbf{a} \cdot \mathbf{b}) + |\mathbf{b}|^2$. Adding these two equations gives

$|\mathbf{a} + \mathbf{b}|^2 + |\mathbf{a} - \mathbf{b}|^2 = 2|\mathbf{a}|^2 + 2|\mathbf{b}|^2$.

EXERCISES 11.4

1. $\mathbf{a} \times \mathbf{b} = \begin{vmatrix} \mathbf{i} & \mathbf{j} & \mathbf{k} \\ 1 & 0 & 1 \\ 0 & 1 & 0 \end{vmatrix} = \begin{vmatrix} 0 & 1 \\ 1 & 0 \end{vmatrix}\mathbf{i} - \begin{vmatrix} 1 & 1 \\ 0 & 0 \end{vmatrix}\mathbf{j} + \begin{vmatrix} 1 & 0 \\ 0 & 1 \end{vmatrix}\mathbf{k} = -\mathbf{i} + \mathbf{k}$

3. $\mathbf{a} \times \mathbf{b} = \begin{vmatrix} \mathbf{i} & \mathbf{j} & \mathbf{k} \\ -2 & 3 & 4 \\ 3 & 0 & 1 \end{vmatrix} = \begin{vmatrix} 3 & 4 \\ 0 & 1 \end{vmatrix}\mathbf{i} - \begin{vmatrix} -2 & 4 \\ 3 & 1 \end{vmatrix}\mathbf{j} + \begin{vmatrix} -2 & 3 \\ 3 & 0 \end{vmatrix}\mathbf{k} = 3\mathbf{i} + 14\mathbf{j} - 9\mathbf{k}$

5. $\mathbf{a} \times \mathbf{b} = \begin{vmatrix} \mathbf{i} & \mathbf{j} & \mathbf{k} \\ 1 & 1 & 1 \\ 1 & 1 & -1 \end{vmatrix} = \begin{vmatrix} 1 & 1 \\ 1 & -1 \end{vmatrix}\mathbf{i} - \begin{vmatrix} 1 & 1 \\ 1 & -1 \end{vmatrix}\mathbf{j} + \begin{vmatrix} 1 & 1 \\ 1 & 1 \end{vmatrix}\mathbf{k} = -2\mathbf{i} + 2\mathbf{j}$

7. $\mathbf{a} \times \mathbf{b} = \begin{vmatrix} \mathbf{i} & \mathbf{j} & \mathbf{k} \\ 2 & 0 & -1 \\ 1 & 2 & 0 \end{vmatrix} = \begin{vmatrix} 0 & -1 \\ 2 & 0 \end{vmatrix}\mathbf{i} - \begin{vmatrix} 2 & -1 \\ 1 & 0 \end{vmatrix}\mathbf{j} + \begin{vmatrix} 2 & 0 \\ 1 & 2 \end{vmatrix}\mathbf{k} = 2\mathbf{i} - \mathbf{j} + 4\mathbf{k}$

9. $\mathbf{a} \times \mathbf{b} = \begin{vmatrix} \mathbf{i} & \mathbf{j} & \mathbf{k} \\ 0 & 1 & 2 \\ 3 & 1 & 0 \end{vmatrix} = \begin{vmatrix} 1 & 2 \\ 1 & 0 \end{vmatrix}\mathbf{i} - \begin{vmatrix} 0 & 2 \\ 3 & 0 \end{vmatrix}\mathbf{j} + \begin{vmatrix} 0 & 1 \\ 3 & 1 \end{vmatrix}\mathbf{k} = -2\mathbf{i} + 6\mathbf{j} - 3\mathbf{k}$

$\mathbf{b} \times \mathbf{a} = \begin{vmatrix} \mathbf{i} & \mathbf{j} & \mathbf{k} \\ 3 & 1 & 0 \\ 0 & 1 & 2 \end{vmatrix} = \begin{vmatrix} 1 & 0 \\ 1 & 2 \end{vmatrix}\mathbf{i} - \begin{vmatrix} 3 & 0 \\ 0 & 2 \end{vmatrix}\mathbf{j} + \begin{vmatrix} 3 & 1 \\ 0 & 1 \end{vmatrix}\mathbf{k} = 2\mathbf{i} - 6\mathbf{j} + 3\mathbf{k}$

(and notice $\mathbf{a} \times \mathbf{b} = -\mathbf{b} \times \mathbf{a}$ here, as we know is always true by Theorem 8.)

11. $\begin{vmatrix} \mathbf{i} & \mathbf{j} & \mathbf{k} \\ 1 & -1 & 1 \\ 0 & 4 & 4 \end{vmatrix} = \begin{vmatrix} -1 & 1 \\ 4 & 4 \end{vmatrix}\mathbf{i} - \begin{vmatrix} 1 & 1 \\ 0 & 4 \end{vmatrix}\mathbf{j} + \begin{vmatrix} 1 & -1 \\ 0 & 4 \end{vmatrix}\mathbf{k} = -8\mathbf{i} - 4\mathbf{j} + 4\mathbf{k}$ and by Theorem 5 this cross product is

orthogonal to both of the original vectors. So two unit vectors orthogonal to both are

$\pm \dfrac{\langle -8, -4, 4 \rangle}{\sqrt{64 + 16 + 16}} = \pm \dfrac{\langle -8, -4, 4 \rangle}{4\sqrt{6}}$ or $\left\langle -\dfrac{2}{\sqrt{6}}, -\dfrac{1}{\sqrt{6}}, \dfrac{1}{\sqrt{6}} \right\rangle$ and $\left\langle \dfrac{2}{\sqrt{6}}, \dfrac{1}{\sqrt{6}}, -\dfrac{1}{\sqrt{6}} \right\rangle$.

13. Let $\mathbf{a} = \langle a_1, a_2, a_3 \rangle$. Then $\mathbf{0} \times \mathbf{a} = \begin{vmatrix} \mathbf{i} & \mathbf{j} & \mathbf{k} \\ 0 & 0 & 0 \\ a_1 & a_2 & a_3 \end{vmatrix} = \begin{vmatrix} 0 & 0 \\ a_2 & a_3 \end{vmatrix} \mathbf{i} - \begin{vmatrix} 0 & 0 \\ a_1 & a_3 \end{vmatrix} \mathbf{j} + \begin{vmatrix} 0 & 0 \\ a_1 & a_2 \end{vmatrix} \mathbf{k} = \mathbf{0},$

$\mathbf{a} \times \mathbf{0} = \begin{vmatrix} \mathbf{i} & \mathbf{j} & \mathbf{k} \\ a_1 & a_2 & a_3 \\ 0 & 0 & 0 \end{vmatrix} = \begin{vmatrix} a_2 & a_3 \\ 0 & 0 \end{vmatrix} \mathbf{i} - \begin{vmatrix} a_1 & a_3 \\ 0 & 0 \end{vmatrix} \mathbf{j} + \begin{vmatrix} a_1 & a_2 \\ 0 & 0 \end{vmatrix} \mathbf{k} = \mathbf{0}.$

15. $\mathbf{a} \times \mathbf{b} = \langle a_2 b_3 - a_3 b_2, a_3 b_1 - a_1 b_3, a_1 b_2 - a_2 b_1 \rangle$

$= \langle (-1)(b_2 a_3 - b_3 a_2), (-1)(b_3 a_1 - b_1 a_3), (-1)(b_1 a_2 - b_2 a_1) \rangle$

$= -\langle b_2 a_3 - b_3 a_2, b_3 a_1 - b_1 a_3, b_1 a_2 - b_2 a_1 \rangle = -\mathbf{b} \times \mathbf{a}$

17. $\mathbf{a} \times (\mathbf{b} + \mathbf{c}) = \mathbf{a} \times \langle b_1 + c_1, b_2 + c_2, b_3 + c_3 \rangle$

$= \langle a_2(b_3 + c_3) - a_3(b_2 + c_2), a_3(b_1 + c_1) - a_1(b_3 + c_3), a_1(b_2 + c_2) - a_2(b_1 + c_1) \rangle$

$= \langle a_2 b_3 + a_2 c_3 - a_3 b_2 - a_3 c_2, a_3 b_1 + a_3 c_1 - a_1 b_3 - a_1 c_3, a_1 b_2 + a_1 c_2 - a_2 b_1 - a_2 c_1 \rangle$

$= \langle (a_2 b_3 - a_3 b_2) + (a_2 c_3 - a_3 c_2), (a_3 b_1 - a_1 b_3) + (a_3 c_1 - a_1 c_3), (a_1 b_2 - a_2 b_1) + (a_1 c_2 - a_2 c_1) \rangle$

$= \langle a_2 b_3 - a_3 b_2, a_3 b_1 - a_1 b_3, a_1 b_2 - a_2 b_1 \rangle + \langle a_2 c_3 - a_3 c_2, a_3 c_1 - a_1 c_3, a_1 c_2 - a_2 c_1 \rangle = (\mathbf{a} \times \mathbf{b}) + (\mathbf{a} \times \mathbf{c})$

19. The vectors corresponding to $\overrightarrow{AB}$ and $\overrightarrow{AD}$ are $\mathbf{a} = \langle 3, -1, 0 \rangle$ and $\mathbf{b} = \langle 2, -2, 0 \rangle$. The area of the parallelogram

with the given vertices is $|\mathbf{a} \times \mathbf{b}| = \left\| \begin{matrix} \mathbf{i} & \mathbf{j} & \mathbf{k} \\ 3 & -1 & 0 \\ 2 & -2 & 0 \end{matrix} \right\| = |(0)\mathbf{i} - (0)\mathbf{j} + (-6 + 2)\mathbf{k}| = |-4\mathbf{k}| = 4.$

21. (a) A vector orthogonal to the plane through the points P, Q and R is a vector orthogonal to both $\overrightarrow{PQ}$ and

$\overrightarrow{PR}$ and thus by Theorem 5 is $\overrightarrow{PQ} \times \overrightarrow{PR}$. Here $\overrightarrow{PQ} = \langle -1, 2, 0 \rangle$ and $\overrightarrow{PR} = \langle -1, 0, 3 \rangle$, so

$\overrightarrow{PQ} \times \overrightarrow{PR} = \langle (2)(3) - (0)(0), (0)(-1) - (-1)(3), (-1)(0) - (2)(-1) \rangle = \langle 6, 3, 2 \rangle$

(or any nonzero scalar multiple of $\langle 6, 3, 2 \rangle$) is the desired vector.

(b) From (a), $\left| \overrightarrow{PQ} \times \overrightarrow{PR} \right| = |\langle 6, 3, 2 \rangle| = \sqrt{36 + 9 + 4} = 7$, so the area of the triangle is $\frac{7}{2}$.

23. (a) $\overrightarrow{PQ} = \langle 1, -1, 1 \rangle$ and $\overrightarrow{PR} = \langle 4, 3, 7 \rangle$ and the desired vector is

$\overrightarrow{PQ} \times \overrightarrow{PR} = \langle (-1)(7) - (1)(3), (1)(4) - (1)(7), (1)(3) - (-1)(4) \rangle = \langle -10, -3, 7 \rangle.$

(b) $\left| \overrightarrow{PQ} \times \overrightarrow{PR} \right| = \sqrt{100 + 9 + 49} = \sqrt{158}$ and the area of the triangle is $\frac{1}{2}\sqrt{158}$.

25. $\mathbf{a} \cdot (\mathbf{b} \times \mathbf{c}) = \begin{vmatrix} 1 & 0 & 6 \\ 2 & 3 & -8 \\ 8 & -5 & 6 \end{vmatrix} = 1 \begin{vmatrix} 3 & -8 \\ -5 & 6 \end{vmatrix} - 0 + 6 \begin{vmatrix} 2 & 3 \\ 8 & -5 \end{vmatrix} = (18 - 40) + 6(-10 - 24) = -226.$ Thus the

volume is $|-226| = 226$ cubic units.

27. $\mathbf{a} = \overrightarrow{PQ} = \langle 1, -1, 2 \rangle$, $\mathbf{b} = \overrightarrow{PR} = \langle 3, 0, 0 \rangle$ and $\mathbf{c} = \overrightarrow{PS} = \langle 2, -2, -3 \rangle.$

$\mathbf{a} \cdot (\mathbf{b} \times \mathbf{c}) = \begin{vmatrix} 1 & -1 & 2 \\ 3 & 0 & 6 \\ 2 & -2 & -3 \end{vmatrix} = 1 \begin{vmatrix} 0 & 6 \\ -2 & -3 \end{vmatrix} - (-1) \begin{vmatrix} 3 & 6 \\ 2 & -3 \end{vmatrix} + 2 \begin{vmatrix} 3 & 0 \\ 2 & -2 \end{vmatrix} = 12 - 21 - 12 = -21,$ and the

volume is 21 cubic units.

29. $\mathbf{a} \cdot (\mathbf{b} \times \mathbf{c}) = \begin{vmatrix} 2 & 3 & 1 \\ 1 & -1 & 0 \\ 7 & 3 & 2 \end{vmatrix} = 2 \begin{vmatrix} -1 & 0 \\ 3 & 2 \end{vmatrix} - 3 \begin{vmatrix} 1 & 0 \\ 7 & 2 \end{vmatrix} + 1 \begin{vmatrix} 1 & -1 \\ 7 & 3 \end{vmatrix} = -4 - 6 + 10 = 0$, which says that the

volume of the parallelepiped determined by $\mathbf{a}$, $\mathbf{b}$ and $\mathbf{c}$ is 0, and thus these three vectors are coplanar.

31. The magnitude of the torque is

$$|\tau| = |\mathbf{r} \times \mathbf{F}| = |\mathbf{r}||\mathbf{F}|\sin\theta = (0.18\,\text{m})(60\,\text{N})\sin(180 - (70 + 10))° = 10.8\sin 100° \approx 10.6\,\text{J}.$$

33. (a)

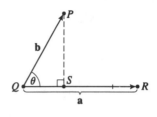

The distance between a point and a line is the length of the perpendicular from the point to the line, here $\left|\overrightarrow{PS}\right| = d$.

But referring to triangle PQS,

$d = \left|\overrightarrow{PS}\right| = \left|\overrightarrow{QP}\right|\sin\theta = |\mathbf{b}|\sin\theta$. But θ is the angle

between $\overrightarrow{QP} = \mathbf{b}$ and $\overrightarrow{QR} = \mathbf{a}$. Thus by Theorem 6,

$\sin\theta = \dfrac{|\mathbf{a} \times \mathbf{b}|}{|\mathbf{a}||\mathbf{b}|}$ and so $d = |\mathbf{b}|\sin\theta = \dfrac{|\mathbf{b}||\mathbf{a} \times \mathbf{b}|}{|\mathbf{a}||\mathbf{b}|} = \dfrac{|\mathbf{a} \times \mathbf{b}|}{|\mathbf{a}|}$.

(b) $\mathbf{a} = \overrightarrow{QR} = \langle -1, -2, -1 \rangle$ and $\mathbf{b} = \overrightarrow{QP} = \langle 1, -5, -7 \rangle$. Then

$\mathbf{a} \times \mathbf{b} = \langle (-2)(-7) - (-1)(-5), (-1)(1) - (-1)(-7), (-1)(-5) - (-2)(1) \rangle = \langle 9, -8, 7 \rangle$. Thus the

distance is $d = \dfrac{|\mathbf{a} \times \mathbf{b}|}{|\mathbf{a}|} = \dfrac{1}{\sqrt{6}}\sqrt{81 + 64 + 49} = \sqrt{\dfrac{194}{6}} = \sqrt{\dfrac{97}{3}}$.

35. $(\mathbf{a} - \mathbf{b}) \times (\mathbf{a} + \mathbf{b}) = (\mathbf{a} - \mathbf{b}) \times \mathbf{a} + (\mathbf{a} - \mathbf{b}) \times \mathbf{b}$ by Theorem 8 #3

$\qquad = \mathbf{a} \times \mathbf{a} + (-\mathbf{b}) \times \mathbf{a} + \mathbf{a} \times \mathbf{b} + (-\mathbf{b}) \times \mathbf{b}$ by Theorem 8 #4

$\qquad = (\mathbf{a} \times \mathbf{a}) - (\mathbf{b} \times \mathbf{a}) + (\mathbf{a} \times \mathbf{b}) - (\mathbf{b} \times \mathbf{b})$ by Theorem 8 #2

$\qquad = \mathbf{0} - (\mathbf{b} \times \mathbf{a}) + (\mathbf{a} \times \mathbf{b}) - \mathbf{0}$ by Example 2

$\qquad = (\mathbf{a} \times \mathbf{b}) + (\mathbf{a} \times \mathbf{b})$ by Theorem 8 #1

$\qquad = 2(\mathbf{a} \times \mathbf{b})$

37. $\mathbf{a} \times (\mathbf{b} \times \mathbf{c}) + \mathbf{b} \times (\mathbf{c} \times \mathbf{a}) + \mathbf{c} \times (\mathbf{a} \times \mathbf{b})$

$\qquad = [(\mathbf{a} \cdot \mathbf{c})\mathbf{b} - (\mathbf{a} \cdot \mathbf{b})\mathbf{c}] + [(\mathbf{b} \cdot \mathbf{a})\mathbf{c} - (\mathbf{b} \cdot \mathbf{c})\mathbf{a}] + [(\mathbf{c} \cdot \mathbf{b})\mathbf{a} - (\mathbf{c} \cdot \mathbf{a})\mathbf{b}]$ (by Exercise 36)

$\qquad = (\mathbf{a} \cdot \mathbf{c})\mathbf{b} - (\mathbf{a} \cdot \mathbf{b})\mathbf{c} + (\mathbf{a} \cdot \mathbf{b})\mathbf{c} - (\mathbf{b} \cdot \mathbf{c})\mathbf{a} + (\mathbf{b} \cdot \mathbf{c})\mathbf{a} - (\mathbf{a} \cdot \mathbf{c})\mathbf{b} = \mathbf{0}$

39. (a) No. If $\mathbf{a} \cdot \mathbf{b} = \mathbf{a} \cdot \mathbf{c}$, then $\mathbf{a} \cdot (\mathbf{b} - \mathbf{c}) = 0$, so $\mathbf{a}$ is perpendicular to $\mathbf{b} - \mathbf{c}$, which can happen if $\mathbf{b} \neq \mathbf{c}$. For example, let $\mathbf{a} = \langle 1, 1, 1 \rangle$, $\mathbf{b} = \langle 1, 0, 0 \rangle$ and $\mathbf{c} = \langle 0, 1, 0 \rangle$.

(b) No. If $\mathbf{a} \times \mathbf{b} = \mathbf{a} \times \mathbf{c}$ then $\mathbf{a} \times (\mathbf{b} - \mathbf{c}) = \mathbf{0}$, which implies that $\mathbf{a}$ is parallel to $\mathbf{b} - \mathbf{c}$ which of course can happen if $\mathbf{b} \neq \mathbf{c}$.

(c) Yes. Since $\mathbf{a} \cdot \mathbf{c} = \mathbf{a} \cdot \mathbf{b}$, $\mathbf{a}$ is perpendicular to $\mathbf{b} - \mathbf{c}$, by part (a). From part (b), $\mathbf{a}$ is also parallel to $\mathbf{b} - \mathbf{c}$. Thus since $\mathbf{a} \neq \mathbf{0}$ but is both parallel and perpendicular to $\mathbf{b} - \mathbf{c}$, we have $\mathbf{b} - \mathbf{c} = \mathbf{0}$ or $\mathbf{b} = \mathbf{c}$.

EXERCISES 11.5

1. $\mathbf{r}_0 = 3\mathbf{i} - \mathbf{j} + 8\mathbf{k}$ and $\mathbf{v} = \mathbf{a}$ so the vector equation is

$\mathbf{r} = (3\mathbf{i} - \mathbf{j} + 8\mathbf{k}) + t(2\mathbf{i} + 3\mathbf{j} + 5\mathbf{k}) = (3 + 2t)\mathbf{i} + (-1 + 3t)\mathbf{j} + (8 + 5t)\mathbf{k}$, and the parametric equations are

$x = 3 + 2t, y = -1 + 3t, z = 8 + 5t$.

3. $\mathbf{r} = (\mathbf{j} + 2\mathbf{k}) + t(6\mathbf{i} + 3\mathbf{j} + 2\mathbf{k}) = (6t)\mathbf{i} + (1 + 3t)\mathbf{j} + (2 + 2t)\mathbf{k}$ is the vector equation, while

$x = 6t, y = 1 + 3t, z = 2 + 2t$ are the parametric equations.

5. The parallel vector is $\mathbf{v} = \langle 6 - 2, 0 - 1, 3 - 8 \rangle = \langle 4, -1, -5 \rangle$ so the direction numbers are $a = 4, b = -1$,

$c = -5$. Letting $P_0 = (2, 1, 8)$, the parametric equations are $x = 2 + 4t, y = 1 - t, z = 8 - 5t$ and symmetric

equations are $\dfrac{x - 2}{4} = \dfrac{y - 1}{-1} = \dfrac{z - 8}{-5}$.

7. $\mathbf{v} = \langle 0, 1, -5 \rangle$ and letting $P_0 = (3, 1, -1)$, the parametric equations are $x = 3, y = 1 + t, z = -1 - 5t$ while

the symmetric equations are $x = 3, y - 1 = \dfrac{z + 1}{-5}$. Notice here that the direction number $a = 0$, so rather than

writing $\dfrac{x - 3}{0}$ in the symmetric equation we write the equation $x = 3$ separately.

9. $\mathbf{v} = \langle \frac{1}{3}, 4, -9 \rangle$ and letting $P_0 = \left(-\frac{1}{3}, 1, 1 \right)$, the parametric equations are $x = -\frac{1}{3} + \frac{1}{3}t, y = 1 + 4t, z = 1 - 9t$

while the symmetric equations are $\dfrac{x + 1/3}{1/3} = \dfrac{y - 1}{4} = \dfrac{z - 1}{-9}$.

11. Direction vectors of the lines are respectively $\mathbf{v}_1 = \langle 6, 9, 12 \rangle$ and $\mathbf{v}_2 = \langle 4, 6, 8 \rangle$ and since $\mathbf{v}_1 = \frac{3}{2}\mathbf{v}_2$ the direction

vectors and thus the lines are parallel.

13. **(a)** A direction vector of the line with given parametric equations is $\mathbf{v} = \langle 2, 3, -7 \rangle$ and the desired parallel line

must also have $\mathbf{v}$ as a direction vector. Here $P_0 = (0, 2, -1)$ so the symmetric equations for the line are

$\dfrac{x}{2} = \dfrac{y - 2}{3} = \dfrac{z + 1}{-7}$.

(b) The line intersects the xy-plane when $z = 0$, so we need $\dfrac{x}{2} = \dfrac{y - 2}{3} = \dfrac{1}{-7}$ or

$x = -\frac{2}{7}, y = \frac{11}{7}$. Thus the point of intersection with the xy-plane is $\left(-\frac{2}{7}, \frac{11}{7}, 0 \right)$. Similarly for the yz- and

xz-planes, we need respectively $x = 0$ and $y = 0$ or $0 = \dfrac{y - 2}{3} = \dfrac{z + 1}{-7}$ and $\dfrac{x}{2} = -\dfrac{2}{3} = \dfrac{z + 1}{-7}$ or $y = 2$,

$z = -1$ and $x = -\frac{4}{3}, z = \frac{11}{3}$. Thus the line intersects the yz-plane at $(0, 2, -1)$ and the xz-plane at

$\left(-\frac{4}{3}, 0, \frac{11}{3} \right)$.

15. The lines aren't parallel since the direction vectors $\langle 2, 4, -3 \rangle$ and $\langle 1, 3, 2 \rangle$ aren't parallel so we check to see if the

lines intersect. The parametric equations of the lines are L_1: $x = 4 + 2t, y = -5 + 4t, z = 1 - 3t$ and L_2:

$x = 2 + s, y = -1 + 3s, z = 2s$. For the lines to intersect we must be able to find one value of t and one value

of s satisfying the following three equations: $4 + 2t = 2 + s, -5 + 4t = -1 + 3s, 1 - 3t = 2s$. Solving the

first two equations we get $t = -5, s = -8$ and checking we see that these values don't satisfy the third equation.

Thus L_1 and L_2 aren't parallel and don't intersect, so they must be skew lines.

17. Since the direction vectors are $v_1 = \langle -6, 9, -3 \rangle$ and $v_2 = \langle 2, -3, 1 \rangle$, we have $v_1 = -3v_2$ so the lines are parallel.

19. Setting $a = 7$, $b = 1$, $c = 4$, $x_0 = 1$, $y_0 = 4$, $z_0 = 5$ in Equation 6 gives $7(x - 1) + 1(y - 4) + 4(z - 5) = 0$ or $7x + y + 4z = 31$ to be the equation of the plane.

21. Setting $a = 15$, $b = 9$, $c = -12$, $x_0 = 1$, $y_0 = 2$, $z_0 = 3$ in Equation 6 gives
$15(x - 1) + 9(y - 2) - 12(z - 3) = 0$ or $5x + 3y - 4z = -1$ to be the equation of the plane.

23. Since the two planes are parallel, they will have the same normal vectors. Thus $n = \langle 1, 1, -1 \rangle$ and the equation of the plane is $1(x - 6) + 1(y - 5) + 1(z - 2) = 0$ or $x + y - z = 13$.

25. The equation is $3(x + 1) - 4(y - 3) - 6(z + 8) = 0$ or $3x - 4y - 6z = 33$.

27. Here the vectors $a = \langle 1, 1, 1 \rangle$ and $b = \langle 1, 2, 3 \rangle$ lie in the plane, so $a \times b$ is a normal vector to the plane. Thus $n = a \times b = \langle 3 - 2, 1 - 3, 2 - 1 \rangle = \langle 1, -2, 1 \rangle$ and the equation of the plane is $x - 2y + z = 0$.

29. $a = \langle -1, -2, -1 \rangle$ and $b = \langle 3, 1, 9 \rangle$ so a normal vector to the plane is
$n = a \times b = \langle -18 + 1, -3 + 9, -1 + 6 \rangle = \langle -17, 6, 5 \rangle$ and the equation of the plane is
$-17(x - 1) + 6(y - 0) + 5(z + 3) = 0$ or $-17x + 6y + 5z = -32$.

31. To find the equation of the plane we must first find two nonparallel vectors in the plane, then their cross product will be a normal vector to the plane. But since the given line lies in the plane, its direction vector $a = \langle 2, -3, -1 \rangle$ is one vector in the plane. To find another nonparallel vector b which lies in the plane pick any point on the line [say $(1, 2, 3)$, found by setting $t = 0$] and let b be the vector connecting this point to the given point in the plane. (But beware; we should first check that the given point is not on the given line. If it were on the line, the plane wouldn't be uniquely determined. What would n then be when we set $n = a \times b$?) Here $b = \langle 0, 4, -7 \rangle$ so $n = a \times b = \langle 21 + 4, 0 + 14, 8 - 0 \rangle = \langle 25, 14, 8 \rangle$ and the equation of the plane is $25(x - 1) + 14(y - 7) + 8(z + 4) = 0$ or $25x + 14y + 8z = 77$.

33. $(0, 0, 0)$ is a point on $x = y = z$. $\langle 1, 1, 1 \rangle$ is the direction of this line, and thus also of the plane.
$\langle 0 - 0, 1 - 0, 2 - 0 \rangle = \langle 0, 1, 2 \rangle$ is also a vector in the plane. Therefore,
$n = \langle 1, 1, 1 \rangle \times \langle 0, 1, 2 \rangle = \langle 2 - 1, -2 + 0, 1 - 0 \rangle = \langle 1, -2, 1 \rangle$. Choosing $(x_0, y_0, z_0) = (0, 0, 0)$, the equation of the plane is, by Equation 7, $x - 2y + z = 1 \cdot 0 - 2 \cdot 0 + 1 \cdot 0 \iff x - 2y + z = 0$.

35. Substituting the parametric equations of the line into the equation of the plane gives
$x + y + z = 1 + t + 2t + 3t = 1 \implies t = 0$. This value of t corresponds to the point of intersection $(1, 0, 0)$, obtained by substitution of $t = 0$ into the equations of the line.

37. Substituting the parametric equations of the line into the equation of the plane gives

$2x + y - z + 5 = 2(1 + 2t) + (-1) - t + 5 = 0 \quad \Rightarrow \quad 3t + 6 = 0 \quad \Rightarrow \quad t = -2$. Therefore, the point of intersection is $x = 1 + 2(-2) = -3$, $y = -1$ and $z = -2$ and the point of intersection is $(-3, -1, -2)$.

39. Setting $x = 0$, we see that $(0, 1, 0)$ satisfies the equations of both planes, so that they do in fact have a line of intersection. $\mathbf{v} = \mathbf{n_1} \times \mathbf{n_2} = \langle 1, 1, 1 \rangle \times \langle 1, 0, 1 \rangle = \langle 1, 0, -1 \rangle$ is the direction of this line. Therefore, direction numbers of the intersecting line are $1, 0, -1$.

41. The normal vectors to the planes are respectively $\mathbf{n_1} = \langle 1, 0, 1 \rangle$ and $\mathbf{n_2} = \langle 0, 1, 1 \rangle$. Thus the normal vectors and consequently the planes aren't parallel. Furthermore $\mathbf{n_1} \cdot \mathbf{n_2} = 1 \neq 0$ so the planes aren't perpendicular. Letting θ be the angle between the two planes, we have $\cos \theta = \dfrac{\mathbf{n_1} \cdot \mathbf{n_2}}{|\mathbf{n_1}||\mathbf{n_2}|} = \dfrac{1}{\sqrt{2}\sqrt{2}} = \dfrac{1}{2}$ and $\theta = \cos^{-1} \frac{1}{2} = 60°$.

43. The respective normals are $\mathbf{n_1} = \langle 1, 4, -3 \rangle$ and $\mathbf{n_2} = \langle -3, 6, 7 \rangle$ so the normals (and thus the planes) fail to be parallel. But $\mathbf{n_1} \cdot \mathbf{n_2} = -3 + 24 - 21 = 0$ so the normals and thus the planes are perpendicular.

45. The normals are $\mathbf{n_1} = \langle 2, 4, -2 \rangle$ and $\mathbf{n_2} = \langle -3, -6, 3 \rangle$ respectively. So $\mathbf{n_1} = -\frac{3}{2}\mathbf{n_2}$. The normals, and hence the planes, are parallel. The planes are not the same because $-\frac{3}{2}(1) \neq 10$.

47. (a) To find a point on the line of intersection, set one of the variables equal to a constant, say $z = 0$. (This will only work if the line of intersection crosses the xy-plane, otherwise try setting x or y equal to 0.) Then the equations of the planes reduce to $x + y = 2$ and $3x - 4y = 6$. Solving these two equations gives $x = 2$, $y = 0$. So a point on the line of intersection is $(2, 0, 0)$. The direction of the line is

$\mathbf{v} = \mathbf{n_1} \times \mathbf{n_2} = \langle 5 - 4, -3 - 5, -4 - 3 \rangle = \langle 1, -8, -7 \rangle$, and symmetric equations for the line are

$x - 2 = \dfrac{y}{-8} = \dfrac{z}{-7}$.

(b) The angle between the planes satisfies $\cos \theta = \dfrac{\mathbf{n_1} \cdot \mathbf{n_2}}{|\mathbf{n_1}||\mathbf{n_2}|} = \dfrac{3 - 4 - 5}{\sqrt{3}\sqrt{50}} = -\dfrac{\sqrt{6}}{5}$. Therefore

$\theta = \cos^{-1}\left(-\frac{\sqrt{6}}{5}\right) \approx 119°$ (or 61°).

49. Setting $x = 0$, the equations of the two planes become $z = y$ and $5y + z = -1$, which intersect at $y = -\frac{1}{6}$ and $z = -\frac{1}{6}$. Thus we can choose $(x_0, y_0, z_0) = \left(0, -\frac{1}{6}, -\frac{1}{6}\right)$. The direction of the line of intersection is

$\mathbf{v} = \mathbf{n_1} \times \mathbf{n_2} = \langle 2, -5, -1 \rangle \times \langle 1, 1, -1 \rangle = \langle 6, 1, 7 \rangle$. Parametric equations for this line are, by Equation 2,

$x = 6t$, $y = -\frac{1}{6} + t$, $z = -\frac{1}{6} + 7t$.

51. The plane contains all perpendicular bisectors of the line segment joining $(1, 1, 0)$ and $(0, 1, 1)$. All of these bisectors pass through the midpoint of this segment $\left(\dfrac{1}{2}, \dfrac{1+1}{2}, \dfrac{1}{2}\right) = \left(\dfrac{1}{2}, 1, \dfrac{1}{2}\right)$. The direction of this line segment $\langle 1 - 0, 1 - 1, 0 - 1 \rangle = \langle 1, 0, -1 \rangle$ is perpendicular to the plane so that we can choose this to be $\mathbf{n}$. Therefore the equation of the plane is $1\left(x - \frac{1}{2}\right) + 0(y - 1) - 1\left(z - \frac{1}{2}\right) = 0 \quad \Leftrightarrow \quad x = z$.

53. A direction vector for the line of intersection is $\mathbf{a} = \mathbf{n}_1 \times \mathbf{n}_2 = \langle 1, 1, -1 \rangle \times \langle 2, -1, 3 \rangle = \langle 2, -5, -3 \rangle$ and $\mathbf{a}$ is parallel to the desired plane. Another vector parallel to the plane is the vector connecting any point on the line of intersection to the given point $(-1, 2, 1)$ in the plane. Setting $x = 0$, the equation of the planes reduce to $y - z = 2$ and $-y + 3z = 1$ with simultaneous solution $y = \frac{7}{2}$ and $z = \frac{3}{2}$. So a point on the line is $\left(0, \frac{7}{2}, \frac{3}{2} \right)$ and another vector parallel to the plane is $\left\langle -1, -\frac{3}{2}, -\frac{1}{2} \right\rangle$. Then a normal vector to the plane is $\mathbf{n} = \langle 2, -5, -3 \rangle \times \left\langle -1, -\frac{3}{2}, -\frac{1}{2} \right\rangle = \langle -2, 4, -8 \rangle$ and an equation of the plane is $-2(x + 1) + 4(y - 2) - 8(z - 1) = 0$ or $x - 2y + 4z = -1$.

55. The plane contains the points $(a, 0, 0)$, $(0, b, 0)$ and $(0, 0, c)$. Thus the vectors $\mathbf{a} = \langle -a, b, 0 \rangle$ and $\mathbf{b} = \langle -a, 0, c \rangle$ lie in the plane and $\mathbf{n} = \mathbf{a} \times \mathbf{b} = \langle bc - 0, 0 + ac, 0 + ab \rangle = \langle bc, ac, ab \rangle$ is a normal vector to the plane. The equation of the plane is therefore $bcx + acy + abz = abc + 0 + 0$ or $bcx + acy + abz = abc$. Notice that if $a \neq 0$, $b \neq 0$ and $c \neq 0$ then we can rewrite the equation as $\frac{x}{a} + \frac{y}{b} + \frac{z}{c} = 1$. This is a good equation to remember!

57. Two vectors which are perpendicular to the required line are the normal of the given plane, $\langle 1, 1, 1 \rangle$, and a direction vector for the given line, $\langle 1, -1, 2 \rangle$. So a direction vector for the required line is $\langle 1, 1, 1 \rangle \times \langle 1, -1, 2 \rangle = \langle 3, -1, -2 \rangle$. So L is given by $\langle x, y, z \rangle = \langle 0, 1, 2 \rangle + t \langle 3, -1, -2 \rangle$ or parametric equations $x = 3t$, $y = 1 - t$, $z = 2 - 2t$.

59. Let P_i have normal vector $\mathbf{n}_i$, then $\mathbf{n}_1 = \langle 4, -2, 6 \rangle$, $\mathbf{n}_2 = \langle 4, -2, -2 \rangle$, $\mathbf{n}_3 = \langle -6, 3, -9 \rangle$, $\mathbf{n}_4 = \langle 2, -1, -1 \rangle$. Then $\mathbf{n}_1 = \frac{-2}{3} \mathbf{n}_3$, so $\mathbf{n}_1$ and $\mathbf{n}_3$ are parallel, and hence P_1 and P_3 are parallel, similarly P_2 and P_4 are parallel because $\mathbf{n}_2 = 2\mathbf{n}_4$. However $\mathbf{n}_1$ and $\mathbf{n}_2$ are not parallel. $\left(0, 0, \frac{1}{2} \right)$ lies on P_1, but not on P_3, so they are not the same plane, but both P_2 and P_4 contain the point $(0, 0, -3)$, so they are identical.

61. Let $Q = (2, 2, 0)$ and $R = (3, -1, 5)$, points on the line corresponding to $t = 0$ and $t = 1$. Let $P = (1, 2, 3)$, then $\mathbf{a} = \overrightarrow{QR} = \langle 1, -3, 5 \rangle$, $\mathbf{b} = \overrightarrow{QP} = \langle -1, 0, 3 \rangle$. The distance is
$$d = \frac{|\mathbf{a} \times \mathbf{b}|}{|\mathbf{a}|} = \frac{|\langle 1, -3, 5 \rangle \times \langle -1, 0, 3 \rangle|}{|\langle 1, -3, 5 \rangle|} = \frac{|\langle -9, -8, -3 \rangle|}{|\langle 1, -3, 5 \rangle|} = \frac{\sqrt{9^2 + 8^2 + 3^2}}{\sqrt{1^2 + 3^2 + 5^2}} = \frac{\sqrt{154}}{\sqrt{35}} = \sqrt{\frac{22}{5}}$$

63. $D = \dfrac{1}{\sqrt{1 + 4 + 4}} [(1)(2) + (-2)(8) + (-2)(5) - 1] = \dfrac{25}{3}$

65. Put $y = z = 0$ in the equation of the first plane, to get the point $(-1, 0, 0)$ on the plane. Because the planes are parallel, the distance, D, between them is the distance from $(-1, 0, 0)$ to the second plane. Using the formula from Example 8, $D = \dfrac{|3(-1) + 6(0) - 3(0) - 4|}{\sqrt{3^2 + 6^2 + (-3)^2}} = \dfrac{7}{3\sqrt{6}}$.

67. The distance between two parallel planes is the same as the distance between a point on one of the planes and the other plane. Let $P_0 = (x_0, y_0, z_0)$ be a point on the plane given by $ax + by + cz = d_1$. Then $ax_0 + by_0 + cz_0 = d_1$ and the distance between P_0 and the plane given by $ax + by + cz = d_2$ is

$$D = \frac{1}{\sqrt{a^2 + b^2 + c^2}} |ax_0 + by_0 + cz_0 - d_2| = \frac{|d_1 - d_2|}{\sqrt{a^2 + b^2 + c^2}}.$$

69. $L_1 \colon x = y = z \;\Rightarrow\; x = y$ (1). $L_2 \colon x + 1 = y/2 = z/3 \;\Rightarrow\; x + 1 = y/2$ (2). The solution of (1) and (2) is $x = y = -2$. However, when $x = -2$, $x = z \;\Rightarrow\; z = -2$, but $x + 1 = z/3 \;\Rightarrow\; z = -3$, a contradiction. Hence the lines do not intersect. For L_1, $\mathbf{v}_1 = \langle 1, 1, 1 \rangle$, and for L_2, $\mathbf{v}_2 = \langle 1, 2, 3 \rangle$, so the lines are not parallel. Thus the lines are skew lines. If two lines are skew, they can be viewed as lying in two parallel planes and so the distance between the skew lines would be the same as the distance between these parallel planes. The common normal vector to the planes must be perpendicular to both $\langle 1, 1, 1 \rangle$ and $\langle 1, 2, 3 \rangle$, the direction vectors of the two lines. So set $\mathbf{n} = \langle 1, 1, 1 \rangle \times \langle 1, 2, 3 \rangle = \langle 3 - 2, -3 + 1, 2 - 1 \rangle = \langle 1, -2, 1 \rangle$. From above, we know that $(-2, -2, -2)$ and $(-2, -2, -3)$ are points of L_1 and L_2 respectively. So in the notation of Equation 7, $d_1 = 1(-2) - 2(-2) + 1(-2) = 0$ and $d_2 = 1(-2) - 2(-2) + 1(-3) = -1$. By Exercise 67, the distance between these two skew lines is $D = \dfrac{|0 - (-1)|}{\sqrt{1 + 4 + 1}} = \dfrac{1}{\sqrt{6}}$.

Alternate solution (without reference to planes): A vector which is perpendicular to both of the lines is $\mathbf{n} = \langle 1, 1, 1 \rangle \times \langle 1, 2, 3 \rangle = \langle 1, -2, 1 \rangle$. Pick any point on each of the lines, say $(-2, -2, -2)$ and $(-2, -2, -3)$, and form the vector $\mathbf{b} = \langle 0, 0, 1 \rangle$ connecting the two points. The distance between the two skew lines is the absolute value of the scalar projection of $\mathbf{b}$ along $\mathbf{n}$, that is, $D = \dfrac{|\mathbf{n} \cdot \mathbf{b}|}{|\mathbf{n}|} = \dfrac{|1 \cdot 0 - 2 \cdot 0 + 1 \cdot 1|}{\sqrt{1 + 4 + 1}} = \dfrac{1}{\sqrt{6}}.$

71. If $a \neq 0$ then $ax + by + cz = d \;\Rightarrow\; ax - d + by + cz = 0 \;\Rightarrow\; a(x - d/a) + b(y - 0) + c(z - 0) = 0$ which by (6) is the scalar equation of the plane through the point $(d/a, 0, 0)$ with normal vector $\langle a, b, c \rangle$. Similarly if $b \neq 0$ (or if $c \neq 0$) the equation of the plane can be rewritten as $a(x - 0) + b(y - d/b) + c(z - 0) = 0$ [or as $a(x - 0) + b(y - 0) + c(z - d/c) = 0$] which by (6) is the scalar equation of a plane through the point $(0, d/b, 0)$ [or the point $(0, 0, d/c)$] with normal vector $\langle a, b, c \rangle$.

EXERCISES 11.6

1. The trace in any plane $x = k$ is given by
$z^2 - y^2 = 1 - k^2$, $x = k$ whose graph is a
hyperbola. The trace in any plane $y = k$ is
the circle given by $x^2 + z^2 = 1 + k^2$, $y = k$,
and the trace in any plane $z = k$ is the
hyperbola given by $x^2 - y^2 = 1 - k^2$, $z = k$.
Thus the surface is a hyperboloid of one sheet
with axis the y-axis.

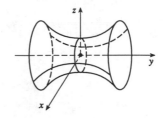

3. Traces: $x = k$, $9y^2 + 36z^2 = 36 - 4k^2$, an
ellipse for $|k| < 3$; $y = k$, $4x^2 + 36z^2 = 36 - 9k^2$,
an ellipse for $|k| < 2$;
$z = k$, $4x^2 + 9y^2 = 36(1 - k^2)$,
an ellipse for $|k| < 1$. Thus the surface is
an ellipsoid with center at the origin and axes
along the x-, y- and z-axes.

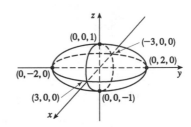

5. Traces: $x = k$, $4z^2 - y^2 = 1 + k^2$,
a hyperbola; $y = k$, $4z^2 - x^2 = 1 + k^2$,
a hyperbola; $z = k$, $-x^2 - y^2 = 1 - 4k^2$
or $x^2 + y^2 = 4k^2 - 1$, a circle for $k > \frac{1}{2}$ or
$k < -\frac{1}{2}$. Thus the surface is a hyperboloid
of two sheets with axis the z-axis.

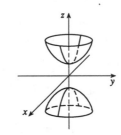

7. Traces: $x = k$, $z = y^2$, a parabola;
$y = k$, $z = k^2$, a line; $z = k$,
$y^2 = k$ or $y = \pm\sqrt{k}$, two parallel
lines for $k > 0$. Thus the surface
is a parabolic cylinder opening upward.

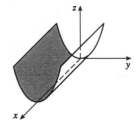

9. Traces: $x = k$, $y^2 = k^2 + z^2$ or $y^2 - z^2 = k^2$, a hyperbola
for $k \neq 0$ and two intersecting lines for $k = 0$;
$y = k$, $x^2 + z^2 = k^2$, a circle for $k \neq 0$;
$z = k$, $y^2 = x^2 + k^2$ or $y^2 - x^2 = k^2$, a hyperbola for $k \neq 0$
and two intersecting lines for $k = 0$. Thus the surface is
a cone (right circular) with axis the y-axis and vertex the origin.

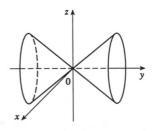

11. Traces: $x = k$, $k^2 + 4z^2 - y = 0$ or $y - k^2 = 4z^2$, a parabola;

 $y = k$, $x^2 + 4z^2 = k$, an ellipse for $k > 0$;

 $z = k$, $x^2 + 4k^2 - y = 0$ or $y - 4k^2 = x^2$, a parabola.

 Thus the surface is an elliptic paraboloid with axis the y-axis

 and vertex the origin.

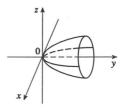

13. Traces: $x = k$ $\Rightarrow$ $\dfrac{y^2}{9} + \dfrac{z^2}{1} = 1$, ellipses;

 $y = k$, $|k| \le 3$ $\Rightarrow$ $9z^2 = 9 - k^2$ $\Rightarrow$ $z = \pm\sqrt{1 - (k^2/9)}$,

 pairs of lines; $z = k$, $|k| \le 1$ $\Rightarrow$ $y^2 = 9(1 - k^2)$ $\Rightarrow$

 $y = \pm 3\sqrt{1 - k^2}$, pairs of lines. This is the equation of

 an elliptic cylinder, centered at the origin, whose axis is the x-axis.

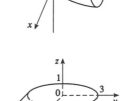

15. Traces: $x = k$ $\Rightarrow$ $y = z^2 - k^2$, parabolas;

 $y = k$ $\Rightarrow$ $k = z^2 - x^2$, hyperbolas on the

 z-axis for $k > 0$, and hyperbolas on the x-axis

 for $k < 0$; $z = k$ $\Rightarrow$ $y = k^2 - x^2$, parabolas.

 Thus, $\dfrac{y}{1} = \dfrac{z^2}{1^2} - \dfrac{x^2}{1^2}$ is a hyperbolic paraboloid.

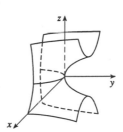

17. This is the equation of an ellipsoid: $x^2 + 4y^2 + 9z^2 = x^2 + \dfrac{y^2}{(1/2)^2} + \dfrac{z^2}{(1/3)^2} = 1$, with x-intercepts ± 1,

 y-intercepts $\pm\frac{1}{2}$ and z-intercepts $\pm\frac{1}{3}$. So the major axis is the x-axis and the only possible graph is VII.

19. This is the equation of a hyperboloid of one sheet, with $a = b = c = 1$. Since the minus sign is in front of the y

 term, the axis of the hyperboloid is the y-axis, hence the correct graph is II.

21. There are no real values of x and z that satisfy this equation for $y < 0$, so this surface does not extend to the left

 of the xz-plane. The surface intersects the plane $y = k > 0$ in an ellipse. Notice that y occurs to the first power

 whereas x and z occur to the second power. So the surface is an elliptic paraboloid with axis the y-axis. Its

 graph is VI.

23. This surface is a cylinder because the variable y is missing from the equation. The intersection of the surface and

 the xz-plane is an ellipse. So the graph is VIII.

25. $z^2 = 3x^2 + 4y^2 - 12$ or $3x^2 + 4y^2 - z^2 = 12$

 or $\dfrac{x^2}{4} + \dfrac{y^2}{3} - \dfrac{z^2}{12} = 1$

 or $\dfrac{x^2}{2^2} + \dfrac{y^2}{\left(\sqrt{3}\right)^2} - \dfrac{z^2}{\left(\sqrt{12}\right)^2} = 1$

 represents a hyperboloid of one sheet with

 axis the z-axis.

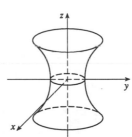

27. $z = x^2 + y^2 + 1$ or $z - 1 = x^2 + y^2$,

a circular paraboloid with axis the z-axis

and vertex $(0, 0, 1)$.

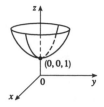

29. Completing the square in all three variables gives

$$(x + 2)^2 + (y - 3)^2 - 4(z + 1)^2 = 13 + 9 \text{ or}$$

$$\frac{(x + 2)^2}{\left(\sqrt{22}\right)^2} + \frac{(y - 3)^2}{\left(\sqrt{22}\right)^2} - \frac{(z + 1)^2}{\left(\frac{1}{2}\sqrt{22}\right)^2} = 1,$$

a hyperboloid of one sheet with center $(-2, 3, -1)$

and axis the vertical line $y = 3$, $x = -2$.

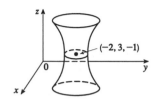

31. $x^2 + 4y^2 = 100$ or $\dfrac{x^2}{10^2} + \dfrac{y^2}{5^2} = 1$,

an elliptic cylinder with axis the z-axis.

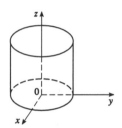

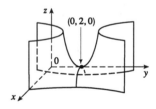

33. Completing the square in y gives

$$x^2 - (y - 2)^2 + z = 4 - 4 = 0$$

or $z = (y - 2)^2 - x^2$, a

hyperbolic paraboloid with

center at $(0, 2, 0)$.

35.

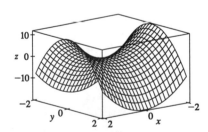

37.

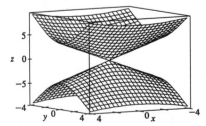
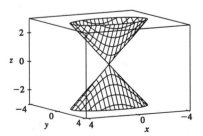

To restrict the z-range as in the second graph, we can use the option view=-2..2 in Maple's plot3d

command, or PlotRange -> {-2,2} in Mathematica's Plot3D command.

39.

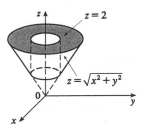

41. The surface is a paraboloid of revolution (circular paraboloid) with vertex at the origin, axis the y-axis and opens to the right. Thus the trace in the yz-plane is also a parabola: $y = z^2$, $x = 0$. The equation is $y = x^2 + z^2$.

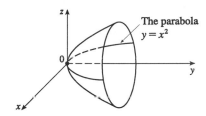

43. Let $P = (x, y, z)$ be an arbitrary point equidistant from $(-1, 0, 0)$ and the plane $x = 1$. Then the distance from P to $(-1, 0, 0)$ is $\sqrt{(x+1)^2 + y^2 + z^2}$ and the distance from P to the plane $x = 1$ is $|x - 1|/\sqrt{1^2} = |x - 1|$ (by the formula in Example 11.5.8). So $|x - 1| = \sqrt{(x+1)^2 + y^2 + z^2}$ $\Leftrightarrow$ $(x - 1)^2 = (x + 1)^2 + y^2 + z^2$ $\Leftrightarrow$ $x^2 - 2x + 1 = x^2 + 2x + 1 + y^2 + z^2$ $\Leftrightarrow$ $-4x = y^2 + z^2$. Thus the collection of all such points P is a circular paraboloid with vertex at the origin, axis the x-axis, which opens in the negative direction.

45. If (a, b, c) satisfies $z = y^2 - x^2$, then $c = b^2 - a^2$. L_1: $x = a + t$, $y = b + t$, $z = c + 2(b - a)t$, L_2: $x = a + t$, $y = b - t$, $z = c - 2(b + a)t$. Substitute the parametric equations of L_1 into the equation of the hyperbolic paraboloid in order to find the points of intersection: $z = y^2 - x^2$ $\Rightarrow$ $c + 2(b - a)t = (b + t)^2 - (a + t)^2 = b^2 - a^2 + 2(b - a)t$ $\Rightarrow$ $c = b^2 - a^2$. As this is true for all values of t, L_1 lies on $z = y^2 - x^2$. Performing similar operations with L_2 gives: $z = y^2 - x^2$ $\Rightarrow$ $c - 2(b + a)t = (b - t)^2 - (a + t)^2 = b^2 - a^2 - 2(b + a)t$ $\Rightarrow$ $c = b^2 - a^2$. This tells us that all of L_2 also lies on $z = y^2 - x^2$.

47. The curve of intersection looks like a bent ellipse. The projection of this onto the xy-plane is the set of x- and y-coordinates of points that satisfy $x^2 + y^2 = z = 1 - y^2$. That is, points in the xy-plane that satisfy $x^2 + y^2 = 1 - y^2$ $\Leftrightarrow$ $x^2 + 2y^2 = 1$ $\Leftrightarrow$ $x^2 + \dfrac{y^2}{(1/\sqrt{2})^2} = 1$, which is the equation of an ellipse.

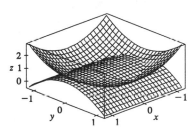

EXERCISES 11.7

1. $x = \cos 4t$, $y = t$, $z = \sin 4t$. At any point (x, y, z) on the curve, $x^2 + z^2 = \cos^2 4t + \sin^2 4t = 1$. So the curve lies on a circular cylinder with axis the y-axis. Since $y = t$, this is a helix. So the graph is V.

3. $x = t$, $y = 1/(1 + t^2)$, $z = t^2$. Note that y and z are positive for all t. The curve passes through $(0, 1, 0)$ when $t = 0$. As $t \to \infty$, $(x, y, z) \to (\infty, 0, \infty)$, and as $t \to -\infty$, $(x, y, z) \to (-\infty, 0, \infty)$. So the graph is I.

5. $x = \cos t$, $y = \sin t$, $z = \sin 5t$. $x^2 + y^2 = \cos^2 t + \sin^2 t = 1$, so the curve lies on a circular cylinder, axis the z-axis. Each of x, y and z is periodic, and at $t = 0$ and $t = 2\pi$ the curve passes through the same point, so the curve repeats itself and the graph is IV.

7. The corresponding parametric equations are $x = t$, $y = -t$, $z = 2t$, which are the parametric equations of a line through the origin and with direction vector $\langle 1, -1, 2 \rangle$.

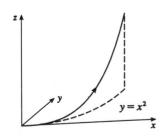

9. The parametric equations give $x^2 + z^2 = \sin^2 t + \cos^2 t = 1$, $y = 3$, which is a circle of radius 1, center $(0, 3, 0)$ in the plane $y = 3$.

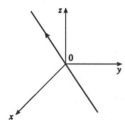

11. Eliminating the parameter t by substituting $z = t$ into $x = t^4 + 1$ gives $x = z^4 + 1$, which is a fourth-degree curve in the xz-plane that opens along the positive x-axis with vertex $(1, 0, 0)$.

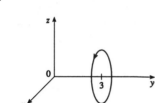

13. The parametric equations are $x = t^2$, $y = t^4$, $z = t^6$. These are positive for $t \neq 0$ and 0 when $t = 0$. So the curve lies entirely in the first quadrant. The projection of the graph onto the xy-plane is $y = x^2$, $y > 0$, a half parabola. On the xz-plane $z = x^3$, $z > 0$, a half cubic, and the yz-plane, $y^3 = z^2$.

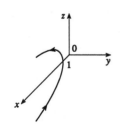

15. If $x = t\cos t$, $y = t\sin t$, and $z = t$, then

$x^2 + y^2 = t^2\cos^2 t + t^2\sin^2 t = t^2 = z^2$,

so the curve lies on the cone $z^2 = x^2 + y^2$.

Thus the curve is a spiral on this cone.

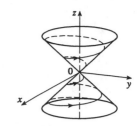

17. $\mathbf{r}(t) = \langle \sin t, \cos t, t^2 \rangle$

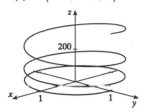

19. $\mathbf{r}(t) = \langle \sqrt{t}, t, t^2 - 2 \rangle$

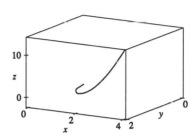

21.

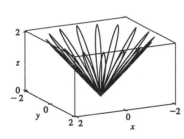

$x = (1 + \cos 16t)\cos t$, $y = (1 + \cos 16t)\sin t$, $z = 1 + \cos 16t$.

At any point on the graph,

$$x^2 + y^2 = (1 + \cos 16t)^2 \cos^2 t + (1 + \cos 16t)^2 \sin^2 t$$

$$= (1 + \cos 16t)^2 = z^2, \text{ so the graph lies on the cone}$$

$x^2 + y^2 = z^2$. From the graph at left, we see that this curve looks

like the projection of a leaved two-dimensional curve onto a cone.

23. $\lim\limits_{t \to 0} \langle t, \cos t, 2 \rangle = \left\langle \lim\limits_{t \to 0} t, \lim\limits_{t \to 0} \cos t, \lim\limits_{t \to 0} 2 \right\rangle = \langle 0, 1, 2 \rangle$

25. $\lim\limits_{t \to 1} \sqrt{t + 3} = 2$, $\lim\limits_{t \to 1} \dfrac{t - 1}{t^2 - 1} = \lim\limits_{t \to 1} \dfrac{1}{t + 1} = \dfrac{1}{2}$, $\lim\limits_{t \to 1} \left(\dfrac{\tan t}{t} \right) = \tan 1$

Thus the given limit equals $\langle 2, \frac{1}{2}, \tan 1 \rangle$.

27. The domain of $\mathbf{r}$ is $\mathbb{R}$ and $\mathbf{r}'(t) = \langle 1, 2t, 3t^2 \rangle$.

29. Since $\tan t$ and $\sec t$ aren't defined for odd multiples of $\frac{\pi}{2}$, the domain of $\mathbf{r}$ is $\{t \mid t \neq (2n + 1)\frac{\pi}{2}, n \text{ an integer}\}$.

$\mathbf{r}'(t) = (\sec^2 t)\mathbf{j} + (\sec t \tan t)\mathbf{k}$.

31. We need $4 - t^2 > 0$ and $1 + t \geq 0$, so the domain of $\mathbf{r}$ is $\{t \mid -1 \leq t < 2\}$.

$\mathbf{r}'(t) = -\dfrac{2t}{4 - t^2}\mathbf{i} + \dfrac{1}{2\sqrt{1 + t}}\mathbf{j} - 12e^{3t}\mathbf{k}$.

33. The domain of $\mathbf{r}$ is $\mathbb{R}$ and $\mathbf{r}'(t) = 0 + \mathbf{b} + 2c t = \mathbf{b} + 2t\mathbf{c}$ by Theorem 5 #1.

35. **(a), (c)**

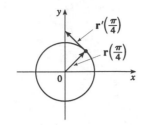

(b) $\mathbf{r}'(t) = \langle -\sin t, \cos t \rangle$

37. Since $(x - 1)^2 = t^2 = y$, the curve is a parabola.

(a), (c)

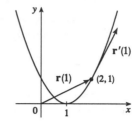

(b) $\mathbf{r}'(t) = \mathbf{i} + 2t\mathbf{j}$

39. $x^{-2} = e^{-2t} = y$, so $y = 1/x^2 > 0$.

(a), (c)

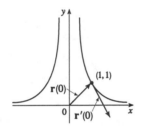

(b) $\mathbf{r}'(t) = e^t\mathbf{i} - 2e^{-2t}\mathbf{j}$

41. $\mathbf{r}'(t) = \langle 2, 6t, 12t^2 \rangle$, $\mathbf{r}(1) = \langle 2, 3, 4 \rangle$, $\mathbf{r}'(1) = \langle 2, 6, 12 \rangle$. Thus

$$\mathbf{T}(1) = \frac{\mathbf{r}'(1)}{|\mathbf{r}'(1)|} = \frac{1}{\sqrt{188}}\langle 2, 6, 12 \rangle = \left\langle \frac{1}{\sqrt{46}}, \frac{3}{\sqrt{46}}, \frac{6}{\sqrt{46}} \right\rangle.$$

43. $\mathbf{r}'(t) = \mathbf{i} + 2\cos t\,\mathbf{j} - 3\sin t\,\mathbf{k}$, $\mathbf{r}'\left(\frac{\pi}{6}\right) = \mathbf{i} + \sqrt{3}\mathbf{j} - \frac{3}{2}\mathbf{k}$. Thus

$\mathbf{T}\left(\frac{\pi}{6}\right) = \frac{1}{\sqrt{25/4}}\left(\mathbf{i} + \sqrt{3}\mathbf{j} - \frac{3}{2}\mathbf{k}\right) = \frac{2}{5}\mathbf{i} + \frac{2\sqrt{3}}{5}\mathbf{j} - \frac{3}{5}\mathbf{k}.$

45. The vector equation of the curve is $\mathbf{r}(t) = t\mathbf{i} + t^2\mathbf{j} + t^3\mathbf{k}$, so $\mathbf{r}'(t) = \mathbf{i} + 2t\mathbf{j} + 3t^2\mathbf{k}$. At the point $(1, 1, 1)$, $t = 1$, so the tangent vector here is $\mathbf{i} + 2\mathbf{j} + 3\mathbf{k}$. The tangent line goes through the point $(1, 1, 1)$ and has direction vector $\mathbf{i} + 2\mathbf{j} + 3\mathbf{k}$. Thus parametric equations are $x = 1 + t$, $y = 1 + 2t$, $z = 1 + 3t$.

47. $\mathbf{r}(t) = \langle t\cos 2\pi t, t\sin 2\pi t, 4t \rangle$, $\mathbf{r}'(t) = \langle \cos 2\pi t - 2\pi t\sin 2\pi t, \sin 2\pi t + 2\pi t\cos 2\pi t, 4 \rangle$. At $\left(0, \frac{1}{4}, 1\right)$, $t = \frac{1}{4}$ and $\mathbf{r}'\left(\frac{1}{4}\right) = \langle 0 - \frac{\pi}{2}, 1 + 0, 4 \rangle = \langle -\frac{\pi}{2}, 1, 4 \rangle$. Thus the parametric equations of the tangent line are $x = -\frac{\pi}{2}t$, $y = \frac{1}{4} + t$, $z = 1 + 4t$.

49. $r(t) = \langle t, \sqrt{2}\cos t, \sqrt{2}\sin t\rangle$,

$r'(t) = \langle 1, -\sqrt{2}\sin t, \sqrt{2}\cos t\rangle$.

At $\left(\frac{\pi}{4}, 1, 1\right)$, $t = \frac{\pi}{4}$ and

$r'\left(\frac{\pi}{4}\right) = \langle 1, -1, 1\rangle$. Thus the

parametric equations of the tangent

line are $x = \frac{\pi}{4} + t$, $y = 1 - t$, $z = 1 + t$.

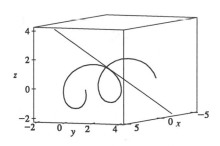

51. The angle of intersection of the two curves is the angle between the two tangent vectors to the curves at the point of intersection. Since $r_1'(t) = \langle 1, 2t, 3t^2\rangle$ and $t = 0$ at $(0, 0, 0)$, $r_1'(0) = \langle 1, 0, 0\rangle$ is a tangent vector to r_1 at $(0, 0, 0)$. Also $r_2' = \langle \cos t, 2\cos t, 1\rangle$ so $r_2'(0) = \langle 1, 2, 1\rangle$ is a tangent vector to r_2 at $(0, 0, 0)$. If θ is the angle between these two tangent vectors, then $\cos\theta = \frac{1}{\sqrt{1}\sqrt{6}}\langle 1, 0, 0\rangle \cdot \langle 1, 2, 1\rangle = \frac{1}{\sqrt{6}}$ and $\theta = \cos^{-1}\frac{1}{\sqrt{6}} \approx 66°$.

53. If the point $(1, 4, 0)$ lies on the curve, then $1 - 3t = y = 4 \Rightarrow t = -1$. The point which corresponds to $t = -1$ is $(t^2, 1 - 3t, 1 + t^3) = ((-1)^2, 1 - 3(-1), 1 + (-1)^3) = (1, 4, 0)$. For the next point, $-8 = y = 1 - 3t \Rightarrow t = 3$. The point which corresponds to $t = 3$ is $(3^2, 1 - 3(3), 1 + 3^3) = (9, -8, 28)$. If the last point is on the curve, then $7 = y = 1 - 3t \Rightarrow t = -2$. But the point for $t = -2$ is $((-2)^2, 1 - 3(-2), 1 + (-2)^3) = (4, 7, -7) \neq (4, 7, -6)$. So this point is not on the curve.

55.

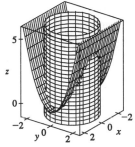 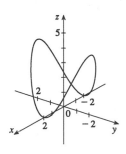

$x = 2\cos t \Rightarrow y^2 = 4 - 4\cos^2 t = 4\sin^2 t \Rightarrow y = \pm 2\sin t$ and $z = x^2 = 4\cos^2 t$. We choose the $+$ sign, so parametric equations for the curve of intersection are $x = 2\cos t$, $y = 2\sin t$, $z = 4\cos^2 t$.

57. $\int_0^1 (t\mathbf{i} + t^2\mathbf{j} + t^3\mathbf{k})dt = \left(\int_0^1 t\,dt\right)\mathbf{i} + \left(\int_0^1 t^2\,dt\right)\mathbf{j} + \left(\int_0^1 t^3\,dt\right)\mathbf{k} = \left[\frac{t^2}{2}\right]_0^1\mathbf{i} + \left[\frac{t^3}{3}\right]_0^1\mathbf{j} + \left[\frac{t^4}{4}\right]_0^1\mathbf{k} = \frac{1}{2}\mathbf{i} + \frac{1}{3}\mathbf{j} + \frac{1}{4}\mathbf{k}$

59. $\int_0^{\pi/4} (\cos 2t\,\mathbf{i} + \sin 2t\,\mathbf{j} + t\sin t\,\mathbf{k})dt = \left[\frac{1}{2}\sin 2t\,\mathbf{i} - \frac{1}{2}\cos 2t\,\mathbf{j}\right]_0^{\pi/4} + \left[[-t\cos t]_0^{\pi/4} + \int_0^{\pi/4}\cos t\,dt\right]\mathbf{k}$

$= \frac{1}{2}\mathbf{i} + \frac{1}{2}\mathbf{j} + \left[-\frac{\pi}{4}\cos\frac{\pi}{4} + \sin\frac{\pi}{4}\right]\mathbf{k} = \frac{1}{2}\mathbf{i} + \frac{1}{2}\mathbf{j} + \frac{1}{\sqrt{2}}\left(1 - \frac{\pi}{4}\right)\mathbf{k} = \frac{1}{2}\mathbf{i} + \frac{1}{2}\mathbf{j} + \frac{4 - \pi}{4\sqrt{2}}\mathbf{k}$

61. $r(t) = \frac{t^3}{3}\mathbf{i} + t^4\mathbf{j} - \frac{t^3}{3}\mathbf{k} + \mathbf{c}$ where $\mathbf{c}$ is a constant vector. But $\mathbf{j} = r(0) = (0)\mathbf{i} + (0)\mathbf{j} - (0)\mathbf{k} + \mathbf{c}$. Thus $\mathbf{c} = \mathbf{j}$ and

$r(t) = \frac{t^3}{3}\mathbf{i} + (t^4 + 1)\mathbf{j} - \frac{t^3}{3}\mathbf{k}$.

63. Let $\mathbf{u}(t) = \langle u_1(t), u_2(t), u_3(t) \rangle$ and $\mathbf{v}(t) = \langle v_1(t), v_2(t), v_3(t) \rangle$. In each part of this problem the basic procedure is to use Equation 1 and then analyze the individual component functions using the limit properties we have already developed for real-valued functions.

(a) $\lim\limits_{t \to a} \mathbf{u}(t) + \lim\limits_{t \to a} \mathbf{v}(t) = \left\langle \lim\limits_{t \to a} u_1(t), \lim\limits_{t \to a} u_2(t), \lim\limits_{t \to a} u_3(t) \right\rangle + \left\langle \lim\limits_{t \to a} v_1(t), \lim\limits_{t \to a} v_2(t), \lim\limits_{t \to a} v_3(t) \right\rangle$ and the limits of

these component functions must each exist since the vector functions both possess limits as $t \to a$. Then

adding the two vectors and using the addition property of limits for real-valued functions we have that

$$\lim_{t \to a} \mathbf{u}(t) + \lim_{t \to a} \mathbf{v}(t) = \left\langle \lim_{t \to a} u_1(t) + \lim_{t \to a} v_1(t), \lim_{t \to a} u_2(t) + \lim_{t \to a} v_2(t), \lim_{t \to a} u_3(t) + \lim_{t \to a} v_3(t) \right\rangle$$

$$= \left\langle \lim_{t \to a}[u_1(t) + v_1(t)], \lim_{t \to a}[u_2(t) + v_2(t)], \lim_{t \to a}[u_3(t) + v_3(t)] \right\rangle$$

$$= \lim_{t \to a} \langle u_1(t) + v_1(t), u_2(t) + v_2(t), u_3(t) + v_3(t) \rangle \quad \text{[using (1) backward]} \quad = \lim_{t \to a}[\mathbf{u}(t) + \mathbf{v}(t)].$$

(b) $\lim\limits_{t \to a} c\mathbf{u}(t) = \lim\limits_{t \to a} \langle cu_1(t), cu_2(t), cu_3(t) \rangle = \left\langle \lim\limits_{t \to a} cu_1(t), \lim\limits_{t \to a} cu_2(t), \lim\limits_{t \to a} cu_3(t) \right\rangle$

$$= \left\langle c \lim_{t \to a} u_1(t), c \lim_{t \to a} u_2(t), c \lim_{t \to a} u_3(t) \right\rangle = c \left\langle \lim_{t \to a} u_1(t), \lim_{t \to a} u_2(t), \lim_{t \to a} u_3(t) \right\rangle$$

$$= c \lim_{t \to a} \langle u_1(t), u_2(t), u_3(t) \rangle = c \lim_{t \to a} \mathbf{u}(t)$$

(c) $\lim\limits_{t \to a} \mathbf{u}(t) \cdot \lim\limits_{t \to a} \mathbf{v}(t) = \left\langle \lim\limits_{t \to a} u_1(t), \lim\limits_{t \to a} u_2(t), \lim\limits_{t \to a} u_3(t) \right\rangle \cdot \left\langle \lim\limits_{t \to a} v_1(t), \lim\limits_{t \to a} v_2(t), \lim\limits_{t \to a} v_3(t) \right\rangle$

$$= \left[\lim_{t \to a} u_1(t)\right]\left[\lim_{t \to a} v_1(t)\right] + \left[\lim_{t \to a} u_2(t)\right]\left[\lim_{t \to a} v_2(t)\right] + \left[\lim_{t \to a} u_3(t)\right]\left[\lim_{t \to a} v_3(t)\right]$$

$$= \lim_{t \to a} u_1(t)v_1(t) + \lim_{t \to a} u_2(t)v_2(t) + \lim_{t \to a} u_3(t)v_3(t)$$

$$= \lim_{t \to a} [u_1(t)v_1(t) + u_2(t)v_2(t) + u_3(t)v_3(t)] = \lim_{t \to a} [\mathbf{u}(t) \cdot \mathbf{v}(t)]$$

(d) $\lim\limits_{t \to a} \mathbf{u}(t) \times \lim\limits_{t \to a} \mathbf{v}(t) = \left\langle \lim\limits_{t \to a} u_1(t), \lim\limits_{t \to a} u_2(t), \lim\limits_{t \to a} u_3(t) \right\rangle \times \left\langle \lim\limits_{t \to a} v_1(t), \lim\limits_{t \to a} v_2(t), \lim\limits_{t \to a} v_3(t) \right\rangle$

$$= \left\langle \left[\lim_{t \to a} u_2(t)\right]\left[\lim_{t \to a} v_3(t)\right] - \left[\lim_{t \to a} u_3(t)\right]\left[\lim_{t \to a} v_2(t)\right], \left[\lim_{t \to a} u_3(t)\right]\left[\lim_{t \to a} v_1(t)\right] - \left[\lim_{t \to a} u_1(t)\right]\left[\lim_{t \to a} v_3(t)\right], \right.$$

$$\left. \left[\lim_{t \to a} u_1(t)\right]\left[\lim_{t \to a} v_2(t)\right] - \left[\lim_{t \to a} u_2(t)\right]\left[\lim_{t \to a} v_1(t)\right] \right\rangle$$

$$= \left\langle \lim_{t \to a}[u_2(t)v_3(t) - u_3(t)v_2(t)], \lim_{t \to a}[u_3(t)v_1(t) - u_1(t)v_3(t)], \lim_{t \to a}[u_1(t)v_2(t) - u_2(t)v_1(t)] \right\rangle$$

$$= \lim_{t \to a} \langle u_2(t)v_3(t) - u_3(t)v_2(t), u_3(t)v_1(t) - u_1(t)v_3(t), u_1(t)v_2(t) - u_2(t)v_1(t) \rangle = \lim_{t \to a} [\mathbf{u}(t) \times \mathbf{v}(t)]$$

For Exercises 65 and 67, let $\mathbf{u}(t) = \langle u_1(t), u_2(t), u_3(t)\rangle$ *and* $\mathbf{v}(t) = \langle v_1(t), v_2(t), v_3(t)\rangle$. *In these exercises, the procedure is to apply Theorem 4 so the corresponding properties of derivatives of real-valued functions can be used.*

65. $\dfrac{d}{dt}[\mathbf{u}(t) + \mathbf{v}(t)] = \dfrac{d}{dt}\langle u_1(t) + v_1(t), u_2(t) + v_2(t), u_3(t) + v_3(t)\rangle$

$$= \left\langle \frac{d}{dt}[u_1(t) + v_1(t)], \frac{d}{dt}[u_2(t) + v_2(t)], \frac{d}{dt}[u_3(t) + v_3(t)]\right\rangle$$

$$= \langle u_1'(t) + v_1'(t), u_2'(t) + v_2'(t), u_3'(t) + v_3'(t)\rangle$$

$$= \langle u_1'(t), u_2'(t), u_3'(t)\rangle + \langle v_1'(t), v_2'(t), v_3'(t)\rangle = \mathbf{u}'(t) + \mathbf{v}'(t).$$

67. $\dfrac{d}{dt}[\mathbf{u}(t) \times \mathbf{v}(t)] = \dfrac{d}{dt}\langle u_2(t)v_3(t) - u_3(t)v_2(t), u_3(t)v_1(t) - u_1(t)v_3(t), u_1(t)v_2(t) - u_2(t)v_1(t)\rangle$

$$= \langle u_2'v_3(t) + u_2(t)v_3'(t) - u_3'(t)v_2(t) - u_3(t)v_2'(t),$$
$$u_3'(t)v_1(t) + u_3(t)v_1'(t) - u_1'(t)v_3(t) - u_1(t)v_3'(t),$$
$$u_1'(t)v_2(t) + u_1(t)v_2'(t) - u_2'(t)v_1(t) - u_2(t)v_1'(t)\rangle$$

$$= \langle u_2'(t)v_3(t) - u_3'(t)v_2(t), u_3'(t)v_1(t) - u_1'(t)v_3(t), u_1'(t)v_2(t) - u_2'(t)v_1(t)\rangle$$
$$+ \langle u_2(t)v_3'(t) - u_3(t)v_2'(t), u_3(t)v_1'(t) - u_1(t)v_3'(t), u_1(t)v_2'(t) - u_2(t)v_1'(t)\rangle$$

$$= \mathbf{u}'(t) \times \mathbf{v}(t) + \mathbf{u}(t) \times \mathbf{v}'(t)$$

Alternate Solution: Let $\mathbf{r}(t) = \mathbf{u}(t) \times \mathbf{v}(t)$, then
$\mathbf{r}(t + h) - \mathbf{r}(t) = [\mathbf{u}(t + h) \times \mathbf{v}(t + h)] - [\mathbf{u}(t) \times \mathbf{v}(t)]$

$$= [\mathbf{u}(t + h) \times \mathbf{v}(t + h)] - [\mathbf{u}(t) \times \mathbf{v}(t)] + [\mathbf{u}(t + h) \times \mathbf{v}(t)] - [\mathbf{u}(t + h) \times \mathbf{v}(t)]$$

$$= \mathbf{u}(t + h) \times [\mathbf{v}(t + h) - \mathbf{v}(t)] + [\mathbf{u}(t + h) - \mathbf{u}(t)] \times \mathbf{v}(t).$$

(Be careful of the order of the cross product.) Dividing through by h and taking the limit as $h \to 0$ we have

$$\mathbf{r}'(t) = \lim_{h\to 0}\frac{\mathbf{u}(t + h) \times [\mathbf{v}(t + h) - \mathbf{v}(t)]}{h} + \lim_{h\to 0}\frac{[\mathbf{u}(t + h) - \mathbf{u}(t)] \times \mathbf{v}(t)}{h}$$

$$= \mathbf{u}(t) \times \mathbf{v}'(t) + \mathbf{u}'(t) \times \mathbf{v}(t) \text{ by Exercise 63(a) and Definition 3.}$$

69. $D_t[\mathbf{u}(t) \cdot \mathbf{v}(t)] = \mathbf{u}'(t) \cdot \mathbf{v}(t) + \mathbf{u}(t) \cdot \mathbf{v}'(t)$ by Theorem 5 #4

$$= \left(-4t\mathbf{j} + 9t^2\mathbf{k}\right) \cdot (t\mathbf{i} + \cos t\,\mathbf{j} + \sin t\,\mathbf{k}) + \left(\mathbf{i} - 2t^2\mathbf{j} + 3t^3\mathbf{k}\right) \cdot (\mathbf{i} - \sin t\,\mathbf{j} + \cos t\,\mathbf{k})$$

$$= -4t\cos t + 9t^2\sin t + 1 + 2t^2\sin t + 3t^3\cos t = 1 - 4t\cos t + 11t^2\sin t + 3t^3\cos t$$

71. $\dfrac{d}{dt}[\mathbf{r}(t) \times \mathbf{r}'(t)] = \mathbf{r}'(t) \times \mathbf{r}'(t) + \mathbf{r}(t) \times \mathbf{r}''(t)$ by Equation 5 #5. But $\mathbf{r}'(t) \times \mathbf{r}'(t) = \mathbf{0}$ by Example 11.4.2. Thus

$\dfrac{d}{dt}[\mathbf{r}(t) \times \mathbf{r}'(t)] = \mathbf{r}(t) \times \mathbf{r}''(t).$

73. $\dfrac{d}{dt}|\mathbf{r}(t)| = \dfrac{d}{dt}[\mathbf{r}(t) \cdot \mathbf{r}(t)]^{1/2} = \tfrac{1}{2}[\mathbf{r}(t) \cdot \mathbf{r}(t)]^{-1/2}[2\mathbf{r}(t) \cdot \mathbf{r}'(t)] = \dfrac{\mathbf{r}(t) \cdot \mathbf{r}'(t)}{|\mathbf{r}(t)|}$

75. Since $\mathbf{u}(t) = \mathbf{r}(t) \cdot [\mathbf{r}'(t) \times \mathbf{r}''(t)]$,

$\mathbf{u}'(t) = \mathbf{r}'(t) \cdot [\mathbf{r}'(t) \times \mathbf{r}''(t)] + \mathbf{r}(t) \cdot \dfrac{d}{dt}[\mathbf{r}'(t) \times \mathbf{r}''(t)]$

$$= 0 + \mathbf{r}(t) \cdot [\mathbf{r}''(t) \times \mathbf{r}''(t) + \mathbf{r}'(t) \times \mathbf{r}'''(t)] \qquad \text{[since } \mathbf{r}'(t) \perp \mathbf{r}'(t) \times \mathbf{r}''(t)]$$

$$= \mathbf{r}(t) \cdot [\mathbf{r}'(t) \times \mathbf{r}'''(t)] \qquad \text{[since } \mathbf{r}''(t) \times \mathbf{r}''(t) = \mathbf{0}]$$

EXERCISES 11.8

1. $\mathbf{r}'(t) = \langle 2, 3\cos t, -3\sin t \rangle$, $|\mathbf{r}'(t)| = \sqrt{4 + 9\cos^2 t + 9\sin^2 t} = \sqrt{13}$

$L = \int_a^b \sqrt{13}\, dt = \sqrt{13}(b - a)$

3. $\mathbf{r}'(t) = \langle 6, 6\sqrt{2}t, 6t^2 \rangle$, $|\mathbf{r}'(t)| = 6\sqrt{1 + 2t^2 + t^4} = 6(1 + t^2)$

$L = \int_0^1 6(1 + t^2)\, dt = \left[\dfrac{6(t + t^3)}{3}\right]_0^1 = \dfrac{24}{3} = 8$

5. The point $(2, 4, 8)$ corresponds to $t = 2$, so by Equation 2, $L = \int_0^2 \sqrt{(1)^2 + (2t)^2 + (3t^2)^2}\, dt$.

If $f(t) = \sqrt{1 + 4t^2 + 9t^4}$, then Simpson's Rule gives

$L \approx \dfrac{2 - 0}{10 \cdot 3}[f(0) + 4f(0.2) + 2f(0.4) + \cdots + 4f(1.8) + f(2)] \approx 9.5706$.

7. $\mathbf{r}'(t) = e^t(\cos t + \sin t)\mathbf{i} + e^t(\cos t - \sin t)\mathbf{j}$,

$ds/dt = |\mathbf{r}'(t)| = e^t\sqrt{(\cos t + \sin t)^2 + (\cos t - \sin t)^2} = e^t\sqrt{2\cos^2 t + 2\sin^2 t} = \sqrt{2}e^t$

$s(t) = \int_0^t |\mathbf{r}'(u)|\, du = \int_0^t \sqrt{2}e^u\, du = \sqrt{2}(e^t - 1) \quad \Rightarrow \quad \tfrac{1}{\sqrt{2}}s + 1 = e^t \quad \Rightarrow \quad t(s) = \ln\left(\tfrac{1}{\sqrt{2}}s + 1\right)$.

Therefore, $\mathbf{r}(t(s)) = \left(\tfrac{1}{\sqrt{2}}s + 1\right)\left[\sin\left(\ln\left(\tfrac{1}{\sqrt{2}}s + 1\right)\right)\mathbf{i} + \cos\left(\ln\left(\tfrac{1}{\sqrt{2}}s + 1\right)\right)\mathbf{j}\right]$.

9. $|\mathbf{r}'(t)| = \sqrt{(3\cos t)^2 + 16 + (-3\sin t)^2} = \sqrt{9 + 16} = 5$ and $s(t) = \int_0^t |\mathbf{r}'(u)|\, du = \int_0^t 5\, du = 5t \quad \Rightarrow$

$t(s) = \tfrac{1}{5}s$. Therefore, $\mathbf{r}(t(s)) = 3\sin\left(\tfrac{1}{5}s\right)\mathbf{i} + \tfrac{4}{5}s\mathbf{j} + 3\cos\left(\tfrac{1}{5}s\right)\mathbf{k}$.

11. **(a)** $\mathbf{T}(t) = \dfrac{\mathbf{r}'(t)}{|\mathbf{r}'(t)|} = \dfrac{1}{\sqrt{16 + 9}}\langle 4\cos 4t, 3, -4\sin 4t\rangle = \dfrac{1}{5}\langle 4\cos 4t, 3, -4\sin 4t\rangle$

$\mathbf{N}(t) = \dfrac{\mathbf{T}'(t)}{|\mathbf{T}'(t)|} = \dfrac{5}{16 \cdot 5}\langle -16\sin 4t, 0, -16\cos 4t\rangle = \langle -\sin 4t, 0, -\cos 4t\rangle$

(b) $\kappa(t) = \dfrac{|\mathbf{T}'(t)|}{|\mathbf{r}'(t)|} = \dfrac{16}{5 \cdot 5} = \dfrac{16}{25}$

13. **(a)** $\mathbf{T}(t) = \dfrac{\mathbf{r}'(t)}{|\mathbf{r}'(t)|} = \dfrac{1}{\sqrt{2\sin^2 t + 2\cos^2 t}}\langle -\sqrt{2}\sin t, \cos t, \cos t\rangle = \tfrac{1}{\sqrt{2}}\langle -\sqrt{2}\sin t, \cos t, \cos t\rangle$

$\mathbf{N}(t) = \dfrac{\mathbf{T}'(t)}{|\mathbf{T}'(t)|} = \dfrac{1}{\sqrt{2\cos^2 t + 2\sin^2 t}}\langle -\sqrt{2}\cos t, -\sin t, -\sin t\rangle = \tfrac{1}{\sqrt{2}}\langle -\sqrt{2}\cos t, -\sin t, -\sin t\rangle$

(b) $\kappa(t) = \dfrac{|\mathbf{T}'(t)|}{|\mathbf{r}'(t)|} = \dfrac{1}{\sqrt{2}}$

15. $\mathbf{r}'(t) = \mathbf{j} - 2t\mathbf{k}$, $\mathbf{r}''(t) = -2\mathbf{k}$, $|\mathbf{r}'(t)|^3 = (4t^2 + 1)^{3/2}$, $|\mathbf{r}'(t) \times \mathbf{r}''(t)| = |-2\mathbf{i}| = 2$,

$\kappa(t) = \dfrac{|\mathbf{r}'(t) \times \mathbf{r}''(t)|}{|\mathbf{r}'(t)|^3} = \dfrac{2}{(4t^2 + 1)^{3/2}}$

17. $\mathbf{r}'(t) = \langle 6t^2, -6t, 6 \rangle$, $\mathbf{r}''(t) = \langle 12t, -6, 0 \rangle$, $|\mathbf{r}'(t)|^3 = 6^3(t^4 + t^2 + 1)^{3/2}$,

$|\mathbf{r}'(t) \times \mathbf{r}''(t)| = |36\langle 1, 2t, t^2 \rangle| = 36\sqrt{1 + 4t^2 + t^4}$,

$$\kappa(t) = \frac{|\mathbf{r}'(t) \times \mathbf{r}''(t)|}{|\mathbf{r}'(t)|^3} = \frac{36\sqrt{1 + 4t^2 + t^4}}{6^3(t^4 + t^2 + 1)^{3/2}} = \frac{\sqrt{1 + 4t^2 + t^4}}{6(t^4 + t^2 + 1)^{3/2}}$$

19. $\mathbf{r}'(t) = \langle \cos t, -\sin t, \cos t \rangle$, $\mathbf{r}''(t) = \langle -\sin t, -\cos t, -\sin t \rangle$, $|\mathbf{r}'(t)|^3 = \left(\sqrt{\cos^2 t + 1} \right)^3$,

$|\mathbf{r}'(t) \times \mathbf{r}''(t)| = |\langle 1, 0, -1 \rangle| = \sqrt{2}$, $\kappa(t) = \dfrac{|\mathbf{r}'(t) \times \mathbf{r}''(t)|}{|\mathbf{r}'(t)|^3} = \dfrac{\sqrt{2}}{(1 + \cos^2 t)^{3/2}}$

21. $f'(x) = 3x^2$, $f''(x) = 6x$, $\kappa(x) = \dfrac{|f''(x)|}{\left[1 + (f'(x))^2 \right]^{3/2}} = \dfrac{6|x|}{[1 + 9x^4]^{3/2}}$

23. $y' = \cos x$, $y'' = -\sin x$, $\kappa(x) = \dfrac{|y''(x)|}{\left[1 + (y'(x))^2 \right]^{3/2}} = \dfrac{|\sin x|}{[1 + \cos^2 x]^{3/2}}$

25. Since $y' = y'' = e^x$, the curvature is $\kappa(x) = \dfrac{|y''(x)|}{\left[1 + (y'(x))^2 \right]^{3/2}} = \dfrac{e^x}{(1 + e^{2x})^{3/2}} = e^x [1 + e^{2x}]^{-3/2} \quad \Rightarrow$

$\kappa'(x) = e^x[1 + e^{2x}]^{-3/2} + e^x \left(-\frac{3}{2} \right) [1 + e^{2x}]^{-5/2} (2e^{2x}) = e^x \dfrac{1 + e^{2x} - 3e^{2x}}{(1 + e^{2x})^{5/2}} = e^x \dfrac{1 - 2e^{2x}}{(1 + e^{2x})^{5/2}}$. Then when

$\kappa'(x) = 0$, we must have $1 - 2e^{2x} = 0$ or $e^{2x} = \frac{1}{2}$ or $x = -\frac{1}{2} \ln 2$. And since $1 - 2e^{2x} > 0$ for $x < -\frac{1}{2} \ln 2$ and

$1 - 2e^{2x} < 0$ for $x > -\frac{1}{2} \ln 2$, the maximum curvature is attained at the point

$\left(-\frac{1}{2} \ln 2, e^{(-\ln 2)/2} \right) = \left(-\frac{1}{2} \ln 2, \frac{1}{\sqrt{2}} \right)$.

27. $y = x^4 \quad \Rightarrow \quad y' = 4x^3$, $y'' = 12x^2$, and

$\kappa(x) = \dfrac{|y''|}{\left[1 + (y')^2 \right]^{3/2}} = \dfrac{12x^2}{(1 + 16x^6)^{3/2}}$.

The appearance of the two humps in this graph is perhaps a little

surprising, but it is explained by the fact that $y = x^4$ is very flat

around the origin, and so here the curvature is zero.

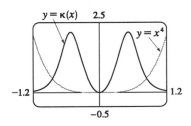

29. $\kappa(t) = \dfrac{|\dot{x}\ddot{y} - \ddot{x}\dot{y}|}{\left(\dot{x}^2 + \dot{y}^2 \right)^{3/2}} = \dfrac{|(3t^2)(2) - (6t)(2t)|}{(9t^4 + 4t^2)^{3/2}} = \dfrac{6t^2}{(t^2)^{3/2}(9t^2 + 4)^{3/2}} = \dfrac{6t^2}{|t|^3(9t^2 + 4)^{3/2}} = \dfrac{6}{|t|(9t^2 + 4)^{3/2}}$

31. $\left(1, \frac{2}{3}, 1 \right)$ corresponds to $t = 1$. $\mathbf{T}(t) = \dfrac{\mathbf{r}'(t)}{|\mathbf{r}'(t)|} = \dfrac{\langle 2t, 2t^2, 1 \rangle}{\sqrt{4t^2 + 4t^4 + 1}} = \dfrac{\langle 2t, 2t^2, 1 \rangle}{2t^2 + 1}$, so $\mathbf{T}(1) = \langle \frac{2}{3}, \frac{2}{3}, \frac{1}{3} \rangle$.

$\mathbf{T}'(t) = -4t(2t^2 + 1)^{-2}\langle 2t, 2t^2, 1 \rangle + (2t^2 + 1)^{-1}\langle 2, 4t, 0 \rangle$ (By Theorem 11.7.5)

$\qquad = (2t^2 + 1)^{-2}\langle -8t^2 + 4t^2 + 2, -8t^3 + 8t^3 + 4t, -4t \rangle = 2(2t^2 + 1)^{-2}\langle 1 - 2t^2, 2t, -2t \rangle$

$\mathbf{N}(t) = \dfrac{\mathbf{T}'(t)}{|\mathbf{T}'(t)|} = \dfrac{2(2t^2 + 1)^{-2}\langle 1 - 2t^2, 2t, -2t \rangle}{2(2t^2 + 1)^{-2}\sqrt{(1 - 2t^2)^2 + (2t)^2 + (-2t)^2}} = \dfrac{\langle 1 - 2t^2, 2t, -2t \rangle}{\sqrt{1 - 4t^2 + 4t^4 + 8t^2}} = \dfrac{\langle 1 - 2t^2, 2t, -2t \rangle}{1 + 2t^2}$

$\mathbf{N}(1) = \langle -\frac{1}{3}, \frac{2}{3}, -\frac{2}{3} \rangle$ and $\mathbf{B}(1) = \mathbf{T}(1) \times \mathbf{N}(1) = \langle -\frac{4}{9} - \frac{2}{9}, -(-\frac{4}{9} + \frac{1}{9}), \frac{4}{9} + \frac{2}{9} \rangle = \langle -\frac{2}{3}, \frac{1}{3}, \frac{2}{3} \rangle$.

33. $t = \pi$ corresponds to $(0, \pi, -2)$. $\mathbf{T}(t) = \dfrac{\mathbf{r}'(t)}{|\mathbf{r}'(t)|} = \dfrac{\langle 6\cos 3t, 1, -6\sin 3t \rangle}{\sqrt{36\cos^2 3t + 1 + 36\sin^2 3t}} = \dfrac{1}{\sqrt{37}}\langle 6\cos 3t, 1, -6\sin 3t \rangle$.

$\mathbf{T}(\pi) = \dfrac{1}{\sqrt{37}}\langle -6, 1, 0 \rangle$ is a normal vector for the normal plane, and so $\langle -6, 1, 0 \rangle$ is also normal. Thus an

equation for the plane is $-6(x - 0) + 1(y - \pi) + 0(z + 2) = 0$ or $y - 6x = \pi$.

$\mathbf{T}'(t) = \dfrac{1}{\sqrt{37}}\langle -18\sin 3t, 0, -18\cos 3t \rangle \quad \Rightarrow \quad |\mathbf{T}'(t)| = \dfrac{\sqrt{18^2 \sin^2 3t + 18^2 \cos^2 3t}}{\sqrt{37}} = \dfrac{18}{\sqrt{37}} \quad \Rightarrow$

$\mathbf{N}(t) = \langle -\sin 3t, 0, -\cos 3t \rangle$. So $\mathbf{B}(\pi) = \dfrac{1}{\sqrt{37}}\langle -6, 1, 0 \rangle \times \langle 0, 0, 1 \rangle = \dfrac{1}{\sqrt{37}}\langle 1, 6, 0 \rangle$. Since $\mathbf{B}(\pi)$ is a normal to

the osculating plane, so is $\langle 1, 6, 0 \rangle$ and an equation for the plane is $1(x - 0) + 6(y - \pi) + 0(z + 2) = 0$ or

$x + 6y = 6\pi$.

35. The ellipse is given by the parametric equations $x = 2\cos t$, $y = 3\sin t$, so using the result from Exercise 28,

$\kappa(t) = \dfrac{|\dot{x}\ddot{y} - \ddot{x}\dot{y}|}{[\dot{x}^2 + \dot{y}^2]^{3/2}} = \dfrac{|(-2\sin t)(-3\sin t) - (3\cos t)(-2\cos t)|}{(4\sin^2 t + 9\cos^2 t)^{3/2}} = \dfrac{6}{(4\sin^2 t + 9\cos^2 t)^{3/2}}$.

At $(2, 0)$, $t = 0$. Now $\kappa(0) = \frac{6}{27} = \frac{2}{9}$, so the radius of

the osculating circle is $1/\kappa(0) = \frac{9}{2}$ and its center is $\left(-\frac{5}{2}, 0\right)$.

Its equation is therefore $\left(x + \frac{5}{2}\right)^2 + y^2 = \frac{81}{4}$.

At $(0, 3)$, $t = \frac{\pi}{2}$, and $\kappa\left(\frac{\pi}{2}\right) = \frac{6}{8} = \frac{3}{4}$. So the radius of

the osculating circle is $\frac{4}{3}$ and its center is $\left(0, \frac{5}{3}\right)$.

Hence its equation is $x^2 + \left(y - \frac{5}{3}\right)^2 = \frac{16}{9}$.

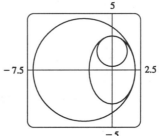

37. The tangent vector is normal to the normal plane, and the vector $\langle 6, 6, -8 \rangle$ is normal to the given plane. But

$\mathbf{T}(t) \parallel \mathbf{r}'(t)$ and $\langle 6, 6, -8 \rangle \parallel \langle 3, 3, -4 \rangle$, so we need to find t such that $\mathbf{r}'(t) \parallel \langle 3, 3, -4 \rangle$. $\mathbf{r}(t) = \langle t^3, 3t, t^4 \rangle$

$\Rightarrow \quad \mathbf{r}'(t) = \langle 3t^2, 3, 4t^3 \rangle \parallel \langle 3, 3, -4 \rangle$ when $t = -1$. So the planes are parallel at the point

$\mathbf{r}(-1) = (-1, -3, 1)$.

39. $\kappa = \left| \dfrac{d\mathbf{T}}{ds} \right| = \left| \dfrac{d\mathbf{T}/dt}{ds/dt} \right| = \dfrac{|d\mathbf{T}/dt|}{ds/dt}$ and $\mathbf{N} = \dfrac{d\mathbf{T}/dt}{|d\mathbf{T}/dt|}$, so $\kappa \mathbf{N} = \dfrac{\left|\dfrac{d\mathbf{T}}{dt}\right|\dfrac{d\mathbf{T}}{dt}}{\left|\dfrac{d\mathbf{T}}{dt}\right|\dfrac{ds}{dt}} = \dfrac{d\mathbf{T}/dt}{ds/dt} = \dfrac{d\mathbf{T}}{ds}$ by the Chain Rule.

41. (a) $|\mathbf{B}| = 1 \quad \Rightarrow \quad \mathbf{B} \cdot \mathbf{B} = 1 \quad \Rightarrow \quad \dfrac{d}{ds}(\mathbf{B} \cdot \mathbf{B}) = 0 \quad \Rightarrow \quad 2\dfrac{d\mathbf{B}}{ds} \cdot \mathbf{B} = 0 \quad \Rightarrow \quad \dfrac{d\mathbf{B}}{ds} \perp \mathbf{B}$

(b) $\mathbf{B} = \mathbf{T} \times \mathbf{N} \quad \Rightarrow \quad \dfrac{d\mathbf{B}}{ds} = \dfrac{d}{ds}(\mathbf{T} \times \mathbf{N}) = \dfrac{d}{dt}(\mathbf{T} \times \mathbf{N})\dfrac{1}{ds/dt} = \dfrac{d}{dt}(\mathbf{T} \times \mathbf{N})\dfrac{1}{|\mathbf{r}'(t)|}$

$= [(\mathbf{T}' \times \mathbf{N}) + (\mathbf{T} \times \mathbf{N}')]\dfrac{1}{|\mathbf{r}'(t)|} = \left[\left(\mathbf{T}' \times \dfrac{\mathbf{T}'}{|\mathbf{T}'|}\right) + (\mathbf{T} \times \mathbf{N}')\right]\dfrac{1}{|\mathbf{r}'(t)|} = \dfrac{\mathbf{T} \times \mathbf{N}'}{|\mathbf{r}'(t)|} \quad \Rightarrow \quad \dfrac{d\mathbf{B}}{ds} \perp \mathbf{T}$.

(c) $\mathbf{B} = \mathbf{T} \times \mathbf{N} \quad \Rightarrow \quad \mathbf{T} \perp \mathbf{N}, \mathbf{B} \perp \mathbf{T}$ and $\mathbf{B} \perp \mathbf{N}$. So $\mathbf{B}, \mathbf{T}$ and $\mathbf{N}$ form an orthogonal set of vectors in the

three-dimensional space $\mathbb{R}^3$, which makes them a basis for this space. From parts (a) and (b), $d\mathbf{B}/ds$ is

perpendicular to both $\mathbf{B}$ and $\mathbf{T}$, so $d\mathbf{B}/ds$ is parallel to $\mathbf{N}$. Therefore, $d\mathbf{B}/ds = -\tau(s)\mathbf{N}$, where $\tau(s)$ is a

scalar.

43. Let $r_1(t)$, $a \le t \le b$ and $r_2(u)$, $\alpha \le u \le \beta$ be two parametrizations of a piecewise smooth curve C where $t = g(u)$, $g'(u) > 0$ (so that as u increases, so does t, and conversely,) $g(\alpha) = a$, and $g(\beta) = b$. Then by (3), $L_1 = \int_a^b |r_1'(t)| dt = \int_\alpha^\beta |r_1'(g(u))| |g'(u)| du = \int_\alpha^\beta |r_1'(g(u))| g'(u) | du$ using the substitution $t = g(u)$ and noting $g'(u) > 0$. But since $r_1(t)$ and $r_2(u)$ are parametrizations of the same curve C and $t = g(u)$ where g is a strictly increasing function, $r_1(g(u)) = r_2(u)$ for each $u \in [\alpha, \beta]$. Thus $L_1 = \int_\alpha^\beta |r_1'(g(u)) g'(u)| du$

$$= \int_\alpha^\beta \left| \frac{d}{du} r_1(g(u)) \right| du = \int_\alpha^\beta |r_2'(u)| du = L_2. \text{ So the arc length is independent of parametrization.}$$

45. (a) $r' = s'T \Rightarrow r'' = s''T + s'T' = s''T + s' \dfrac{dT}{ds} s' = s''T + \kappa(s')^2 N$ by the first Serret-Frenet formula.

(b) Using part (a), we have

$$r' \times r'' = (s'T) \times \left[s''T + \kappa(s')^2 N \right] = [(s'T) \times (s''T)] + \left[(s'T) \times \left(\kappa(s')^2 N \right) \right] \quad \text{(By Theorem 11.4.8)}$$

$$= (s's'')(T \times T) + \kappa(s')^3 (T \times N) = 0 + \kappa(s')^3 B = \kappa(s')^3 B.$$

(c) Using part (a), we have

$$r''' = \left[s''T + \kappa(s')^2 N \right]' = s'''T + s''T' + \kappa'(s')^2 N + 2\kappa s' s'' N + \kappa(s')^2 N'$$

$$= s'''T + s'' \frac{dT}{ds} s' + \kappa'(s')^2 N + 2\kappa s' s'' N + \kappa(s')^2 \frac{dN}{ds} s'$$

$$= s'''T + s'' s' \kappa N + \kappa'(s')^2 N + 2\kappa s' s'' N + \kappa(s')^3 (-\kappa T + \tau B) \quad \text{(by the second formula)}$$

$$= \left[s''' - \kappa^2 (s')^3 \right] T + \left[3\kappa s' s'' + \kappa'(s')^2 \right] N + \kappa \tau (s')^3 B.$$

(d) Using parts (b) and (c) and the facts that $B \cdot T = 0$, $B \cdot N = 0$, and $B \cdot B = 1$, we get

$$\frac{(r' \times r'') \cdot r'''}{|r' \times r''|^2} = \frac{\kappa(s')^3 B \cdot \left\{ \left[s''' - \kappa^2(s')^3 \right] T + \left[3\kappa s' s'' + \kappa'(s')^2 \right] N + \kappa \tau(s')^3 B \right\}}{\left| \kappa(s')^3 B \right|^2} = \frac{\kappa(s')^3 \kappa \tau(s')^3}{\left[\kappa(s')^3 \right]^2} = \tau.$$

47. $r = \langle t, \frac{1}{2}t^2, \frac{1}{3}t^3 \rangle \Rightarrow r' = \langle 1, t, t^2 \rangle$, $r'' = \langle 0, 1, 2t \rangle$, $r''' = \langle 0, 0, 2 \rangle \Rightarrow r' \times r'' = \langle t^2, -2t, 1 \rangle \Rightarrow$

$$\tau = \frac{(r' \times r'') \cdot r'''}{|r' \times r''|^2} = \frac{\langle t^2, -2t, 1 \rangle \cdot \langle 0, 0, 2 \rangle}{t^4 + 4t^2 + 1} = \frac{2}{t^4 + 4t^2 + 1}$$

49. For one helix the vector equation is $r(t) = \langle 10 \cos t, 10 \sin t, 34t/(2\pi) \rangle$ because the radius of each helix is 10 angstroms and z increases by 34 angstroms for each increase of 2π in t. Therefore

$$L = \int_0^{2\pi \cdot 2.9 \times 10^8} |r'(t)| dt = \int_0^{2\pi \cdot 2.9 \times 10^8} \sqrt{(-10 \sin t)^2 + (10 \cos t)^2 + \left(\frac{34}{2\pi} \right)^2} \, dt$$

$$= \int_0^{2\pi \cdot 2.9 \times 10^8} \sqrt{100 + \left(\frac{34}{2\pi} \right)^2} \, dt = \sqrt{100 + \left(\frac{34}{2\pi} \right)^2} (2\pi) 2.9 \times 10^8 \approx 2.07 \times 10^{10} \text{ angstroms} \approx 2 \text{ m}.$$

This is the approximate length of each helix in a DNA molecule.

EXERCISES 11.9

1. $\mathbf{r}(t) = \langle t^2 - 1, t \rangle \quad \Rightarrow$
$\mathbf{v}(t) = \mathbf{r}'(t) = \langle 2t, 1 \rangle,$
$\mathbf{a}(t) = \mathbf{r}''(t) = \langle 2, 0 \rangle,$
$|\mathbf{v}(t)| = \sqrt{4t^2 + 1}$

At $t = 1$:
$\mathbf{v}(1) = \langle 2, 1 \rangle$
$\mathbf{a}(1) = \langle 2, 0 \rangle$

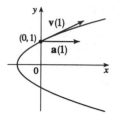

3. $\mathbf{r}(t) = e^t\mathbf{i} + e^{-t}\mathbf{j} \quad \Rightarrow$
$\mathbf{v}(t) = e^t\mathbf{i} - e^{-t}\mathbf{j},$
$\mathbf{a}(t) = e^t\mathbf{i} + e^{-t}\mathbf{j},$
$|\mathbf{v}(t)| = \sqrt{e^{2t} + e^{-2t}} = e^{-t}\sqrt{e^{4t} + 1}$
Since $x = e^t$, $t = \ln x$ and $y = e^{-t} = e^{-\ln x} = 1/x$, and
$x > 0, y > 0$.

At $t = 0$:
$\mathbf{v}(0) = \mathbf{i} - \mathbf{j},$
$\mathbf{a}(0) = \mathbf{i} + \mathbf{j}$

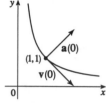

5. $\mathbf{r}(t) = \langle \sin t, t, \cos t \rangle \quad \Rightarrow$
$\mathbf{v}(t) = \langle \cos t, 1, -\sin t \rangle, \mathbf{v}(0) = \langle 1, 1, 0 \rangle$
$\mathbf{a}(t) = \langle -\sin t, 0, -\cos t \rangle, \mathbf{a}(0) = \langle 0, 0, -1 \rangle$
$|\mathbf{v}(t)| = \sqrt{\cos^2 t + 1 + \sin^2 t} = \sqrt{2}$
Since $x^2 + z^2 = 1$, $y = t$, the path of the
particle is a helix about the y-axis.

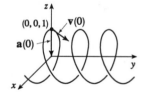

7. $\mathbf{r}(t) = \langle t^3, t^2 + 1, t^3 - 1 \rangle \quad \Rightarrow \quad \mathbf{v}(t) = \langle 3t^2, 2t, 3t^2 \rangle, \mathbf{a}(t) = \langle 6t, 2, 6t \rangle,$
$|\mathbf{v}(t)| = \sqrt{9t^4 + 4t^2 + 9t^4} = \sqrt{18t^4 + 4t^2} = |t|\sqrt{18t^2 + 4}$

9. $\mathbf{r}(t) = \langle 1/t, 1, t^2 \rangle \quad \Rightarrow \quad \mathbf{v}(t) = \langle -t^{-2}, 0, 2t \rangle, \mathbf{a}(t) = \langle 2t^{-3}, 0, 2 \rangle, |\mathbf{v}(t)| = \sqrt{t^{-4} + 4t^2} = \dfrac{1}{t^2}\sqrt{4t^6 + 1}$

11. $\mathbf{r}(t) = e^t\langle \cos t, \sin t, t \rangle \quad \Rightarrow$
$\mathbf{v}(t) = e^t\langle \cos t, \sin t, t \rangle + e^t\langle -\sin t, \cos t, 1 \rangle = e^t\langle \cos t - \sin t, \sin t + \cos t, t + 1 \rangle$
$\mathbf{a}(t) = e^t\langle \cos t - \sin t - \sin t - \cos t, \sin t + \cos t + \cos t - \sin t, t + 1 + 1 \rangle = e^t\langle -2\sin t, 2\cos t, t + 2 \rangle$
$|\mathbf{v}(t)| = e^t\sqrt{\cos^2 t + \sin^2 t - 2\cos t \sin t + \sin^2 t + \cos^2 t + 2\sin t \cos t + t^2 + 2t + 1} = e^t\sqrt{t^2 + 2t + 3}$

13. $\mathbf{a}(t) = \mathbf{k} \quad \Rightarrow \quad \mathbf{v}(t) = \int \mathbf{k}\,dt = t\mathbf{k} + \mathbf{c}_1$ and $\mathbf{i} - \mathbf{j} = \mathbf{v}(0) = 0\mathbf{k} + \mathbf{c}_1$, so $\mathbf{c}_1 = \mathbf{i} - \mathbf{j}$ and $\mathbf{v}(t) = \mathbf{i} - \mathbf{j} + t\mathbf{k}$.
$\mathbf{r}(t) = \int(\mathbf{i} - \mathbf{j} + t\mathbf{k})dt = t\mathbf{i} - t\mathbf{j} + \frac{1}{2}t^2\mathbf{k} + \mathbf{c}_2$. But $\mathbf{0} = \mathbf{r}(0) = \mathbf{0} + \mathbf{c}_2$, so $\mathbf{c}_2 = \mathbf{0}$ and $\mathbf{r}(t) = t\mathbf{i} - t\mathbf{j} + \frac{1}{2}t^2\mathbf{k}$.

15. (a) $\mathbf{a}(t) = \mathbf{i} + 2\mathbf{j} + 2t\mathbf{k} \quad \Rightarrow$

$\mathbf{v}(t) = \int (\mathbf{i} + 2\mathbf{j} + 2t\mathbf{k})dt = t\mathbf{i} + 2t\mathbf{j} + t^2\mathbf{k} + \mathbf{c}_1,$

and $\mathbf{0} = \mathbf{v}(0) = \mathbf{0} + \mathbf{c}_1$, so $\mathbf{c}_1 = \mathbf{0}$ and $\mathbf{v}(t) = t\mathbf{i} + 2t\mathbf{j} + t^2\mathbf{k}$.

$\mathbf{r}(t) = \int (t\mathbf{i} + 2t\mathbf{j} + t^2\mathbf{k})dt = \frac{1}{2}t^2\mathbf{i} + t^2\mathbf{j} + \frac{1}{3}t^3\mathbf{k} + \mathbf{c}_2.$

But $\mathbf{i} + \mathbf{k} = \mathbf{r}(0) = \mathbf{0} + \mathbf{c}_2$, so $\mathbf{c}_2 = \mathbf{i} + \mathbf{k}$ and

$\mathbf{r}(t) = \left(1 + \frac{1}{2}t^2\right)\mathbf{i} + t^2\mathbf{j} + \left(1 + \frac{1}{3}t^3\right)\mathbf{k}.$

(b)

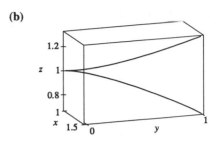

17. $\mathbf{r}(t) = \langle t^2, 5t, t^2 - 16t \rangle \quad \Rightarrow \quad \mathbf{v}(t) = \langle 2t, 5, 2t - 16 \rangle,$

$|\mathbf{v}(t)| = \sqrt{4t^2 + 25 + 4t^2 - 64t + 256} = \sqrt{8t^2 - 64t + 281}$ and

$\dfrac{d}{dt}|\mathbf{v}(t)| = \frac{1}{2}(8t^2 - 64t + 281)^{-1/2}(16t - 64)$. This is zero if and only if the numerator is zero, that is,

$16t - 64 = 0$ or $t = 4$. Since $\dfrac{d}{dt}|\mathbf{v}(t)| < 0$ for $t < 4$ and $\dfrac{d}{dt}|\mathbf{v}(t)| > 0$ for $t > 4$, the minimum speed of $\sqrt{153}$ is

attained at $t = 4$ units of time.

19. $|\mathbf{F}(t)| = 20$ N in the direction of the positive z-axis, so $\mathbf{F}(t) = 20\mathbf{k}$. Also $m = 4$ kg, $\mathbf{r}(0) = \mathbf{0}$ and $\mathbf{v}(0) = \mathbf{i} - \mathbf{j}$.

Since $20\mathbf{k} = \mathbf{F}(t) = 4\mathbf{a}(t)$, $\mathbf{a}(t) = 5\mathbf{k}$. Then $\mathbf{v}(t) = 5t\mathbf{k} + \mathbf{c}_1$ where $\mathbf{c}_1 = \mathbf{i} - \mathbf{j}$ so $\mathbf{v}(t) = \mathbf{i} - \mathbf{j} + 5t\mathbf{k}$ and the

speed is $|\mathbf{v}(t)| = \sqrt{1 + 1 + 25t^2} = \sqrt{25t^2 + 2}$. Also $\mathbf{r}(t) = t\mathbf{i} - t\mathbf{j} + \frac{5}{2}t^2\mathbf{k} + \mathbf{c}_2$ and $\mathbf{0} = \mathbf{r}(0)$, so $\mathbf{c}_2 = \mathbf{0}$ and

$\mathbf{r}(t) = t\mathbf{i} - t\mathbf{j} + \frac{5}{2}t^2\mathbf{k}.$

21. $|\mathbf{v}(0)| = 500$ m/s and since the angle of elevation is $30°$, the direction of the velocity is $\frac{1}{2}\left(\sqrt{3}\mathbf{i} + \mathbf{j}\right)$. Thus

$\mathbf{v}(0) = 250\left(\sqrt{3}\mathbf{i} + \mathbf{j}\right)$ and if we set up the axes so the projectile starts at the origin, then $\mathbf{r}(0) = \mathbf{0}$. Ignoring air

resistance, the only force is that due to gravity so $\mathbf{F}(t) = -mg\mathbf{j}$ where $g \approx 9.8$ m/s^2. Thus $\mathbf{a}(t) = -g\mathbf{j}$ and

$\mathbf{v}(t) = -gt\mathbf{j} + \mathbf{c}_1$. But $250\left(\sqrt{3}\mathbf{i} + \mathbf{j}\right) = \mathbf{v}(0) = \mathbf{c}_1$, so $\mathbf{v}(t) = 250\sqrt{3}\mathbf{i} + (250 - gt)\mathbf{j}$ and

$\mathbf{r}(t) = 250\sqrt{3}t\mathbf{i} + \left(250t - \frac{1}{2}gt^2\right)\mathbf{j} + \mathbf{c}_2$ where $\mathbf{0} = \mathbf{r}(0) = \mathbf{c}_2$. Thus $\mathbf{r}(t) = 250\sqrt{3}t\mathbf{i} + \left(250t - \frac{1}{2}gt^2\right)\mathbf{j}$.

(a) Setting $250t - \frac{1}{2}gt^2 = 0$ gives $t = 0$ or $t = 500/g \approx 51.0$ s. So the range is $\left(250\sqrt{3}\right)(500/g) \approx 22$ km.

(b) $0 = \dfrac{d}{dt}\left(250t - \frac{1}{2}gt^2\right) = 250 - gt$ implies the maximum height is attained when $t = 250/g \approx 25.5$ s and

thus the maximum height is $(250)(250/g) - g(250/g)^2\frac{1}{2} = (250)^2/(2g) \approx 3.2$ km.

(c) From part (a), impact occurs at $t = 500/g \approx 51.0$. Thus the velocity at impact is

$\mathbf{v}(500/g) = 250\sqrt{3}\mathbf{i} + [250 - g(500/g)]\mathbf{j} = 250\sqrt{3}\mathbf{i} - 250\mathbf{j}$ and the speed is

$|\mathbf{v}(500/g)| = 250\sqrt{3 + 1} = 500$ m/s.

23. As in Example 5, $\mathbf{r}(t) = (v_0\cos 45°)t\mathbf{i} + \left[(v_0\sin 45°)t - \frac{1}{2}gt^2\right]\mathbf{j} = \frac{1}{2}\left[v_0\sqrt{2}t\mathbf{i} + \left(v_0\sqrt{2}t - gt^2\right)\right]\mathbf{j}$. Then the

ball lands at $t = \dfrac{v_0\sqrt{2}}{g}$ s. Now since it lands 90 m away, $90 = \frac{1}{2}v_0\sqrt{2}\dfrac{v_0\sqrt{2}}{g}$ or $v_0^2 = 90g$ and the initial velocity

is $v_0 = \sqrt{90g} \approx 30$ m/s.

25. From (4), $x = (v_0 \cos \alpha)t$ or $t = \dfrac{x}{v_0 \cos \alpha}$. Thus $y = (v_0 \sin \alpha)\dfrac{x}{v_0 \cos \alpha} - \dfrac{g}{2}\left(\dfrac{x}{v_0 \cos \alpha}\right)^2$

$= (\tan \alpha)x - \dfrac{g}{2v_0^2 \cos^2 \alpha}x^2$. Thus the trajectory is a parabola. Continuing by completing the square, we see that

$$y - \dfrac{(\tan^2 \alpha)v_0^2 \cos^2 \alpha}{2g} = -\dfrac{g}{2v_0^2 \cos^2 \alpha}\left[x - \dfrac{(\tan \alpha)v_0^2(\cos^2 \alpha)}{g}\right]^2 \quad \text{or}$$

$$y - \dfrac{v_0^2 \sin^2 \alpha}{2g} = -\dfrac{g}{2v_0^2 \cos^2 \alpha}\left(x - \dfrac{v_0^2 \sin \alpha \cos \alpha}{g}\right)^2. \quad \text{Thus the vertex of the parabola lies at}$$

$\left(\dfrac{v_0^2 \sin \alpha \cos \alpha}{g}, \dfrac{v_0^2 \sin^2 \alpha}{2g}\right)$, so the maximum height is $y = \dfrac{v_0^2 \sin^2 \alpha}{2g}$.

27. $\mathbf{r}'(t) = (1 - \cos t)\mathbf{i} + (\sin t)\mathbf{j}$, $|\mathbf{r}'(t)| = \sqrt{1 - 2\cos t + 1} = \sqrt{2(1 - \cos t)}$, $\mathbf{r}''(t) = (\sin t)\mathbf{i} + (\cos t)\mathbf{j}$. Thus

$a_T = \dfrac{\sin t}{\sqrt{2(1 - \cos t)}}$ and

$a_N = \dfrac{|(\cos t - \cos^2 t - \sin^2 t)\mathbf{k}|}{\sqrt{2(1 - \cos t)}} = \dfrac{\sqrt{[(\cos t) - 1]^2}}{\sqrt{2}\sqrt{1 - \cos t}} = \dfrac{1}{\sqrt{2}}\sqrt{\dfrac{(1 - \cos t)^2}{1 - \cos t}} = \dfrac{\sqrt{1 - \cos t}}{\sqrt{2}}$.

29. $\mathbf{r}'(t) = 3t^2\mathbf{i} + 2t\mathbf{j} + \mathbf{k}$, $|\mathbf{r}'(t)| = \sqrt{9t^4 + 4t^2 + 1}$, $\mathbf{r}''(t) = 6t\mathbf{i} + 2\mathbf{j}$. Thus $a_T = \dfrac{18t^3 + 4t}{\sqrt{9t^4 + 4t^2 + 1}}$ and

$a_N = \dfrac{|-2\mathbf{i} + 6t\mathbf{j} + (6t^2 - 12t^2)\mathbf{k}|}{\sqrt{9t^4 + 4t^2 + 1}} = \dfrac{\sqrt{4 + 36t^2 + 36t^4}}{\sqrt{9t^4 + 4t^2 + 1}} = \dfrac{2\sqrt{9t^4 + 9t^2 + 1}}{\sqrt{9t^4 + 4t^2 + 1}}$.

31. $\mathbf{r}'(t) = e^t\mathbf{i} + \sqrt{2}\mathbf{j} - e^{-t}\mathbf{k}$, $|\mathbf{r}(t)| = \sqrt{e^{2t} + 2 + e^{-2t}} = e^t + e^{-t}$, $\mathbf{r}''(t) = e^t\mathbf{i} + e^{-t}\mathbf{k}$. Then

$a_T = \dfrac{e^{2t} - e^{-2t}}{e^t + e^{-t}} = e^t - e^{-t} = 2\sinh t$ and

$a_N = \dfrac{\left|\sqrt{2}e^{-t}\mathbf{i} - 2\mathbf{j} - \sqrt{2}e^t\mathbf{k}\right|}{e^t + e^{-t}} = \dfrac{\sqrt{2(e^{-2t} + 2 + e^{2t})}}{e^t + e^{-t}} = \sqrt{2}\dfrac{e^t + e^{-t}}{e^t + e^{-t}} = \sqrt{2}$.

33. If the engines are turned off at time t, then the spacecraft will continue to travel in the direction of $\mathbf{v}(t)$, so we

need a t such that for some scalar $s > 0$, $\mathbf{r}(t) + s\mathbf{v}(t) = \langle 6, 4, 9 \rangle$. $\mathbf{v}(t) = \mathbf{r}'(t) = \mathbf{i} + \dfrac{1}{t}\mathbf{j} + \dfrac{8t}{(t^2 + 1)^2}\mathbf{k} \quad \Rightarrow$

$\mathbf{r}(t) + s\mathbf{v}(t) = \left\langle 3 + t + s, 2 + \ln t + \dfrac{s}{t}, 7 - \dfrac{4}{t^2 + 1} + \dfrac{8st}{(t^2 + 1)^2}\right\rangle \quad \Rightarrow \quad 3 + t + s = 6 \quad \Rightarrow \quad s = 3 - t$, so

$7 - \dfrac{4}{t^2 + 1} + \dfrac{8(3 - t)t}{(t^2 + 1)^2} = 9 \quad \Leftrightarrow \quad \dfrac{24t - 12t^2 - 4}{(t^2 + 1)^2} = 2 \quad \Leftrightarrow \quad t^4 + 8t^2 - 12t + 3 = 0$. It is easily seen that

$t = 1$ is a root of this polynomial. Also $2 + \ln 1 + \dfrac{3 - 1}{1} = 4$, so $t = 1$ is the desired solution.

35. With $\mathbf{r} = (r \cos \theta)\mathbf{i} + (r \sin \theta)\mathbf{j}$ and $\mathbf{h} = \alpha\mathbf{k}$ where $\alpha > 0$,

    (a) $\mathbf{h} = \mathbf{r} \times \mathbf{r}' = [(r \cos \theta)\mathbf{i} + (r \sin \theta)\mathbf{j}] \times \left[\left(r' \cos \theta - r \sin \theta \dfrac{d\theta}{dt}\right)\mathbf{i} + \left(r' \sin \theta + r \cos \theta \dfrac{d\theta}{dt}\right)\mathbf{j}\right]$

    $= \left[rr' \cos \theta \sin \theta + r^2 \cos^2\theta \dfrac{d\theta}{dt} - rr' \cos \theta \sin \theta + r^2 \sin^2\theta \dfrac{d\theta}{dt}\right]\mathbf{k} = r^2 \dfrac{d\theta}{dt}\mathbf{k}$

    (b) Since $\mathbf{h} = \alpha\mathbf{k}$, $\alpha > 0$, $\alpha = |\mathbf{h}|$. But by (a), $\alpha = |\mathbf{h}| = r^2\dfrac{d\theta}{dt}$.

(c) $A(t) = \dfrac{1}{2} \displaystyle\int_{\theta_0}^{\theta} |\mathbf{r}|^2 \, d\theta = \dfrac{1}{2} \int_{t_0}^{t} r^2 \dfrac{d\theta}{dt} \, dt$ in polar coordinates. Thus, by the Fundamental Theorem of Calculus,

$\dfrac{dA}{dt} = \dfrac{r^2}{2} \dfrac{d\theta}{dt}$.

(d) $\dfrac{dA}{dt} = \dfrac{r^2}{2} \dfrac{d\theta}{dt} = \dfrac{h}{2} = $ constant since $\mathbf{h}$ is a constant vector and $h = |\mathbf{h}|$.

37.   From Exercise 36, $T^2 = \dfrac{4\pi^2}{GM} a^3$. $T \approx 365.25 \text{ days} \times 24 \cdot 60^2 \dfrac{\text{seconds}}{\text{day}} \approx 3.1558 \times 10^7$ seconds. Therefore

$a^3 = \dfrac{GMT^2}{4\pi^2} \approx \dfrac{(6.67 \times 10^{-11})(1.99 \times 10^{30})(3.1558 \times 10^7)^2}{4\pi^2} \approx 3.348 \times 10^{33} \text{ m}^3 \quad \Rightarrow \quad a \approx 1.496 \times 10^{11} \text{ m}.$

Thus, the length of the major axis of the earth's orbit (that is, $2a$) is approximately

$2.99 \times 10^{11} \text{ m} = 2.99 \times 10^8 \text{ km}.$

EXERCISES 11.10

1. $r = 3$, $\theta = \frac{\pi}{2}$, $z = 1$ so $x = 0$, $y = 3$ and the point is $(0, 3, 1)$.

3. $x = 2\cos\frac{4\pi}{3} = -1$, $z = 8$, $y = 2\sin\frac{4\pi}{3} = -\sqrt{3}$ so the point is $\left(-1, -\sqrt{3}, 8\right)$.

5. $x = 3\cos 0 = 3$, $y = 3\sin 0 = 0$ and $z = -6$ so the point is $(3, 0, -6)$.

7. $r^2 = (-1)^2 + (0)^2 = 1$ so $r = 1$; $z = 0$; $\tan\theta = 0$ so $\theta = 0$ or π. But $x = -1$ so $\theta = \pi$ and the point is $(1, \pi, 0)$.

9. $r^2 = 4$ so $r = 2$, $\tan\theta = \frac{1}{\sqrt{3}}$ so $\theta = \frac{\pi}{6}$ and $z = 4$. Thus the point in cylindrical coordinates is $\left(2, \frac{\pi}{6}, 4\right)$.

11. $r = \sqrt{4^2 + 4^2} = 4\sqrt{2}$; $z = 4$; $\tan\theta = \frac{4}{4}$, so $\theta = \frac{\pi}{4}$ or $\theta = \frac{5\pi}{4}$, but both x and y are positive, so $\theta = \frac{\pi}{4}$ and the point is $\left(4\sqrt{2}, \frac{\pi}{4}, 4\right)$.

13. $x = (1)\sin 0 \cos 0 = 0$, $y = (1)\sin 0 \sin 0 = 0$, $z = (1)\cos 0 = 1$ so the point in rectangular coordinates is $(0, 0, 1)$.

15. $x = \sin\frac{\pi}{6}\cos\frac{\pi}{6} = \frac{\sqrt{3}}{4}$, $y = \sin\frac{\pi}{6}\sin\frac{\pi}{6} = \frac{1}{4}$ and $z = \cos\frac{\pi}{6} = \frac{\sqrt{3}}{2}$ so the point is $\left(\frac{\sqrt{3}}{4}, \frac{1}{4}, \frac{\sqrt{3}}{2}\right)$.

17. $x = 4\sin\frac{\pi}{6}\cos\frac{\pi}{4} = 4\left(\frac{1}{2}\right)\frac{1}{\sqrt{2}} = \sqrt{2}$, $y = 4\sin\frac{\pi}{6}\sin\frac{\pi}{4} = \sqrt{2}$ and $z = 4\cos\frac{\pi}{6} = 4\left(\frac{\sqrt{3}}{2}\right) = 2\sqrt{3}$ so the point is $\left(\sqrt{2}, \sqrt{2}, 2\sqrt{3}\right)$.

19. $\rho = \sqrt{9 + 0 + 0} = 3$, $\cos\phi = \frac{0}{3} = 0$ so $\phi = \frac{\pi}{2}$ and $\cos\theta = \dfrac{-3}{3\sin\frac{\pi}{2}} = -1$ so $\theta = \pi$ and the spherical

coordinates are $\left(3, \pi, \frac{\pi}{2}\right)$.

21. $\rho = \sqrt{3 + 1} = 2$, $\cos\phi = \frac{1}{2}$ so $\phi = \frac{\pi}{3}$ and $\cos\theta = \dfrac{\sqrt{3}}{2\sin\frac{\pi}{3}} = \dfrac{\sqrt{3}\cdot 2}{2\cdot\sqrt{3}} = 1$ so $\theta = 0$ and the point is $\left(2, 0, \frac{\pi}{3}\right)$.

Note: It is also apparent that $\theta = 0$ since the point is in the xz-plane and $x > 0$.

23. $\rho = \sqrt{1 + 1 + 2} = 2$; $z = -\sqrt{2} = 2\cos\phi$, so $\cos\phi = -\frac{1}{\sqrt{2}}$ which implies that $\phi = \frac{3\pi}{4}$;

$\cos\theta = \dfrac{1}{2\sin\frac{3\pi}{4}} = \dfrac{1}{\sqrt{2}}$, so $\theta = \frac{\pi}{4}$ or $\theta = \frac{7\pi}{4}$, but $x > 0$ and $y < 0$ so $\theta = \frac{7\pi}{4}$. Thus the point is $\left(2, \frac{7\pi}{4}, \frac{3\pi}{4}\right)$.

25. $\rho = \sqrt{r^2 + z^2} = \sqrt{2 + 0} = \sqrt{2}$; $\theta = \frac{\pi}{4}$; $z = 0 = \sqrt{2}\cos\phi$ so $\phi = \frac{\pi}{2}$ and the point is $\left(\sqrt{2}, \frac{\pi}{4}, \frac{\pi}{2}\right)$.

27. $\rho = \sqrt{r^2 + z^2} = \sqrt{4^2 + 4^2} = 4\sqrt{2}$; $\theta = \frac{\pi}{3}$; $z = 4 = 4\sqrt{2}\cos\phi$ so $\cos\phi = \frac{1}{\sqrt{2}}$ $\Rightarrow$ $\phi = \frac{\pi}{4}$ and the point is $\left(4\sqrt{2}, \frac{\pi}{3}, \frac{\pi}{4}\right)$.

29. $z = 2\cos 0 = 2$, $r = \sqrt{\rho^2 - z^2} = \sqrt{2^2 - 2^2} = 0$, (or $r = 2\sin 0 = 0$), $\theta = 0$ and the point is $(0, 0, 2)$.

31. $z = 8\cos\frac{\pi}{2} = 0$, $r = 8\sin\frac{\pi}{2} = 8$, $\theta = \frac{\pi}{6}$ and the point is $\left(8, \frac{\pi}{6}, 0\right)$.

33. Since $r = 3$, $x^2 + y^2 = 9$ and the surface is a cylinder of radius 3 and axis the z-axis.

35. Since $\phi = \frac{\pi}{3}$, the surface is one frustum of the right circular cone with vertex at the origin and axis the positive z-axis.

37. $z = r^2 = x^2 + y^2$ so the surface is a circular paraboloid with vertex at the origin and axis the positive z-axis.

39. $2 = \rho\cos\phi = z$ is a plane through the point $(0, 0, 2)$ and parallel to the xy-plane.

41. Since $\phi = 0$, $x = 0$ and $y = 0$ while $z = \rho \geq 0$. Thus the locus is the positive z-axis including the origin.

43. $r = 2\cos\theta$ $\Rightarrow$ $r^2 = x^2 + y^2 = 2r\cos\theta = 2x$ $\Leftrightarrow$ $(x - 1)^2 + y^2 = 1$ which is the equation of a circular cylinder of radius 1, whose axis is the vertical line $x = 1$, $y = 0$, $z = z$.

45. Since $r^2 + z^2 = 25$ and $r^2 = x^2 + y^2$, we have $x^2 + y^2 + z^2 = 25$, a sphere of radius 5 and center at the origin.

47. Since $x^2 = \rho^2\sin^2\phi\cos^2\theta$ and $z^2 = \rho^2\cos^2\phi$, the equation of the surface in rectangular coordinates is $x^2 + z^2 = 4$. Thus the surface is a right circular cylinder of radius 2 about the y-axis.

49. Since $r^2 - r = 0$, $r = 0$ or $r = 1$. But $x^2 + y^2 = r^2$. Thus the surface consists of the right circular cylinder of radius 1 and axis the z-axis along with the surface given by $x^2 + y^2 = 0$, that is, the z-axis.

51. (a) $r^2 = x^2 + y^2$, so $r^2 + z^2 = 16$.

(b) $\rho^2 = x^2 + y^2 + z^2$, so $\rho^2 = 16$ or $\rho = 4$.

53. (a) $r\cos\theta + 2r\sin\theta + 3z = 6$

(b) $\rho\sin\phi\cos\theta + 2\rho\sin\phi\sin\theta + 3\rho\cos\phi = 6$ or $\rho(\sin\phi\cos\theta + 2\sin\phi\sin\theta + 3\cos\phi) = 6$.

55. (a) $r^2(\cos^2\theta - \sin^2\theta) - 2z^2 = 4$ or $2z^2 = r^2\cos 2\theta - 4$.

(b) $\rho^2(\sin^2\phi\cos^2\theta - \sin^2\phi\sin^2\theta - 2\cos^2\phi) = 4$ or $\rho^2(\sin^2\phi\cos 2\theta - 2\cos^2\phi) = 4$.

57. (a) $r^2 = 2r\sin\theta$ or $r = 2\sin\theta$.

(b) $\rho^2\sin^2\phi(\cos^2\theta + \sin^2\theta) = 2\rho\sin\phi\sin\theta$ or $\rho\sin^2\phi = 2\sin\phi\sin\theta$ or $\rho\sin\phi = 2\sin\theta$.

59. $z = r^2 = x^2 + y^2$ is a circular paraboloid with vertex $(0, 0, 0)$, opening upward. $z = 2 - r^2 \Rightarrow$

$z - 2 = -(x^2 + y^2)$ is a circular paraboloid with vertex $(0, 0, 2)$ opening downward. Thus $r^2 \le z \le 2 - r^2$ is

the solid region enclosed by these two surfaces.

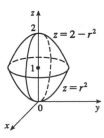

61. $-\frac{\pi}{2} \le \theta \le \frac{\pi}{2}$ restricts the solid to the four octants in which x is positive. $\rho = \sec \phi \Rightarrow \rho \cos \phi = z = 1$,

which is the equation of a horizontal plane. $0 \le \phi \le \frac{\pi}{6}$ describes a cone, opening upward. So the solid lies

above the cone $\phi = \frac{\pi}{6}$ and below the plane $z = 1$.

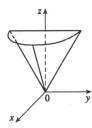

63. $z \ge \sqrt{x^2 + y^2}$ because the solid lies above the cone. Squaring both sides of this inequality gives $z^2 \ge x^2 + y^2$

$\Rightarrow 2z^2 \ge x^2 + y^2 + z^2 = \rho^2 \Rightarrow z^2 = \rho^2 \cos^2\phi \ge \frac{1}{2}\rho^2 \Rightarrow \cos^2\phi \ge \frac{1}{2}$. The cone opens upward so

that the inequality is $\cos \phi \ge \frac{1}{\sqrt{2}}$, or equivalently $0 \le \phi \le \frac{\pi}{4}$. In spherical coordinates the sphere

$z = x^2 + y^2 + z^2$ is $\rho \cos \phi = \rho^2 \Rightarrow \rho = \cos \phi$. $0 \le \rho \le \cos \phi$ because the solid lies below the sphere. The

solid can therefore be described as the region in spherical coordinates satisfying $0 \le \rho \le \cos \phi, 0 \le \phi \le \frac{\pi}{4}$.

65. In cylindrical coordinates, the equation of the cylinder is $r = 3$, $0 \le z \le 10$. The hemisphere is the upper part of

the sphere radius 3, center $(0, 0, 10)$, equation $r^2 + (z - 10)^2 = 3^2$, $z \ge 10$. In Maple, we can use either the

`coords=cylindrical` option in a regular `plot` command, or the `plots[cylinderplot]` command.

In Mathematica, we can use `ParametricPlot3d`.

REVIEW EXERCISES FOR CHAPTER 11

1. By Theorem 11.3.2 #2, this is true.

3. True. If θ is the angle between $\mathbf{u}$ and $\mathbf{v}$, then by Theorem 11.4.6 $|\mathbf{u} \times \mathbf{v}| = |\mathbf{u}||\mathbf{v}|\sin\theta = |\mathbf{v}||\mathbf{u}|\sin\theta = |\mathbf{v} \times \mathbf{u}|$.
    (Or, by Theorem 11.4.8, $|\mathbf{u} \times \mathbf{v}| = |-\mathbf{v} \times \mathbf{u}| = |-1||\mathbf{v} \times \mathbf{u}| = |\mathbf{v} \times \mathbf{u}|$.)

5. Theorem 11.4.8 #2 tells us that this is true.

7. This is true by Theorem 11.4.8 #5.

9. This is true by Theorem 11.4.5.

11. If $|\mathbf{u}| = 1$, $|\mathbf{v}| = 1$ and θ is the angle between these two vectors (so $0 \leq \theta \leq \pi$), then by Theorem 11.4.6
    $|\mathbf{u} \times \mathbf{v}| = |\mathbf{u}||\mathbf{v}|\sin\theta = \sin\theta$, which is equal to 1 if and only if $\theta = \frac{\pi}{2}$ (that is, the two vectors are orthogonal).
    Therefore, the assertion that the cross product of two unit vectors is a unit vector is false.

13. This is false because by 11.6.7, $\dfrac{x^2}{1} + \dfrac{y^2}{1} = 1$ is the equation of a circular cylinder.

15. $|AB| = \sqrt{9 + 16 + 144} = \sqrt{169} = 13$, $|BC| = \sqrt{1 + 1 + 36} = \sqrt{38}$, $|CA| = \sqrt{4 + 25 + 36} = \sqrt{65}$

17. Completing the squares gives $(x + 2)^2 + (y + 3)^2 + (z - 5)^2 = -2 + 4 + 9 + 25 = 36$. Thus the circle is
    centered at $(-2, -3, 5)$ and has radius 6.

19. $6\mathbf{a} - 5\mathbf{c} = (6 - 0)\mathbf{i} + (6 - 5)\mathbf{j} + (-12 + 25)\mathbf{k} = 6\mathbf{i} + \mathbf{j} + 13\mathbf{k}$

21. $\mathbf{a} \cdot \mathbf{b} = (1)(3) + (1)(-2) + (-2)(1) = -1$

23. $\mathbf{b} \times \mathbf{c} = \begin{vmatrix} \mathbf{i} & \mathbf{j} & \mathbf{k} \\ 3 & -2 & 1 \\ 0 & 1 & -5 \end{vmatrix} = 9\mathbf{i} + 15\mathbf{j} + 3\mathbf{k}$, $|\mathbf{b} \times \mathbf{c}| = 3\sqrt{9 + 25 + 1} = 3\sqrt{35}$

25. $\mathbf{c} \times \mathbf{c} = \mathbf{0}$ for any $\mathbf{c}$.

27. $\cos\theta = \dfrac{\mathbf{a} \cdot \mathbf{b}}{|\mathbf{a}||\mathbf{b}|} = \dfrac{-1}{\sqrt{6}\sqrt{14}} = \dfrac{-1}{2\sqrt{21}}$ and $\theta = \cos^{-1}\frac{-1}{2\sqrt{21}} \approx 96°$.

29. The scalar projection is $\text{comp}_{\mathbf{a}}\,\mathbf{b} = |\mathbf{b}|\cos\theta = \mathbf{a} \cdot \mathbf{b}/|\mathbf{a}| = -\frac{1}{\sqrt{6}}$.

31. We need $4x + 3x - 28 = 0$ or $x = 4$.

33. (a) $(\mathbf{u} \times \mathbf{v}) \cdot \mathbf{w} = \mathbf{u} \cdot (\mathbf{v} \times \mathbf{w}) = 2$

    (b) $\mathbf{u} \cdot (\mathbf{w} \times \mathbf{v}) = \mathbf{u} \cdot [-(\mathbf{v} \times \mathbf{w})] = -\mathbf{u} \cdot (\mathbf{v} \times \mathbf{w}) = -2$

    (c) $\mathbf{v} \cdot (\mathbf{u} \times \mathbf{w}) = (\mathbf{v} \times \mathbf{u}) \cdot \mathbf{w} = -(\mathbf{u} \times \mathbf{v}) \cdot \mathbf{w} = -2$

    (d) $(\mathbf{u} \times \mathbf{v}) \cdot \mathbf{v} = \mathbf{u} \cdot (\mathbf{v} \times \mathbf{v}) = \mathbf{u} \cdot \mathbf{0} = 0$

35. Determine the vectors $\overrightarrow{PQ} = \langle a_1, a_2, a_3 \rangle$ and $\overrightarrow{PR} = \langle b_1, b_2, b_3 \rangle$. If there is a scalar t such that $\langle a_1, a_2, a_3 \rangle = t\langle b_1, b_2, b_3 \rangle$, then the vectors are parallel and the points must all lie on the same line. Alternatively, if $\overrightarrow{PQ} \times \overrightarrow{PR} = \mathbf{0}$, then $\overrightarrow{PQ}$ and $\overrightarrow{PR}$ are parallel, so P, Q, and R are collinear. Thirdly, an algebraic method is to determine the equation of the line joining two of the points, and then check whether or not the third point is on that line.

37. For simplicity, consider a unit cube positioned with its back left corner at the origin. Vector representations of the diagonals joining the points $(0,0,0)$ to $(1,1,1)$ and $(1,0,0)$ to $(0,1,1)$ are $\langle 1,1,1 \rangle$ and $\langle -1,1,1 \rangle$ respectively. Let θ be the angle between these two vectors.
$$\langle 1,1,1 \rangle \cdot \langle -1,1,1 \rangle = -1+1+1 = 1 = |\langle 1,1,1 \rangle||\langle -1,1,1 \rangle|\cos\theta = 3\cos\theta \quad \Rightarrow \quad \cos\theta = \tfrac{1}{3} \quad \Rightarrow$$
$\theta = \cos^{-1}\tfrac{1}{3} \approx 71°$ (or $109°$).

39. $\overrightarrow{AB} = \langle 1,0,-1 \rangle$, $\overrightarrow{AC} = \langle 0,4,3 \rangle$, so

 (a) a vector perpendicular to the plane is $\overrightarrow{AB} \times \overrightarrow{AC} = \langle 0+4, -(3+0), 4-0 \rangle = \langle 4,-3,4 \rangle$.
 (b) $\tfrac{1}{2}\left|\overrightarrow{AB} \times \overrightarrow{AC}\right| = \tfrac{1}{2}\sqrt{16+9+16} = \dfrac{\sqrt{41}}{2}$

41. Let F_1 be the magnitude of the force directed $20°$ away from the direction of shore, and let F_2 be the magnitude of the other force. Separating these forces into components parallel to the direction of the resultant force and perpendicular to it gives $F_1\cos 20° + F_2\cos 30° = 255$ (1), and $F_1\sin 20° - F_2\sin 30° = 0 \quad \Rightarrow$
$F_1 = F_2\dfrac{\sin 30°}{\sin 20°}$ (2). Substituting (2) into (1) gives $F_2(\sin 30° \cot 20° + \cos 30°) = 255 \quad \Rightarrow \quad F_2 \approx 114\,\text{N}$.
Substituting this into (2) gives $F_1 \approx 166\,\text{N}$.

43. $x = 1 + 2t$, $y = 2 - t$, $z = 4 + 3t$

45. $\mathbf{v} = \langle 4,-3,5 \rangle$ so $x = 1 + 4t$, $y = -3t$, $z = 1 + 5t$.

47. $(x+4) + 2(y-1) + 5(z-2) = 0$ or $x + 2y + 5z = 8$.

49. Substitution of the parametric equations into the equation of the plane gives
$2x - y + z = 2(2-t) - (1+3t) + 4t = 2 \quad \Rightarrow \quad -t+3 = 2 \quad \Rightarrow \quad t = 1$. When $t = 1$, the parametric equations give $x = 2 - 1 = 1$, $y = 1 + 3 = 4$ and $z = 4$. Therefore, the point of intersection is $(1, 4, 4)$.

51. Since the direction vectors $\langle 2,3,4 \rangle$ and $\langle 6,-1,2 \rangle$ aren't parallel, neither are the lines. For the lines to intersect, the three equations $1 + 2t = -1 + 6s$, $2 + 3t = 3 - s$, $3 + 4t = -5 + 2s$ must be satisfied simultaneously. Solving the first two equations gives $t = \tfrac{1}{5}$, $s = \tfrac{2}{5}$ and checking we see these values don't satisfy the third equation. Thus the lines aren't parallel and they don't intersect, so they must be skew.

53. By Exercise 11.5.67, $D = \dfrac{|2-24|}{\sqrt{26}} = \dfrac{22}{\sqrt{26}}$.

55. A plane through the x-axis intersecting the yz-plane in the line $y = z$, $x = 0$.

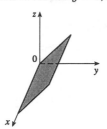

57. A circular paraboloid with vertex the origin and axis the y-axis.

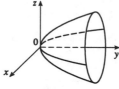

59. A (right elliptical) cone with vertex at the origin and axis the x-axis.

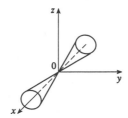

61. A hyperboloid of two sheets with axis the y-axis. For $|y| > 2$, traces parallel to the xz-plane are circles.

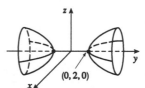

63. $4x^2 + y^2 = 16 \quad \Leftrightarrow \quad \dfrac{x^2}{4} + \dfrac{y^2}{16} = 1$. The equation of the ellipsoid is $\dfrac{x^2}{4} + \dfrac{y^2}{16} + \dfrac{z^2}{c^2} = 1$, since the horizontal trace in the plane $z = 0$ must be the original ellipse. The traces of the ellipsoid in the yz-plane must be circles since the surface is obtained by rotation about the x-axis. Therefore, $c^2 = 16$ and the equation of the ellipsoid is

$$\dfrac{x^2}{4} + \dfrac{y^2}{16} + \dfrac{z^2}{16} = 1 \quad \Leftrightarrow \quad 4x^2 + y^2 + z^2 = 16.$$

65. **(a)** Since $x = 2$ and $y^2 + z^2 = 1$, the curve is a circle in the plane $x = 2$ with center $(2, 0, 0)$ and radius 1.

(b) $\mathbf{r}'(t) = \cos t\,\mathbf{j} - \sin t\,\mathbf{k} \quad \Rightarrow$

$\mathbf{r}''(t) = -\sin t\,\mathbf{j} - \cos t\,\mathbf{k}$

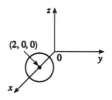

67. $\int_0^1 \left[(t + t^2)\mathbf{i} + (2 + t^3)\mathbf{j} + t^4\mathbf{k}\right] dt = \left[\left(\tfrac{1}{2}t^2 + \tfrac{1}{3}t^3\right)\mathbf{i} + \left(2t + \tfrac{1}{4}t^4\right)\mathbf{j} + \left(\tfrac{1}{5}t^5\right)\mathbf{k}\right]_0^1 = \tfrac{5}{6}\mathbf{i} + \tfrac{9}{4}\mathbf{j} + \tfrac{1}{5}\mathbf{k}$

69. $t = 1$ at $(1, 4, 2)$ and $t = 4$ at $(2, 1, 17)$, so

$$L = \int_1^4 \sqrt{\dfrac{1}{4t} + \dfrac{16}{t^4} + 4t^2}\, dt$$

$$\approx \dfrac{4 - 1}{3 \cdot 4}\left[\sqrt{\dfrac{1}{4} + 16 + 4} + 4 \cdot \sqrt{\dfrac{1}{4 \cdot 7/4} + \dfrac{16}{(7/4)^4} + 4\left(\dfrac{7}{4}\right)^2} + 2 \cdot \sqrt{\dfrac{1}{4 \cdot 10/4} + \dfrac{16}{(10/4)^4} + 4\left(\dfrac{10}{4}\right)^2}\right.$$

$$\left. + 4 \cdot \sqrt{\dfrac{1}{4 \cdot 13/4} + \dfrac{16}{(13/4)^4} + 4\left(\dfrac{13}{4}\right)^2} + \sqrt{\dfrac{1}{4 \cdot 4} + \dfrac{16}{4^4} + 4 \cdot 4^2}\right]$$

$$\approx 15.9241.$$

71. The angle of intersection of the two curves, θ, is the angle between their respective tangents at the point of intersection. For both curves the point $(1, 0, 0)$ occurs when $t = 0$. $\mathbf{r}_1'(t) = -\sin t\,\mathbf{i} + \cos t\,\mathbf{j} + \mathbf{k}$ $\Rightarrow$ $\mathbf{r}_1'(0) = \mathbf{j} + \mathbf{k}$ and $\mathbf{r}_2'(t) = \mathbf{i} + 2t\mathbf{j} + 3t^2\mathbf{k}$ $\Rightarrow$ $\mathbf{r}_2'(0) = \mathbf{i}$. $\mathbf{r}_1'(0) \cdot \mathbf{r}_2'(0) = (\mathbf{j} + \mathbf{k}) \cdot \mathbf{i} = 0$. Therefore, the curves intersect at right angles to each other, that is, $\theta = \frac{\pi}{2}$.

73. **(a)** $\mathbf{T}(t) = \dfrac{\langle t^2, t, 1 \rangle}{\sqrt{t^4 + t^2 + 1}}$

(b) $\mathbf{T}'(t) = -\frac{1}{2}\left(t^4 + t^2 + 1\right)^{-3/2}(4t^3 + 2t)\langle t^2, t, 1 \rangle + \left(t^4 + t^2 + 1\right)^{-1/2}\langle 2t, 1, 0 \rangle$

$= \dfrac{-2t^3 - t}{(t^4 + t^2 + 1)^{3/2}}\langle t^2, t, 1 \rangle + \dfrac{1}{(t^4 + t^2 + 1)^{1/2}}\langle 2t, 1, 0 \rangle$

$= \dfrac{\langle -2t^5 - t^3, -2t^4 - t^2, -2t^3 - t \rangle + \langle 2t^5 + 2t^3 + 2t, t^4 + t^2 + 1, 0 \rangle}{(t^4 + t^2 + 1)^{3/2}} = \dfrac{\langle 2t, -t^4 + 1, -2t^3 - t \rangle}{(t^4 + t^2 + 1)^{3/2}}$

$|\mathbf{T}'(t)| = \dfrac{\sqrt{4t^2 + t^8 - 2t^4 + 1 + 4t^6 + 4t^4 + t^2}}{(t^4 + t^2 + 1)^{3/2}} = \dfrac{\sqrt{t^8 + 4t^6 + 2t^4 + 5t^2}}{(t^4 + t^2 + 1)^{3/2}}$, and

$\mathbf{N}(t) = \dfrac{\langle 2t, 1 - t^4, -2t^3 - t \rangle}{\sqrt{t^8 + 4t^6 + 2t^4 + 5t^2}}$.

(c) $\kappa(t) = \dfrac{|\mathbf{T}'(t)|}{|\mathbf{r}'(t)|} = \dfrac{\sqrt{t^8 + 4t^6 + 2t^4 + 5t^2}}{(t^4 + t^2 + 1)^2}$

75. $y' = 4x^3$, $y'' = 12x^2$ and $\kappa(x) = |12x^2|/[1 + 16x^6]^{3/2}$ so $\kappa(1) = 12/(17)^{3/2}$.

77. $\mathbf{r}(t) = \langle \sin 2t, t, \cos 2t \rangle$ $\Rightarrow$ $\mathbf{r}'(t) = \langle 2\cos 2t, 1, -2\sin 2t \rangle$ $\Rightarrow$ $\mathbf{T}(t) = \frac{1}{\sqrt{5}}\langle 2\cos 2t, 1, -2\sin 2t \rangle$ $\Rightarrow$ $\mathbf{T}'(t) = \frac{1}{\sqrt{5}}\langle -4\sin 2t, 0, -4\cos 2t \rangle$ $\Rightarrow$ $\mathbf{N}(t) = \langle -\sin 2t, 0, -\cos 2t \rangle$. So $\mathbf{N} = \mathbf{N}(\pi) = \langle 0, 0, -1 \rangle$ and $\mathbf{B} = \mathbf{T} \times \mathbf{N} = \frac{1}{\sqrt{5}}\langle -1, 2, 0 \rangle$. So a normal to the osculating plane is $\langle -1, 2, 0 \rangle$ and the equation is

$-1(x - 0) + 2(y - \pi) + 0(z - 1) = 0$ or $x - 2y + 2\pi = 0$.

79. $\mathbf{v}(t) = \int (t\mathbf{i} + \mathbf{j} + t^2\mathbf{k})dt = \frac{1}{2}t^2\mathbf{i} + t\mathbf{j} + \frac{1}{3}t^3\mathbf{k} + \mathbf{c}_1$, but $\mathbf{i} + 2\mathbf{j} + \mathbf{k} = \mathbf{v}(0) = \mathbf{0} + \mathbf{c}_1$ so $\mathbf{c}_1 = \mathbf{i} + 2\mathbf{j} + \mathbf{k}$ and $\mathbf{v}(t) = \left(1 + \frac{1}{2}t^2\right)\mathbf{i} + (2 + t)\mathbf{j} + \left(1 + \frac{1}{3}t^3\right)\mathbf{k}$. $\mathbf{r}(t) = \int \mathbf{v}(t)dt = \left(t + \frac{1}{6}t^3\right)\mathbf{i} + \left(2t + \frac{1}{2}t^2\right)\mathbf{j} + \left(t + \frac{1}{12}t^4\right)\mathbf{k} + \mathbf{c}_2$. But $\mathbf{r}(0) = \mathbf{0}$ so $\mathbf{c}_2 = \mathbf{0}$ and $\mathbf{r}(t) = \left(t + \frac{1}{6}t^3\right)\mathbf{i} + \left(2t + \frac{1}{2}t^2\right)\mathbf{j} + \left(t + \frac{1}{12}t^4\right)\mathbf{k}$.

81. $x = 2\cos\frac{\pi}{6} = \sqrt{3}$, $y = 2\sin\frac{\pi}{6} = 1$, $z = 2$, so in rectangular coordinates the point is $\left(\sqrt{3}, 1, 2\right)$. $\rho = \sqrt{3 + 1 + 4} = 2\sqrt{2}$, $\theta = \frac{\pi}{6}$, and $\cos\phi = z/\rho = \frac{1}{\sqrt{2}}$, so $\phi = \frac{\pi}{4}$ and the spherical coordinates are $\left(2\sqrt{2}, \frac{\pi}{6}, \frac{\pi}{4}\right)$.

83. $x = 4\sin\frac{\pi}{6}\cos\frac{\pi}{3} = 1$, $y = 4\sin\frac{\pi}{6}\sin\frac{\pi}{3} = \sqrt{3}$, $z = 4\cos\frac{\pi}{6} = 2\sqrt{3}$ so in rectangular coordinates the point is $\left(1, \sqrt{3}, 2\sqrt{3}\right)$. $r^2 = x^2 + y^2 = 4$, $r = 2$, so the cylindrical coordinates are $\left(2, \frac{\pi}{3}, 2\sqrt{3}\right)$.

85. A half-plane including the z-axis and intersecting the xy-plane in the half-line $x = y$, $x > 0$

87. Since $\rho = 3\sec\phi$, $\rho\cos\phi = 3$ or $z = 3$. Thus the surface is a plane parallel to the xy-plane and through the point $(0, 0, 3)$.

89. In cylindrical coordinates: $r^2 + z^2 = 4$. In spherical coordinates: $\rho^2 = 4$ or $\rho = 2$.

91. The resulting surface is a paraboloid of revolution with equation $z = 4x^2 + 4y^2$. Changing to cylindrical coordinates we have $z = 4(x^2 + y^2) = 4r^2$.

93. **(a)** Instead of proceeding directly, we use Theorem 11.7.5 #3:

$$\mathbf{r}(t) = t\mathbf{R}(t) \quad \Rightarrow \quad \mathbf{v} = \mathbf{r}'(t) = \mathbf{R}(t) + t\mathbf{R}'(t) = \cos\omega t\,\mathbf{i} + \sin\omega t\,\mathbf{j} + t\mathbf{v}_d$$

(b) Using the same method as in part (a) and starting with $\mathbf{v} = \mathbf{R}(t) + t\mathbf{R}'(t)$, we have

$$\mathbf{a} = \mathbf{v}' = \mathbf{R}'(t) + \mathbf{R}'(t) + t\mathbf{R}''(t) = 2\mathbf{R}'(t) + t\mathbf{R}''(t) = 2\mathbf{v}_d + t\mathbf{a}_d.$$

(c) Here we have $\mathbf{r}(t) = e^{-t}\cos\omega t\,\mathbf{i} + e^{-t}\sin\omega t\,\mathbf{j} = e^{-t}\mathbf{R}(t)$. So, as in parts (a) and (b),

$$\mathbf{v} = \mathbf{r}'(t) = e^{-t}\mathbf{R}'(t) - e^{-t}\mathbf{R}(t) = e^{-t}[\mathbf{R}'(t) - \mathbf{R}(t)] \quad \Rightarrow$$

$$\mathbf{a} = \mathbf{v}' = e^{-t}[\mathbf{R}''(t) - \mathbf{R}'(t)] - e^{-t}[\mathbf{R}'(t) - \mathbf{R}(t)] = e^{-t}[\mathbf{R}''(t) - 2\mathbf{R}'(t) + \mathbf{R}(t)] = e^{-t}\mathbf{a}_d - 2e^{-t}\mathbf{v}_d + e^{-t}\mathbf{R}.$$

Thus the Coriolis acceleration (the "extra" terms not involving $\mathbf{a}_d$) is $-2e^{-t}\mathbf{v}_d + e^{-t}\mathbf{R}$.

APPLICATIONS PLUS (after Chapter 11)

1. **(a)** $\mathbf{r}(t) = R\cos\omega t\,\mathbf{i} + R\sin\omega t\,\mathbf{j} \Rightarrow \mathbf{v} = \mathbf{r}'(t) = -\omega R\sin\omega t\,\mathbf{i} + \omega R\cos\omega t\,\mathbf{j}$, so

$\mathbf{r} = R(\cos\omega t\,\mathbf{i} + \sin\omega t\,\mathbf{j})$ and $\mathbf{v} = \omega R(-\sin\omega t\,\mathbf{i} + \cos\omega t\,\mathbf{j})$.

$\mathbf{v}\cdot\mathbf{r} = \omega R^2(-\cos\omega t\sin\omega t + \sin\omega t\cos\omega t) = 0$, so $\mathbf{v}\perp\mathbf{r}$. Since $\mathbf{r}$ points along a radius of the circle, and

$\mathbf{v}\perp\mathbf{r}$, $\mathbf{v}$ is tangent to the circle. Because it is a velocity vector, $\mathbf{v}$ points in the direction of motion.

(b) In (a), we wrote $\mathbf{v}$ in the form $\omega R\mathbf{u}$, where $\mathbf{u}$ is the unit vector $-\sin\omega t\,\mathbf{i} + \cos\omega t\,\mathbf{j}$. Clearly

$|\mathbf{v}| = \omega R|\mathbf{u}| = \omega R$. At speed ωR, the particle completes one revolution, a distance $2\pi R$, in time

$T = \dfrac{2\pi R}{\omega R} = \dfrac{2\pi}{\omega}$.

(c) $\mathbf{a} = \dfrac{d\mathbf{v}}{dt} = -\omega^2 R\cos\omega t\,\mathbf{i} - \omega^2 R\sin\omega t\,\mathbf{j} = -\omega^2 R(\cos\omega t\,\mathbf{i} + \sin\omega t\,\mathbf{j})$, so $\mathbf{a} = -\omega^2\mathbf{r}$. This shows that $\mathbf{a}$ is

proportional to $\mathbf{r}$ and points in the opposite direction (toward the origin). Also, $|\mathbf{a}| = \omega^2|\mathbf{r}| = \omega^2 R$.

(d) By Newton's Second Law (see Section 11.9 of the text), $\mathbf{F} = m\mathbf{a}$, so

$|\mathbf{F}| = m|\mathbf{a}| = mR\omega^2 - \dfrac{m(\omega R)^2}{R} = \dfrac{m|\mathbf{v}|^2}{R}$.

3. **(a)** The projectile reaches maximum height when $0 = \dfrac{dy}{dt} = \dfrac{d}{dt}\left[(v_0\sin\alpha)t - \tfrac{1}{2}gt^2\right] = v_0\sin\alpha - gt$; that is,

when $t = \dfrac{v_0\sin\alpha}{g}$ and $y = (v_0\sin\alpha)\left(\dfrac{v_0\sin\alpha}{g}\right) - \dfrac{g}{2}\left(\dfrac{v_0\sin\alpha}{g}\right)^2 = \dfrac{v_0^2\sin^2\alpha}{2g}$. This is the maximum

height attained when the projectile is fired with an angle of elevation α. This maximum height is largest

when $\alpha = \dfrac{\pi}{2}$. In that case, $\sin\alpha = 1$ and the maximum height is $v_0^2/(2g)$.

(b) Let $R = v_0^2/g$. We are asked to consider the parabola $x^2 + 2Ry - R^2 = 0$ which can be rewritten as

$y = -\dfrac{1}{2R}x^2 + \dfrac{R}{2}$. The points on or inside this parabola are those for which $-R \le x \le R$ and

$0 \le y \le \dfrac{-1}{2R}x^2 + \dfrac{R}{2}$. When the projectile is fired at angle of elevation α, the points (x,y) along its path

satisfy the relations $x = (v_0\cos\alpha)t$ and $y = (v_0\sin\alpha)t - \tfrac{1}{2}gt^2$, where $0 \le t \le (2v_0\sin\alpha)/g$ (as in

Example 5 in Section 11.9). Thus $|x| \le |(v_0\cos\alpha)(2v_0\sin\alpha)/g| = |(v_0^2/g)\sin 2\alpha| \le |v_0^2/g| = |R|$. This

shows that $-R \le x \le R$.

For t in the specified range, we also have $y = t(v_0\sin\alpha - \tfrac{1}{2}gt) = \tfrac{1}{2}gt\left(\dfrac{2v_0\sin\alpha}{g} - t\right) \ge 0$ and

$y = (v_0\sin\alpha)\dfrac{x}{v_0\cos\alpha} - \dfrac{g}{2}\left(\dfrac{x}{v_0\cos\alpha}\right)^2 = (\tan\alpha)x - \dfrac{g}{2v_0^2\cos^2\alpha}x^2 = \dfrac{-1}{2R\cos^2\alpha}x^2 + (\tan\alpha)x$. Thus

$y - \left(\dfrac{-1}{2R}x^2 + \dfrac{R}{2}\right) = \dfrac{-1}{2R\cos^2\alpha}x^2 + \dfrac{1}{2R}x^2 + (\tan\alpha)x - \dfrac{R}{2} = \dfrac{x^2}{2R}\left(1 - \dfrac{1}{\cos^2\alpha}\right) + (\tan\alpha)x - \dfrac{R}{2}$

$= \dfrac{x^2(1 - \sec^2\alpha) + 2R(\tan\alpha)x - R^2}{2R} = \dfrac{-(\tan^2\alpha)x^2 + 2R(\tan\alpha)x - R^2}{2R} = \dfrac{-[(\tan\alpha)x - R]^2}{2R} \le 0$.

We have shown that every target that can be hit by the projectile lies on or inside the parabola

$y = -\dfrac{1}{2R}x^2 + \dfrac{R}{2}$. Now let (a, b) be any point on or inside that parabola. Then $-R \le a \le R$ and

$0 \le b \le -\dfrac{1}{2R}a^2 + \dfrac{R}{2}$. We seek an angle α such that (a, b) lies in the path of the projectile; that is, we

wish to find an angle α such that $b = \dfrac{-1}{2R\cos^2\alpha}a^2 + (\tan\alpha)a$ or equivalently

$b = \dfrac{-1}{2R}(\tan^2\alpha + 1)a^2 + (\tan\alpha)a$. Rearranging this equation we get

$\dfrac{a^2}{2R}\tan^2\alpha - a\tan\alpha + \left(\dfrac{a^2}{2R} + b\right) = 0$ or $a^2(\tan\alpha)^2 - 2aR(\tan\alpha) + (a^2 + 2bR) = 0$ (★). This

quadratic equation for $\tan\alpha$ has real solutions exactly when the discriminant is nonnegative. Now

$B^2 - 4AC \ge 0 \;\Leftrightarrow\; (-2aR)^2 - 4a^2(a^2 + 2bR) \ge 0 \;\Leftrightarrow\; 4a^2(R^2 - a^2 - 2bR) \ge 0 \;\Leftrightarrow$

$-a^2 - 2bR + R^2 \ge 0 \;\Leftrightarrow\; b \le \dfrac{1}{2R}(R^2 - a^2) \;\Leftrightarrow\; b \le \dfrac{-1}{2R}a^2 + \dfrac{R}{2}$. This condition is satisfied

since (a, b) is on or inside the parabola $y = \dfrac{-1}{2R}x^2 + \dfrac{R}{2}$. It follows that (a, b) lies in the path of the

projectile when $\tan\alpha$ satisfies (★), that is, when

$\tan\alpha = \dfrac{2aR \pm \sqrt{4a^2(R^2 - a^2 - 2bR)}}{2a^2} = \dfrac{R \pm \sqrt{R^2 - 2bR - a^2}}{a}$.

(c)

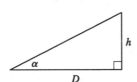

If the gun is pointed at a target with height h at a distance D downrange, then $\tan\alpha = h/D$. When the

projectile reaches a distance D downrange (remember we are assuming that it doesn't hit the ground first,)

we have $D = x = (v_0\cos\alpha)t$, so $t = \dfrac{D}{v_0\cos\alpha}$ and $y = (v_0\sin\alpha)t - \tfrac{1}{2}gt^2 = D\tan\alpha - \dfrac{gD^2}{2v_0^2\cos^2\alpha}$.

Meanwhile, the target, whose x-coordinate is also D, has fallen from height h to height

$h - \tfrac{1}{2}gt^2 = D\tan\alpha - \dfrac{gD^2}{2v_0^2\cos^2\alpha}$. Thus the projectile hits the target.

5. **(a)** $m\dfrac{d^2\mathbf{R}}{dt^2} = -mg\mathbf{j} - k\dfrac{d\mathbf{R}}{dt} \;\Rightarrow\; \dfrac{d}{dt}\left(m\dfrac{d\mathbf{R}}{dt} + k\mathbf{R} + mgt\mathbf{j}\right) = 0 \;\Rightarrow\; m\dfrac{d\mathbf{R}}{dt} + k\mathbf{R} + mgt\mathbf{j} = \mathbf{c}$ (**c** is a

constant vector in the xy-plane). At $t = 0$, this says $m\mathbf{v}(0) + k\mathbf{R}(0) = \mathbf{c}$. Since $\mathbf{v}(0) = \mathbf{v}_0$ and $\mathbf{R}(0) = \mathbf{0}$,

we have $\mathbf{c} = m\mathbf{v}_0$. Therefore $\dfrac{d\mathbf{R}}{dt} + \dfrac{k}{m}\mathbf{R} + gt\mathbf{j} = \mathbf{v}_0$, or $\dfrac{d\mathbf{R}}{dt} + \dfrac{k}{m}\mathbf{R} = \mathbf{v}_0 - gt\mathbf{j}$.

(b) Multiplying by $e^{(k/m)t}$ gives $e^{(k/m)t}\dfrac{d\mathbf{R}}{dt} + \dfrac{k}{m}e^{(k/m)t}\mathbf{R} = e^{(k/m)t}\mathbf{v}_0 - gte^{(k/m)t}\mathbf{j}$ or

$\dfrac{d}{dt}\left(e^{(k/m)t}\mathbf{R}\right) = e^{(k/m)t}\mathbf{v}_0 - gte^{(k/m)t}\mathbf{j}$. Integrating gives

$e^{(k/m)t}\mathbf{R} = \dfrac{m}{k}e^{(k/m)t}\mathbf{v}_0 - \left(\dfrac{mg}{k}te^{(k/m)t} - \dfrac{m^2g}{k^2}e^{(k/m)t}\right)\mathbf{j} + \mathbf{b}$ for some constant vector $\mathbf{b}$. Setting $t = 0$

yields the relation $\mathbf{R}(0) = \dfrac{m}{k}\mathbf{v}_0 + \dfrac{m^2g}{k^2}\mathbf{j} + \mathbf{b}$, so $\mathbf{b} = -\dfrac{m}{k}\mathbf{v}_0 - \dfrac{m^2g}{k^2}\mathbf{j}$. Thus

$e^{(k/m)t}\mathbf{R} = \dfrac{m}{k}\left[e^{(k/m)t} - 1\right]\mathbf{v}_0 - \left[\dfrac{mg}{k}te^{(k/m)t} - \dfrac{m^2g}{k^2}\left(e^{(k/m)t} - 1\right)\right]\mathbf{j}$ and

$\mathbf{R}(t) = \dfrac{m}{k}\left[1 - e^{-(k/m)t}\right]\mathbf{v}_0 + \dfrac{mg}{k}\left[\dfrac{m}{k}\left(1 - e^{-(k/m)t}\right) - t\right]\mathbf{j}.$

7. **(a)** $F(x) = \begin{cases} 1 & \text{if } x \le 0 \\ \sqrt{1 - x^2} & \text{if } 0 < x < \frac{1}{\sqrt{2}} \\ -x + \sqrt{2} & \text{if } x \ge \frac{1}{\sqrt{2}} \end{cases} \Rightarrow F'(x) = \begin{cases} 0 & \text{if } x < 0 \\ -x/\sqrt{1 - x^2} & \text{if } 0 < x < \frac{1}{\sqrt{2}} \\ -1 & \text{if } x > \frac{1}{\sqrt{2}} \end{cases} \Rightarrow$

$F''(x) = \begin{cases} 0 & \text{if } x < 0 \\ -1/(1 - x^2)^{3/2} & \text{if } 0 < x < \frac{1}{\sqrt{2}} \\ 0 & \text{if } x > \frac{1}{\sqrt{2}} \end{cases}$ since

$\dfrac{d}{dx}\left[-x\left(1 - x^2\right)^{-1/2}\right] = -\left(1 - x^2\right)^{-1/2} - x^2\left(1 - x^2\right)^{-3/2} = -\left(1 - x^2\right)^{-3/2}.$

Now $\lim\limits_{x \to 0^+} \sqrt{1 - x^2} = 1 = F(0)$ and $\lim\limits_{x \to (1/\sqrt{2})^-} \sqrt{1 - x^2} = \frac{1}{\sqrt{2}} = F\!\left(\frac{1}{\sqrt{2}}\right)$, so F is continuous. Also,

since $\lim\limits_{x \to 0^+} F'(x) = 0 = \lim\limits_{x \to 0^-} F'(x)$ and $\lim\limits_{x \to (1/\sqrt{2})^-} F'(x) = -1 = \lim\limits_{x \to (1/\sqrt{2})^+} F'(x)$, F' is continuous. But

$\lim\limits_{x \to 0^+} F''(x) = -1 \ne 0 = \lim\limits_{x \to 0^-} F''(x)$, so F'' is not continuous at $x = 0$. $\left(\text{The same is true at } x = \frac{1}{\sqrt{2}}.\right)$

So F does not have continuous curvature.

(b) Set $P(x) = ax^5 + bx^4 + cx^3 + dx^2 + ex + f$. The continuity conditions on P are $P(0) = 0$, $P(1) = 1$,

$P'(0) = 0$ and $P'(1) = 1$. Also the curvature must be continuous. For $x \le 0$ and $x \ge 1$ $\kappa(x) = 0$,

elsewhere $\kappa(x) = \dfrac{|P''(x)|}{\left(1 + [P'(x)]^2\right)^{3/2}}$, so we need $P''(0) = 0$ and $P''(1) = 0$. The conditions

$P(0) = P'(0) = P''(0) = 0$ imply that $d = e = f = 0$.

The other conditions imply that $a + b + c = 1$,

$5a + 4b + 3c = 1$, and $10a + 6b + 3c = 0$. From these, we

find that $a = 3$, $b = -8$, and $c = 6$. Therefore

$P(x) = 3x^5 - 8x^4 + 6x^3$. Since there was no solution with $a = 0$,

this could not have been done with a polynomial of degree 4.

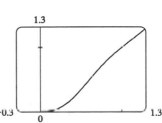

CHAPTER TWELVE

EXERCISES 12.1

1. **(a)** $f(2,1) = 4 - 1 + 4(2)(1) - 7(2) + 10 = 7$

   (b) $f(-3,5) = 9 - 25 + 4(-3)(5) + 21 + 10 = -45$

   (c) $f(x+h,y) = (x+h)^2 - y^2 + 4(x+h)y - 7(x+h) + 10$
   $$= x^2 + 2xh + h^2 - y^2 + 4xy + 4hy - 7x - 7h + 10$$

   (d) $f(x,y+k) = x^2 - (y+k)^2 + 4x(y+k) - 7x + 10 = x^2 - y^2 - 2ky - k^2 + 4xy + 4xk - 7x + 10$

   (e) $f(x,x) = x^2 - x^2 + 4x^2 - 7x + 10 = 4x^2 - 7x + 10$

3. **(a)** $F(1,1) = \dfrac{3(1)(1)}{1+2} = 1$

   (b) $F(-1,2) = \dfrac{3(-1)(2)}{1+8} = -\dfrac{2}{3}$

   (c) $F(t,1) = \dfrac{3t}{t^2+2}$

   (d) $F(-1,y) = \dfrac{3(-1)y}{1+2y^2} = -\dfrac{3y}{1+2y^2}$

   (e) $F(x,x^2) = \dfrac{3xx^2}{x^2 + 2(x^2)^2} = \dfrac{3x^3}{x^2 + 2x^4} = \dfrac{3x}{1+2x^2}$

5. $D = \mathbb{R}^2$ and the range is $\mathbb{R}$.

7. $x + y \neq 0$ so $D = \{(x,y) \mid x + y \neq 0\}$. Since $2/(x+y)$ can't be zero, the range is $\{z \mid z \neq 0\}$.

9. $D = \mathbb{R}^2$ since the exponential function is defined everywhere and the range is $\{z \mid z > 0\}$.

11. For the logarithmic function to be defined, we need $x - y + z > 0$. Thus $D = \{(x,y,z) \mid x + z > y\}$ and the range is $\mathbb{R}$.

13. $D = \mathbb{R}^3$ and the range is $\mathbb{R}$.

15. $y - 2x \geq 0$ so $D = \{(x,y) \mid y \geq 2x\}$.

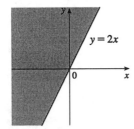

17. $x + 2y \neq 0$ and $9 - x^2 - y^2 \geq 0$, so
    $$D = \{(x,y) \mid y \neq -\tfrac{1}{2}x \text{ and } x^2 + y^2 \leq 9\}.$$

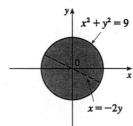

19. $D = \{(x, y) \mid x^2 + y \geq 0\} = \{(x, y) \mid y \geq -x^2\}$

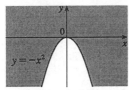

21. $D = \{(x, y) \mid xy > 1\}$

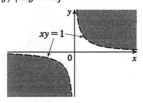

23. $D = \{(x, y) \mid y \neq \frac{\pi}{2} + n\pi, n \text{ an integer}\}$

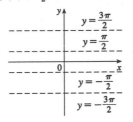

25. $D = \{(x, y) \mid -1 \leq x + y \leq 1\}$
$= \{(x, y) \mid -1 - x \leq y \leq 1 - x\}$

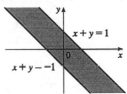

27. Since $\sin y > 0$ implies $2n\pi < y < (2n+1)\pi$, n an integer, $D = \{(x, y) \mid x > 0$ and $2n\pi < y < (2n+1)\pi, n$ an integer$\}$.

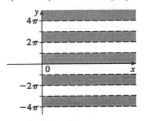

29. $D = \{(x, y, z) \mid x^2 + y^2 + z^2 \leq 1\}$ (the points inside or on the sphere of radius 1, center the origin).

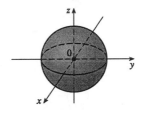

31. $z = 3$, a horizontal plane through the point $(0, 0, 3)$.

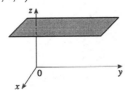

33. $z = 1 - x - y$ or $x + y + z = 1$, a plane with intercepts 1, 1, and 1.

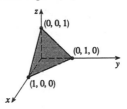

35. $z = x^2 + 9y^2$, an elliptic paraboloid with vertex the origin.

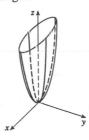

37. $z = \sqrt{x^2 + y^2}$, the top half of a right circular cone.

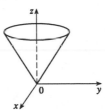

39. $z = y^2 - x^2$, a hyperbolic paraboloid.

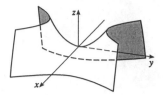

41. $z = 1 - x^2$, a parabolic cylinder.

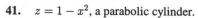
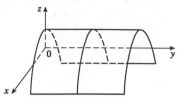

43. The level curves are $xy = k$. For $k = 0$ the curves are the coordinate axis; if $k > 0$, they are hyperbolas in the first and third quadrants; if $k < 0$, they are hyperbolas in the second and fourth quadrants.

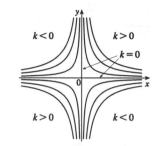

45. $k = x^2 + 9y^2$, a family of ellipses with major axis the x-axis. (Or, if $k = 0$, the origin.)

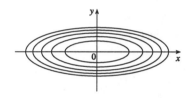

47. $k = x/y$ is a family of lines without the point $(0, 0)$.

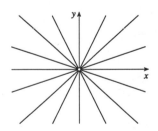

49. $k = \sqrt{x + y}$ or for $x + y \geq 0$, $k^2 = x + y$, or $y = -x + k^2$

Note: $k \geq 0$ since $k = \sqrt{x + y}$.

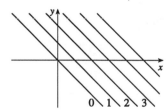

51. $k = x - y^2$, or $x - k = y^2$, a family of parabolas with vertex $(k, 0)$.

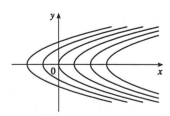

53. $k = x + 3y + 5z$ is a family of parallel planes with normal vector $\langle 1, 3, 5 \rangle$.

55. $k = x^2 - y^2 + z^2$ are the equations of the level surfaces. For $k = 0$, the surface is a right circular cone with vertex the origin and axis the y-axis. For $k > 0$, we have a family of hyperboloids of one sheet with axis the y-axis. For $k < 0$, we have a family of hyperboloids of two sheets with axis the y-axis.

57. The isothermals are given by

$k = 100/(1 + x^2 + 2y^2)$ or

$x^2 + 2y^2 = (100 - k)/k$

$(0 < k \leq 100)$, a family of ellipses.

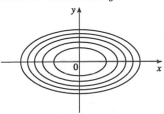

59. (a) B   *Reasons:* This function is constant on any circle centered at the origin, a

   (b) III   description which matches only B and III.

61. (a) F   *Reasons:* $f(x, y) \to \infty$ as $(x, y) \to (0, 0)$, a condition satisfied only by F and V.

   (b) V

63. (a) D   *Reasons:* This function is periodic in both x and y, with period 2π in each variable.

   (b) IV

65. $f(x, y) = x^3 + y^3$

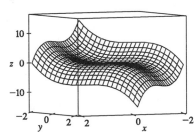 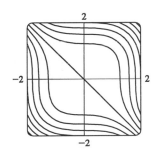

Note that the function is 0 along the line $y = -x$.

67. $f(x, y) = xy^2 - x^3$

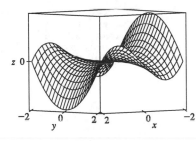 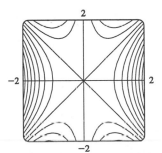

The cross-sections parallel to the yz-plane (such as the left-front trace in the graph above) are parabolas; those parallel to the xz-plane (such as the right-front trace) are cubic curves. The surface is called a monkey saddle because a monkey sitting on the surface near the origin has places for both legs and tail to rest.

69. $f(x, y) = e^{ax^2+by^2}$. We start with the case $a = b = 1$. This gives a graph whose level curves are circles centered at the origin.

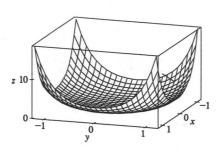

$$a = 1, b = 1$$

If we increase the ratio a/b, the level curves become ellipses whose eccentricity increases as a/b increases.

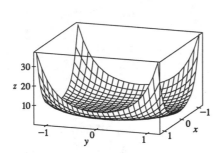

$$a = 1.5, b = 1$$

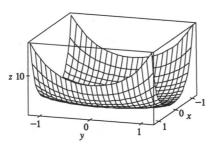

$$a = 1.5, b = 0.5$$

If one of a and b is 0, the graph is a cylinder. Note that in general, if we interchange a and b, the graph is rotated by 90° about the z-axis.

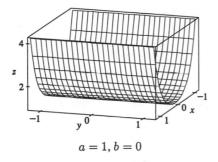

$$a = 1, b = 0$$

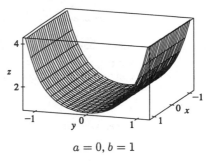

$$a = 0, b = 1$$

If a is positive and b is negative, the graph is saddle-shaped near the point $(0, 0, 1)$.

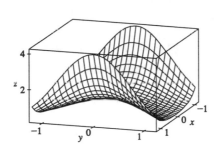

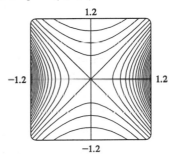

$$a = 1, b = -1$$

If a and b are both negative, the graph has a bump at the origin, which gets narrower in the x-direction as a decreases, and narrower in the y-direction as b decreases.

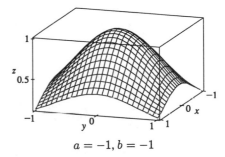

$$a = -1, b = -1$$

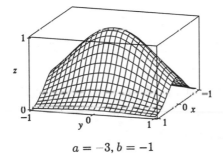

$$a = -3, b = -1$$

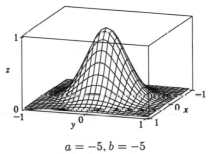

$$a = -5, b = -5$$

EXERCISES 12.2

1. The function is a polynomial, so the limit equals $(2^2)(3^2) - 2(2)(3^5) + 3(3) = -927$.

3. Since this is a rational function defined at $(0,0)$, the limit equals $(0 + 0 - 5)/(2 - 0) = -\frac{5}{2}$.

5. The product of two functions continuous at (π, π), so the limit equals $\pi \sin[(\pi + \pi)/4] = \pi$.

7. Let $f(x, y) = (x - y)/(x^2 + y^2)$. First approach $(0,0)$ along the x-axis. Then $f(x, 0) = x/x^2 = 1/x$ and $\lim_{x \to 0} f(x, 0)$ doesn't exist. Thus $\lim_{(x,y) \to (0,0)} f(x, y)$ doesn't exist.

9. Let $f(x, y) = 8x^2 y^2/(x^4 + y^4)$. Approaching $(0,0)$ along the x-axis gives $f(x, y) \to 0$ as $(x, y) \to (0,0)$ along the x-axis. Approaching $(0,0)$ along the line $y = x$, $f(x, x) = 8x^4/2x^4 = 4$ for $x \neq 0$, so along this line $f(x, y) \to 4$ as $(x, y) \to (0,0)$. Thus the limit doesn't exist.

11. Let $f(x, y) = 2xy/(x^2 + 2y^2)$. As $(x, y) \to (0,0)$ along the x-axis, $f(x, y) \to 0$. But as $(x, y) \to (0,0)$ along the line $y = x$, $f(x, x) = 2x^2/3x^2$, so $f(x, y) \to \frac{2}{3}$ as $(x, y) \to (0,0)$ along this line. So the limit doesn't exist.

13. We can show that the limit along any line through $(0,0)$ is 0 and that the limits along the paths $x = y^2$ and $y = x^2$ are also 0. So we suspect that the limit exists and equals 0. Let $\epsilon > 0$ be given. We need to find $\delta > 0$ such that $\left| xy/\sqrt{x^2 + y^2} - 0 \right| < \epsilon$ whenever $0 < \sqrt{x^2 + y^2} < \delta$ or $|xy|/\sqrt{x^2 + y^2} < \epsilon$ whenever $0 < \sqrt{x^2 + y^2} < \delta$. But $|x| = \sqrt{x^2} \leq \sqrt{x^2 + y^2}$ so $|xy|/\sqrt{x^2 + y^2} \leq |y| = \sqrt{y^2} \leq \sqrt{x^2 + y^2}$. Thus choose $\delta = \epsilon$ and let $0 < \sqrt{x^2 + y^2} < \delta = \epsilon$, then $\left| xy/\sqrt{x^2 + y^2} - 0 \right| \leq \sqrt{x^2 + y^2} < \delta = \epsilon$. Hence by definition, $\lim_{(x,y) \to (0,0)} xy/\sqrt{x^2 + y^2} = 0$.

    Or: Use the Squeeze Theorem. $0 \leq \left| \dfrac{xy}{\sqrt{x^2 + y^2}} \right| \leq |x|$ since $|y| \leq \sqrt{x^2 + y^2}$, and $|x| \to 0$ as $(x, y) \to (0,0)$.

15. Let $f(x, y) = \dfrac{2x^2 y}{x^4 + y^2}$. Then $f(x, 0) = 0$ for $x \neq 0$, so $f(x, y) \to 0$ as $(x, y) \to (0,0)$ along the x-axis. But $f(x, x^2) = \dfrac{2x^4}{2x^4} = 1$ for $x \neq 0$, so $f(x, y) \to 1$ as $(x, y) \to (0,0)$ along the parabola $y = x^2$. Thus the limit doesn't exist.

17. $\lim_{(x,y) \to (0,0)} \dfrac{x^2 + y^2}{\sqrt{x^2 + y^2 + 1} - 1} = \lim_{(x,y) \to (0,0)} \dfrac{(x^2 + y^2)(\sqrt{x^2 + y^2 + 1} + 1)}{x^2 + y^2} = \lim_{(x,y) \to (0,0)} \left[\sqrt{x^2 + y^2 + 1} + 1 \right] = 2$

19. Let $f(x, y) = (xy - x)/(x^2 + y^2 - 2y + 1)$. Then $f(0, y) = 0$ for $y \neq 1$, so $f(x, y) \to 0$ as $(x, y) \to (0, 1)$ along the y-axis. But $f(x, x + 1) = x(x + 1 - 1)/(x^2 + (x + 1 - 1)^2) = \frac{1}{2}$ for $x \neq 0$ so $f(x, y) \to 1/2$ as $(x, y) \to (0, 1)$ along the line $y = x + 1$. Thus the limit doesn't exist.

21. $\lim_{(x,y,z) \to (1,2,3)} \dfrac{xz^2 - y^2 z}{xyz - 1} = \dfrac{1 \cdot 3^2 - 2^2 \cdot 3}{1 \cdot 2 \cdot 3 - 1} = -\dfrac{3}{5}$ since the function is continuous at $(1, 2, 3)$.

23. Let $f(x, y, z) = (x^2 - y^2 - z^2)/(x^2 + y^2 + z^2)$. Then $f(x, 0, 0) = 1$ for $x \neq 0$ and $f(0, y, 0) = -1$ for $y \neq 0$, so as $(x, y, z) \rightarrow (0, 0, 0)$ along the x-axis, $f(x, y, z) \rightarrow 1$ but as $(x, y, z) \rightarrow (0, 0, 0)$ along the y-axis, $f(x, y, z) \rightarrow -1$. Thus the limit doesn't exist.

25. Let $f(x, y, z) = \dfrac{xy + yz^2 + xz^2}{x^2 + y^2 + z^4}$. Then $f(x, 0, 0) = 0/x^2 = 0$ for $x \neq 0$, so as $(x, y, z) \rightarrow (0, 0, 0)$ along the x-axis, $f(x, y, z) \rightarrow 0$. But $f(x, x, 0) = x^2/(2x^2) = \frac{1}{2}$ for $x \neq 0$, so as $(x, y, z) \rightarrow (0, 0, 0)$ along the line $y = x, z = 0$, $f(x, y, z) \rightarrow \frac{1}{2}$. Thus the limit does not exist.

27.

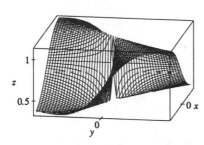

From the ridges on the graph, we see that as $(x, y) \rightarrow (0, 0)$ along the lines under the two ridges, $f(x, y)$ approaches different values. So the limit does not exist.

29. $h(x, y) = g(f(x, y)) = g(x^4 + x^2y^2 + y^4) = e^{-(x^4 + x^2y^2 + y^4)}\cos(x^4 + x^2y^2 + y^4)$. Since f is a polynomial it is continuous throughout $\mathbb{R}^2$ and g is the product of two functions, both of which are continuous on $\mathbb{R}$, h is continuous on $\mathbb{R}^2$ by Theorem 5.

31. $h(x, y) = g(f(x, y)) = (2x + 3y - 6)^2 + \sqrt{2x + 3y - 6}$. Since f is a polynomial, it is continuous on $\mathbb{R}^2$ and g is continuous on its domain $\{t \mid t \geq 0\}$. Thus h is continuous on its domain $D = \{(x, y) \mid 2x + 3y - 6 \geq 0\}$ $= \{(x, y) \mid y \geq -\frac{2}{3}x + 2\}$ which consists of all points on and above right of the line $y = -\frac{2}{3}x + 2$.

33. $F(x, y)$ is a rational function and thus is continuous on its domain $D = \{(x, y) \mid x^2 + y^2 - 1 \neq 0\}$, that is, F is continuous except on the circle $x^2 + y^2 = 1$.

35. $F(x, y) = g(f(x, y))$ where $f(x, y) = x^4 - y^4$, a polynomial so continuous on $\mathbb{R}^2$ and $g(t) = \tan t$, continuous on its domain $\{t \mid t \neq (2n + 1)\frac{\pi}{2}, n \text{ an integer}\}$. Thus F is continuous on its domain $D = \{(x, y) \mid x^4 - y^4 \neq (2n + 1)\frac{\pi}{2}, n \text{ an integer}\}$.

37. $G(x, y) = g(x, y)f(x, y)$ where $g(x, y) = e^{xy}$ and $f(x, y) = \sin(x + y)$ both of which are continuous on $\mathbb{R}^2$. Thus G is continuous on $\mathbb{R}^2$.

39. $G(x, y) = g_1(f_1(x, y)) - g_2(f_2(x, y))$ where $f_1(x, y) = x + y$ and $f_2(x, y) = x - y$ both of which are polynomials so continuous on $\mathbb{R}^2$ and $g_1(t) = \sqrt{t}$, $g_2(s) = \sqrt{s}$ both of which are continuous on their respective domains $\{t \mid t > 0\}$ and $\{s \mid s \geq 0\}$. Thus $g_1 \circ f_1$ is continuous on its domain $D_1 = \{(x, y) \mid x + y \geq 0\} = \{(x, y) \mid y \geq -x\}$ and $g_2 \circ f_2$ is continuous on its domain $D_2 = \{(x, y) \mid x - y \geq 0\} = \{(x, y) \mid y \leq x\}$. Then G, being the difference of these two composite functions, is continuous on its domain $D = D_1 \cap D_2 = \{(x, y) \mid -x \leq y \leq x\} = \{(x, y) \mid |y| \leq x\}$.

41. $f(x, y, z) = xg(f(y, z))$ where $f(y, z) = yz$, continuous on $\mathbb{R}^2$ and $g(t) = \ln t$, continuous on its domain $\{t \mid t > 0\}$. Since $h(x) = x$ is continuous on $\mathbb{R}$, $f(x, y, z)$ is continuous on its domain $D = \{(x, y, z) \mid yz > 0\}$.

In Exercises 43-46 each f is a piecewise defined function where each first piece is a rational function defined everywhere except at the origin. Thus each f is continuous on $\mathbb{R}^2$ except possibly at the origin. So for each we need only check $\lim_{(x,y)\to(0,0)} f(x, y)$.

43. Letting $z = \sqrt{2}x$, $\lim_{(x,y)\to(0,0)} \dfrac{2x^2 - y^2}{2x^2 + y^2} = \lim_{(z,y)\to(0,0)} \dfrac{z^2 - y^2}{z^2 + y^2}$ which doesn't exist by Example 1. Thus f is not continuous at $(0, 0)$ and the largest set on which f is continuous is $\{(x, y) \mid (x, y) \neq (0, 0)\}$.

45. Since $x^2 \leq 2x^2 + y^2$, we have $|x^2 y^3 / (2x^2 + y^2)| \leq |y^3|$. We know that $|y^3| \to 0$ as $(x, y) \to (0, 0)$. So, by the Squeeze Theorem, $\lim_{(x,y)\to(0,0)} f(x, y) = \lim_{(x,y)\to(0,0)} \dfrac{x^2 y^3}{2x^2 + y^2} = 0$. But $f(0, 0) = 1$, so f is discontinuous at $(0, 0)$.

For $(x, y) \neq (0, 0)$, $f(x, y)$ is equal to a rational function and is therefore continuous. Therefore f is continuous on the set $\{(x, y) \mid (x, y) \neq (0, 0)\}$.

47. $\lim_{(x,y)\to(0,0)} \dfrac{x^3 + y^3}{x^2 + y^2} = \lim_{r\to 0^+} \dfrac{(r\cos\theta)^3 + (r\sin\theta)^3}{r^2} = \lim_{r\to 0^+} \left[r\cos^3\theta + r\sin^3\theta \right] = 0$

49. $\lim_{(x,y,z)\to(0,0,0)} \dfrac{xyz}{x^2 + y^2 + z^2} = \lim_{\rho\to 0^+} \dfrac{(\rho\sin\phi\cos\theta)(\rho\sin\phi\sin\theta)(\rho\cos\phi)}{\rho} = \lim_{\rho\to 0^+} \left(\rho^2 \sin^2\phi \cos\phi \sin\theta \cos\theta \right) = 0$

51. **(a)** Let $\epsilon > 0$ be given. We need to find $\delta > 0$ such that $|x - a| < \epsilon$ whenever
$0 < \sqrt{(x-a)^2 + (y-b)^2} < \delta$. But $|x - a| = \sqrt{(x-a)^2} \leq \sqrt{(x-a)^2 + (y-b)^2}$. Thus setting $\delta = \epsilon$ and letting $0 < \sqrt{(x-a)^2 + (y-b)^2} < \delta$, we have $|x - a| \leq \sqrt{(x-a)^2 + (y-b)^2} < \delta = \epsilon$. Hence, by Definition 1, $\lim_{(x,y)\to(a,b)} x = a$.

(b) The argument is the same as in (a) with the roles of x and y interchanged.

(c) Let $\epsilon > 0$ be given and set $\delta = \epsilon$. Then $|f(x, y) - L| = |c - c| = 0 \leq \sqrt{(x-a)^2 + (y-b)^2} < \delta = \epsilon$ whenever $0 < \sqrt{(x-a)^2 + (y-b)^2} < \delta$. Thus by Definition 1, $\lim_{(x,y)\to(a,b)} c = c$.

53. Since $|\mathbf{x} - \mathbf{a}|^2 = |\mathbf{x}|^2 + |\mathbf{a}|^2 - 2|\mathbf{x}||\mathbf{a}|\cos\theta \geq |\mathbf{x}|^2 + |\mathbf{a}|^2 - 2|\mathbf{x}||\mathbf{a}| = (|\mathbf{x}| - |\mathbf{a}|)^2$, we have $||\mathbf{x}| - |\mathbf{a}|| \leq |\mathbf{x} - \mathbf{a}|$. Let $\epsilon > 0$ be given and set $\delta = \epsilon$. Then whenever $0 < |\mathbf{x} - \mathbf{a}| < \delta$, $||\mathbf{x}| - |\mathbf{a}|| \leq |\mathbf{x} - \mathbf{a}| < \delta = \epsilon$. Hence $\lim_{\mathbf{x}\to\mathbf{a}} |\mathbf{x}| = |\mathbf{a}|$ and $f(\mathbf{x}) = |\mathbf{x}|$ is continuous on $\mathbb{R}^n$.

EXERCISES 12.3

1. $f(x, y) = 16 - 4x^2 - y^2 \Rightarrow f_x(x, y) = -8x$ and $f_y(x, y) = -2y \Rightarrow f_x(1, 2) = -8$ and

$f_y(1, 2) = -4$. The graph of f is the paraboloid $z = 16 - 4x^2 - y^2$ and the vertical plane $y = 2$ intersects it in

the parabola $z = 12 - 4x^2$, $y = 2$, (the curve C_1 in the first figure). The slope of the tangent line to this

parabola at $(1, 2, 8)$ is $f_x(1, 2) = -8$.

Similarly the plane $x = 1$ intersects

the paraboloid in the parabola

$z = 12 - y^2$, $x = 1$, (the curve C_2

in the second figure) and the slope

of the tangent line at $(1, 2, 8)$

is $f_y(1, 2) = -4$.

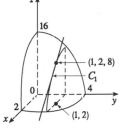
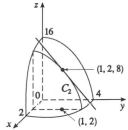

3. $f(x, y) = x^3 y^5 \Rightarrow f_x(x, y) = 3x^2 y^5$, $f_x(3, -1) = -27$

5. $f(x, y) = xe^{-y} + 3y \Rightarrow \dfrac{\partial f}{\partial y} = x(-1)e^{-y} + 3, \dfrac{\partial f}{\partial y}(1, 0) = -1 + 3 = 2$

7. $z = \dfrac{x^3 + y^3}{x^2 + y^2} \Rightarrow \dfrac{\partial z}{\partial x} = \dfrac{3x^2(x^2 + y^2) - (x^3 + y^3)(2x)}{(x^2 + y^2)^2} = \dfrac{x^4 + 3x^2y^2 - 2xy^3}{(x^2 + y^2)^2}$,

$\dfrac{\partial z}{\partial y} = \dfrac{3y^2(x^2 + y^2) - (x^3 + y^3)(2y)}{(x^2 + y^2)^2} = \dfrac{3x^2y^2 + y^4 - 2yx^3}{(x^2 + y^2)^2}$.

9. $xy + yz = xz \Rightarrow \dfrac{\partial}{\partial x}(xy + yz) = \dfrac{\partial}{\partial x}(xz) \Leftrightarrow y + y\dfrac{\partial z}{\partial x} = z + x\dfrac{\partial z}{\partial x} \Leftrightarrow (y - x)\dfrac{\partial z}{\partial x} = z - y$, so

$\dfrac{\partial z}{\partial x} = \dfrac{z - y}{y - x}$. $\dfrac{\partial}{\partial y}(xy + yz) = \dfrac{\partial}{\partial y}(xz) \Leftrightarrow x + z + y\dfrac{\partial z}{\partial y} = x\dfrac{\partial z}{\partial y} \Leftrightarrow (y - x)\dfrac{\partial z}{\partial y} = -(x + z)$, so

$\dfrac{\partial z}{\partial y} = \dfrac{x + z}{x - y}$.

11. $x^2 + y^2 - z^2 = 2x(y + z) \Leftrightarrow \dfrac{\partial}{\partial x}(x^2 + y^2 - z^2) = \dfrac{\partial}{\partial x}[2x(y + z)] \Leftrightarrow$

$2x - 2z\dfrac{\partial z}{\partial x} = 2(y + z) + 2x\dfrac{\partial z}{\partial x} \Leftrightarrow 2(x + z)\dfrac{\partial z}{\partial x} = 2(x - y - z)$, so $\dfrac{\partial z}{\partial x} = \dfrac{x - y - z}{x + z}$.

$\dfrac{\partial}{\partial y}(x^2 + y^2 - z^2) = \dfrac{\partial}{\partial y}[2x(y + z)] \Leftrightarrow 2y - 2z\dfrac{\partial z}{\partial y} = 2x\left(1 + \dfrac{\partial z}{\partial y}\right) \Leftrightarrow 2(x + z)\dfrac{\partial z}{\partial y} = 2(y - x)$, so

$\dfrac{\partial z}{\partial y} = \dfrac{y - x}{x + z}$.

13. $f(x, y, z) = xyz \Rightarrow f_y(x, y, z) = xz$, so $f_y(0, 1, 2) = 0$.

15. $u = xy + yz + zx \Rightarrow u_x = y + z, u_y = x + z, u_z = y + x$

17. $f(x, y) = x^3 y^5 - 2x^2 y + x \Rightarrow f_x(x, y) = 3x^2 y^5 - 4xy + 1, f_y(x, y) = 5x^3 y^4 - 2x^2$

19. $f(x, y) = x^4 + x^2 y^2 + y^4 \Rightarrow f_x(x, y) = 4x^3 + 2xy^2, f_y(x, y) = 2x^2 y + 4y^3$

21. $f(x, y) = \dfrac{x - y}{x + y}$ $\Rightarrow$ $f_x(x, y) = \dfrac{(1)(x + y) - (x - y)(1)}{(x + y)^2} = \dfrac{2y}{(x + y)^2}$,

$f_y(x, y) = \dfrac{(-1)(x + y) - (x - y)(1)}{(x + y)^2} = -\dfrac{2x}{(x + y)^2}$

23. $f(x, y) = e^x \tan(x - y)$ $\Rightarrow$ $f_x(x, y) = e^x \tan(x - y) + e^x \sec^2(x - y) = e^x[\tan(x - y) + \sec^2(x - y)]$,

$f_y(x, y) = e^x[\sec^2(x - y)](-1) = -e^x \sec^2(x - y)$

25. $f(u, v) = \tan^{-1}\left(\dfrac{u}{v}\right)$ $\Rightarrow$ $f_u(u, v) = \dfrac{1}{1 + (u/v)^2}\left(\dfrac{1}{v}\right) = \dfrac{1}{v}\left(\dfrac{v^2}{u^2 + v^2}\right) = \dfrac{v}{u^2 + v^2}$,

$f_v(u, v) = \dfrac{1}{1 + (u/v)^2}\left(-\dfrac{u}{v^2}\right) = -\dfrac{u}{v^2}\left(\dfrac{v^2}{u^2 + v^2}\right) = -\dfrac{u}{u^2 + v^2}$

27. $g(x, y) = y \tan(x^2 y^3)$ $\Rightarrow$ $g_x(x, y) = [y \sec^2(x^2 y^3)](2xy^3) = 2xy^4 \sec^2(x^2 y^3)$,

$g_y(x, y) = \tan(x^2 y^3) + [y \sec^2(x^2 y^3)](3x^2 y^2) = \tan(x^2 y^3) + 3x^2 y^3 \sec^2(x^2 y^3)$

29. $z = \ln\left(x + \sqrt{x^2 + y^2}\right)$ $\Rightarrow$

$\dfrac{\partial z}{\partial x} = \dfrac{1}{x + \sqrt{x^2 + y^2}}\left[1 + \tfrac{1}{2}(x^2 + y^2)^{-1/2}(2x)\right] = \dfrac{\left(\sqrt{x^2 + y^2} + x\right)/\sqrt{x^2 + y^2}}{\left(x + \sqrt{x^2 + y^2}\right)} = \dfrac{1}{\sqrt{x^2 + y^2}}$,

$\dfrac{\partial z}{\partial y} = \dfrac{1}{x + \sqrt{x^2 + y^2}}\left(\dfrac{1}{2}\right)(x^2 + y^2)^{-1/2}(2y) = \dfrac{y}{x\sqrt{x^2 + y^2} + x^2 + y^2}$

31. $f(x, y) = \displaystyle\int_x^y e^{t^2}\, dt$. By the Fundamental Theorem of Calculus, Part I, $\dfrac{d}{dx}\displaystyle\int_a^x f(t)dt = f(x)$ for f continuous.

Thus $f_x(x, y) = \dfrac{\partial}{\partial x}\displaystyle\int_x^y e^{t^2}\, dt = \dfrac{\partial}{\partial x}\left(-\displaystyle\int_y^x e^{t^2}\, dt\right) = -e^{x^2}$ and $f_y(x, y) = \dfrac{\partial}{\partial y}\displaystyle\int_x^y e^{t^2}\, dt = e^{y^2}$.

33. $f(x, y, z) = x^2 y z^3 + xy - z$ $\Rightarrow$ $f_x(x, y, z) = 2xyz^3 + y$, $f_y(x, y, z) = x^2 z^3 + x$, $f_z(x, y, z) = 3x^2 y z^2 - 1$

35. $f(x, y, z) = x^{yz}$ $\Rightarrow$ $f_x(x, y, z) = yzx^{yz-1}$. By Theorem 6.4.7 or 6.4*.3 (or 3.4.5, in the Early

Transcendentals version), $f_y(x, y, z) = x^{yz}\ln(x^z) = zx^{yz}\ln x$ and by symmetry $f_z(x, y, z) = yx^{yz}\ln x$.

37. $u = z\sin\left(\dfrac{y}{x + z}\right)$ $\Rightarrow$ $u_x = z\cos\left(\dfrac{y}{x + z}\right)[-y(x + z)^{-2}] = \dfrac{-yz}{(x + z)^2}\cos\left(\dfrac{y}{x + z}\right)$,

$u_y = z\cos\left(\dfrac{y}{x + z}\right)\left(\dfrac{1}{x + z}\right) = \dfrac{z}{x + z}\cos\left(\dfrac{y}{x + z}\right)$,

$u_z = \sin\left(\dfrac{y}{x + z}\right) + z\cos\left(\dfrac{y}{x + z}\right)[-y(x + z)^{-2}] = \sin\left(\dfrac{y}{x + z}\right) - \dfrac{yz}{(x + z)^2}\cos\left(\dfrac{y}{x + z}\right)$

39. $u = xy^2 z^3 \ln(x + 2y + 3z)$ $\Rightarrow$

$u_x = y^2 z^3 \ln(x + 2y + 3z) + xy^2 z^3\left(\dfrac{1}{x + 2y + 3z}\right) = y^2 z^3\left[\ln(x + 2y + 3z) + \dfrac{x}{x + 2y + 3z}\right]$,

$u_y = 2xyz^3 \ln(x + 2y + 3z) + xy^2 z^3\left(\dfrac{1}{x + 2y + 3z}\right)(2) = 2xyz^3\left[\ln(x + 2y + 3z) + \dfrac{y}{x + 2y + 3z}\right]$, and by

symmetry, $u_z = 3xy^2 z^2\left[\ln(x + 2y + 3z) + \dfrac{z}{x + 2y + 3z}\right]$.

41. $f(x, y, z, t) = \dfrac{x - y}{z - t}$ $\Rightarrow$ $f_x(x, y, z, t) = \dfrac{1}{z - t}$, $f_y(x, y, z, t) = -\dfrac{1}{z - t}$,

$f_z(x, y, z, t) = (x - y)(-1)(z - t)^{-2} = \dfrac{y - x}{(z - t)^2}$, and $f_t(x, y, z, t) = (x - y)(-1)(z - t)^{-2}(-1) = \dfrac{x - y}{(z - t)^2}$.

43. $u = \sqrt{x_1^2 + x_2^2 + \cdots + x_n^2}$. For each $i = 1, \ldots, n$,

$u_{x_i} = \tfrac{1}{2}(x_1^2 + x_2^2 + \cdots + x_n^2)^{-1/2}(2x_i) = \dfrac{x_i}{\sqrt{x_1^2 + x_2^2 + \cdots + x_n^2}}.$

45. $f(x, y) = x^2 - xy + 2y^2$ $\Rightarrow$

$f_x(x, y) = \lim_{h \to 0} \dfrac{f(x + h, y) - f(x, y)}{h} = \lim_{h \to 0} \dfrac{(x + h)^2 - (x + h)y + 2y^2 - (x^2 - xy + 2y^2)}{h}$

$= \lim_{h \to 0} \dfrac{h(2x - y + h)}{h} = \lim_{h \to 0}(2x - y + h) = 2x - y,$

$f_y(x, y) = \lim_{h \to 0} \dfrac{f(x, y + h) - f(x, y)}{h} = \lim_{h \to 0} \dfrac{x^2 - x(y + h) + 2(y + h)^2 - (x^2 - xy + 2y^2)}{h}$

$= \lim_{h \to 0} \dfrac{h(4y - x + 2h)}{h} = \lim_{h \to 0}(4y - x + 2h) = 4y - x$

47. $f(x, y) = x^2 + y^2 + x^2 y$ $\Rightarrow$ $f_x = 2x + 2xy$, $f_y = 2y + x^2$

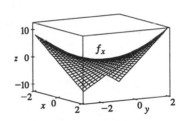

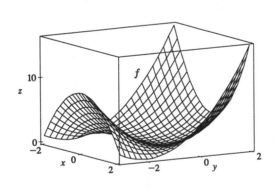

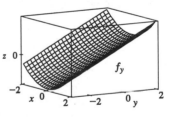

Note that the traces of f in planes parallel to the xz-plane are parabolas which open downward for $y < -1$ and upward for $y > -1$, and the traces of f_x in these planes are straight lines, which have negative slopes for $y < -1$ and positive slopes for $y > -1$. The traces of f in planes parallel to the yz-plane are parabolas which always open upward, and the traces of f_y in these planes are straight lines with positive slopes.

49. First of all, if we start at the point $(3, -3)$ and move in the positive y-direction, we see that both b and c decrease, while a increases. Both b and c have a low point at about $(3, -1.5)$, while a is 0 at this point. So a is definitely the graph of f_y, and one of b and c is the graph of f. To see which is which, we start at the point $(-3, -1.5)$ and move in the positive x-direction. b traces out a line with negative slope, while c traces out a parabola opening downward. This tells us that b is the x-derivative of c. So c is the graph of f, b is the graph of f_x, and a is the graph of f_y.

51. $z = f(x) + g(y) \quad \Rightarrow \quad \dfrac{\partial z}{\partial x} = f'(x), \dfrac{\partial z}{\partial y} = g'(y)$

53. $z = f(x + y)$. Let $u = x + y$. Then $\dfrac{\partial z}{\partial x} = \dfrac{df}{du}\dfrac{\partial u}{\partial x} = \dfrac{df}{d(x + y)} = f'(x + y)$,

$\dfrac{\partial z}{\partial y} = \dfrac{df}{du}\dfrac{\partial u}{\partial y} = \dfrac{df}{d(x + y)} = f'(x + y)$.

55. $z = f\left(\dfrac{x}{y}\right)$. Let $u = \dfrac{x}{y}$. Then $\dfrac{\partial u}{\partial x} = \dfrac{1}{y}$ and $\dfrac{\partial u}{\partial y} = -\dfrac{x}{y^2}$. Hence $\dfrac{\partial z}{\partial x} = \dfrac{df}{du}\dfrac{\partial u}{\partial x} = \dfrac{df/d(x/y)}{y} = \dfrac{f'(x/y)}{y}$ and

$\dfrac{\partial z}{\partial y} = \dfrac{df}{d(x/y)}\left(-\dfrac{x}{y^2}\right) = -x\left[\dfrac{df/d(x/y)}{y^2}\right] = -\dfrac{xf'(x/y)}{y^2}$.

57. $f(x, y) = x^2 y + x\sqrt{y} \quad \Rightarrow \quad f_x = 2xy + \sqrt{y},\ f_y = x^2 + \dfrac{x}{2\sqrt{y}}$. Thus $f_{xx} = 2y$, $f_{xy} = 2x + \dfrac{1}{2\sqrt{y}}$,

$f_{yx} = 2x + \dfrac{1}{2\sqrt{y}}$ and $f_{yy} = -\dfrac{x}{4y^{3/2}}$.

59. $z = (x^2 + y^2)^{3/2} \quad \Rightarrow \quad z_x = \frac{3}{2}(x^2 + y^2)^{1/2}(2x) = 3x(x^2 + y^2)^{1/2}$ and $z_y = 3y(x^2 + y^2)^{1/2}$. Thus

$z_{xx} = 3(x^2 + y^2)^{1/2} + 3x(x^2 + y^2)^{-1/2}(\frac{1}{2})(2x) = \dfrac{3(x^2 + y^2) + 3x^2}{\sqrt{x^2 + y^2}} = \dfrac{3(2x^2 + y^2)}{\sqrt{x^2 + y^2}}$ and

$z_{xy} = 3x(\frac{1}{2})(x^2 + y^2)^{-1/2}(2y) = \dfrac{3xy}{\sqrt{x^2 + y^2}}$. By symmetry $z_{yx} = \dfrac{3xy}{\sqrt{x^2 + y^2}}$ and $z_{yy} = \dfrac{3(x^2 + 2y^2)}{\sqrt{x^2 + y^2}}$.

61. $z = t\sin^{-1}x \quad \Rightarrow \quad z_x = t\dfrac{1}{\sqrt{1 - (\sqrt{x})^2}}(\frac{1}{2})x^{-1/2} = \dfrac{t}{2\sqrt{x - x^2}}$, $z_t = \sin^{-1}\sqrt{x}$. Thus

$z_{xx} = \frac{1}{2}t(-\frac{1}{2})(x - x^2)^{-3/2}(1 - 2x) = \dfrac{t(2x - 1)}{4(x - x^2)^{3/2}}$, $z_{xt} = \dfrac{1}{2\sqrt{x - x^2}}$,

$z_{tx} = \dfrac{1}{\sqrt{1 - (\sqrt{x})^2}}(\frac{1}{2}x^{-1/2}) = \dfrac{1}{2\sqrt{x - x^2}}$, and $z_{tt} = 0$.

63. $u = x^5 y^4 - 3x^2 y^3 + 2x^2 \quad \Rightarrow \quad u_x = 5x^4 y^4 - 6xy^3 + 4x$, $u_{xy} = 20x^4 y^3 - 18xy^2$ and

$u_y = 4x^5 y^3 - 9x^2 y^2$, $u_{yx} = 20x^4 y^3 - 18xy^2$. Thus $u_{xy} = u_{yx}$.

65. $u = \sin^{-1}(xy^2) \quad \Rightarrow \quad u_x = \dfrac{1}{\sqrt{1 - (xy^2)^2}}(y^2) = y^2\sqrt{1 - x^2 y^4}$,

$u_{xy} = 2y(1 - x^2 y^4)^{-1/2} + y^2(-\frac{1}{2})(1 - x^2 y^4)^{-3/2}(-4x^2 y^3) = \dfrac{2y(1 - x^2 y^4) + 2x^2 y^5}{(1 - x^2 y^4)^{3/2}} = \dfrac{2y}{(1 - x^2 y^4)^{3/2}}$, and

$u_y = \dfrac{1}{\sqrt{1 - (xy^2)^2}}(2xy) = \dfrac{2xy}{\sqrt{1 - x^2 y^4}}$, $u_{yx} = \dfrac{2y\sqrt{1 - x^2 y^4} - 2xy(\frac{1}{2})(1 - x^2 y^4)^{-1/2}(-2xy^4)}{1 - x^2 y^4}$

$= \dfrac{2y - 2x^2 y^5 + 2x^2 y^5}{(1 - x^2 y^4)^{3/2}} = \dfrac{2y}{(1 - x^2 y^4)^{3/2}}$.

Thus $u_{xy} = u_{yx}$.

67. $f(x, y) = x^2y^3 - 2x^4y \Rightarrow f_x = 2xy^3 - 8x^3y, f_{xx} = 2y^3 - 24x^2y, f_{xxx} - -48xy$

69. $f(x, y, z) = x^5 + x^4y^4z^3 + yz^2 \Rightarrow f_x = 5x^4 + 4x^3y^4z^3, f_{xy} = 16x^3y^3z^3$, and $f_{xyz} = 48x^3y^3z^2$

71. $z = x \sin y \Rightarrow \dfrac{\partial z}{\partial x} = \sin y, \dfrac{\partial^2 z}{\partial y \partial x} = \cos y$, and $\dfrac{\partial^3 z}{\partial y^2 \partial x} = -\sin y.$

73. $u = \ln(x + 2y^2 + 3z^3) \Rightarrow \dfrac{\partial u}{\partial z} = \dfrac{1}{x + 2y^2 + 3z^3}(9z^2) = \dfrac{9z^2}{x + 2y^2 + 3z^3}$,

$\dfrac{\partial^2 u}{\partial y \partial z} = -9z^2(x + 2y^2 + 3z^3)^{-2}(4y) = -\dfrac{36yz^2}{(x + 2y^2 + 3z^3)^2}$, and $\dfrac{\partial^3 u}{\partial x \partial y \partial z} = \dfrac{72yz^2}{(x + 2y^2 + 3z^3)^3}.$

75. $u = e^{-\alpha^2 k^2 t} \sin kx \Rightarrow u_x = ke^{-\alpha^2 k^2 t} \cos kx, u_{xx} = -k^2 e^{-\alpha^2 k^2 t} \sin kx$, and $u_t = -\alpha^2 k^2 e^{-\alpha^2 k^2 t} \sin kx.$ Thus

$\alpha^2 u_{xx} = u_t.$

77. $u = \dfrac{1}{\sqrt{x^2 + y^2 + z^2}} \Rightarrow u_x = \left(-\frac{1}{2}\right)(x^2 + y^2 + z^2)^{-3/2}(2x) = x(x^2 + y^2 + z^2)^{-3/2}$ and

$u_{xx} = -(x^2 + y^2 + z^2)^{-3/2} - x\left(-\frac{3}{2}\right)(x^2 + y^2 + z^2)^{-5/2}(2x) = \dfrac{2x^2 - y^2 - z^2}{(x^2 + y^2 + z^2)^{5/2}}.$ By symmetry,

$u_{yy} = \dfrac{2y^2 - x^2 - z^2}{(x^2 + y^2 + z^2)^{5/2}}$ and $u_{zz} = \dfrac{2z^2 - x^2 - y^2}{(x^2 + y^2 + z^2)^{5/2}}.$ Thus

$u_{xx} + u_{yy} + u_{zz} = \dfrac{2x^2 - y^2 - z^2 + 2y^2 - x^2 - z^2 + 2z^2 - x^2 - y^2}{(x^2 + y^2 + z^2)^{5/2}} = 0.$

79. Let $v = x + at, w = x - at.$ Then $u_t = \dfrac{\partial[f(v) + g(w)]}{\partial t} = \dfrac{df(v)}{dv}\dfrac{\partial v}{\partial t} + \dfrac{dg(w)}{dw}\dfrac{\partial w}{\partial t} = af'(v) - ag'(w)$ and

$u_{tt} = \dfrac{\partial[af'(v) - ag'(w)]}{\partial t} = a[af''(v) + ag''(w)] = a^2(f''(v) + g''(w)).$ Similarly by using the Chain Rule we

have $u_x = f'(v) + g'(w)$ and $u_{xx} = f''(v) + g''(w).$ Thus $u_{tt} = a^2 u_{xx}.$

81. $z_x = e^y + ye^x, z_{xx} = ye^x, \partial^3 z / \partial x^3 = ye^x.$ By symmetry $z_y = xe^y + e^x, z_{yy} = xe^y$ and $\partial^3 z / \partial y^3 = xe^y.$ Then

$\partial^3 z / \partial x \partial y^2 = e^y$ and $\partial^3 z / \partial x^2 \partial y = e^x.$ Thus $z = xe^y + ye^x$ satisfies the given partial differential equation.

83. $f(x_1, \ldots, x_n) = (x_1^2 + \cdots + x_n^2)^{(2-n)/2} \Rightarrow \dfrac{\partial f}{\partial x_i} = \left(1 - \dfrac{n}{2}\right)2x_i(x_1^2 + \cdots + x_n^2)^{-n/2}, 1 \le i \le n \Rightarrow$

$\dfrac{\partial^2 f}{\partial x_i^2} = 2\left(1 - \dfrac{n}{2}\right)(x_1^2 + \cdots + x_n^2)^{-n/2} - (2n)\left(1 - \dfrac{n}{2}\right)(x_i^2)(x_1^2 + \cdots + x_n^2)^{-(2+n)/2}, 1 \le i \le n.$ Therefore

$\dfrac{\partial^2 f}{\partial x_1^2} + \cdots + \dfrac{\partial^2 f}{\partial x_n^2} = \displaystyle\sum_{i=1}^{n}\left[(2-n)(x_1^2 + \cdots + x_n^2)^{-n/2} - n(2-n)(x_i^2)(x_1^2 + \cdots + x_n^2)^{-(2+n)/2}\right]$

$= n(2-n)(x_1^2 + \cdots + x_n^2)^{-n/2} - n(2-n)(x_1^2 + \cdots + x_n^2)(x_1^2 + \cdots + x_n^2)^{-(2+n)/2}$

$= n(2-n)(x_1^2 + \cdots + x_n^2)^{-n/2} - n(2-n)(x_1^2 + \cdots + x_n^2)^{-n/2} = 0.$

85. By the Chain Rule, taking the partial derivative of both sides with respect to R_1 gives

$\dfrac{\partial R^{-1}}{\partial R}\dfrac{\partial R}{\partial R_1} = \dfrac{\partial[(1/R_1) + (1/R_2) + (1/R_3)]}{\partial R_1}$ or $-R^{-2}\dfrac{\partial R}{\partial R_1} = -R_1^{-2}.$ Thus $\dfrac{\partial R}{\partial R_1} = \dfrac{R^2}{R_1^2}.$

87. $\dfrac{\partial K}{\partial m} = \frac{1}{2}V^2, \dfrac{\partial K}{\partial V} = mV, \dfrac{\partial^2 K}{\partial V^2} = m.$ Thus $\dfrac{\partial K}{\partial m} \cdot \dfrac{\partial^2 K}{\partial V^2} = \frac{1}{2}V^2 m = K.$

89. $f_x(x, y) = x + 4y \Rightarrow f_{xy}(x, y) = 4$ and $f_y(x, y) = 3x - y \Rightarrow f_{yx}(x, y) = 3$. Since f_{xy} and f_{yx} are continuous everywhere but $f_{xy}(x, y) \neq f_{yx}(x, y)$, Clairaut's Theorem implies that such a function $f(x, y)$ does not exist.

91. By the geometry of partial derivatives, the slope of the tangent line is $f_x(1, 2)$. By implicit differentiation of $4x^2 + 2y^2 + z^2 = 16$, we get $8x + 2z(\partial z/\partial x) = 0 \Rightarrow \partial z/\partial x = -4x/z$, so when $x = 1$ and $z = 2$ we have $\partial z/\partial x = -2$. So the slope is $f_x(1, 2) = -2$. Thus the tangent line is given by $z - 2 = -2(x - 1)$, $y = 2$. Taking the parameter to be $t = x - 1$, we can write parametric equations for this line: $x = 1 + t$, $y = 2$, $z = 2 - 2t$.

93. By Clairaut's Theorem, $f_{xyy} = (f_{xy})_y = (f_{yx})_y = f_{yxy} = (f_y)_{xy} = (f_y)_{yx} = f_{yyx}$.

95. Let $g(x) = f(x, 0) = x(x^2)^{-3/2}e^0 = x|x|^{-3}$. But we are using the point $(1, 0)$, so near $(1, 0)$, $g(x) = x^{-2}$. Then $g'(x) = -2x^{-3}$ and $g'(1) = -2$, so using (1) we have $f_x(1, 0) = g'(1) = -2$.

97. (a)

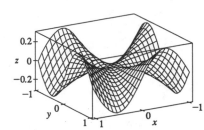

(b) For $(x, y) \neq (0, 0)$,

$$f_x(x, y) = \frac{(3x^2y - y^3)(x^2 + y^2) - (x^3y - xy^3)(2x)}{(x^2 + y^2)^2}$$

$$= \frac{x^4y + 4x^2y^3 - y^5}{(x^2 + y^2)^2}$$

and by symmetry

$$f_y(x, y) = \frac{x^5 - 4x^3y^2 - xy^4}{(x^2 + y^2)^2}.$$

(c) $f_x(0, 0) = \lim\limits_{h \to 0} \dfrac{f(h, 0) - f(0, 0)}{h} = \lim\limits_{h \to 0} \dfrac{(0/h^2) - 0}{h} = 0$ and $f_y(0, 0) = \lim\limits_{h \to 0} \dfrac{f(0, h) - f(0, 0)}{h} = 0$.

(d) Using (3), $f_{xy}(0, 0) = \dfrac{\partial f_x}{\partial y} = \lim\limits_{h \to 0} \dfrac{f_x(0, h) - f_x(0, 0)}{h} = \lim\limits_{h \to 0} \dfrac{(-h^5 - 0)/h^4}{h} = -1$ while by (2),

$$f_{yx}(0, 0) = \frac{\partial f_y}{\partial x} = \lim\limits_{h \to 0} \frac{f_y(h, 0) - f_y(0, 0)}{h} = \lim\limits_{h \to 0} \frac{h^5/h^4}{h} = 1.$$

(e) For $(x, y) \neq (0, 0)$, we use a CAS to compute that $f_{xy}(x, y) = \dfrac{x^6 + 9x^4y^2 - 4x^2y^4 + 4y^6}{(x^2 + y^2)^3}$. Now as

$(x, y) \to (0, 0)$ along the x-axis, $f_{xy}(x, y) \to 1$ while as $(x, y) \to (0, 0)$ along the y-axis, $f_{xy}(x, y) \to 4$.

Thus f_{xy} isn't continuous at $(0, 0)$ and Clairaut's Theorem doesn't apply, so there is no contradiction. The graphs of f_{xy} and f_{yx} are identical except at the origin, where we observe the discontinuity.

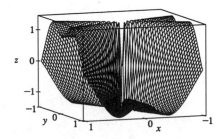

EXERCISES 12.4

1. $z = f(x, y) = x^2 + 4y^2 \Rightarrow f_x(x, y) = 2x, f_y(x, y) = 8y, f_x(2, 1) = 4, f_y(2, 1) = 8$. Thus the equation of
the tangent plane is $z - 8 = 4(x - 2) + 8(y - 1)$ or $4x + 8y - z = 8$.

3. $z = f(x, y) = 5 + (x - 1)^2 + (y + 2)^2 \Rightarrow f_x(x, y) = 2(x - 1), f_y(x, y) = 2(y + 2), f_x(2, 0) = 2,$
$f_y(2, 0) = 4$ and the equation is $z - 10 = 2(x - 2) + 4y$ or $2x + 4y - z = -6$.

5. $z = f(x, y) = \ln(2x + y) \Rightarrow f_x(x, y) = \dfrac{2}{2x + y}, f_y(x, y) = \dfrac{1}{2x + y}, f_x(-1, 3) = 2, f_y(-1, 3) = 1$. Thus
the equation of the tangent plane is $z = 2(x + 1) + (y - 3)$ or $2x + y - z = 1$.

7. $z = f(x, y) = xy$, so $f_x(-1, 2) = 2, f_y(-1, 2) = -1$
and the equation of the tangent plane is
$z + 2 = 2(x + 1) + (-1)(y - 2)$ or
$2x - y - z = -2$.

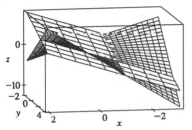

9. $f(x, y) = e^{-(x^2 + y^2)/15}(\sin^2 x + \cos^2 y)$. A CAS gives
$f_x = -\frac{2}{15} e^{-(x^2 + y^2)/15}(x \sin^2 x + x \cos^2 y - 15 \sin x \cos x)$ and
$f_y = -\frac{2}{15} e^{-(x^2 + y^2)/15}(y \sin^2 x + y \cos^2 y + 15 \sin y \cos y)$.
We use the CAS to evaluate these at $(2, 3)$, and then
substitute the results into Equation 2 in order to plot
the tangent plane.

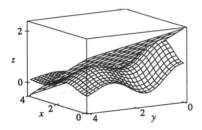

11. $z = x^2 y^3 \Rightarrow dz = \dfrac{\partial z}{\partial x} dx + \dfrac{\partial z}{\partial y} dy = 2xy^3 dx + 3x^2 y^2 dy$

13. $u = e^x \cos xy \Rightarrow du = e^x(\cos xy - y \sin xy)dx - (xe^x \sin xy)dy$

15. $w = x^2 y + y^2 z \Rightarrow dw = \dfrac{\partial w}{\partial x} dx + \dfrac{\partial w}{\partial y} dy + \dfrac{\partial w}{\partial z} dz = 2xy\, dx + (x^2 + 2yz)dy + y^2 dz$

17. $w = \ln \sqrt{x^2 + y^2 + z^2} \Rightarrow$
$$dw = \left(\frac{1}{2}\right) \frac{2x(x^2 + y^2 + z^2)^{-1/2} dx + 2y(x^2 + y^2 + z^2)^{-1/2} dy + 2z(x^2 + y^2 + z^2)^{-1/2} dz}{(x^2 + y^2 + z^2)^{1/2}} = \frac{x\, dx + y\, dy + z\, dz}{x^2 + y^2 + z^2}$$

19. $\Delta x = 0.05, \Delta y = 0.1, z = 5x^2 + y^2, z_x = 10x, z_y = 2y$. Thus when $x = 1, y = 2$,
$dz = (10)(0.05) + (4)(0.1) = 0.9$, while $\Delta z = f(1.05, 2.1) - f(1, 2) = 5(1.05)^2 + (2.1)^2 - 5 - 4 = 0.9225$.

21. $f(x, y) = \sqrt{20 - x^2 - 7y^2} \Rightarrow f_x = -\dfrac{x}{\sqrt{20 - x^2 - 7y^2}}$ and $f_y = -\dfrac{7y}{\sqrt{20 - x^2 - 7y^2}}$. Since
$f(2, 1) = \sqrt{20 - 4 - 7} = 3$, we set $(a, b) = (2, 1)$. Then $\Delta x = -0.05, \Delta y = 0.08$. Thus
$f(1.95, 1.08) \approx f(2, 1) + dz = 3 + (-\frac{2}{3})(-0.05) + (-\frac{7}{3})(0.08) = 2.84\overline{6}$.

23. $f(x, y, z) = x^2 y^3 z^4 \Rightarrow f_x = 2xy^3 z^4$, $f_y = 3x^2 y^2 z^4$ and $f_z = 4x^2 y^3 z^3$. Since $f(1, 1, 3) = 81$, we set $(a, b, c) = (1, 1, 3)$. Then $\Delta x = 0.05$, $\Delta y = -0.1$, $\Delta z = 0.01$, and so
$f(1.05, 0.9, 3.01) \approx f(1, 1, 3) + dw = 81 + (162)(0.05) + (243)(-0.1) + (108)(0.01) = 65.88$.

25. Let $w = f(x, y, z) = x\sqrt{y - z^3} \Rightarrow f_x = \sqrt{y - z^3}$, $f_y = \dfrac{x}{2\sqrt{y - z^3}}$, and $f_z = -\dfrac{3xz^2}{2\sqrt{y - z^3}}$. Then
$f(9, 10, 1) = 27$, so we set $(a, b, c) = (9, 10, 1)$. Then $\Delta x = -0.06$, $\Delta y = -0.01$, and $\Delta z = 0.01$. Thus
$8.94\sqrt{9.99 - (1.01)^3} \approx 27 + (3)(-0.06) + \frac{9}{6}(-0.01) + \left(-\frac{27}{6}\right)(0.01) = 26.76$.

27. Let $z = f(x, y) = \sqrt{x}\,e^y \Rightarrow f_x = e^y/(2\sqrt{x})$, $f_y = \sqrt{x}\,e^y$. Now $f(1, 0) = 1$, so we set $(a, b) = (1, 0)$,
$\Delta x = -0.01$, $\Delta y = 0.02$. Thus $\sqrt{0.99}\,e^{0.02} \approx 1 + \frac{1}{2}(-0.01) + 1(0.02) = 1.015$.

29. $dA = \dfrac{\partial A}{\partial x}\,dx + \dfrac{\partial A}{\partial y}\,dy = y\,dx + x\,dy$ and $|\Delta x| \le 0.1$, $|\Delta y| \le 0.1$. Thus the maximum error in the area is
about $dA = 24(0.1) + 30(0.1) = 5.4\,\text{cm}^2$.

31. The volume of a can is $V = \pi r^2 h$ and $\Delta V \approx dV$ is an estimate of the amount of tin. Here
$dV = 2\pi rh\,dr + \pi r^2\,dh$, so $\Delta V \approx dV = 2\pi(48)(-0.04) + \pi(16)(-0.08) \approx -16.08\,\text{cm}^3$. Thus the amount of
tin is about $16\,\text{cm}^3$.

33. The area of the rectangle is $A = xy$, and $\Delta A \approx dA$ is an estimate of the area of paint in the stripe. Here
$dA = y\,dx + x\,dy$, so $\Delta A \approx dA = (100)\left(\frac{1}{2}\right) + (200)\left(\frac{1}{2}\right) = 150\,\text{ft}^2$. Thus there are approximately $150\,\text{ft}^2$ of
paint in the stripe.

35. By the Chain Rule, taking partial derivatives of both sides with respect to R_1 gives
$$\frac{\partial R^{-1}}{\partial R}\frac{\partial R}{\partial R_1} = \frac{\partial[(1/R_1) + (1/R_2) + (1/R_3)]}{\partial R_1} \Rightarrow -R^{-2}\frac{\partial R}{\partial R_1} = -R_1^{-2} \Rightarrow \frac{\partial R}{\partial R_1} = \frac{R^2}{R_1^2}, \text{ and by}$$
symmetry, $\dfrac{\partial R}{\partial R_2} = \dfrac{R^2}{R_2^2}$, $\dfrac{\partial R}{\partial R_3} = \dfrac{R^2}{R_3^2}$. When $R_1 = 25$, $R_2 = 40$ and $R_3 = 50$, $\dfrac{1}{R} = \dfrac{17}{200} \Leftrightarrow R = \dfrac{200}{17}$ ohms.
Since the possible error for each R_i is 0.5%, the maximum error of R is attained by setting $\Delta R_i = 0.005 R_i$. So
$$\Delta R \approx dR = \frac{\partial R}{\partial R_1}\Delta R_1 + \frac{\partial R}{\partial R_2}\Delta R_2 + \frac{\partial R}{\partial R_3}\Delta R_3 = (0.005)R^2\left[\frac{1}{R_1} + \frac{1}{R_2} + \frac{1}{R_3}\right] = (0.005)R = \frac{1}{17} \approx 0.059\,\text{ohms}.$$

37. $\Delta z = f(a + \Delta x, b + \Delta y) - f(a, b) = (a + \Delta x)^2 + (b + \Delta y)^2 - (a^2 + b^2)$
$= a^2 + 2a\Delta x + (\Delta x)^2 + b^2 + 2b\Delta y + (\Delta y)^2 - a^2 - b^2 = 2a\Delta x + (\Delta x)^2 + 2b\Delta y + (\Delta y)^2$.
But $f_x(a, b) = 2a$ and $f_y(a, b) = 2b$ and so $\Delta z = f_x(a, b)\Delta x + f_y(a, b)\Delta y + \Delta x\Delta x + \Delta y\Delta y$, which is
Definition 12 with $\epsilon_1 = \Delta x$ and $\epsilon_2 = \Delta y$. Hence f is differentiable.

39. To show that f is continuous at (a, b) we need to show that $\displaystyle\lim_{(x,y)\to(a,b)} f(x, y) = f(a, b)$ or equivalently
$\displaystyle\lim_{(\Delta x, \Delta y)\to(0,0)} f(a + \Delta x, b + \Delta y) = f(a, b)$. Since f is differentiable at (a, b),
$f(a + \Delta x, b + \Delta y) - f(a, b) = \Delta z = f_x(a, b)\Delta x + f_y(a, b)\Delta y + \epsilon_1\Delta x + \epsilon_2\Delta y$, where ϵ_1 and $\epsilon_2 \to 0$ as
$(\Delta x, \Delta y) \to (0, 0)$. Thus $f(a + \Delta x, b + \Delta y) = f(a, b) + f_x(a, b)\Delta x + f_y(a, b)\Delta y + \epsilon_1\Delta x + \epsilon_2\Delta y$. Taking
the limit of both sides as $(\Delta x, \Delta y) \to (0, 0)$ gives $\displaystyle\lim_{(\Delta x, \Delta y)\to(0,0)} f(a + \Delta x, b + \Delta y) = f(a, b)$. Thus f is
continuous at (a, b).

EXERCISES 12.5

1. $z = x^2 + y^2$, $x = t^3$, $y = 1 + t^2$ $\Rightarrow$ $\dfrac{dz}{dt} = 2x\dfrac{dx}{dt} + 2y\dfrac{dy}{dt} = (2t^3)(3t^2) + 2(1 + t^2)(2t) = 6t^5 + 4t^3 + 4t$

3. $z = \ln(x + y^2)$, $x = \sqrt{1 + t}$, $y = 1 + \sqrt{t}$ $\Rightarrow$

$\dfrac{dz}{dt} = \dfrac{1}{(x + y^2)}\dfrac{1}{2\sqrt{1+t}} + \dfrac{1}{(x + y^2)}2y\dfrac{1}{2\sqrt{t}} = \dfrac{1}{\sqrt{1+t}+1+\sqrt{t}}\left(\dfrac{1}{2\sqrt{1+t}} + \dfrac{1+\sqrt{t}}{\sqrt{t}}\right)$

5. $w = xy^2z^3$, $x = \sin t$, $y = \cos t$, $z = 1 + e^{2t}$ $\Rightarrow$ $\dfrac{dw}{dt} = y^2z^3(\cos t) + 2xyz^3(-\sin t) + 3xy^2z^2(2e^{2t})$

7. $z = x^2 \sin y$, $x = s^2 + t^2$, $y = 2st$ $\Rightarrow$ $\dfrac{\partial z}{\partial s} = (2x \sin y)(2s) + (x^2 \cos y)(2t) = 4sx \sin y + 2tx^2 \cos y$,

$\dfrac{\partial z}{\partial t} = (2x \sin y)(2t) + (x^2 \cos y)(2s) = 4xt \sin y + 2sx^2 \cos y$

9. $z = x^2 - 3x^2y^3$, $x = se^t$, $y = se^{-t}$ $\Rightarrow$

$\partial z/\partial s = (2x - 6xy^3)(e^t) + (-9x^2y^2)(e^{-t}) = (2x - 6xy^3)e^t - 9x^2y^2e^{-t}$,

$\partial z/\partial t = (2x - 6xy^3)(se^t) + (-9x^2y^2)(-se^{-t}) = (2x - 6xy^3)se^t + 9x^2y^2se^{-t}$

11. $z = 2^{x-3y}$, $x = s^2t$, $y = st^2$ $\Rightarrow$ $\partial z/\partial s = (z \ln 2)(2st) + z(-3 \ln 2)(t^2) = (2^{x-3y} \ln 2)(2st - 3t^2)$,

$\partial z/\partial t = (z \ln 2)(s^2) + z(-3 \ln 2)(2st) = (2^{x-3y} \ln 2)(s^2 - 6st)$

13. $u = f(x, y)$, $x = x(r, s, t)$, $y = y(r, s, t)$ $\Rightarrow$

$\dfrac{\partial u}{\partial r} = \dfrac{\partial u}{\partial x}\dfrac{\partial x}{\partial r} + \dfrac{\partial u}{\partial y}\dfrac{\partial y}{\partial r}$, $\dfrac{\partial u}{\partial s} = \dfrac{\partial u}{\partial x}\dfrac{\partial x}{\partial s} + \dfrac{\partial u}{\partial y}\dfrac{\partial y}{\partial s}$, $\dfrac{\partial u}{\partial t} = \dfrac{\partial u}{\partial x}\dfrac{\partial x}{\partial t} + \dfrac{\partial u}{\partial y}\dfrac{\partial y}{\partial t}$

15. $v = f(p, q, r)$, $p = p(x, y, z)$, $q = q(x, y, z)$, $r = r(x, y, z)$ $\Rightarrow$

$\dfrac{\partial v}{\partial x} = \dfrac{\partial v}{\partial p}\dfrac{\partial p}{\partial x} + \dfrac{\partial v}{\partial q}\dfrac{\partial q}{\partial x} + \dfrac{\partial v}{\partial r}\dfrac{\partial r}{\partial x}$, $\dfrac{\partial v}{\partial y} = \dfrac{\partial v}{\partial p}\dfrac{\partial p}{\partial y} + \dfrac{\partial v}{\partial q}\dfrac{\partial q}{\partial y} + \dfrac{\partial v}{\partial r}\dfrac{\partial r}{\partial y}$, $\dfrac{\partial v}{\partial z} = \dfrac{\partial v}{\partial p}\dfrac{\partial p}{\partial z} + \dfrac{\partial v}{\partial q}\dfrac{\partial q}{\partial z} + \dfrac{\partial v}{\partial r}\dfrac{\partial r}{\partial z}$

17. $w = x^2 + y^2 + z^2$, $x = st$, $y = s \cos t$, $z = s \sin t$ $\Rightarrow$

$\dfrac{\partial w}{\partial s} = \dfrac{\partial w}{\partial x}\dfrac{\partial x}{\partial s} + \dfrac{\partial w}{\partial y}\dfrac{\partial y}{\partial s} + \dfrac{\partial w}{\partial z}\dfrac{\partial z}{\partial s} = 2xt + 2y \cos t + 2z \sin t$. When $s = 1$, $t = 0$, we have $x = 0$, $y = 1$ and

$z = 0$, so $\partial w/\partial s = 2 \cos 0 = 2$. Similarly $\partial w/\partial t = 2xs + 2y(-s \sin t) + 2z(s \cos t) = 0 + (-2)\sin 0 + 0 = 0$,

when $s = 1$, $t = 0$.

19. $z = y^2 \tan x$, $x = t^2uv$, $y = u + tv^2$ $\Rightarrow$

$\partial z/\partial t = (y^2 \sec^2 x)2tuv + (2y \tan x)v^2$, $\partial z/\partial u = (y^2 \sec^2 x)t^2v + 2y \tan x$,

$\partial z/\partial v = (y^2 \sec^2 x)t^2u + (2y \tan x)2tv$. When $t = 2$, $u = 1$ and $v = 0$, we have $x = 0$, $y = 1$, so $\partial z/\partial t = 0$,

$\partial z/\partial u = 0$, $\partial z/\partial v = 4$.

21. $u = \dfrac{x+y}{y+z}$, $x = p+r+t$, $y = p-r+t$, $z = p+r-t$ $\Rightarrow$

$$\frac{\partial u}{\partial p} = \frac{1}{y+z} + \frac{(y+z)-(x+y)}{(y+z)^2} - \frac{x+y}{(y+z)^2} = \frac{(y+z)+(z-x)-(x+y)}{(y+z)^2} = 2\frac{z-x}{(y+z)^2} = 2\frac{-2t}{4p^2} = -\frac{t}{p^2},$$

$$\frac{\partial u}{\partial r} = \frac{1}{y+z} + \frac{z-x}{(y+z)^2}(-1) - \frac{x+y}{(y+z)^2} = 0, \text{ and}$$

$$\frac{\partial u}{\partial t} = \frac{1}{y+z} + \frac{z-x}{(y+z)^2} + \frac{x+y}{(y+z)^2} = 2\frac{y+z}{(y+z)^2} = \frac{2}{2p} = \frac{1}{p}.$$

23. $x^2 - xy + y^3 = 8$, so let $F(x,y) = x^2 - xy + y^3 - 8 = 0$. Then $\dfrac{dy}{dx} = -\dfrac{F_x}{F_y} = \dfrac{-(2x-y)}{-x+3y^2} = \dfrac{y-2x}{3y^2-x}$.

25. $2y^2 + \sqrt[3]{xy} = 3x^2 + 17$, so let $F(x,y) = 2y^2 + \sqrt[3]{xy} - 3x^2 - 17 = 0$. Then

$$\frac{dy}{dx} = -\frac{y/\left[3(xy)^{2/3}\right] - 6x}{4y + x/\left[3(xy)^{2/3}\right]} = \frac{18x - x^{-2/3}y^{1/3}}{12y + x^{1/3}y^{-2/3}}.$$

27. Let $F(x,y,z) = xy + yz - xz = 0$. Then $\dfrac{\partial z}{\partial x} = -\dfrac{F_x}{F_z} = -\dfrac{y-z}{y-x} = \dfrac{z-y}{y-x}$, $\dfrac{\partial z}{\partial y} = -\dfrac{F_y}{F_z} = -\dfrac{x+z}{y-x} = \dfrac{x+z}{x-y}$.

29. $x^2 + y^2 - z^2 = 2x(y+z)$. Let $F(x,y,z) = x^2 + y^2 - z^2 - 2x(y+z) = 0$. Then

$$\frac{\partial z}{\partial x} = -\frac{F_x}{F_z} = -\frac{2x-2y-2z}{-2z-2x} = \frac{x-y-z}{z+x}, \quad \frac{\partial z}{\partial y} = -\frac{F_y}{F_z} = -\frac{2y-2x}{-2z-2x} = \frac{y-x}{z+x}.$$

31. Let $F(x,y,z) = xe^y + yz + ze^x = 0$. Then $\dfrac{\partial z}{\partial x} = -\dfrac{F_x}{F_z} = -\dfrac{e^y + ze^x}{y+e^x}$, $\dfrac{\partial z}{\partial y} = -\dfrac{F_y}{F_z} = -\dfrac{xe^y + z}{y+e^x}$.

33. $dr/dt = -1.2$, $dh/dt = 3$, $V = \pi r^2 h$ and $dV/dt = 2\pi rh(dr/dt) + \pi r^2(dh/dt)$. Thus when $r = 80$ and $h = 150$, $dV/dt = (-28{,}800)\pi + (19{,}200)\pi = -9600\pi \text{ cm}^3/\text{s}$.

35. **(a)** $V = \ell wh$, so by the Chain Rule, $\dfrac{dV}{dt} = \dfrac{\partial V}{\partial \ell}\dfrac{d\ell}{dt} + \dfrac{\partial V}{\partial w}\dfrac{dw}{dt} + \dfrac{\partial V}{\partial h}\dfrac{dh}{dt} = wh\dfrac{d\ell}{dt} + \ell h\dfrac{dw}{dt} + \ell w\dfrac{dh}{dt}$

$$= 2\cdot 2\cdot 2 + 1\cdot 2\cdot 2 + 1\cdot 2\cdot(-3) = 6\,\text{m}^3/\text{s}.$$

(b) $S = 2(\ell w + \ell h + wh)$, so by the Chain Rule,

$$\frac{dS}{dt} = \frac{\partial S}{\partial \ell}\frac{d\ell}{dt} + \frac{\partial S}{\partial w}\frac{dw}{dt} + \frac{\partial S}{\partial h}\frac{dh}{dt} = 2(w+h)\frac{d\ell}{dt} + 2(\ell+h)\frac{dw}{dt} + 2(\ell+w)\frac{dh}{dt}$$

$$= 2(2+2)2 + 2(1+2)2 + 2(1+2)(-3) = 10\,\text{m}^2/\text{s}.$$

(c) $L^2 = \ell^2 + w^2 + h^2 \Rightarrow 2L\dfrac{dL}{dt} = 2\ell\dfrac{d\ell}{dt} + 2w\dfrac{dw}{dt} + 2h\dfrac{dh}{dt} = 2(1)(2) + 2(2)(2) + 2(2)(-3) = 0$

$$\Rightarrow \quad dL/dt = 0\,\text{m}/\text{s}.$$

37. $\dfrac{dP}{dt} = 0.05$, $\dfrac{dT}{dt} = 0.15$, $V = 8.31\dfrac{T}{P}$ and $\dfrac{dV}{dt} = \dfrac{8.31}{P}\dfrac{dT}{dt} - 8.31\dfrac{T}{P^2}\dfrac{dP}{dt}$. Thus when $P = 20$ and $T = 320°$,

$$\frac{dV}{dt} = 8.31\left[\frac{0.15}{20} - \frac{(0.05)(320)}{400}\right] \approx -0.27\,\text{L/s}.$$

39. (a) Using the Chain Rule, $\dfrac{\partial z}{\partial r} = \dfrac{\partial z}{\partial x}\cos\theta + \dfrac{\partial z}{\partial y}\sin\theta$, $\dfrac{\partial z}{\partial \theta} = \dfrac{\partial z}{\partial x}(-r\sin\theta) + \dfrac{\partial z}{\partial y}r\cos\theta$.

(b) $\left(\dfrac{\partial z}{\partial r}\right)^2 = \left(\dfrac{\partial z}{\partial x}\right)^2\cos^2\theta + 2\dfrac{\partial z}{\partial x}\dfrac{\partial z}{\partial y}\cos\theta\sin\theta + \left(\dfrac{\partial z}{\partial y}\right)^2\sin^2\theta$,

$\left(\dfrac{\partial z}{\partial \theta}\right)^2 = \left(\dfrac{\partial z}{\partial x}\right)^2 r^2\sin^2\theta - 2\dfrac{\partial z}{\partial x}\dfrac{\partial z}{\partial y}r^2\cos\theta\sin\theta + \left(\dfrac{\partial z}{\partial y}\right)^2 r^2\cos^2\theta$. Thus

$\left(\dfrac{\partial z}{\partial r}\right)^2 + \dfrac{1}{r^2}\left(\dfrac{\partial z}{\partial \theta}\right)^2 = \left[\left(\dfrac{\partial z}{\partial x}\right)^2 + \left(\dfrac{\partial z}{\partial y}\right)^2\right](\cos^2\theta + \sin^2\theta) = \left(\dfrac{\partial z}{\partial x}\right)^2 + \left(\dfrac{\partial z}{\partial y}\right)^2$.

41. Let $u = x - y$. Then $\dfrac{\partial z}{\partial x} = \dfrac{dz}{du}\dfrac{\partial u}{\partial x} = \dfrac{dz}{du}$ and $\dfrac{\partial z}{\partial y} = \dfrac{dz}{du}(-1)$. Thus $\dfrac{\partial z}{\partial x} + \dfrac{\partial z}{\partial y} = 0$.

43. Let $u = x + at$, $v = x - at$. Then $z = f(u) + g(v)$, so $\partial z/\partial u = f'(u)$ and $\partial z/\partial v = g'(v)$. Thus

$\dfrac{\partial z}{\partial t} = \dfrac{\partial z}{\partial u}\dfrac{\partial u}{\partial t} + \dfrac{\partial z}{\partial v}\dfrac{\partial v}{\partial t} = af'(u) - ag'(v)$ and $\dfrac{\partial^2 z}{\partial t^2} = a\dfrac{\partial}{\partial t}[f'(u) - g'(v)] = a\left(\dfrac{df'(u)}{du}\dfrac{\partial u}{\partial t} - \dfrac{dg'(v)}{dv}\dfrac{\partial v}{\partial t}\right)$

$= a^2 f''(u) + a^2 g''(v)$. Similarly $\dfrac{\partial z}{\partial x} = f'(u) + g'(v)$ and $\dfrac{\partial^2 z}{\partial x^2} = f''(u) + g''(v)$. Thus $\dfrac{\partial^2 z}{\partial t^2} = a^2\dfrac{\partial^2 z}{\partial x^2}$.

45. $\dfrac{\partial z}{\partial s} = \dfrac{\partial z}{\partial x}2s + \dfrac{\partial z}{\partial y}2r$. Then

$\dfrac{\partial^2 z}{\partial r\partial s} = \dfrac{\partial}{\partial r}\left(\dfrac{\partial z}{\partial x}2s\right) + \dfrac{\partial}{\partial r}\left(\dfrac{\partial z}{\partial y}2r\right)$

$= \dfrac{\partial^2 z}{\partial x^2}\dfrac{\partial x}{\partial r}2s + \dfrac{\partial}{\partial y}\left(\dfrac{\partial z}{\partial x}\right)\dfrac{\partial y}{\partial r}2s + \dfrac{\partial z}{\partial x}\dfrac{\partial}{\partial r}(2s) + \dfrac{\partial^2 z}{\partial y^2}\dfrac{\partial y}{\partial r}2r + \dfrac{\partial}{\partial x}\left(\dfrac{\partial z}{\partial y}\right)\dfrac{\partial x}{\partial r}2r + \dfrac{\partial z}{\partial y}2$

$= 4rs\dfrac{\partial^2 z}{\partial x^2} + \dfrac{\partial^2 z}{\partial y\partial x}4s^2 + 0 + 4rs\dfrac{\partial^2 z}{\partial y^2} + \dfrac{\partial^2 z}{\partial x\partial y}4r^2 + 2\dfrac{\partial z}{\partial y}$.

And by the continuity of the partials, $\dfrac{\partial^2 z}{\partial r\partial s} = 4rs\dfrac{\partial^2 z}{\partial x^2} + 4rs\dfrac{\partial^2 z}{\partial y^2} + (4r^2 + 4s^2)\dfrac{\partial^2 z}{\partial x\partial y} + 2\dfrac{\partial z}{\partial y}$.

47. $\dfrac{\partial z}{\partial r} = \dfrac{\partial z}{\partial x}\cos\theta + \dfrac{\partial z}{\partial y}\sin\theta$ and $\dfrac{\partial z}{\partial \theta} = -\dfrac{\partial z}{\partial x}r\sin\theta + \dfrac{\partial z}{\partial y}r\cos\theta$. Then

$\dfrac{\partial^2 z}{\partial r^2} = \cos\theta\left(\dfrac{\partial^2 z}{\partial x^2}\cos\theta + \dfrac{\partial^2 z}{\partial y\partial x}\sin\theta\right) + \sin\theta\left(\dfrac{\partial^2 z}{\partial y^2}\sin\theta + \dfrac{\partial^2 z}{\partial x\partial y}\cos\theta\right)$

$= \cos^2\theta\dfrac{\partial^2 z}{\partial x^2} + 2\cos\theta\sin\theta\dfrac{\partial^2 z}{\partial x\partial y} + \sin^2\theta\dfrac{\partial^2 z}{\partial y^2}$ and

$\dfrac{\partial^2 z}{\partial \theta^2} = -r\cos\theta\dfrac{\partial z}{\partial x} + (-r\sin\theta)\left(\dfrac{\partial^2 z}{\partial x^2}(-r\sin\theta) + \dfrac{\partial^2 z}{\partial y\partial x}r\cos\theta\right)$

$-r\sin\theta\dfrac{\partial z}{\partial y} + r\cos\theta\left(\dfrac{\partial^2 z}{\partial y^2}r\cos\theta + \dfrac{\partial^2 z}{\partial x\partial y}(-r\sin\theta)\right)$

$= -r\cos\theta\dfrac{\partial z}{\partial x} - r\sin\theta\dfrac{\partial z}{\partial y} + r^2\sin^2\theta\dfrac{\partial^2 z}{\partial x^2} - 2r^2\cos\theta\sin\theta\dfrac{\partial^2 z}{\partial x\partial y} + r^2\cos^2\theta\dfrac{\partial^2 z}{\partial y^2}$. Thus

$\dfrac{\partial^2 z}{\partial r^2} + \dfrac{1}{r^2}\dfrac{\partial^2 z}{\partial \theta^2} + \dfrac{1}{r}\dfrac{\partial z}{\partial r} = (\cos^2\theta + \sin^2\theta)\dfrac{\partial^2 z}{\partial x^2} + (\sin^2\theta + \cos^2\theta)\dfrac{\partial^2 z}{\partial y^2} - \dfrac{1}{r}\cos\theta\dfrac{\partial z}{\partial x}$

$-\dfrac{1}{r}\sin\theta\dfrac{\partial z}{\partial y} + \dfrac{1}{r}\left(\cos\theta\dfrac{\partial z}{\partial x} + \sin\theta\dfrac{\partial z}{\partial y}\right) = \dfrac{\partial^2 z}{\partial x^2} + \dfrac{\partial^2 z}{\partial y^2}$ as desired.

49. Differentiating both sides of $f(tx, ty) = t^n f(x, y)$ with respect to t using the Chain Rule, we get

$$\frac{\partial}{\partial t} f(tx, ty) = \frac{\partial}{\partial t} [t^n f(x, y)] \quad \Leftrightarrow \quad \frac{\partial}{\partial(tx)} f(tx, ty) \cdot \frac{\partial(tx)}{\partial t} + \frac{\partial}{\partial(ty)} f(tx, ty) \cdot \frac{\partial(ty)}{\partial t}$$

$$= x \frac{\partial}{\partial(tx)} f(tx, ty) + y \frac{\partial}{\partial(ty)} f(tx, ty) = nt^{n-1} f(x, y). \text{ Setting } t = 1: \ x \frac{\partial}{\partial x} f(x, y) + y \frac{\partial}{\partial y} f(x, y) = nf(x, y).$$

51. Differentiating both sides of $f(tx, ty) = t^n f(x, y)$ with respect to x using the Chain Rule, we get

$$\frac{\partial}{\partial x} f(tx, ty) = \frac{\partial}{\partial x} [t^n f(x, y)] \quad \Leftrightarrow \quad \frac{\partial}{\partial(tx)} f(tx, ty) \cdot \frac{\partial(tx)}{\partial x} + \frac{\partial}{\partial(ty)} f(tx, ty) \cdot \frac{\partial(ty)}{\partial x} = t^n \frac{\partial}{\partial x} f(x, y) \quad \Leftrightarrow$$

$$tf_x(tx, ty) = t^n f_x(x, y). \text{ Thus } f_x(tx, ty) = t^{n-1} f_x(x, y).$$

EXERCISES 12.6

1. $f(x, y) = x^2 y^3 + 2x^4 y \ \Rightarrow \ D_{\mathbf{u}} f(x, y) = (2xy^3 + 8x^3 y) \cos \frac{\pi}{3} + (3x^2 y^2 + 2x^4) \sin \frac{\pi}{3}$. Thus
$D_{\mathbf{u}} f(1, -2) = (-16 - 16)(\frac{1}{2}) + (12 + 2)(\frac{\sqrt{3}}{2}) = 7\sqrt{3} - 16$.

3. $f(x, y) = y^x \ \Rightarrow \ D_{\mathbf{u}} f(x, y) = (y^x \ln y) \cos \frac{\pi}{2} + (xy^{x-1}) \sin \frac{\pi}{2} = xy^{x-1}$. Thus $D_{\mathbf{u}} f(1, 2) = (1)(2)^{1-1} = 1$.

5. $f(x, y) = x^3 - 4x^2 y + y^2$

    (a) $\nabla f(x, y) = f_x \mathbf{i} + f_y \mathbf{j} = (3x^2 - 8xy)\mathbf{i} + (2y - 4x^2)\mathbf{j}$

    (b) $\nabla f(0, -1) = -2\mathbf{j}$

    (c) $\nabla f(0, -1) \cdot \mathbf{u} = -\frac{8}{5}$

7. $f(x, y, z) = xy^2 z^3$

    (a) $\nabla f(x, y, z) = f_x \mathbf{i} + f_y \mathbf{j} + f_z \mathbf{k} = y^2 z^3 \mathbf{i} + 2xyz^3 \mathbf{j} + 3xy^2 z^2 \mathbf{k}$

    (b) $\nabla f(1, -2, 1) = 4\mathbf{i} - 4\mathbf{j} + 12\mathbf{k}$

    (c) $\nabla f(1, -2, 1) \cdot \mathbf{u} = \frac{1}{\sqrt{3}}(20) = \frac{20}{\sqrt{3}}$

9. $f(x, y) = \sqrt{x - y} \ \Rightarrow \ \nabla f(x, y) = \left\langle \frac{1}{2}(x - y)^{-1/2}, -\frac{1}{2}(x - y)^{-1/2} \right\rangle, \nabla f(5, 1) = \left\langle \frac{1}{4}, -\frac{1}{4} \right\rangle$, and a unit vector
in the direction of $\mathbf{v}$ is $\mathbf{u} = \left\langle \frac{12}{13}, \frac{5}{13} \right\rangle$, so $D_{\mathbf{u}} f(5, 1) = \nabla f(5, 1) \cdot \mathbf{u} = \frac{12}{52} - \frac{5}{52} = \frac{7}{52}$.

11. $g(x, y) = xe^{xy} \ \Rightarrow \ \nabla g(x, y) = \left\langle e^{xy}(1 + xy), x^2 e^{xy} \right\rangle, \nabla g(-3, 0) = \langle 1, 9 \rangle, \mathbf{u} = \left\langle \frac{2}{\sqrt{13}}, \frac{3}{\sqrt{13}} \right\rangle$ and
$D_{\mathbf{u}} g(-3, 0) = \frac{2}{\sqrt{13}} + \frac{27}{\sqrt{13}} = \frac{29}{\sqrt{13}}$.

13. $f(x, y, z) = \sqrt{xyz} \ \Rightarrow \ \nabla f(x, y, z) = \frac{1}{2}(xyz)^{-1/2} \langle yz, xz, xy \rangle, \nabla f(2, 4, 2) = \langle 1, \frac{1}{2}, 1 \rangle, \mathbf{u} = \left\langle \frac{2}{3}, \frac{1}{3}, -\frac{2}{3} \right\rangle$ and
$D_{\mathbf{u}} f(2, 4, 2) = 1 \cdot \frac{2}{3} + \frac{1}{2} \cdot \frac{1}{3} + 1 \left(-\frac{2}{3} \right) = \frac{1}{6}$.

15. $g(x, y, z) = x \tan^{-1}(y/z) \quad \Rightarrow \quad \nabla g(x, y, z) = \langle \tan^{-1}(y/z), xz/(y^2 + z^2), -xy/(y^2 + z^2) \rangle$,

$\nabla g(1, 2, -2) = \langle -\frac{\pi}{4}, -\frac{1}{4}, -\frac{1}{4} \rangle$, $\mathbf{u} = \frac{1}{\sqrt{3}} \langle 1, 1, -1 \rangle$ and $D_{\mathbf{u}} g(1, 2, -2) = \frac{(-\pi)(1)}{4\sqrt{3}} + \frac{(-1)(1)}{4\sqrt{3}} + \frac{(-1)(-1)}{4\sqrt{3}} = -\frac{\pi}{4\sqrt{3}}$.

17. $f(x, y) = xe^{-y} + 3y \quad \Rightarrow \quad \nabla f(x, y) = \langle e^{-y}, 3 - xe^{-y} \rangle$, $\nabla f(1, 0) = \langle 1, 2 \rangle$ is the direction and the maximum rate is $|\nabla f(1, 0)| = \sqrt{5}$.

19. $f(x, y) = \sqrt{x^2 + 2y} \quad \Rightarrow \quad \nabla f(x, y) = \left\langle \dfrac{x}{\sqrt{x^2 + 2y}}, \dfrac{1}{\sqrt{x^2 + 2y}} \right\rangle$. Thus the maximum rate of change is

$|\nabla f(4, 10)| = \frac{\sqrt{17}}{6}$ in the direction $\langle \frac{2}{3}, \frac{1}{6} \rangle$ or $\langle 4, 1 \rangle$.

21. $f(x, y) = \cos(3x + 2y) \quad \Rightarrow \quad \nabla f(x, y) = \langle -3\sin(3x + 2y), -2\sin(3x + 2y) \rangle$, so the maximum rate of

change is $|\nabla f(\frac{\pi}{6}, -\frac{\pi}{8})| = \sqrt{\frac{13}{2}}$ in the direction $\langle -\frac{3\sqrt{2}}{2}, -\sqrt{2} \rangle$ or $\langle -3, -2 \rangle$.

23. As in the proof of Theorem 15, $D_{\mathbf{u}} f = |\nabla f| \cos \theta$. Since the minimum value of $\cos \theta$ is -1 occurring when $\theta = \pi$, the minimum value of $D_{\mathbf{u}} f$ is $-|\nabla f|$ occurring when $\theta = \pi$, that is when $\mathbf{u}$ is in the opposite direction of ∇f (assuming $\nabla f \neq \mathbf{0}$).

25. $T = \dfrac{k}{\sqrt{x^2 + y^2 + z^2}}$ and $120 = T(1, 2, 2) = \dfrac{k}{3}$ so $k = 360$.

   (a) $\mathbf{u} = \dfrac{\langle 1, -1, 1 \rangle}{\sqrt{3}}$,

$$D_{\mathbf{u}} T(1, 2, 2) = \nabla T(1, 2, 2) \cdot \mathbf{u} = \left[-360(x^2 + y^2 + z^2)^{-3/2} \langle x, y, z \rangle \right]_{(1,2,2)} \cdot \mathbf{u}$$

$$= -\frac{40}{3} \langle 1, 2, 2 \rangle \cdot \frac{1}{\sqrt{3}} \langle 1, -1, 1 \rangle = -\frac{40}{3\sqrt{3}}$$

   (b) From (a), $\nabla T = -360(x^2 + y^2 + z^2)^{-3/2} \langle x, y, z \rangle$, and since $\langle x, y, z \rangle$ is the position vector of the point (x, y, z), the vector $-\langle x, y, z \rangle$, and thus ∇T, always points toward the origin.

27. $\nabla V(x, y, z) = \langle 10x - 3y + yz, xz - 3x, xy \rangle$, $\nabla V(3, 4, 5) = \langle 38, 6, 12 \rangle$

   (a) $D_{\mathbf{u}} V(3, 4, 5) = \langle 38, 6, 12 \rangle \cdot \frac{1}{\sqrt{3}} \langle 1, 1, -1 \rangle = \frac{32}{\sqrt{3}}$

   (b) $\langle 38, 6, 12 \rangle$ or $\langle 19, 3, 6 \rangle$.

   (c) $|\nabla V(x, y, z)| = \sqrt{(10x - 3y + yz)^2 + (xz - 3x)^2 + x^2 y^2}$ in general or at $(3, 4, 5)$ is

      $|\nabla V(3, 4, 5)| = \sqrt{1624} = 2\sqrt{406}$.

29. A unit vector in the direction of $\overrightarrow{AB}$ is $\mathbf{i}$ and a unit vector in the direction of $\overrightarrow{AC}$ is $\mathbf{j}$. Thus

$D_{\overrightarrow{AB}} f(1, 3) = f_x(1, 3) = 3$ and $D_{\overrightarrow{AC}} f(1, 3) = f_y(1, 3) = 26$. Therefore

$\nabla f(1, 3) = \langle f_x(1, 3), f_y(1, 3) \rangle = \langle 3, 26 \rangle$ and by definition $D_{\overrightarrow{AD}} f(1, 3) = \nabla f \cdot \mathbf{u}$ where $\mathbf{u}$ is a unit vector in the

direction of $\overrightarrow{AD}$, which is $\langle \frac{5}{13}, \frac{12}{13} \rangle$. Therefore $D_{\overrightarrow{AD}} f(1, 3) = \langle 3, 26 \rangle \cdot \langle \frac{5}{13}, \frac{12}{13} \rangle = 3 \cdot \frac{5}{13} + 26 \cdot \frac{12}{13} = \frac{327}{13}$.

31. $\nabla(au + bv) = \left\langle \dfrac{\partial(au + bv)}{\partial x}, \dfrac{\partial(au + bv)}{\partial y} \right\rangle = \left\langle a \dfrac{\partial u}{\partial x} + b \dfrac{\partial v}{\partial x}, a \dfrac{\partial u}{\partial y} + b \dfrac{\partial v}{\partial y} \right\rangle = a \left\langle \dfrac{\partial u}{\partial x}, \dfrac{\partial u}{\partial y} \right\rangle + b \left\langle \dfrac{\partial v}{\partial x}, \dfrac{\partial v}{\partial y} \right\rangle$

$= a \nabla u + b \nabla v$

33. $\nabla\left(\dfrac{u}{v}\right) = \left\langle \dfrac{v\dfrac{\partial u}{\partial x} - u\dfrac{\partial v}{\partial x}}{v^2},\ \dfrac{v\dfrac{\partial u}{\partial y} - u\dfrac{\partial v}{\partial y}}{v^2} \right\rangle = \dfrac{v\left\langle \dfrac{\partial u}{\partial x}, \dfrac{\partial u}{\partial y}\right\rangle - u\left\langle \dfrac{\partial v}{\partial x}, \dfrac{\partial v}{\partial y}\right\rangle}{v^2} = \dfrac{v\nabla u - u\nabla v}{v^2}$

35. $F(x, y, z) = 4x^2 + y^2 + z^2,\ \nabla F(2, 2, 2) = \langle 16, 4, 4\rangle$, so

    (a)   the equation of the tangent plane is $16x + 4y + 4z = 48$ or $4x + y + z = 12$, and

    (b)   the normal line is given by $\dfrac{x - 2}{16} = \dfrac{y - 2}{4} = \dfrac{z - 2}{4}$ or $\dfrac{x - 2}{4} = y - 2 = z - 2$.

37. $\nabla F(x, y, z) = \langle 2x - 2y + 4z, 2y - 2x, -2z + 4x\rangle,\ \nabla F(1, 0, 1) = \langle 6, -2, 2\rangle$

    (a) $6x - 2y + 2z = 8$ or $3x - y + z = 4$ **(b)** $\dfrac{x - 1}{3} = -y = z - 1$

39. $F(x, y, z) = -z + xe^y \cos z,\ \nabla F(x, y, z) = \langle e^y \cos z, xe^y \cos z, -1 - xe^y \sin z\rangle,\ \nabla F(1, 0, 0) = \langle 1, 1, -1\rangle$

    (a) $x + y - z = 1$ **(b)** $x - 1 = y = -z$

41. $\nabla F(x, y, z) = \langle y + z, x + z, y + x\rangle$,

    $\nabla F(1, 1, 1) = \langle 2, 2, 2\rangle$, so the equation of the tangent plane

    is $2x + 2y + 2z = 6$ or $x + y + z = 3$, and the normal line

    is given by $x - 1 = y - 1 = z - 1$ or $x = y = z$.

43. $\nabla f(x, y) = \langle 2x, 8y\rangle,\ \nabla f(2, 1) = \langle 4, 8\rangle$.

    The tangent line has equation

    $\nabla f(2, 1) \cdot \langle x - 2, y - 1\rangle = 0 \quad\Rightarrow$

    $4(x - 2) + 8(y - 1) = 0$, which

    simplifies to $x + 2y = 4$.

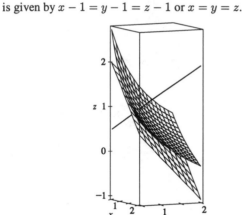

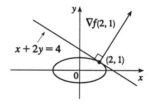

45. $\nabla F(x_0, y_0, z_0) = \left\langle \dfrac{2x_0}{a^2}, \dfrac{2y_0}{b^2}, \dfrac{2z_0}{c^2}\right\rangle$. Thus an equation of the tangent plane at (x_0, y_0, z_0) is

$\dfrac{2x_0}{a^2}x + \dfrac{2y_0}{b^2}y + \dfrac{2z_0}{c^2}z = 2\left(\dfrac{x_0^2}{a^2} + \dfrac{y_0^2}{b^2} + \dfrac{z_0^2}{c^2}\right) = 2(1) = 2$ since (x_0, y_0, z_0) is a point on the ellipsoid. Hence

$\dfrac{x_0}{a^2}x + \dfrac{y_0}{b^2}y + \dfrac{z_0}{c^2}z = 1$ is an equation of the tangent plane.

47. $\nabla F(x_0, y_0, z_0) = \left\langle \dfrac{2x_0}{a^2}, \dfrac{2y_0}{b^2}, \dfrac{-1}{c}\right\rangle$, so an equation of the tangent plane is

$\dfrac{2x_0}{a^2}x + \dfrac{2y_0}{b^2}y - \dfrac{1}{c}z = \dfrac{2x_0^2}{a^2} + \dfrac{2y_0^2}{b^2} - \dfrac{z_0}{c}$ or $\dfrac{2x_0}{a^2}x + \dfrac{2y_0}{b^2}y = \dfrac{z}{c} + 2\left(\dfrac{x_0^2}{a^2} + \dfrac{y_0^2}{b^2}\right) - \dfrac{z_0}{c}$. But $\dfrac{z_0}{c} = \dfrac{x_0^2}{a^2} + \dfrac{y_0^2}{b^2}$, so

the equation can be written as $\dfrac{2x_0}{a^2}x + \dfrac{2y_0}{b^2}y = \dfrac{z + z_0}{c}$.

49. $\nabla f(x_0, y_0, z_0) = \langle 2x_0, -2y_0, 4z_0 \rangle$ and the given line has direction numbers $2, 4, 6$, so

$\langle 2x_0, -2y_0, 4z_0 \rangle = k\langle 2, 4, 6 \rangle$ or $x_0 = k$, $y_0 = -2k$ and $z_0 = \frac{3}{2}k$. But $x_0^2 - y_0^2 + 2z_0^2 = 1$ or
$\left(1 - 4 + \frac{9}{2}\right)k^2 = 1$, so $k = \pm\sqrt{\frac{2}{3}} = \pm\frac{\sqrt{6}}{3}$ and there are two such points: $\left(\pm\frac{\sqrt{6}}{3}, \mp\frac{2\sqrt{6}}{3}, \pm\frac{\sqrt{6}}{2}\right)$.

51. Let (x_0, y_0, z_0) be a point on the cone [other than $(0, 0, 0)$]. Then the equation of the tangent plane to the cone at this point is $2x_0 x + 2y_0 y - 2z_0 z = 2(x_0^2 + y_0^2 - z_0^2)$. But $x_0^2 + y_0^2 = z_0^2$ so the tangent plane is given by $x_0 x + y_0 y - z_0 z = 0$, a plane which always contains the origin.

53. Let (x_0, y_0, z_0) be a point on the surface. Then the equation of the tangent plane at the point is

$$\frac{x}{2\sqrt{x_0}} + \frac{y}{2\sqrt{y_0}} + \frac{z}{2\sqrt{z_0}} = \frac{\sqrt{x_0} + \sqrt{y_0} + \sqrt{z_0}}{2}.$$ But $\sqrt{x_0} + \sqrt{y_0} + \sqrt{z_0} = \sqrt{c}$, so the equation is

$\frac{x}{\sqrt{x_0}} + \frac{y}{\sqrt{y_0}} + \frac{z}{\sqrt{z_0}} = \sqrt{c}$. The x-, y-, and z-intercepts are $\sqrt{cx_0}$, $\sqrt{cy_0}$ and $\sqrt{cz_0}$ respectively. (The

x-intercept is found by setting $y = z = 0$ and solving the resulting equation for x, and the y- and z-intercepts are found similarly.) So the sum of the intercepts is $\sqrt{c}\left(\sqrt{x_0} + \sqrt{y_0} + \sqrt{z_0}\right) = c$, a constant.

55. If $f(x, y, z) = z - x^2 - y^2$ and $g(x, y, z) = 4x^2 + y^2 + z^2$, then the tangent line is perpendicular to both ∇f and ∇g at $(-1, 1, 2)$. The vector $\mathbf{v} = \nabla f \times \nabla g$ will therefore be parallel to the tangent line. We have:
$\nabla f(x, y, z) = \langle -2x, -2y, 1 \rangle \quad \Rightarrow \quad \nabla f(-1, 1, 2) = \langle 2, -2, 1 \rangle$, and $\nabla g(x, y, z) = \langle 8x, 2y, 2z \rangle \quad \Rightarrow$

$\nabla g(-1, 1, 2) = \langle -8, 2, 4 \rangle$. Hence $\mathbf{v} = \nabla f \times \nabla g = \begin{vmatrix} \mathbf{i} & \mathbf{j} & \mathbf{k} \\ 2 & -2 & 1 \\ -8 & 2 & 4 \end{vmatrix} = -10\mathbf{i} - 16\mathbf{j} - 12\mathbf{k}$. Parametric equations

are: $x = -1 - 10t$, $y = 1 - 16t$, $z = 2 - 12t$.

57. (a) The direction of the normal line of F is given by ∇F, and that of G by ∇G. Assuming that
$\nabla F \neq 0 \neq \nabla G$, the two normal lines are be perpendicular at P if $\nabla F \cdot \nabla G = 0$ at $P \quad \Leftrightarrow$
$\langle \partial F/\partial x, \partial F/\partial y, \partial F/\partial z \rangle \cdot \langle \partial G/\partial x, \partial G/\partial y, \partial G/\partial z \rangle = 0$ at $P \quad \Leftrightarrow \quad F_x G_x + F_y G_y + F_z G_z = 0$ at P.

(b) Here $F = x^2 + y^2 - z^2$ and $G = x^2 + y^2 + z^2 - r^2$, so
$\nabla F \cdot \nabla G = \langle 2x, 2y, -2z \rangle \cdot \langle 2x, 2y, 2z \rangle = 4x^2 + 4y^2 - 4z^2 = 4F = 0$, since the point $\langle x, y, z \rangle$ lies on the graph of $F = 0$. To see that this is true without using calculus, note that $G = 0$ is the equation of a sphere centered at the origin and $F = 0$ is the equation of a right circular cone with vertex at the origin (which is generated by lines through the origin). At any point of intersection, the sphere's normal line (which passes through the origin) lies on the cone, and thus is perpendicular to the cone's normal line. So the surfaces with equations $F = 0$ and $G = 0$ are everywhere orthogonal.

59. Let $\mathbf{u} = \langle a, b \rangle$ and $\mathbf{v} = \langle c, d \rangle$. Then we know that at the given point, $D_{\mathbf{u}}f = \nabla f \cdot \mathbf{u} = af_x + bf_y$ and $D_{\mathbf{v}}f = \nabla f \cdot \mathbf{v} = cf_x + df_y$. But these are just two linear equations in the two unknowns f_x and f_y, and since $\mathbf{u}$ and $\mathbf{v}$ are not parallel, we can solve the equations to find $\nabla f = \langle f_x, f_y \rangle$ at the given point. In fact,
$$\nabla f = \left\langle \frac{dD_{\mathbf{u}}f - bD_{\mathbf{v}}f}{ad - bc}, \frac{aD_{\mathbf{v}}f - cD_{\mathbf{u}}f}{ad - bc} \right\rangle.$$

EXERCISES 12.7

1. $f(x, y) = x^2 + y^2 + 4x - 6y \Rightarrow f_x = 2x + 4, f_y = 2y - 6,$
$f_{xx} = f_{yy} = 2, f_{xy} = 0.$ Then $f_x = 0$ and $f_y = 0$ implies
$(x, y) = (-2, 3)$ and $D(-2, 3) = 4 > 0,$ so $f(-2, 3) = -13$ is
a local minimum.

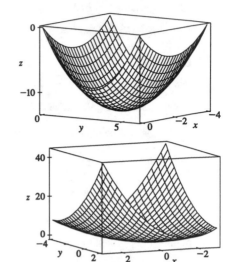

3. $f(x, y) = 2x^2 + y^2 + 2xy + 2x + 2y \Rightarrow$
$f_x = 4x + 2y + 2, f_y = 2y + 2x + 2,$
$f_{xx} = 4, f_{yy} = 2, f_{xy} = 2.$
Then $f_x = 0$ and $f_y = 0$ implies $2x = 0,$ so the critical point is
$(0, -1).$ $D(0, -1) = 8 - 4 > 0,$ so $f(0, -1) = -1$ is a
local minimum.

5. $f(x, y) = x^2 + y^2 + x^2 y + 4 \Rightarrow f_x = 2x + 2xy, f_y = 2y + x^2,$
$f_{xx} = 2 + 2y, f_{yy} = 2, f_{xy} = 2x.$ Then $f_y = 0$ implies $y = -\frac{1}{2}x^2,$
substituting into $f_x = 0$ gives $2x - x^3 = 0$ so $x = 0$ or $x = \pm\sqrt{2}.$
Thus the critical points are $(0, 0), \left(\sqrt{2}, -1\right)$ and $\left(-\sqrt{2}, -1\right).$
Now $D(0, 0) = 4, D\left(\sqrt{2}, -1\right) = -8 = D\left(-\sqrt{2}, -1\right),$
$f_{xx}(0, 0) = 2, f_{xx}\left(\pm\sqrt{2}, -1\right) = 0.$ Thus $f(0, 0) = 4$ is
a local minimum and $\left(\pm\sqrt{2}, -1\right)$ are saddle points.

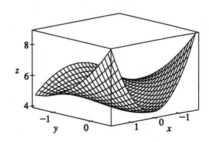

7. $f(x, y) = x^3 - 3xy + y^3 \Rightarrow f_x = 3x^2 - 3y, f_y = 3y^2 - 3x,$
$f_{xx} = 6x, f_{yy} = 6y, f_{xy} = -3.$ Then $f_x = 0$ implies $x^2 = y$ and
substituting into $f_y = 0$ gives $x = 0$ or $x = 1.$ Thus the
critical points are $(0, 0)$ and $(1, 1).$ Now $D(0, 0) = -9 < 0$ so
$(0, 0)$ is a saddle point and $D(1, 1) = 36 - 9 > 0$ while
$f_{xx}(1, 1) = 6$ so $f(1, 1) = -1$ is a local minimum.

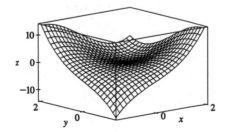

9. $f(x, y) = xy - 2x - y \Rightarrow f_x = y - 2, f_y = x - 1,$
$f_{xx} = f_{yy} = 0, f_{xy} = 1$ and the only critical point is $(1, 2).$
Now $D(1, 2) = -1$ so $(1, 2)$ is a saddle point and f has no
local maximum or minimum.

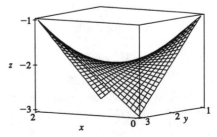

11. $f(x,y) = \dfrac{x^2y^2 - 8x + y}{xy}$ $\Rightarrow$ $f_x = y - x^{-2}$, $f_y = x + 8y^{-2}$, $f_{xx} = 2x^{-3}$, $f_{yy} = -16y^{-3}$ and $f_{xy} = 1$. Then $f_x = 0$ implies $y = x^{-2}$, substituting into $f_y = 0$ gives $x + 8x^4 = 0$ so $x = 0$ or $x = -\frac{1}{2}$ but $(0, y)$ is not in the domain of f. Thus the only critical point is $\left(-\frac{1}{2}, 4\right)$. Then $f_{xx}\left(-\frac{1}{2}, 4\right) = -16$ and $D\left(-\frac{1}{2}, 4\right) = 4 - 1 > 0$ so $f\left(-\frac{1}{2}, 4\right) = -6$ is a local maximum.

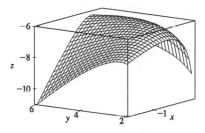 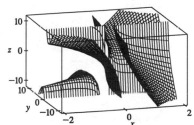

13. $f(x,y) = e^x \cos y$ $\Rightarrow$ $f_x = e^x \cos y$, $f_y = -e^x \sin y$.
Now $f_x = 0$ implies $\cos y = 0$ or $y = \frac{\pi}{2} + n\pi$ for n an integer.
But $\sin\left(\frac{\pi}{2} + n\pi\right) \neq 0$, so there are no critical points.

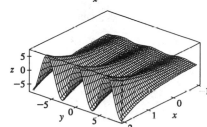

15. $f(x,y) = x \sin y$ $\Rightarrow$ $f_x = \sin y$, $f_y = x \cos y$, $f_{xx} = 0$, $f_{yy} = -x \sin y$ and $f_{xy} = \cos y$. Then $f_x = 0$ if and only if $y = n\pi$, n an integer, and substituting into $f_y = 0$ requires $x = 0$ for each of these y-values. Thus the critical points are $(0, n\pi)$, n an integer. But $D(0, n\pi) = -\cos^2(n\pi) < 0$, so so each critical point is a saddle point.

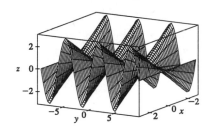

17. $f(x,y) = 3x^2y + y^3 - 3x^2 - 3y^2 + 2$

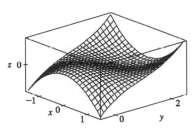 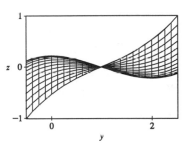 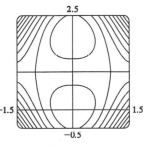

From the graphs, it appears that f has a local maximum $f(0,0) \approx 2$ and a local minimum $f(0,2) \approx -2$. There appear to be saddle points near $(\pm 1, 1)$. $f_x = 6xy - 6x$, $f_y = 3x^2 + 3y^2 - 6y$. Then $f_x = 0$ implies $x = 0$ or $y = 1$ and when $x = 0$, $f_y = 0$ implies $y = 0$ or $y = 2$; when $y = 1$, $f_y = 0$ implies $x^2 = 1$ or $x = \pm 1$. Thus the critical points are $(0,0)$, $(0,2)$, $(\pm 1, 1)$. Now $f_{xx} = 6y - 6$, $f_{yy} = 6y - 6$ and $f_{xy} = 6x$, so $D(0,0) = D(0,2) = 36 > 0$ while $D(\pm 1, 1) = -36 < 0$ and $f_{xx}(0,0) = -6$, $f_{xx}(0,2) = 6$. Hence $(\pm 1, 1)$ are saddle points while $f(0,0) = 2$ is a local maximum and $f(0,2) = -2$ is a local minimum.

19. $f(x, y) = \sin x + \sin y + \sin(x + y)$, $0 \le x \le 2\pi$, $0 \le y \le 2\pi$

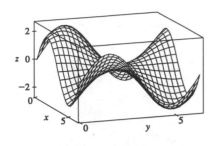
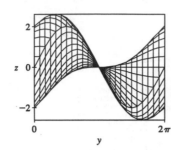
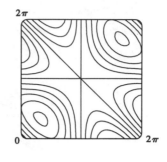

From the graphs it appears that f has a local maximum at about $(1, 1)$ with value ≈ 2.6, a local minimum at about $(5, 5)$ with value ≈ -2.6, and a saddle point at about $(3, 3)$.

$f_x = \cos x + \cos(x + y)$, $f_y = \cos y + \cos(x + y)$, $f_{xx} = -\sin x - \sin(x + y)$, $f_{yy} = -\sin y - \sin(x + y)$, $f_{xy} = -\sin(x + y)$. Setting $f_x = 0$ and $f_y = 0$ and subtracting gives $\cos x - \cos y = 0$ or $\cos x = \cos y$. Thus $x = y$ or $x = 2\pi - y$. If $x = y$, $f_x = 0$ becomes $\cos x + \cos 2x = 0$ or $2\cos^2 x + \cos x - 1 = 0$, a quadratic in $\cos x$. Thus $\cos x = -1$ or $\frac{1}{2}$ and $x = \pi$, $\frac{\pi}{3}$, or $\frac{5\pi}{3}$, yielding the critical points (π, π), $\left(\frac{\pi}{3}, \frac{\pi}{3}\right)$ and $\left(\frac{5\pi}{3}, \frac{5\pi}{3}\right)$. Similarly if $x = 2\pi - y$, $f_x = 0$ becomes $(\cos x) + 1 = 0$ and the resulting critical point is (π, π). Now $D(x, y) = \sin x \sin y + \sin x \sin(x + y) + \sin y \sin(x + y)$. So $D(\pi, \pi) = 0$ and (3) doesn't apply. $D\left(\frac{\pi}{3}, \frac{\pi}{3}\right) = \frac{9}{4} > 0$ and $f_{xx}\left(\frac{\pi}{3}, \frac{\pi}{3}\right) < 0$ so $f\left(\frac{\pi}{3}, \frac{\pi}{3}\right) = \frac{3\sqrt{3}}{2}$ is a local maximum while $D\left(\frac{5\pi}{3}, \frac{5\pi}{3}\right) = \frac{9}{4} > 0$ and $f_{xx}\left(\frac{5\pi}{3}, \frac{5\pi}{3}\right) > 0$, so $f\left(\frac{5\pi}{3}, \frac{5\pi}{3}\right) = -\frac{3\sqrt{3}}{2}$ is a local minimum.

21. $f(x, y) = x^4 - 5x^2 + y^2 + 3x + 2$ $\Rightarrow$ $f_x(x, y) = 4x^3 - 10x + 3$ and $f_y(x, y) = 2y$. $f_y = 0$ $\Rightarrow$ $y = 0$, and the graph of f_x shows that the roots of $f_x = 0$ are approximately $x = -1.714, 0.312$ and 1.402. (Alternatively, we could have used a calculator or a CAS to find these roots.) So to three decimal places, the critical points are $(-1.714, 0)$, $(1.402, 0)$, and $(0.312, 0)$. Now since $f_{xx} = 12x^2 - 10$, $f_{xy} = 0$, $f_{yy} = 2$, and $D = 24x^2 - 20$, we have $D(-1.714, 0) > 0$, $f_{xx}(-1.714, 0) > 0$, $D(1.402, 0) > 0$, $f_{xx}(1.402, 0) > 0$, and $D(0.312, 0) < 0$. Therefore $f(-1.714, 0) \approx -9.200$ and $f(1.402, 0) \approx 0.242$ are local minima, and $(0.312, 0)$ is a saddle point. The lowest point on the graph is approximately $(-1.714, 0, -9.200)$.

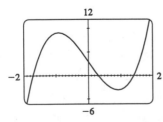
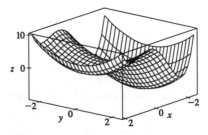

23. $f(x,y) = 2x + 4x^2 - y^2 + 2xy^2 - x^4 - y^4 \Rightarrow f_x(x,y) = 2 + 8x + 2y^2 - 4x^3$,
$f_y(x,y) = -2y + 4xy - 4y^3$. Now $f_y = 0 \Leftrightarrow 2y(2y^2 - 2x + 1) = 0 \Leftrightarrow y = 0$ or $y^2 = x - \frac{1}{2}$. The
first of these implies that $f_x = -4x^3 + 8x + 2$, and the second implies that
$f_x = 2 + 8x + 2\left(x - \frac{1}{2}\right) - 4x^3 = -4x^3 + 10x + 1$. From the graphs, we see that the first possibility for f_x has
roots at approximately -1.267, -0.259, and 1.526, and the second has a root at approximately 1.629 (the
negative roots do not give critical points, since $y^2 = x - \frac{1}{2}$ must be positive). So to three decimal places, f has
critical points at $(-1.267, 0)$, $(-0.259, 0)$, $(1.526, 0)$, and $(1.629, \pm 1.063)$. Now since $f_{xx} = 8 - 12x^2$,
$f_{xy} = 4y$, $f_{yy} = 4x - 12y^2$, and $D = (8 - 12x^2)(4x - 12y^2) - 16y^2$, we have $D(-1.267, 0) > 0$,
$f_{xx}(-1.267, 0) > 0$, $D(-0.259, 0) < 0$, $D(1.526, 0) < 0$, $D(1.629, \pm 1.063) > 0$, and $f_{xx}(1.629, \pm 1.063) < 0$.
Therefore, to three decimal places, $f(-1.267, 0) \approx 1.310$ and $f(1.629, \pm 1.063) \approx 8.105$ are local maxima, and
$(-0.259, 0)$ and $(1.526, 0)$ are saddle points. The highest points on the graph are approximately
$(1.629, \pm 1.063, 8.105)$.

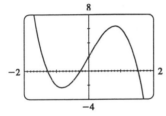
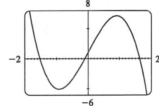
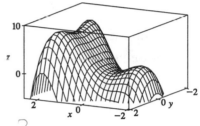

25. Since f is a polynomial it is continuous on D, so an absolute maximum
and minimum exist. Here $f_x = -3$, $f_y = 4$ so there are no critical points
inside D. Thus the absolute extrema must both occur on the boundary.
Along L_1, $y = 0$ and $f(x, 0) = 5 - 3x$, a decreasing function in x, so
the maximum value is $f(0, 0) = 5$ and the minimum value is
$f(4, 0) = -7$. Along L_2, $x = 4$ and $f(4, y) = -7 + 4y$, an increasing
function in y, so the minimum value is $f(4, 0) = -7$ and the maximum

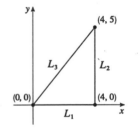

value is $f(4, 5) = 13$. Along L_3, $y = \frac{5}{4}x$ and $f\left(x, \frac{5}{4}x\right) = 5 + 2x$, an increasing function in x, so the minimum
value is $f(0, 0) = 5$ and the maximum value is $f(4, 5) = 13$. Thus the absolute minimum of f on D is
$f(4, 0) = -7$ and the absolute maximum is $f(4, 5) = 13$.

27. In Exercise 5, we found the critical points of f; only $(0, 0)$ with
$f(0, 0) = 4$ is in D. On L_1: $y = -1$, $f(x, -1) = 5$, a constant.
On L_2: $x = 1$, $f(1, y) = y^2 + y + 5$, a quadratic in y which attains
its maximum at $(1, 1)$, $f(1, 1) = 7$ and its minimum at $\left(1, -\frac{1}{2}\right)$,
$f\left(1, -\frac{1}{2}\right) = \frac{17}{4}$. On L_3: $f(x, 1) = 2x^2 + 5$ which attains its maximum

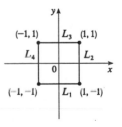

at $(-1, 1)$ and $(1, 1)$ with $f(\pm 1, 1) = 7$ and its minimum at $(0, 1)$,
$f(0, 1) = 5$. On L_4: $f(-1, y) = y^2 + y + 5$ with maximum at $(-1, 1)$, $f(-1, 1) = 7$ and minimum at
$\left(-1, -\frac{1}{2}\right)$, $f\left(-1, -\frac{1}{2}\right) = \frac{17}{4}$. Thus the absolute maximum is attained at both $(\pm 1, 1)$ with $f(\pm 1, 1) = 7$ and the
absolute minimum on D is attained at $(0, 0)$ with $f(0, 0) = 4$.

29. $f_x(x,y) = y - 1$ and $f_y(x,y) = x - 1$ and so the critical point is

$(1,1)$ (in D), where $f(1,1) = 0$. Along L_1: $y = 4$, so

$f(x,4) = 1 + 4x - x - 4 = 3x - 3$, $-2 \leq x \leq 2$, which is an

increasing function and has a maximum value when $x = 2$

where $f(2,4) = 3$ and a minimum of $f(-2,4) = -9$. Along

L_2: $y = x^2$, so let $g(x) = f(x,x^2) = x^3 - x^2 - x + 1$. Then

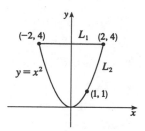

$g'(x) = 3x^2 - 2x - 1 = 0 \Leftrightarrow x = -\frac{1}{3}$ or $x = 1$. $f\left(-\frac{1}{3}, \frac{1}{9}\right) = \frac{32}{27}$ and $f(1,1) = 0$. As a result, the absolute

maximum and minimum values of f on D are $f(2,4) = 3$ and $f(-2,4) = -9$.

31. $f_x(x,y) = 6x^2$ and $f_y(x,y) = 4y^3$. And so $f_x = 0$ and $f_y = 0$ only occur when $x = y = 0$. Hence, the only

critical point inside the disk is at $x = y = 0$ where $f(0,0) = 0$. Now on the circle $x^2 + y^2 = 1$, $y^2 = 1 - x^2$ so

let $g(x) = f(x,y) = 2x^3 + (1 - x^2)^2 = x^4 + 2x^3 - 2x^2 + 1$, $-1 \leq x \leq 1$. Then $g'(x) = 4x^3 + 6x^2 - 4x = 0$

$\Rightarrow \quad x = 0, -2$, or $\frac{1}{2}$. $f(0, \pm 1) = g(0) = 1$, $f\left(\frac{1}{2}, \pm\frac{\sqrt{3}}{2}\right) = g\left(\frac{1}{2}\right) = \frac{13}{16}$, and $(-2, -3)$ is not in D. Checking

the endpoints, we get $f(-1,0) = g(-1) = -2$ and $f(1,0) = g(1) = 2$. Thus the absolute maximum and

minimum of f on D are $f(1,0) = 2$ and $f(-1,0) = -2$.

Another method: On the boundary $x^2 + y^2 = 1$ we can write $x = \cos\theta$, $y = \sin\theta$, so

$f(\cos\theta, \sin\theta) = 2\cos^3\theta + \sin^4\theta$, $0 \leq \theta \leq 2\pi$.

33. $f(x,y) = -(x^2 - 1)^2 - (x^2 y - x - 1)^2 \quad \Rightarrow \quad f_x(x,y) = -2(x^2 - 1)(2x) - 2(x^2 y - x - 1)(2xy - 1)$ and

$f_y(x,y) = -2(x^2 y - x - 1)x^2$. Setting $f_y(x,y) = 0$ gives either $x = 0$ or $x^2 y - x - 1 = 0$. There are no

critical points for $x = 0$, since $f_x(0,y) = -2$, so we set $x^2 y - x - 1 = 0 \quad \Leftrightarrow \quad y = \frac{x+1}{x^2}$ ($x \neq 0$), so

$f_x\left(x, \frac{x+1}{x^2}\right) = -2(x^2 - 1)(2x) - 2\left(x^2 \frac{x+1}{x^2} - x - 1\right)\left(2x\frac{x+1}{x^2} - 1\right) = -4x(x^2 - 1)$. Therefore

$f_x(x,y) = f_y(x,y) = 0$ at the points $(1,2)$ and $(-1,0)$. To classify these critical points, we calculate

$f_{xx}(x,y) = -12x^2 - 12x^2 y^2 + 12xy + 4y + 2$,

$f_{yy}(x,y) = -2x^4$, and $f_{xy}(x,y) = -8x^3 y + 6x^2 + 4x$.

In order to use the Second Derivatives Test we calculate

$D(-1,0) = f_{xx}(-1,0)f_{yy}(-1,0) - [f_{xy}(-1,0)]^2 = 16 > 0$,

$f_{xx}(-1,0) = -10 < 0$, $D(1,2) = 16 > 0$, and

$f_{xx}(1,2) = -26 < 0$, so both $(-1,0)$ and $(1,2)$ give local maxima.

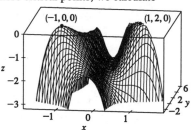

35. Here $d = \sqrt{x^2 + y^2 + z^2}$ with $z = \frac{1}{3}(4 - x - 2y)$. So minimize $d^2 = x^2 + y^2 + \frac{1}{9}(4 - x - 2y)^2 = f(x,y)$.

Then $f_x = 2x - \frac{2}{9}(4 - x - 2y) = \frac{1}{9}(20x + 4y - 8)$, $f_y = \frac{1}{9}(26y + 4x - 16)$. Solving the equations

$20x + 4y - 8 = 0$ and $26y + 4x - 16 = 0$, we get $x = \frac{2}{7}$, $y = \frac{4}{7}$, so the only critical point is $\left(\frac{2}{7}, \frac{4}{7}\right)$. Since the

absolute minimum has to occur at a critical point, the point $\left(\frac{2}{7}, \frac{4}{7}, \frac{6}{7}\right)$ is is the closest point to the origin

$\left[\text{or check } D\left(\frac{2}{7}, \frac{4}{7}\right)\right]$.

37. $d = \sqrt{(x-2)^2 + (y+2)^2 + (z-3)^2}$, where $z = \frac{1}{3}(6x + 4y - 2)$, so we minimize

$d^2 = f(x,y) = (x-2)^2 + (y+2)^2 + \left(2x + \frac{4}{3}y - \frac{11}{3}\right)^2$. Then $f_x = 10x + \frac{16}{3}y - \frac{56}{3}$ and $f_y = \frac{50}{9}y + \frac{16}{3}x - \frac{52}{9}$.

Solving $50y + 48x = 52$ and $16y + 30x = 56$ simultaneously gives $x = \frac{164}{61}$, $y = -\frac{94}{61}$. The absolute minimum

must occur at a critical point. Thus $d^2 = \left(\frac{42}{61}\right)^2 + \left(\frac{28}{61}\right)^2 + \left(-\frac{21}{61}\right)^2$ or $d = \frac{7}{\sqrt{61}}$.

39. Minimize $d^2 = x^2 + y^2 + z^2 = x^2 + y^2 + xy + 1$. Then $f_x = 2x + y$, $f_y = 2y + x$ so the critical point is $(0,0)$

and $D(0,0) = 4 - 1 > 0$ with $f_{xx}(0,0) = 2$ so this is a minimum. Thus $z^2 = 1$ or $z = \pm 1$ and the points on

the surface are $(0, 0, \pm 1)$.

41. $x + y + z = 100$, so maximize $f(x,y) = xy(100 - x - y)$. $f_x = 100y - 2xy - y^2$, $f_y = 100x - x^2 - 2xy$,

$f_{xx} = -2y$, $f_{yy} = -2x$, $f_{xy} = 100 - 2x - 2y$. Then $f_x = 0$ implies $y = 0$ or $y = 100 - 2x$. Substituting $y = 0$

into $f_y = 0$ gives $x = 0$ or $x = 100$ and substituting $y = 100 - 2x$ into $f_y = 0$ gives $3x^2 - 100x = 0$ so $x = 0$

or $\frac{100}{3}$. Thus the critical points are $(0,0)$, $(100,0)$, $(0,100)$ and $\left(\frac{100}{3}, \frac{100}{3}\right)$.

$D(0,0) = D(100,0) = D(0,100) = -10,000$ while $D\left(\frac{100}{3}, \frac{100}{3}\right) = \frac{10,000}{3}$ and $f_{xx}\left(\frac{100}{3}, \frac{100}{3}\right) = -\frac{200}{3} < 0$. Thus

$(0,0)$, $(100,0)$ and $(0,100)$ are saddle points whereas $f\left(\frac{100}{3}, \frac{100}{3}\right)$ is a local maximum. Thus the numbers are

$x = y = z = \frac{100}{3}$.

43. Maximize $f(x,y) = xy(36 - 9x^2 - 36y^2)^{1/2}/2$ with (x,y,z) in first octant. Then

$f_x = \dfrac{y(36 - 9x^2 - 36y^2)^{1/2}}{2} + \dfrac{-9x^2 y(36 - 9x^2 - 36y^2)^{-1/2}}{2} = \dfrac{(36y - 18x^2 y - 36y^3)}{2(36 - 9x^2 - 36y^2)^{1/2}}$ and

$f_y = \dfrac{36x - 9x^3 - 72xy^2}{2(36 - 9x^2 - 36y^2)^{1/2}}$. Setting $f_x = 0$ gives $y = 0$ or $y^2 = \dfrac{2 - x^2}{2}$ but $y > 0$, so only the latter solution

applies. Substituting this y into $f_y = 0$ gives $x^2 = \frac{4}{3}$ or $x = \frac{2}{\sqrt{3}}$, $y = \frac{1}{\sqrt{3}}$ and then $z^2 = (36 - 12 - 12)/4 = 3$.

That this gives a maximum volume follows from the geometry. This maximum volume is

$V = (2x)(2y)(2z) = 8\left(\frac{2}{\sqrt{3}}\right)\left(\frac{1}{\sqrt{3}}\right)\left(\sqrt{3}\right) = \frac{16}{\sqrt{3}}$.

45. Maximize $f(x,y) = \dfrac{xy}{3}(6 - x - 2y)$, then the maximum volume is $V = xyz$.

$f_x = \frac{1}{3}(6y - 2xy - y^2) = \frac{y}{3}(6 - 2x - 2y)$ and $f_y = \frac{x}{3}(6 - x - 4y)$. Setting $f_x = 0$ and $f_y = 0$ gives the

critical point $(2,1)$ which geometrically must yield a maximum. Thus the volume of the largest such box is

$V = (2)(1)\left(\frac{2}{3}\right) = \frac{4}{3}$.

47. Let the dimensions be x, y, and z; then $4x + 4y + 4z = c$ and the volume is

$V = xyz = xy\left[\frac{1}{4}c - x - y\right] = \frac{1}{4}cxy - x^2 y - xy^2$, $x > 0$, $y > 0$. Then $V_x = \frac{1}{4}cy - 2xy - y^2$ and

$V_y = \frac{1}{4}cx - x^2 - 2xy$, so $V_x = 0 = V_y$ when $2x + y = \frac{1}{4}c$ and $x + 2y = \frac{1}{4}c$. Solving, we get $x = \frac{1}{12}c$, $y = \frac{1}{12}c$

and $z = \frac{1}{4}c - x - y = \frac{1}{12}c$. From the geometrical nature of the problem, this critical point must give an absolute

maximum. Thus the box is a cube with edge length $\frac{1}{12}c$.

49. Let the dimensions be x, y and z, then minimize $xy + 2(xz + yz)$ if $xyz = 32{,}000$ m^3. Then

$f(x, y) = xy + [64{,}000(x + y)/xy] = xy + 64{,}000(x^{-1} + y^{-1})$, $f_x = y - 64{,}000x^{-2}$, $f_y = x - 64{,}000y^{-2}$.

And $f_x = 0$ implies $y = 64{,}000/x^2$; substituting into $f_y = 0$ implies $x^3 = 64{,}000$ or $x = 40$ and then $y = 40$.

Now $D(x, y) = [(2)(64{,}000)]^2 x^{-3} y^{-3} - 1 > 0$ for $(40, 40)$ and $f_{xx}(40, 40) > 0$ so this is indeed a minimum.

Thus the dimensions of the box are $x = y = 40$ cm, $z = 20$ cm.

51. Note that here the variables are m and b, and $f(m, b) = \sum_{i=1}^{n} [y_i - (mx_i + b)]^2$. Then

$f_m = \sum_{i=1}^{n} -2x_i[y_i - (mx_i + b)] = 0$ implies $\sum_{i=1}^{n} (x_i y_i - mx_i^2 - bx_i) = 0$ or $\sum_{i=1}^{n} x_i y_i = m\sum_{i=1}^{n} x_i^2 + b\sum_{i=1}^{n} x_i$ and

$f_b = \sum_{i=1}^{n} -2[y_i - (mx_i + b)] = 0$ implies $\sum_{i=1}^{n} y_i = m\sum_{i=1}^{n} x_i + \sum_{i=1}^{n} b = m\left(\sum_{i=1}^{n} x_i\right) + nb$. Thus we have the

two desired equations. Now $f_{mm} = \sum_{i=1}^{n} 2x_i^2$, $f_{bb} = \sum_{i=1}^{n} 2 = 2n$ and $f_{mb} = \sum_{i=1}^{n} 2x_i$. And $f_{mm}(m, b) > 0$ always

and $D(m, b) = 4n\left(\sum_{i=1}^{n} x_i^2\right) - 4\left(\sum_{i=1}^{n} x_i\right)^2 = 4\left[n\left(\sum_{i=1}^{n} x_i^2\right) - \left(\sum_{i=1}^{n} x_i\right)^2\right] > 0$ always so the solutions of

these two equations do indeed minimize $\sum_{i=1}^{n} d_i^2$.

53.

x_i	y_i	x_i^2	$x_i y_i$
69	138	4761	9522
65	127	4225	8255
71	178	5041	12,638
73	185	5329	13,505
68	141	4624	9588
63	122	3969	7686
70	158	4900	11,060
67	135	4489	9045
69	145	4761	10,005
70	162	4900	11,340
685	1491	46,999	102,644

From the table, $685m + 10b = 1491$ and

$46{,}999m + 685b = 102{,}644$.

Solving simultaneously gives $m = \frac{1021}{153} \approx 6.67$

and $b = 149.1 - 68.5\left(\frac{1021}{153}\right) \approx -308.01$.

Thus a 6 ft boy is predicted to weigh about

$(6.67)(72) - 308.01$ or 172.2 lbs.

EXERCISES 12.8

1. $f(x,y) = x^2 - y^2$, $g(x,y) = x^2 + y^2 = 1$ $\Rightarrow$ $\nabla f = \langle 2x, -2y \rangle$, $\lambda \nabla g = \langle 2\lambda x, 2\lambda y \rangle$. Then $2x = 2\lambda x$ implies $x = 0$ or $\lambda = 1$. If $x = 0$, then $x^2 + y^2 = 1$ implies $y = \pm 1$ and if $\lambda = 1$, then $-2y = 2\lambda y$ implies $y = 0$ and thus $x = \pm 1$. Thus the possible points for the extrema of f are $(\pm 1, 0)$, $(0, \pm 1)$. But $f(\pm 1, 0) = 1$ while $f(0, \pm 1) = -1$ so the maximum value of f on $x^2 + y^2 = 1$ is $f(\pm 1, 0) = 1$ and the minimum value is $f(0, \pm 1) = -1$.

3. $f(x,y) = xy$, $g(x,y) = 9x^2 + y^2 = 4$ $\Rightarrow$ $\nabla f = \langle y, x \rangle$, $\lambda \nabla g = \langle 18\lambda x, 2\lambda y \rangle$. Then $y = 18\lambda x$ implies $(x,y) = (0,0)$ or $\lambda = y/18x$ and $x = 2\lambda y$ implies $(x,y) = (0,0)$ or $\lambda = \dfrac{x}{2y}$. Thus $(x,y) = (0,0)$ or

$\dfrac{y}{18x} = \dfrac{x}{2y}$ implies $y^2 = 9x^2$. Now $(x,y) = (0,0)$ doesn't satisfy $g(x,y) = 4$, and when $y^2 = 9x^2$, $g(x,y) = 4$

implies $x^2 = \frac{2}{9}$ or $x = \pm \frac{\sqrt{2}}{3}$. Hence the possible points are $\left(\pm \frac{\sqrt{2}}{3}, \sqrt{2} \right)$, $\left(\pm \frac{\sqrt{2}}{3}, -\sqrt{2} \right)$ and the maximum

value of f on the ellipse is $f\left(\frac{\sqrt{2}}{3}, \sqrt{2} \right) = f\left(-\frac{\sqrt{2}}{3}, -\sqrt{2} \right) = \frac{2}{3}$ while the minimum value is

$f\left(-\frac{\sqrt{2}}{3}, \sqrt{2} \right) = f\left(\frac{\sqrt{2}}{3}, -\sqrt{2} \right) = -\frac{2}{3}$.

5. $f(x,y,z) = x + 3y + 5z$, $g(x,y,z) = x^2 + y^2 + z^2 = 1$ $\Rightarrow$ $\nabla f = \langle 1, 3, 5 \rangle$, $\lambda \nabla g = \langle 2\lambda x, 2\lambda y, 2\lambda z \rangle$.

Then $\nabla f = \lambda \nabla g$ implies $\lambda = \dfrac{1}{2x} = \dfrac{3}{2y} = \dfrac{5}{2z}$ so $x = \frac{1}{5}z$, $y = \frac{3}{5}z$. Then $x^2 + y^2 + z^2 = 1$ implies

$\frac{1}{25}z^2 + \frac{9}{25}z^2 + z^2 = 1$ or $z = \pm\sqrt{\frac{5}{7}}$. Thus the possible points are $\left(\pm\frac{1}{\sqrt{35}}, \pm\frac{3}{\sqrt{35}}, \pm\frac{5}{\sqrt{35}} \right)$ with the maximum

being $f\left(\frac{1}{\sqrt{35}}, \frac{3}{\sqrt{35}}, \frac{5}{\sqrt{35}} \right) = \sqrt{35}$ and the minimum being $f\left(-\frac{1}{\sqrt{35}}, -\frac{3}{\sqrt{35}}, -\frac{5}{\sqrt{35}} \right) = -\sqrt{35}$.

7. $f(x,y,z) = xyz$, $g(x,y,z) = x^2 + 2y^2 + 3z^2 = 6$ $\Rightarrow$ $\nabla f = \langle yz, xz, xy \rangle$, $\lambda \nabla g = \langle 2\lambda x, 4\lambda y, 6\lambda z \rangle$. Then $\nabla f = \lambda \nabla g$ implies $\lambda = (yz)/(2x) = (xz)/(4y) = (xy)/(6z)$ or $x^2 = 2y^2$ and $z^2 = \frac{2}{3}y^2$. Thus $x^2 + 2y^2 + 3z^2 = 6$ implies $6y^2 = 6$ or $y = \pm 1$. Then the possible points are $\left(\sqrt{2}, \pm 1, \sqrt{\frac{2}{3}} \right)$,

$\left(\sqrt{2}, \pm 1, -\sqrt{\frac{2}{3}} \right)$, $\left(-\sqrt{2}, \pm 1, \sqrt{\frac{2}{3}} \right)$, $\left(-\sqrt{2}, \pm 1, -\sqrt{\frac{2}{3}} \right)$. And the maximum value of f on the ellipsoid is

$\frac{2}{\sqrt{3}}$, occurring when all coordinates are positive or exactly two are negative and the minimum is $-\frac{2}{\sqrt{3}}$ occurring

when 1 or 3 of the coordinates are negative.

9. $f(x,y,z) = x^2 + y^2 + z^2$, $g(x,y,z) = x^4 + y^4 + z^4 = 1$ $\Rightarrow$ $\nabla f = \langle 2x, 2y, 2z \rangle$,
$\lambda \nabla g = \langle 4\lambda x^3, 4\lambda y^3, 4\lambda z^3 \rangle$.

Case 1: If $x \neq 0$, $y \neq 0$ and $z \neq 0$, then $\nabla f = \lambda \nabla g$ implies $\lambda = 1/(2x^2) = 1/(2y^2) = 1/(2z^2)$ or

$x^2 = y^2 = z^2$ and $3x^4 = 1$ or $x = \pm\frac{1}{\sqrt[4]{3}}$ giving the points $\left(\pm\frac{1}{\sqrt[4]{3}}, \frac{1}{\sqrt[4]{3}}, \frac{1}{\sqrt[4]{3}} \right)$, $\left(\pm\frac{1}{\sqrt[4]{3}}, -\frac{1}{\sqrt[4]{3}}, \frac{1}{\sqrt[4]{3}} \right)$,

$\left(\pm\frac{1}{\sqrt[4]{3}}, \frac{1}{\sqrt[4]{3}}, -\frac{1}{\sqrt[4]{3}} \right)$, $\left(\pm\frac{1}{\sqrt[4]{3}}, -\frac{1}{\sqrt[4]{3}}, -\frac{1}{\sqrt[4]{3}} \right)$ all with an f-value of $\sqrt{3}$.

Case 2: If one of the variables equals zero and the other two are not zero, then the squares of the two nonzero coordinates are equal with common value $\frac{1}{\sqrt{2}}$ and corresponding f value of $\sqrt{2}$.

Case 3: If exactly two of the variables are zero, then the third variable has value ± 1 with the corresponding f value of 1. Thus on $x^4 + y^4 + z^4 = 1$, the maximum value of f is $\sqrt{3}$ and the minimum value is 1.

11. $f(x, y, z, t) = x + y + z + t$, $g(x, y, z, t) = x^2 + y^2 + z^2 + t^2 = 1$ $\Rightarrow$ $\langle 1, 1, 1, 1 \rangle = \langle 2\lambda x, 2\lambda y, 2\lambda z, 2\lambda t \rangle$
so $\lambda = 1/(2x) = 1/(2y) = 1/(2z) = 1/(2t)$ and $x = y = z = t$. But $x^2 + y^2 + z^2 + t^2 = 1$, so the possible
points are $\left(\pm\frac{1}{2}, \pm\frac{1}{2}, \pm\frac{1}{2}, \pm\frac{1}{2}\right)$. Thus the maximum value of f is $f\left(\frac{1}{2}, \frac{1}{2}, \frac{1}{2}, \frac{1}{2}\right) = 2$ and the minimum value is
$f\left(-\frac{1}{2}, -\frac{1}{2}, -\frac{1}{2}, -\frac{1}{2}\right) = -2$.

13. $f(x, y, z) = x + 2y$, $g(x, y, z) = x + y + z = 1$, $h(x, y, z) = y^2 + z^2 = 4$ $\Rightarrow$ $\nabla f = \langle 1, 2, 0 \rangle$,
$\lambda \nabla g = \langle \lambda, \lambda, \lambda \rangle$ and $\mu \nabla h = \langle 0, 2\mu y, 2\mu z \rangle$. Then $1 = \lambda$, $2 = \lambda + 2\mu y$ and $0 = \lambda + 2\mu z$ so $\mu y = \frac{1}{2} = -\mu z$ or
$y = 1/(2\mu)$, $z = -1/(2\mu)$. Thus $x + y + z = 1$ implies $x = 1$ and $y^2 + z^2 = 4$ implies $\mu = \pm\frac{1}{2\sqrt{2}}$. Then the
possible points are $\left(1, \pm\sqrt{2}, \mp\sqrt{2}\right)$ and the maximum value is $f\left(1, \sqrt{2}, -\sqrt{2}\right) = 1 + 2\sqrt{2}$ and the minimum
value is $f\left(1, -\sqrt{2}, \sqrt{2}\right) = 1 - 2\sqrt{2}$.

15. $f(x, y, z) = yz + xy$, $g(x, y, z) = xy = 1$, $h(x, y, z) = y^2 + z^2 = 1$ $\Rightarrow$ $\nabla f = \langle y, x + z, y \rangle$,
$\lambda \nabla g = \langle \lambda y, \lambda x, 0 \rangle$, $\mu \nabla h = \langle 0, 2\mu y, 2\mu z \rangle$. Then $y = \lambda y$ implies $\lambda = 1$ [$y \neq 0$ since $g(x, y, z) = 1$],
$x + z = \lambda x + 2\mu y$ and $y = 2\mu z$. Thus $\mu = z/(2y) = y/(2z)$ or $y^2 = z^2$, and so $y^2 + z^2 = 1$ implies $y = \pm\frac{1}{\sqrt{2}}$,
$z = \pm\frac{1}{\sqrt{2}}$. Then $xy = 1$ implies $x = \pm\sqrt{2}$ and the possible points are
$\left(\pm\sqrt{2}, \pm\frac{1}{\sqrt{2}}, \frac{1}{\sqrt{2}}\right), \left(\pm\sqrt{2}, \pm\frac{1}{\sqrt{2}}, -\frac{1}{\sqrt{2}}\right)$. Hence the maximum of f subject to the constraints is
$f\left(\pm\sqrt{2}, \pm\frac{1}{\sqrt{2}}, \pm\frac{1}{\sqrt{2}}\right) = \frac{3}{2}$ and the minimum is $f\left(\pm\sqrt{2}, \pm\frac{1}{\sqrt{2}}, \mp\frac{1}{\sqrt{2}}\right) = \frac{1}{2}$.

Note: Since $xy = 1$ is one of the constraints we could have solved the problem by solving $f(y, z) = yz + 1$
subject to $y^2 + z^2 = 1$.

17. $f(x, y) = e^{-xy}$. For the interior of the region, we find the critical points: $f_x = -ye^{-xy}$, $f_y = -xe^{-xy}$, so the
only critical point is $(0, 0)$, and $f(0, 0) = 1$. For the boundary, we use Lagrange multipliers.
$g(x, y) = x^2 + 4y^2 = 1$ $\Rightarrow$ $\lambda \nabla g = \langle 2\lambda x, 8\lambda y \rangle$, so setting $\nabla f = \lambda \nabla g$ we get $-ye^{-xy} = 2\lambda x$ and
$-xe^{-xy} = 8\lambda y$. The first of these gives $e^{-xy} = -2\lambda x/y$, and then the second gives $-x(-2\lambda x/y) = 8\lambda y$ $\Rightarrow$
$x^2 = 4y^2$. Solving this last equation with the constraint $x^2 + 4y^2 = 1$ gives $x = \pm\frac{1}{\sqrt{2}}$ and $y = \pm\frac{1}{2\sqrt{2}}$. Now
$f\left(\pm\frac{1}{\sqrt{2}}, \mp\frac{1}{2\sqrt{2}}\right) = e^{1/4} \approx 1.284$ and $f\left(\pm\frac{1}{\sqrt{2}}, \pm\frac{1}{2\sqrt{2}}\right) = e^{-1/4} \approx 0.779$. The former are the maxima on the
region and the latter are the minima.

19. (a) The graphs of $f(x, y) = 3.7$ and $f(x, y) = 350$ seem to be
tangent to the circle, and so 3.7 and 350 are the approximate
minimum and maximum values of the function $f(x, y)$
subject to the constraint $(x - 3)^2 + (y - 3)^2 = 9$.

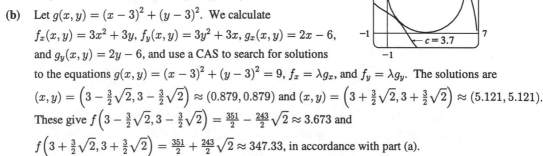

(b) Let $g(x, y) = (x - 3)^2 + (y - 3)^2$. We calculate
$f_x(x, y) = 3x^2 + 3y$, $f_y(x, y) = 3y^2 + 3x$, $g_x(x, y) = 2x - 6$,
and $g_y(x, y) = 2y - 6$, and use a CAS to search for solutions
to the equations $g(x, y) = (x - 3)^2 + (y - 3)^2 = 9$, $f_x = \lambda g_x$, and $f_y = \lambda g_y$. The solutions are
$(x, y) = \left(3 - \frac{3}{2}\sqrt{2}, 3 - \frac{3}{2}\sqrt{2}\right) \approx (0.879, 0.879)$ and $(x, y) = \left(3 + \frac{3}{2}\sqrt{2}, 3 + \frac{3}{2}\sqrt{2}\right) \approx (5.121, 5.121)$.
These give $f\left(3 - \frac{3}{2}\sqrt{2}, 3 - \frac{3}{2}\sqrt{2}\right) = \frac{351}{2} - \frac{243}{2}\sqrt{2} \approx 3.673$ and
$f\left(3 + \frac{3}{2}\sqrt{2}, 3 + \frac{3}{2}\sqrt{2}\right) = \frac{351}{2} + \frac{243}{2}\sqrt{2} \approx 347.33$, in accordance with part (a).

21. $Q(x,y) = Kx^\alpha y^{1-\alpha}$, $g(x,y) = mx + ny = p$ $\Rightarrow$ $\nabla Q = \langle \alpha Kx^{\alpha-1}y^{1-\alpha}, (1-\alpha)Kx^\alpha y^{-\alpha} \rangle$,
$\lambda \nabla g = \langle \lambda m, \lambda n \rangle$. Then $\alpha K(y/x)^{1-\alpha} = \lambda m$ and $(1-\alpha)K(x/y)^\alpha = \lambda n$ and $mx + ny = p$, so
$\alpha K(y/x)^{1-\alpha}/m = (1-\alpha)K(x/y)^\alpha/n$ or $n\alpha/[m(1-\alpha)] = (x/y)^\alpha (x/y)^{1-\alpha}$ or $x = yn\alpha/[m(1-\alpha)]$.
Substituting into $mx + ny = p$ gives $y = p(1-\alpha)/n$ and $x = p\alpha/m$ for the maximum production.

23. Let the sides of the rectangle be x and y. Then $f(x,y) = xy$, $g(x,y) = 2x + 2y = p$ $\Rightarrow$ $\nabla f(x,y) = \langle y, x \rangle$,
$\lambda \nabla g = \langle 2\lambda, 2\lambda \rangle$. Then $\lambda = \frac{1}{2}y = \frac{1}{2}x$ implies $x = y$ and the rectangle with maximum area is a square with side
length $\frac{1}{4}p$.

25. $f(x,y,z) = x^2 + y^2 + z^2$, $g(x,y,z) = x + 2y + 3z = 4$. Then $\nabla f = \langle 2x, 2y, 2z \rangle = \lambda \nabla g = \langle \lambda, 2\lambda, 3\lambda \rangle$ $\Rightarrow$
$x = \frac{1}{2}\lambda$, $y = \lambda$, $z = \frac{3}{2}\lambda$ and $\frac{1}{2}\lambda + 2\lambda + \frac{9}{2}\lambda = 4$ $\Rightarrow$ $\lambda = \frac{4}{7}$. Hence the point closest to the origin is
$\left(\frac{2}{7}, \frac{4}{7}, \frac{6}{7} \right)$.

27. $f(x,y,z) = (x-2)^2 + (y+2)^2 + (z-3)^2$, $g(x,y,z) = 6x + 4y - 3z = 2$ $\Rightarrow$
$\nabla f = \langle 2(x-2), 2(y+2), 2(z-3) \rangle = \lambda \nabla g = \langle 6\lambda, 4\lambda, -3\lambda \rangle$, so $x = 3\lambda + 2$, $y = 2\lambda$, $z = -\frac{3}{2}\lambda + 3$ and
$(18\lambda + 12) + (8\lambda - 8) + \frac{9}{2}\lambda - 9 = 2$ implies $\lambda = \frac{14}{61}$. Thus the shortest distance is
$\sqrt{\left(\frac{42}{61} \right)^2 + \left(\frac{28}{61} \right)^2 + \left(-\frac{21}{61} \right)^2} = \frac{7}{\sqrt{61}}$.

29. $f(x,y,z) = x^2 + y^2 + z^2$, $g(x,y,z) = z^2 - xy - 1 = 0$ $\Rightarrow$ $\nabla f = \langle 2x, 2y, 2z \rangle = \lambda \nabla g = \langle -\lambda y, -\lambda x, 2\lambda z \rangle$.
Then $2z = 2\lambda z$ implies $z = 0$ or $\lambda = 1$. If $z = 0$ then $g(x,y,z) = 1$ implies $xy = -1$ or $x = -1/y$. Thus
$2x = -\lambda y$ and $2y = -\lambda x$ imply $\lambda = 2/y^2 = 2y^2$ or $y = \pm 1$, $x = \pm 1$. If $\lambda = 1$, then $2x = -y$ and $2y = -x$
imply $x = y = 0$, so $z = \pm 1$. Hence the possible points are $(\pm 1, \mp 1, 0)$, $(0, 0, \pm 1)$ and the minimum value of f
is $f(0, 0, \pm 1) = 1$, so the points closest to the origin are $(0, 0, \pm 1)$.

31. $f(x,y,z) = xyz$, $g(x,y,z) = x + y + z = 100$ $\Rightarrow$ $\nabla f = \langle yz, xz, xy \rangle = \lambda \nabla g = \langle \lambda, \lambda, \lambda \rangle$. Then
$\lambda = yz = xz = xy$ implies $x = y = z = \frac{100}{3}$.

33. If the dimensions are $2x$, $2y$ and $2z$, then $f(x,y,z) = 8xyz$ and $g(x,y,z) = 9x^2 + 36y^2 + 4z^2 = 36$ $\Rightarrow$
$\nabla f = \langle 8yz, 8xz, 8xy \rangle = \lambda \nabla g = \langle 18\lambda x, 72\lambda y, 8\lambda z \rangle$. Thus $18\lambda x = 8yz$, $72\lambda y = 8xz$, $8\lambda z = 8xy$ so
$x^2 = 4y^2$, $z^2 = 9y^2$ and $36y^2 + 36y^2 + 36y^2 = 36$ or $y = \frac{1}{\sqrt{3}}$ $(y > 0)$. Thus the volume of the largest such
rectangle is $8 \left(\frac{1}{\sqrt{3}} \right) \left(\frac{2}{\sqrt{3}} \right) \left(\frac{3}{\sqrt{3}} \right) = 16\sqrt{3}$.

35. $f(x,y,z) = xyz$, $g(x,y,z) = x + 2y + 3z = 6$ $\Rightarrow$ $\nabla f = \langle yz, xz, xy \rangle = \lambda \nabla g = \langle \lambda, 2\lambda, 3\lambda \rangle$.
Then $\lambda = yz = \frac{1}{2}xz = \frac{1}{3}xy$ implies $x = 2y$, $z = \frac{2}{3}y$. But $2y + 2y + 2y = 6$ so $y = 1$, $x = 2$, $z = \frac{2}{3}$ and the
volume is $V = \frac{4}{3}$.

37. $f(x,y,z) = xyz$, $g(x,y,z) = 4(x + y + z) = c$ $\Rightarrow$ $\nabla f = \langle yz, xz, xy \rangle$, $\lambda \nabla g = \langle 4\lambda, 4\lambda, 4\lambda \rangle$. Thus
$4\lambda = yz = xz = xy$ or $x = y = z = \frac{1}{12}c$ are the dimensions giving the maximum volume.

39. $f(x,y,z) = xy + 2xz + 2yz$, $g(x,y,z) = xyz = 32,000 \text{ cm}^3$ $\Rightarrow$
$\nabla f = \langle 2z + y, 2z + x, 2(x+y) \rangle = \lambda \nabla g = \langle \lambda yz, \lambda xz, \lambda xy \rangle$. Then (1) $\lambda yz = 2z + y$, (2) $\lambda xz = 2z + x$,
and (3) $\lambda xy = 2(x+y)$. Now (1) $-$ (2) implies $\lambda z(y - x) = y - x$, so $x = y$ or $\lambda = 1/z$. If $\lambda = 1/z$ then (1)
implies $z = 0$ which can't be, so $x = y$. But twice (2) minus (3) together with $x = y$ implies
$\lambda y(2x - y) = (4z + 2y) - 4y$ or $\lambda y(2z - y) = 2(2z - y)$ so $z = y/2$ or $\lambda = 2/y$. If $\lambda = 2/y$ then (3) implies
$y = 0$ which can't be. Thus $x = y = 2z$ and $\frac{1}{2}y^3 = 32,000$ or $y = 40$ and the dimensions which minimize the
volume are $x = y = 40 \text{ cm}$, $z = 20 \text{ cm}$.

41. We need to find the extrema of $f(x, y, z) = x^2 + y^2 + z^2$ subject to the two constraints

$g(x, y, z) = x + y + 2z = 2$ and $h(x, y, z) = x^2 + y^2 - z = 0$. $\nabla f = \langle 2x, 2y, 2z \rangle$, $\lambda \nabla g = \langle \lambda, \lambda, 2\lambda \rangle$ and

$\mu \nabla h = \langle 2\mu x, 2\mu y, -\mu \rangle$. Thus we need (1) $2x = \lambda + 2\mu x$, (2) $2y = \lambda + 2\mu y$, (3) $2z = 2\lambda - \mu$,

(4) $x + y + 2z = 2$, and (5) $x^2 + y^2 - z = 0$. From (1) and (2), $2(x - y) = 2\mu(x - y)$, so if $x \neq y$, $\mu = 1$.

Putting this in (3) gives $2z = 2\lambda - 1$ or $\lambda = z + \frac{1}{2}$, but putting $\mu = 1$ into (1) says $\lambda = 0$. Hence $z + \frac{1}{2} = 0$ or

$z = -\frac{1}{2}$. Then (4) and (5) become $x + y - 3 = 0$ and $x^2 + y^2 + \frac{1}{2} = 0$. The last equation cannot be true, so this

case gives no solution. So we must have $x = y$. Then (4) and (5) become $2x + 2z = 2$ and $2x^2 - z = 0$ which

imply $z = 1 - x$ and $z = 2x^2$. Thus $2x^2 = 1 - x$ or $2x^2 + x - 1 = (2x - 1)(x + 1) = 0$ so $x = \frac{1}{2}$ or $x = -1$.

The two points to check are $\left(\frac{1}{2}, \frac{1}{2}, \frac{1}{2}\right)$ and $(-1, -1, 2)$: $f\left(\frac{1}{2}, \frac{1}{2}, \frac{1}{2}\right) = \frac{3}{4}$ and $f(-1, -1, 2) = 6$. Thus $\left(\frac{1}{2}, \frac{1}{2}, \frac{1}{2}\right)$ is

the point on the ellipse nearest the origin and $(-1, -1, 2)$ is the one farthest from the origin.

REVIEW EXERCISES FOR CHAPTER 12

1. True. $f_y(a, b) = \lim\limits_{h \to 0} \dfrac{f(a, b + h) - f(a, b)}{h}$ from (12.3.3). Let $h = y - b$. As $h \to 0$, $y \to b$. Then by

substituting, we get $f_y(a, b) = \lim\limits_{y \to b} \dfrac{f(a, y) - f(a, b)}{y - b}$.

3. False. $f_{xy} = \partial^2 f / (\partial y \partial x)$.

5. False. See Example 3 in Section 12.2.

7. True. If f has a local minimum and f is differentiable at (a, b) then by (12.7.2), $f_x(a, b) = 0$ and $f_y(a, b) = 0$ so

$\nabla f(a, b) = \langle f_x(a, b), f_y(a, b) \rangle = \langle 0, 0 \rangle = \mathbf{0}$.

9. False. $\nabla f(x, y) = \langle 0, 1/y \rangle$.

10. True. This is simply part (c) of the Second Derivative Test (12.7.3).

11. True. $\nabla f = \langle \cos x, \cos y \rangle$, so $|\nabla f| = \sqrt{\cos^2 x + \cos^2 y}$. But $|\cos \theta| \leq 1$, so $|\nabla f| \leq \sqrt{2}$. Now

$D_{\mathbf{u}} f(x, y) = \nabla f \cdot \mathbf{u} = |\nabla f||\mathbf{u}|\cos \theta$, but $\mathbf{u}$ is a unit vector, so $|D_{\mathbf{u}} f(x, y)| \leq \sqrt{2} \cdot 1 \cdot 1 = \sqrt{2}$.

13. $x \neq 1$ and $x + y + 1 > 0$, so
$D = \{(x, y) \mid y > -x - 1, x \neq 1\}$.

15. $D = \{(x, y) \mid -1 \leq x \leq 1\}$

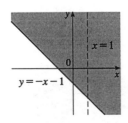

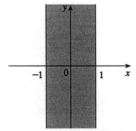

17. $z = f(x, y) = 1 - x^2 - y^2$,

a paraboloid with vertex $(0, 0, 1)$.

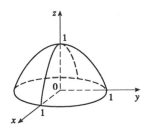

19. Let $k = e^{-c} = e^{-(x^2+y^2)}$ be the

level curves, then $-\ln k = c = x^2 + y^2$,

so we have a family of concentric circles.

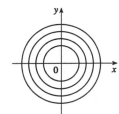

21. Since $0 \le \dfrac{x^2y^2}{x^2 + 2y^2} \le \dfrac{x^2y^2}{x^2 + y^2} \le \dfrac{(x^2 + y^2)(x^2 + y^2)}{x^2 + y^2} = x^2 + y^2$, given $\epsilon > 0$, let $\delta = \sqrt{\epsilon}$. Then whenever

$0 < \sqrt{x^2 + y^2} < \delta$, $\left| \dfrac{x^2y^2}{x^2 + 2y^2} - 0 \right| = \dfrac{x^2y^2}{x^2 + 2y^2} \le x^2 + y^2 < \delta^2 = \epsilon$. Hence $\displaystyle\lim_{(x,y)\to(0,0)} \dfrac{x^2y^2}{x^2 + 2y^2} = 0$.

(Or use the Squeeze Theorem.)

23. $f(x, y) = 3x^4 - x\sqrt{y} \quad \Rightarrow \quad f_x = 12x^3 - \sqrt{y},\ f_y = -\frac{1}{2}xy^{-1/2}$

25. $f(s, t) = e^{2s} \cos \pi t \quad \Rightarrow \quad f_s = 2e^{2s} \cos \pi t,\ f_t = -\pi e^{2s} \sin \pi t$

27. $f(x, y, z) = xy^z \quad \Rightarrow \quad f_x = y^z,\ f_y = xzy^{z-1},\ f_z = xy^z \ln y$

29. $f(x, y) = x^2y^3 - 2x^4 + y^2 \quad \Rightarrow \quad f_x = 2xy^3 - 8x^3,\ f_y = 3x^2y^2 + 2y,\ f_{xx} = 2y^3 - 24x^2,\ f_{yy} = 6x^2y + 2$, and

$f_{xy} = f_{yx} = 6xy^2$.

31. $f(x, y, z) = xy^2z^3 \quad \Rightarrow \quad f_x = y^2z^3,\ f_y = 2xyz^3,\ f_z = 3xy^2z^2,\ f_{xx} = 0,\ f_{yy} = 2xz^3,\ f_{zz} = 6xy^2z$,

$f_{xy} = f_{yx} = 2yz^3,\ f_{xz} = f_{zx} = 3y^2z^2$, and $f_{yz} = f_{zy} = 6xyz^2$.

33. $u = x^y \quad \Rightarrow \quad u_x = yx^{y-1},\ u_y = x^y \ln x$ and $(x/y)u_x + (\ln x)^{-1}u_y = x^y + x^y = 2u$.

35. $z_x(0, 1) = 0,\ z_y(0, 1) = 6$ and the equation of the tangent plane is $z - 5 = 6(y - 1)$ or $z - 6y = -1$.

37. $F(x, y, z) = xy^2z^3,\ \nabla F = \langle y^2z^3, 2xyz^3, 3xy^2z^2 \rangle$ and $\nabla F(3, 2, 1) = \langle 4, 12, 36 \rangle$. Thus the equation of the

tangent plane is $4x + 12y + 36z = 72$ or $x + 3y + 9z = 18$.

39. $F(x, y, z) = x^2 + 2y^2 - 3z^2,\ F_x = 2x,\ F_y = 4y,\ F_z = -6z;\ F_x(3, 2, -1) = 6,\ F_y(3, 2, -1) = 8$,

$F_z(3, 2, -1) = 6$. So the equation of the tangent plane is $6(x - 3) + 8(y - 2) + 6(z + 1) = 0$ or

$3x + 4y + 3z = 14$.

41. $F(x, y, z) = x^2 + y^2 + z^2,\ \nabla F(x_0, y_0, z_0) = \langle 2x_0, 2y_0, 2z_0 \rangle = k\langle 2, 1, -3 \rangle$ or $x_0 = k,\ y_0 = \frac{1}{2}k$ and $z_0 = -\frac{3}{2}k$.

But $x_0^2 + y_0^2 + z_0^2 = 1$, so $\frac{7}{2}k^2 = 1$ and $k = +\sqrt{\frac{2}{7}}$. Hence there are two such points: $\left(\pm\sqrt{\frac{2}{7}}, \pm\frac{1}{\sqrt{14}}, \mp\frac{3}{\sqrt{14}} \right)$.

43. Let $w = f(x, y, z) = x^3\sqrt{y^2 + z^2}$, then $f(2, 3, 4) = 8(5) = 40$ so set $(a, b, c) = (2, 3, 4)$. Then $\Delta x = -0.02$,

$\Delta y = 0.01,\ \Delta z = -0.03,\ f_x = 3x^2\sqrt{y^2 + z^2},\ f_y = yx^3/\sqrt{y^2 + z^2},\ f_z = zx^3/\sqrt{y^2 + z^2}$. Thus

$(1.98)^3\sqrt{(3.01)^2 + (3.97)^2} \approx 40 + (60)(-0.02) + (24/5)(0.01) + (32/5)(-0.03) = 38.656$.

45. $\dfrac{dw}{dt} = \dfrac{1}{2\sqrt{x}}(2e^{2t}) + \dfrac{2y}{z}(3t^2+4) + \dfrac{-y^2}{z^2}(2t) = e^t + \dfrac{2y}{z}(3t^2+4) - 2t\dfrac{y^2}{z^2}$

47. $\dfrac{\partial z}{\partial x} = 2xf'(x^2 - y^2)$, $\dfrac{\partial z}{\partial y} = 1 - 2yf'(x^2 - y^2)$ $\left[\text{where } f' = \dfrac{df}{d(x^2 - y^2)}\right]$. Then

$y\dfrac{\partial z}{\partial x} + x\dfrac{\partial z}{\partial y} = 2xyf'(x^2 - y^2) + x - 2xyf'(x^2 - y^2) = x.$

49. $\dfrac{\partial z}{\partial x} = \dfrac{\partial z}{\partial u}y + \dfrac{\partial z}{\partial v}\dfrac{-y}{x^2}$ and

$\dfrac{\partial^2 z}{\partial x^2} = y\dfrac{\partial}{\partial x}\left(\dfrac{\partial z}{\partial u}\right) + \dfrac{2y}{x^3}\dfrac{\partial z}{\partial v} + \dfrac{-y}{x^2}\dfrac{\partial}{\partial x}\left(\dfrac{\partial z}{\partial v}\right)$

$\qquad = \dfrac{2y}{x^3}\dfrac{\partial z}{\partial v} + y\left(\dfrac{\partial^2 z}{\partial u^2}y + \dfrac{\partial^2 z}{\partial v\partial u}\dfrac{-y}{x^2}\right) + \dfrac{-y}{x^2}\left(\dfrac{\partial^2 z}{\partial v^2}\dfrac{-y}{x^2} + \dfrac{\partial^2 z}{\partial u\partial v}y\right)$

$\qquad = \dfrac{2y}{x^3}\dfrac{\partial z}{\partial v} + y^2\dfrac{\partial^2 z}{\partial u^2} - \dfrac{2y^2}{x^2}\dfrac{\partial^2 z}{\partial u\partial v} + \dfrac{y^2}{x^4}\dfrac{\partial^2 z}{\partial v^2}$. Also $\dfrac{\partial z}{\partial y} = x\dfrac{\partial z}{\partial u} + \dfrac{1}{x}\dfrac{\partial z}{\partial v}$ and

$\dfrac{\partial^2 z}{\partial y^2} = x\dfrac{\partial}{\partial y}\left(\dfrac{\partial z}{\partial u}\right) + \dfrac{1}{x}\dfrac{\partial}{\partial y}\left(\dfrac{\partial z}{\partial v}\right) = x\left(\dfrac{\partial^2 z}{\partial u^2}x + \dfrac{\partial^2 z}{\partial v\partial u}\dfrac{1}{x}\right) + \dfrac{1}{x}\left(\dfrac{\partial^2 z}{\partial v^2}\dfrac{1}{x} + \dfrac{\partial^2 z}{\partial u\partial v}x\right)$

$\qquad = x^2\dfrac{\partial^2 z}{\partial u^2} + 2\dfrac{\partial^2 z}{\partial u\partial v} + \dfrac{1}{x^2}\dfrac{\partial^2 z}{\partial v^2}$. Thus

$x^2\dfrac{\partial^2 z}{\partial x^2} - y^2\dfrac{\partial^2 z}{\partial y^2} = \dfrac{2y}{x}\dfrac{\partial z}{\partial v} + x^2y^2\dfrac{\partial^2 z}{\partial u^2} - 2y^2\dfrac{\partial^2 z}{\partial u\partial v} + \dfrac{y^2}{x^2}\dfrac{\partial^2 z}{\partial v^2} - x^2y^2\dfrac{\partial^2 z}{\partial u^2} - 2y^2\dfrac{\partial^2 z}{\partial u\partial v} - \dfrac{y^2}{x^2}\dfrac{\partial^2 z}{\partial v^2}$

$\qquad = \dfrac{2y}{x}\dfrac{\partial z}{\partial v} - 4y^2\dfrac{\partial^2 z}{\partial u\partial v} = 2v\dfrac{\partial z}{\partial v} - 4uv\dfrac{\partial^2 z}{\partial u\partial v}$ since $y = xv = \dfrac{uv}{y}$ or $y^2 = uv.$

51. $\nabla f = \left\langle z^2\sqrt{y}e^{x\sqrt{y}}, \dfrac{xz^2e^{x\sqrt{y}}}{2\sqrt{y}}, 2ze^{x\sqrt{y}}\right\rangle = ze^{x\sqrt{y}}\left\langle z\sqrt{y}, \dfrac{xz}{2\sqrt{y}}, 2\right\rangle$

53. $\nabla f = \langle 1/\sqrt{x}, -2y\rangle$, $\nabla f(1,5) = \langle 1, -10\rangle$, $\mathbf{u} = \frac{1}{5}\langle 3, -4\rangle$. Then $D_{\mathbf{u}}f(1,5) = \frac{43}{5}.$

55. $\nabla f = \langle 2xy, x^2 + 1/(2\sqrt{y})\rangle$, $|\nabla f(2,1)| = |\langle 4, \frac{9}{2}\rangle|$. Thus the maximum rate of change of f at $(2,1)$ is $\frac{\sqrt{145}}{2}$ in the direction $\langle 4, \frac{9}{2}\rangle$.

57. $\dfrac{x^2 + y^2}{(x-1)^2 + y^2} \to \infty$ as $(x,y) \to (1,0)$ and so $\displaystyle\lim_{(x,y)\to(1,0)} \tan^{-1}\dfrac{x^2+y^2}{(x-1)^2+y^2} = \dfrac{\pi}{2}.$

59. $f(x,y) = x^2 - xy + y^2 + 9x - 6y + 10 \quad\Rightarrow$

$f_x = 2x - y + 9$, $f_y = -x + 2y - 6$,

$f_{xx} = 2 = f_{yy}$, $f_{xy} = -1$. Then $f_x = 0$

and $f_y = 0$ imply $y = 1$, $x = -4$. Thus

the only critical point is $(-4, 1)$ and

$f_{xx}(-4,1) > 0$, $D(-4,1) = 3 > 0$ so

$f(-4,1) = -9$ is a local minimum.

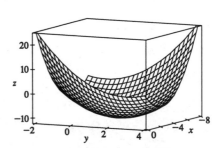

61. $f(x, y) = 3xy - x^2y - xy^2 \quad \Rightarrow \quad f_x = 3y - 2xy - y^2,$

$f_y = 3x - x^2 - 2xy, \; f_{xx} = -2y, \; f_{yy} = -2x,$

$f_{xy} = 3 - 2x - 2y.$ Then $f_x = 0$ implies $y(3 - 2x - y) = 0$

so $y = 0$ or $y = 3 - 2x.$ Substituting into $f_y = 0$ implies

$x(3 - x) = 0$ or $3x(-1 + x) = 0.$ Hence the critical points

are $(0, 0), (3, 0), (0, 3)$ and $(1, 1).$

$D(0, 0) = D(3, 0) = D(0, 3) = -9 < 0$ so $(0, 0), (3, 0),$

and $(0, 3)$ are saddle points. $D(1, 1) = 3 > 0$ and

$f_{xx}(1, 1) = -2 < 0,$ so $f(1, 1) = 1$ is a local maximum.

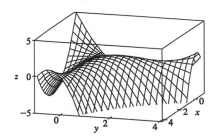

63. First solve inside $D.$ Here $f_x = 4y^2 - 2xy^2 - y^3,$

$f_y = 8xy - 2x^2y - 3xy^2.$ Then $f_x = 0$ implies $y = 0$ or

$y = 4 - 2x,$ but $y = 0$ isn't inside $D.$ Substituting

$y = 4 - 2x$ into $f_y = 0$ implies $x = 0, x = 2$ or $x = 1,$

but $x = 0$ isn't inside $D,$ and when $x = 2, y = 0$ but

$(2, 0)$ isn't inside $D.$ Thus the only critical point inside D is

$(1, 2)$ and $f(1, 2) = 4.$ Secondly we consider the boundary of $D.$

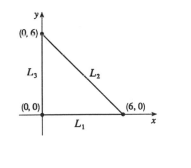

On $L_1, f(x, 0) = 0$ and so $f = 0$ on $L_1.$ On $L_2, x = -y + 6$ and

$f(-y + 6, y) = y^2(6 - y)(-2) = -2(6y^2 - y^3)$ which has critical points at $y = 0$ and $y = 4.$ Then $f(6, 0) = 0$

while $f(2, 4) = -64.$ On $L_3, f(0, y) = 0,$ so $f = 0$ on $L_3.$ Thus on D the absolute maximum of f is

$f(1, 2) = 4$ while the absolute minimum is $f(2, 4) = -64.$

65. $f(x, y) = x^3 - 3x + y^4 - 2y^2$

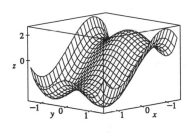
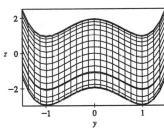
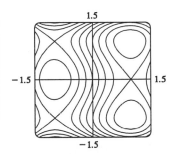

From the graphs, it appears that f has a local maximum $f(-1, 0) \approx 2,$ local minima $f(1, \pm 1) \approx -3,$ and saddle

points at $(-1, \pm 1)$ and $(1, 0).$

To find the exact quantities, we calculate $f_x = 3x^2 - 3 = 0 \quad \Leftrightarrow \quad x = \pm 1$ and $f_y = 4y^3 - 4y = 0 \quad \Leftrightarrow$

$y = 0, \pm 1,$ giving the critical points estimated above. Also $f_{xx} = 6x, f_{xy} = 0, f_{yy} = 12y^2 - 4,$ so using the

Second Derivatives Test, $D(-1, 0) = 24 > 0$ and $f_{xx}(-1, 0) = -6 < 0$ indicating a local maximum

$f(-1, 0) \approx 2; D(1, \pm 1) = 48 > 0$ and $f_{xx}(1, \pm 1) = 6 > 0$ indicating local minima $f(1, \pm 1) = -3;$ and

$D(-1, \pm 1) = -48$ and $D(1, 0) = -24,$ indicating saddle points.

67. $f(x,y) = x^2y$, $g(x,y) = x^2 + y^2 = 1$ $\Rightarrow$ $\nabla f = \langle 2xy, x^2 \rangle = \lambda \nabla g = \langle 2\lambda x, 2\lambda y \rangle$. Then $2xy = 2\lambda x$ and

$x^2 = 2\lambda y$ imply $\lambda = x^2/(2y)$ and $\lambda = y$ if $x \neq 0$ and $y \neq 0$. Hence $x^2 = 2y^2$. Then $x^2 + y^2 = 1$ implies

$3y^2 = 1$ so $y = \pm\frac{1}{\sqrt{3}}$ and $x = \pm\sqrt{\frac{2}{3}}$. [Note if $x = 0$ then $x^2 = 2\lambda y$ implies $y = 0$ and $f(0,0) = 0$.] Thus the

possible points are $\left(\pm\sqrt{\frac{2}{3}}, \pm\frac{1}{\sqrt{3}}\right)$ and the absolute maxima are $f\left(\pm\sqrt{\frac{2}{3}}, \frac{1}{\sqrt{3}}\right) = \frac{2}{3\sqrt{3}}$ while the absolute minima

are $f\left(\pm\sqrt{\frac{2}{3}}, -\frac{1}{\sqrt{3}}\right) = -\frac{2}{3\sqrt{3}}$.

69. $f(x,y,z) = x + y + z$, $g(x,y,z) = 1/x + 1/y + 1/z = 1$ $\Rightarrow$

$\nabla f = \langle 1,1,1 \rangle = \lambda \nabla g = \langle -\lambda x^{-2}, -\lambda y^{-2}, -\lambda z^{-2} \rangle$. Thus $\lambda = -x^2 = -y^2 = -z^2$ or $y = \pm x$, $z = \pm x$.

Substituting into $1/x + 1/y + 1/z = 1$ gives (1) $3/x = 1$ so $x = 3$, or (2) $1/x = 1$ so $x = 1$, or

(3) $-1/x = 1$ so $x = -1$ with the associated points (1) $(3,3,3)$, (2) $(1,1,-1)$ or $(1,-1,1)$, (3) $(-1,1,1)$.

Thus the absolute maximum is $f(3,3,3) = 9$ and the absolute minimum is

$f(1,1,-1) = f(1,-1,1) = f(-1,1,1) = 1$.

71. $f(x,y,z) = x^2 + y^2 + z^2$, $g(x,y,z) = xy^2z^3 = 2$, $\nabla f = \langle 2x, 2y, 2z \rangle = \lambda \nabla g = \langle \lambda y^2z^3, 2\lambda xyz^3, 3\lambda xy^2z^2 \rangle$.

Since $g(x,y,z) = 2$, $x \neq 0$, $y \neq 0$ and $z \neq 0$, so (1) $2x = \lambda y^2z^3$, (2) $1 = \lambda xz^3$, (3) $2 = 3\lambda xy^2z$. Then (2)

and (3) imply $\dfrac{1}{xz^3} = \dfrac{2}{3xy^2z}$ or $y^2 = \frac{2}{3}z^2$ so $y = \pm z\sqrt{\frac{2}{3}}$. Similarly (1) and (3) imply $\dfrac{2x}{y^2z^3} = \dfrac{2}{3xy^2z}$ or

$3x^2 = z^2$ so $x = \pm\frac{1}{\sqrt{3}}z$. But $xy^2z^3 = 2$ so x and z must have the same sign, that is, $x = \frac{1}{\sqrt{3}}z$. Thus

$g(x,y,z) = 2$ implies $\frac{1}{\sqrt{3}}z\left(\frac{2}{3}z^2\right)z^3 = 2$ or $z = \pm3^{1/4}$ and the possible points are $\left(\pm3^{-1/4}, 3^{-1/4}\sqrt{2}, \pm3^{1/4}\right)$,

$\left(\pm3^{-1/4}, -3^{-1/4}\sqrt{2}, \pm3^{1/4}\right)$. However at each of these points f takes on the same value, $2\sqrt{3}$. But $(2,1,1)$

also satisfies $g(x,y,z) = 2$ and $f(2,1,1) = 6 > 2\sqrt{3}$. Thus f has an absolute minimum value of $2\sqrt{3}$ and no

absolute maximum subject to the constraint $g(x,y,z) = 2$.

Alternate Solution: $g(x,y,z) = xy^2z^3 = 2$ implies $y^2 = \dfrac{2}{xz^3}$, so minimize $f(x,z) = x^2 + \dfrac{2}{xz^3} + z^2$. Then

$f_x = 2x - \dfrac{2}{x^2z^3}$, $f_z = -\dfrac{6}{xz^4} + 2z$, $f_{xx} = 2 + \dfrac{4}{x^3z^3}$, $f_{zz} = \dfrac{24}{xz^5} + 2$ and $f_{xz} = \dfrac{6}{x^2z^4}$. Now $f_x = 0$ implies

$2x^3z^3 - 2 = 0$ or $z = 1/x$. Substituting into $f_y = 0$ implies $-6x^3 + 2x^{-1} = 0$ or $x = \sqrt[4]{\frac{1}{3}}$, so the two critical

points are $\left(\pm\sqrt[4]{\frac{1}{3}}, \pm\sqrt[4]{3}\right)$. Then $D\left(\pm\sqrt[4]{\frac{1}{3}}, \pm\sqrt[4]{3}\right) = (2+4)\left(2 + \frac{24}{3}\right) - \left(\frac{6}{\sqrt{3}}\right)^2 > 0$ and

$f_{xx}\left(\pm\sqrt[4]{\frac{1}{3}}, \pm\sqrt[4]{3}\right) = 6 > 0$, so each point is a minimum. Finally, $y^2 = \dfrac{2}{xz^3}$, so the four points closest to the

origin are $\left(\pm\sqrt[4]{\frac{1}{3}}, \frac{\sqrt{2}}{\sqrt[4]{3}}, \pm\sqrt[4]{3}\right)$, $\left(\pm\sqrt[4]{\frac{1}{3}}, -\frac{\sqrt{2}}{\sqrt[4]{3}}, \pm\sqrt[4]{3}\right)$.

73.

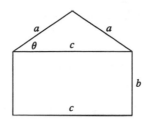

The area of the triangle is $\frac{1}{2}ca\sin\theta$ and the area of the rectangle is bc. Thus, the area of the whole object is

$f(a,b,c) = \frac{1}{2}ca\sin\theta + bc$. The perimeter of the object is $g(a,b,c) = 2a + 2b + c = P$. To simplify $\sin\theta$ in

terms of a, b, and c notice that $a^2\sin^2\theta + \left(\frac{1}{2}c\right)^2 = a^2$ $\Rightarrow$ $\sin\theta = \dfrac{1}{2a}\sqrt{4a^2 - c^2}$. Thus

$f(a,b,c) - \dfrac{c}{4}\sqrt{4a^2 - c^2} + bc$. (Instead of using θ, we could just have used the Pythagorean Theorem.) As a

result, by Lagrange's method, we must find a, b, c, and λ by solving $\nabla f = \lambda\nabla g$ which gives the following

equations: (1) $ca(4a^2 - c^2)^{-1/2} = 2\lambda$, (2) $c = 2\lambda$, (3) $\frac{1}{4}(4a^2 - c^2)^{1/2} - \frac{1}{4}c^2(4a^2 - c^2)^{-1/2} + b = \lambda$, and

(4) $2a + 2b + c = P$. From (2), $\lambda = \frac{1}{2}c$ and so (1) produces $ca(4a^2 - c^2)^{-1/2} = c$ $\Rightarrow$ $(4a^2 - c^2)^{1/2} = a$

$\Rightarrow$ $4a^2 - c^2 = a^2$ $\Rightarrow$ (5) $c = \sqrt{3}a$. Similarly, since $(4a^2 - c^2)^{1/2} = a$ and $\lambda = \frac{1}{2}c$, (3) gives

$\frac{1}{4}a - \dfrac{c^2}{4a} + b = \dfrac{c}{2}$, so from (5), $\dfrac{a}{4} - \dfrac{3a}{4} + b = \dfrac{\sqrt{3}a}{2}$ $\Rightarrow$ $-\dfrac{a}{2} - \dfrac{\sqrt{3}a}{2} = -b$ $\Rightarrow$ (6) $b = \dfrac{a}{2}\left(1 + \sqrt{3}\right)$.

Substituting (5) and (6) into (4) we get: $2a + a\left(1 + \sqrt{3}\right) + \sqrt{3}a = P$ $\Rightarrow$ $3a + 2\sqrt{3}a = P$ $\Rightarrow$

$a = \dfrac{P}{3 + 2\sqrt{3}} = \dfrac{2\sqrt{3} - 3}{3}P$ and thus $b = \dfrac{\left(2\sqrt{3} - 3\right)\left(1 + \sqrt{3}\right)}{6}P = \dfrac{3 - \sqrt{3}}{6}P$ and $c = \left(2 - \sqrt{3}\right)P$.

PROBLEMS PLUS (after Chapter 12)

1. Since three-dimensional situations are often difficult to visualize and work
   with, let us first try to find an analogous problem in two dimensions.
   The analogue of a cube is a square and the analogue of a sphere is a circle.
   Thus a similar problem in two dimensions is the following: if five circles
   with the same radius r are contained in a square of side 1 m so that
   the circles touch each other and four of the circles touch two sides of
   the square, find r.

   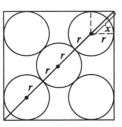

   The diagonal of the square is $\sqrt{2}$. The diagonal is also $4r + 2x$. But x is the diagonal of a smaller square of
   side r. Therefore $x = \sqrt{2}r \quad \Rightarrow \quad \sqrt{2} = 4r + 2x = 4r + 2\sqrt{2}r = \left(4 + 2\sqrt{2}\right)r \quad \Rightarrow \quad r = \dfrac{\sqrt{2}}{4 + 2\sqrt{2}}$.

   Let us use these ideas to solve the original three-dimensional problem. The diagonal of the cube is
   $\sqrt{1^2 + 1^2 + 1^2} = \sqrt{3}$. The diagonal of the cube is also $4r + 2x$ where x is the diagonal of a smaller cube with
   edge r. Therefore $x = \sqrt{r^2 + r^2 + r^2} = \sqrt{3}r \quad \Rightarrow \quad \sqrt{3} = 4r + 2x = 4r + 2\sqrt{3}r = \left(4 + 2\sqrt{3}\right)r$.

   Therefore $r = \dfrac{\sqrt{3}}{4 + 2\sqrt{3}} = \dfrac{2\sqrt{3} - 3}{2}$. The radius of each ball is $\left(\sqrt{3} - \tfrac{3}{2}\right)$ m.

3. We introduce a coordinate system, as shown. Recall that
   the area of the parallelogram spanned by two vectors is
   equal to the length of their cross product, so since

   $\mathbf{u} \times \mathbf{v} = \langle -q, r, 0 \rangle \times \langle -q, 0, p \rangle = \langle pr, pq, qr \rangle$, we have

   $|\mathbf{u} \times \mathbf{v}| = \sqrt{(pr)^2 + (pq)^2 + (qr)^2}$, and therefore

   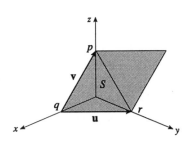

   $$D^2 = \left(\tfrac{1}{2}|\mathbf{u} \times \mathbf{v}|\right)^2 = \tfrac{1}{4}\left[(pr)^2 + (pq)^2 + (qr)^2\right]$$
   $$= \left(\tfrac{1}{2}pr\right)^2 + \left(\tfrac{1}{2}pq\right)^2 + \left(\tfrac{1}{2}qr\right)^2 = A^2 + B^2 + C^2.$$

   Another Method: We draw a line from S perpendicular
   to QR, as shown. Now $D = \tfrac{1}{2}ch$, so $D^2 = \tfrac{1}{4}c^2h^2$.

   Substituting $h^2 = p^2 + k^2$, we get

   $D^2 = \tfrac{1}{4}c^2(p^2 + k^2) = \tfrac{1}{4}c^2p^2 + \tfrac{1}{4}c^2k^2$. But $C = \tfrac{1}{2}ck$, so

   $D^2 = \tfrac{1}{4}c^2p^2 + C^2$. Now substituting $c^2 = q^2 + r^2$ gives

   $D^2 = \tfrac{1}{4}p^2q^2 + \tfrac{1}{4}q^2r^2 + C^2 = A^2 + B^2 + C^2.$

   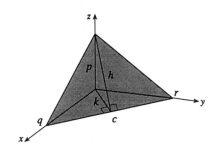

5.

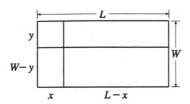

The areas of the smaller rectangles are $A_1 = xy$, $A_2 = (L-x)y$, $A_3 = (L-x)(W-y)$, $A_4 = x(W-y)$. For $0 \le x \le L$, $0 \le y \le W$, let

$$f(x,y) = A_1^2 + A_2^2 + A_3^2 + A_4^2 = x^2y^2 + (L-x)^2y^2 + (L-x)^2(W-y)^2 + x^2(W-y)^2$$
$$= \left[x^2 + (L-x)^2 \right] \left[y^2 + (W-y)^2 \right].$$

Then we need to find the maximum and minimum values of $f(x,y)$. Here

$$f_x(x,y) = [2x - 2(L-x)]\left[y^2 + (W-y)^2 \right] = 0 \quad \Rightarrow \quad 4x - 2L = 0 \text{ or } x = \tfrac{1}{2}L, \text{ and}$$

$$f_y(x,y) = \left[x^2 + (L-x)^2 \right][2y - 2(W-y)] = 0 \quad \Rightarrow \quad 4y - 2W = 0 \text{ or } y = W/2. \text{ Also}$$

$$f_{xx} = 4\left[y^2 + (W-y)^2 \right], \ f_{yy} = 4\left[x^2 + (L-x)^2 \right], \text{ and } f_{xy} = (4x - 2L)(4y - 2W). \text{ Then}$$

$$D = 16\left[y^2 + (W-y)^2 \right]\left[x^2 + (L-x)^2 \right] - (4x - 2L)^2(4y - 2W)^2. \text{ Thus when } x = \tfrac{1}{2}L \text{ and } y = \tfrac{1}{2}W, \ D > 0$$

and $f_{xx} = 2W^2 > 0$. Thus a minimum of f occurs at $\left(\tfrac{1}{2}L, \tfrac{1}{2}W \right)$ and this minimum value is

$f\left(\tfrac{1}{2}L, \tfrac{1}{2}W \right) = \tfrac{1}{4}L^2W^2$. There are no other critical points, so the maximum must occur on the boundary. Now

along the width of the rectangle let $g(y) = f(0,y) = f(L,y) = L^2\left[y^2 + (W-y)^2 \right], 0 \le y \le W$. Then

$g'(y) = L^2[2y - 2(W-y)] = 0 \quad \Leftrightarrow \quad y = \tfrac{1}{2}W$. And $g\left(\tfrac{1}{2} \right) = \tfrac{1}{2}L^2W^2$. Checking the endpoints we get

$g(0) = g(W) = L^2W^2$. Along the length of the rectangle let

$h(x) = f(x,0) = f(x,W) = W^2\left[x^2 + (L-x)^2 \right], 0 \le x \le L$. By symmetry $h'(x) = 0 \quad \Leftrightarrow \quad x = \tfrac{1}{2}L$ and

$h\left(\tfrac{1}{2}L \right) = \tfrac{1}{2}L^2W^2$. At the endpoints we have $h(0) = h(L) = L^2W^2$. Therefore L^2W^2 is the maximum value of

f. This maximum value of f occurs when the "cutting" lines correspond to sides of the rectangle.

7. Let $g(x,y) = xf\left(\dfrac{y}{x} \right)$. Then $g_x(x,y) = f\left(\dfrac{y}{x} \right) + xf'\left(\dfrac{y}{x} \right)\left(-\dfrac{y}{x^2} \right) = f\left(\dfrac{y}{x} \right) - \dfrac{y}{x}f'\left(\dfrac{y}{x} \right)$ and

$g_y(x,y) = xf'\left(\dfrac{y}{x} \right)\left(\dfrac{1}{x} \right) = f'\left(\dfrac{y}{x} \right)$. Thus the tangent plane at (x_0, y_0, z_0) on the surface has equation

$$z - x_0f\left(\dfrac{y_0}{x_0} \right) = \left[f\left(\dfrac{y_0}{x_0} \right) - y_0x_0^{-1}f'\left(\dfrac{y_0}{x_0} \right) \right](x - x_0) + f'\left(\dfrac{y_0}{x_0} \right)(y - y_0) \quad \Rightarrow$$

$$\left[f\left(\dfrac{y_0}{x_0} \right) - y_0x_0^{-1}f'\left(\dfrac{y_0}{x_0} \right) \right]x + \left[f'\left(\dfrac{y_0}{x_0} \right) \right]y - z = 0. \text{ But any plane whose equation is of the form}$$

$ax + by + cz = 0$ passes through the origin. Thus the origin is the common point of intersection.

9. **(a)** $x = r\cos\theta$, $y = r\sin\theta$, $z = z$. Then $\dfrac{\partial u}{\partial r} = \dfrac{\partial u}{\partial x}\dfrac{\partial x}{\partial r} + \dfrac{\partial u}{\partial y}\dfrac{\partial y}{\partial r} + \dfrac{\partial u}{\partial z}\dfrac{\partial z}{\partial r} = \dfrac{\partial u}{\partial x}\cos\theta + \dfrac{\partial u}{\partial y}\sin\theta$ and

$$\frac{\partial^2 u}{\partial r^2} = \cos\theta\left[\frac{\partial^2 u}{\partial x^2}\frac{\partial x}{\partial r} + \frac{\partial^2 u}{\partial y\partial x}\frac{\partial y}{\partial r} + \frac{\partial^2 u}{\partial z\partial x}\frac{\partial z}{\partial r}\right] + \sin\theta\left[\frac{\partial^2 u}{\partial y^2}\frac{\partial y}{\partial r} + \frac{\partial^2 u}{\partial x\partial y}\frac{\partial x}{\partial r} + \frac{\partial^2 u}{\partial z\partial y}\frac{\partial z}{\partial r}\right]$$

$$= \frac{\partial^2 u}{\partial x^2}\cos^2\theta + \frac{\partial^2 u}{\partial y^2}\sin^2\theta + 2\frac{\partial^2 u}{\partial y\partial x}\cos\theta\sin\theta.$$

Similarly $\dfrac{\partial u}{\partial \theta} = -\dfrac{\partial u}{\partial x}r\sin\theta + \dfrac{\partial u}{\partial y}r\cos\theta$ and

$$\frac{\partial^2 u}{\partial \theta^2} = \frac{\partial^2 u}{\partial x^2}r^2\sin^2\theta + \frac{\partial^2 u}{\partial y^2}r^2\cos^2\theta - 2\frac{\partial^2 u}{\partial y\partial x}r^2\sin\theta\cos\theta - \frac{\partial u}{\partial x}r\cos\theta - \frac{\partial u}{\partial y}r\sin\theta. \text{ So}$$

$$\frac{\partial^2 u}{\partial r^2} + \frac{1}{r}\frac{\partial u}{\partial r} + \frac{1}{r^2}\frac{\partial^2 u}{\partial \theta^2} + \frac{\partial^2 u}{\partial z^2}$$

$$= \frac{\partial^2 u}{\partial x^2}\cos^2\theta + \frac{\partial^2 u}{\partial y^2}\sin^2\theta + 2\frac{\partial^2 u}{\partial y\partial x}\cos\theta\sin\theta + \frac{\partial u}{\partial x}\frac{\cos\theta}{r} + \frac{\partial u}{\partial y}\frac{\sin\theta}{r}$$

$$+ \frac{\partial^2 u}{\partial x^2}\sin^2\theta + \frac{\partial^2 u}{\partial y^2}\cos^2\theta - 2\frac{\partial^2 u}{\partial y\partial x}\sin\theta\cos\theta - \frac{\partial u}{\partial x}\frac{\cos\theta}{r} - \frac{\partial u}{\partial y}\frac{\sin\theta}{r} + \frac{\partial^2 u}{\partial z^2}$$

$$= \frac{\partial^2 u}{\partial x^2} + \frac{\partial^2 u}{\partial y^2} + \frac{\partial^2 u}{\partial z^2}.$$

(b) $x = \rho\sin\phi\cos\theta$, $y = \rho\sin\phi\sin\theta$, $z = \rho\cos\phi$. Then

$$\frac{\partial u}{\partial \rho} = \frac{\partial u}{\partial x}\frac{\partial x}{\partial \rho} + \frac{\partial u}{\partial y}\frac{\partial y}{\partial \rho} + \frac{\partial u}{\partial z}\frac{\partial z}{\partial \rho} = \frac{\partial u}{\partial x}\sin\phi\cos\theta + \frac{\partial u}{\partial y}\sin\phi\sin\theta + \frac{\partial u}{\partial z}\cos\phi, \text{ and}$$

$$\frac{\partial^2 u}{\partial \rho^2} = \sin\phi\cos\theta\left[\frac{\partial^2 u}{\partial x^2}\frac{\partial x}{\partial \rho} + \frac{\partial^2 u}{\partial y\partial x}\frac{\partial y}{\partial \rho} + \frac{\partial^2 u}{\partial z\partial x}\frac{\partial z}{\partial \rho}\right] + \sin\phi\sin\theta\left[\frac{\partial^2 u}{\partial y^2}\frac{\partial y}{\partial \rho} + \frac{\partial^2 u}{\partial x\partial y}\frac{\partial x}{\partial \rho} + \frac{\partial^2 u}{\partial z\partial y}\frac{\partial z}{\partial \rho}\right]$$

$$+ \cos\phi\left[\frac{\partial^2 u}{\partial z^2}\frac{\partial z}{\partial \rho} + \frac{\partial^2 u}{\partial x\partial z}\frac{\partial x}{\partial \rho} + \frac{\partial^2 u}{\partial y\partial z}\frac{\partial y}{\partial \rho}\right]$$

$$= 2\frac{\partial^2 u}{\partial y\partial x}\sin^2\phi\sin\theta\cos\theta + 2\frac{\partial^2 u}{\partial z\partial x}\sin\phi\cos\phi\cos\theta + 2\frac{\partial^2 u}{\partial y\partial z}\sin\phi\cos\phi\sin\theta$$

$$+ \frac{\partial^2 u}{\partial x^2}\sin^2\phi\cos^2\theta + \frac{\partial^2 u}{\partial y^2}\sin^2\phi\sin^2\theta + \frac{\partial^2 u}{\partial z^2}\cos^2\phi.$$

Similarly $\dfrac{\partial u}{\partial \phi} = \dfrac{\partial u}{\partial x}\rho\cos\phi\cos\theta + \dfrac{\partial u}{\partial y}\rho\cos\phi\sin\theta - \dfrac{\partial u}{\partial z}\rho\sin\phi$, and

$$\frac{\partial^2 u}{\partial \phi^2} = 2\frac{\partial^2 u}{\partial y\partial x}\rho^2\cos^2\phi\sin\theta\cos\theta - 2\frac{\partial^2 u}{\partial x\partial z}\rho^2\sin\phi\cos\phi\cos\theta$$

$$- 2\frac{\partial^2 u}{\partial y\partial z}\rho^2\sin\phi\cos\phi\sin\theta + \frac{\partial^2 u}{\partial x^2}\rho^2\cos^2\phi\cos^2\theta + \frac{\partial^2 u}{\partial y^2}\rho^2\cos^2\phi\sin^2\theta$$

$$+ \frac{\partial^2 u}{\partial z^2}\rho^2\sin^2\phi - \frac{\partial u}{\partial x}\rho\sin\phi\cos\theta - \frac{\partial u}{\partial y}\rho\sin\phi\sin\theta - \frac{\partial u}{\partial z}\rho\cos\phi.$$

And $\dfrac{\partial u}{\partial \theta} = -\dfrac{\partial u}{\partial x}\rho\sin\phi\sin\theta + \dfrac{\partial u}{\partial y}\rho\sin\phi\cos\theta$, while

$$\frac{\partial^2 u}{\partial \theta^2} = -2\frac{\partial^2 u}{\partial y\partial x}\rho^2\sin^2\phi\cos\theta\sin\theta + \frac{\partial^2 u}{\partial x^2}\rho^2\sin^2\phi\sin^2\theta$$

$$+ \frac{\partial^2 u}{\partial y^2}\rho^2\sin^2\phi\cos^2\theta - \frac{\partial u}{\partial x}\rho\sin\phi\cos\theta - \frac{\partial u}{\partial y}\rho\sin\phi\sin\theta.$$

Therefore

$$-\frac{\partial^2 u}{\partial \rho^2} + \frac{2}{\rho}\frac{\partial u}{\partial \rho} + \frac{\cot \phi}{\rho^2}\frac{\partial u}{\partial \phi} + \frac{1}{\rho^2}\frac{\partial^2 u}{\partial \phi^2} + \frac{1}{\rho^2 \sin^2 \phi}\frac{\partial^2 u}{\partial \theta^2}$$

$$= \frac{\partial^2 u}{\partial x^2}\left[(\sin^2 \phi \cos^2 \theta) + (\cos^2 \phi \cos^2 \theta) + \sin^2 \theta\right]$$

$$+ \frac{\partial^2 u}{\partial y^2}\left[(\sin^2 \phi \sin^2 \theta) + (\cos^2 \phi \sin^2 \theta) + \cos^2 \theta\right] + \frac{\partial^2 u}{\partial z^2}\left[\cos^2 \phi + \sin^2 \phi\right]$$

$$+ \frac{\partial u}{\partial x}\left[\frac{2\sin^2 \phi \cos \theta + \cos^2 \phi \cos \theta - \sin^2 \phi \cos \theta - \cos \theta}{\rho \sin \phi}\right]$$

$$+ \frac{\partial u}{\partial y}\left[\frac{2\sin^2 \phi \sin \theta + \cos^2 \phi \sin \theta - \sin^2 \phi \sin \theta - \sin \theta}{\rho \sin \phi}\right].$$

But $2\sin^2 \phi \cos \theta + \cos^2 \phi \cos \theta - \sin^2 \phi \cos \theta - \cos \theta = (\sin^2 \phi + \cos^2 \phi - 1)\cos \theta = 0$ and similarly the

coefficient of $\partial u/\partial y$ is zero. Also

$\sin^2 \phi \cos^2 \theta + \cos^2 \phi \cos^2 \theta + \sin^2 \theta = \cos^2 \theta(\sin^2 \phi + \cos^2 \phi) + \sin^2 \theta = 1$, and similarly the coefficient of

$\partial^2 u/\partial y^2$ is 1. So Laplace's Equation in spherical coordinates is as stated.

11. At $(x_1, y_1, 0)$ the equations of the tangent planes to $z = f(x, y)$ and $z = g(x, y)$ are

$P_1: z - f(x_1, y_1) = f_x(x_1, y_1)(x - x_1) + f_y(x_1, y_1)(y - y_1)$ and

$P_2: z - g(x_1, y_1) = g_x(x_1, y_1)(x - x_1) + g_y(x_1, y_1)(y - y_1)$, respectively. P_1 intersects the xy-plane in the

line given by $f_x(x_1, y_1)(x - x_1) + f_y(x_1, y_1)(y - y_1) = -f(x_1, y_1)$, $z = 0$; and P_2 intersects the xy-plane in the

line given by $g_x(x_1, y_1)(x - x_1) + g_y(x_1, y_1)(y - y_1) = -g(x_1, y_1)$, $z = 0$. The point $(x_2, y_2, 0)$ is the point of

intersection of these two lines, since $(x_2, y_2, 0)$ is the point where the line of intersection of the two tangent

planes intersects the xy-plane. Thus (x_2, y_2) is the solution of the simultaneous equations

$f_x(x_1, y_1)(x_2 - x_1) + f_y(x_1, y_1)(y_2 - y_1) = -f(x_1, y_1)$ and

$g_x(x_1, y_1)(x_2 - x_1) + g_y(x_1, y_1)(y_2 - y_1) = -g(x_1, y_1)$.

For simplicity, rewrite $f_x(x_1, y_1)$ as f_x and similarly for f_y, g_x, g_y, f and g and solve the equations

$(f_x)(x_2 - x_1) + (f_y)(y_2 - y_1) = -f$ and $(g_x)(x_2 - x_1) + (g_y)(y_2 - y_1) = -g$ simultaneously for $(x_2 - x_1)$ and

$(y_2 - y_1)$. Then $y_2 - y_1 = \dfrac{gf_x - fg_x}{g_x f_y - f_x g_y}$ or $y_2 = y_1 - \dfrac{gf_x - fg_x}{f_x g_y - g_x f_y}$ and

$(f_x)(x_2 - x_1) + \dfrac{(f_y)(gf_x - fg_x)}{g_x f_y - f_x g_y} = -f$ so

$x_2 - x_1 = \dfrac{-f - [(f_y)(gf_x - fg_x)/(g_x f_y - f_x g_y)]}{f_x} = \dfrac{fg_y - f_y g}{g_x f_y - f_x g_y}$. Hence $x_2 = x_1 - \dfrac{fg_y - f_y g}{f_x g_y - g_x f_y}$.

PROBLEMS PLUS

13. Since we are minimizing the area of the ellipse, and the circle lies above the x-axis, the ellipse will intersect the circle for only one value of y. This y-value must satisfy both the equation of the circle and the equation of the ellipse. Now $\dfrac{x^2}{a^2} + \dfrac{y^2}{b^2} = 1 \;\Rightarrow\; x^2 = \dfrac{a^2}{b^2}(b^2 - y^2)$. Substituting

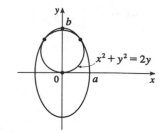

into the equation of the circle gives $\dfrac{a^2}{b^2}(b^2 - y^2) + y^2 - 2y = 0 \;\Rightarrow$

$\left(\dfrac{b^2 - a^2}{b^2}\right)y^2 - 2y + a^2 = 0$. In order for there to be only one solution to this quadratic equation, the

discriminant must be 0, so $4 - 4a^2\dfrac{b^2 - a^2}{b^2} = 0 \;\Rightarrow\; b^2 - a^2b^2 + a^4 = 0$. The area of the ellipse is

$A(a, b) = \pi ab$, and we minimize this function subject to the constraint $g(a, b) = b^2 - a^2b^2 + a^4 = 0$.

Now $\nabla A = \lambda \nabla g \;\Leftrightarrow\; \pi b = \lambda(4a^3 - 2ab^2), \; \pi a = \lambda(2b - 2ba^2) \;\Rightarrow\;$ (1) $\lambda = \dfrac{\pi b}{2a(2a^2 - b^2)}$,

(2) $\lambda = \dfrac{\pi a}{2b(1 - a^2)}$, (3) $b^2 - a^2b^2 + a^4 = 0$. Comparing (1) and (2) gives $\dfrac{\pi b}{2a(2a^2 - b^2)} = \dfrac{\pi a}{2b(1 - a^2)} \;\Rightarrow$

$2\pi b^2 = 4\pi a^4 \;\Leftrightarrow\; a^2 = \tfrac{1}{\sqrt{2}}b$. Substitute this into (3) to get $b = \dfrac{3}{\sqrt{2}} \;\Rightarrow\; a = \sqrt{\tfrac{3}{2}}$.

15. (a) We find the line of intersection L as in Example 11.5.7(b). Observe that the point $(-1, c, c)$ lies on both planes. Now since L lies in both planes, it is perpendicular to both of the normal vectors $\mathbf{n}_1$ and $\mathbf{n}_2$, and

thus parallel to their cross product $\mathbf{n}_1 \times \mathbf{n}_2 = \begin{vmatrix} \mathbf{i} & \mathbf{j} & \mathbf{k} \\ c & 1 & 1 \\ 1 & -c & c \end{vmatrix} = \langle 2c, -c^2 + 1, -c^2 - 1 \rangle$. So the symmetric

equations of L can be written as $\dfrac{x + 1}{-2c} = \dfrac{y - c}{c^2 - 1} = \dfrac{z - c}{c^2 + 1}$, provided that $c \neq 0, \pm 1$.

If $c = 0$, then the two planes are given by $y + z = 0$ and $x = -1$, so the symmetric equations of L are

$x = -1, y = -z$. If $c = -1$, then the two planes are given by $-x + y + z = -1$ and $x + y + z = -1$,

and they intersect in the line $x = 0, y = -z - 1$. If $c = 1$, then the two planes are given by $x + y + z = 1$

and $x - y + z = 1$, and they intersect in the line $y = 0, x = 1 - z$.

(b) If we set $z = t$ in the symmetric equations and solve for x and y separately, we get

$x + 1 = \dfrac{(t - c)(-2c)}{c^2 + 1}, \; y - c = \dfrac{(t - c)(c^2 - 1)}{c^2 + 1} \;\Rightarrow\; x = \dfrac{-2ct + (c^2 - 1)}{c^2 + 1}, \; y = \dfrac{(c^2 - 1)t + 2c}{c^2 + 1}$.

Eliminating c from these equations, we have $x^2 + y^2 = t^2 + 1$. So the curve traced out by L in the plane

$z = t$ is a circle with center at $(0, 0, t)$ and radius $\sqrt{t^2 + 1}$.

(c) The area of a horizontal cross-section of the solid is $A(z) = \pi(z^2 + 1)$, so

$V = \int_0^1 A(z)\,dz = \pi\left[\tfrac{1}{3}z^3 + z\right]_0^1 = \tfrac{4\pi}{3}$.

CHAPTER THIRTEEN

EXERCISES 13.1

1. **(a)** $\sum_{i=1}^{2}\sum_{j=1}^{2} f(x_{ij}^*, y_{ij}^*)\Delta A_{ij} = \frac{1}{2}\left[f\left(0,\frac{3}{2}\right) + f(0,2) + f\left(1,\frac{3}{2}\right) + f(1,2)\right]$

$$= \frac{1}{2}\left[\left(-\frac{27}{4}\right) + (-12) + \left(1 - \frac{27}{4}\right) + (1 - 12)\right] = \frac{1}{2}\left(-\frac{71}{2}\right) = -17.75$$

(b) $\frac{1}{2}\left[f\left(1,\frac{3}{2}\right) + f(1,2) + f\left(2,\frac{3}{2}\right) + f(2,2)\right] = \frac{1}{2}\left[-\frac{23}{4} + (-11) + \left(-\frac{19}{4}\right) + (-10)\right] = \frac{1}{2}\left(-\frac{63}{2}\right) = -15.75$

(c) $\frac{1}{2}\left[f(0,1) + f\left(0,\frac{3}{2}\right) + f(1,1) + f\left(1,\frac{3}{2}\right)\right] = \frac{1}{2}\left[-3 - \frac{27}{4} - 2 - \frac{23}{4}\right] = -8.75$

(d) $\frac{1}{2}\left[f(1,1) + f\left(1,\frac{3}{2}\right) + f(2,1) + f\left(2,\frac{3}{2}\right)\right] = \frac{1}{2}\left[-2 - \frac{23}{4} - 1 - \frac{19}{4}\right] = -6.75$

3.

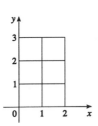

$\Delta A_{ij} = 1$ for $i = 1, 2, j = 1, 2, 3$.

$\iint_R(x^2 + 4y)dA \approx (1)\big[f(1,1) + f(1,2) + f(1,3)$

$\qquad\qquad\qquad\qquad + f(2,1) + f(2,2) + f(2,3)\big]$

$\qquad\qquad = 5 + 9 + 13 + 8 + 12 + 16 = 63$

$\|P\| = \sqrt{1+1} = \sqrt{2}$

5.

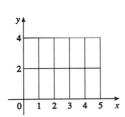

$\Delta A_{ij} = 2$ for $i = 1, 2, 3, 4, 5, j = 1, 2$.

$\iint_R(xy - y^2)dA$

$\approx (2)\big[f\left(\frac{1}{2},1\right) + f\left(\frac{1}{2},3\right) + f\left(\frac{3}{2},1\right) + f\left(\frac{3}{2},3\right) + f\left(\frac{5}{2},1\right)$

$\quad + f\left(\frac{5}{2},3\right) + f\left(\frac{7}{2},1\right) + f\left(\frac{7}{2},3\right) + f\left(\frac{9}{2},1\right) + f\left(\frac{9}{2},3\right)\big]$

$= (2)\big[-\frac{1}{2} + \left(-\frac{15}{2}\right) + \frac{1}{2} + \left(-\frac{9}{2}\right) + \frac{3}{2}$

$\qquad + \left(-\frac{3}{2}\right) + \frac{5}{2} + \frac{3}{2} + \frac{7}{2} + \frac{9}{2}\big] = 0$

and $\|P\| = \sqrt{1+4} = \sqrt{5}$.

7.

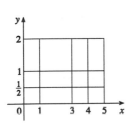

$\iint_R(x^2 - y^2)dA \approx \frac{1}{2}f\left(0,\frac{1}{2}\right) + \frac{1}{2}f(0,1) + 1f(0,2) + 1f\left(1,\frac{1}{2}\right)$

$\qquad + 1f(1,1) + 2f(1,2) + \frac{1}{2}f\left(3,\frac{1}{2}\right) + \frac{1}{2}f(3,1)$

$\qquad + 1f(3,2) + \frac{1}{2}f\left(4,\frac{1}{2}\right) + \frac{1}{2}f(4,1) + 1f(4,2)$

$= \frac{1}{2}\left(-\frac{1}{4}\right) + \frac{1}{2}(-1) + 1(-4) + 1\left(\frac{3}{4}\right) + 1(0) + 2(-3)$

$\qquad + \frac{1}{2}\left(\frac{35}{4}\right) + \frac{1}{2}(8) + (1)(5) + \frac{1}{2}\left(\frac{63}{4}\right) + \frac{1}{2}(15) + 1(12)$

$= \frac{1}{2}\left(\frac{185}{4}\right) + \frac{55}{4} - 6 = \frac{247}{8}$,

and the length of the longest diagonal is $\|P\| = \sqrt{1+4} = \sqrt{5}$.

9. The values of $f(x,y) = \sqrt{52 - x^2 - y^2}$ get smaller as we move further from the origin, so on any of the subrectangles in the problem, the function will have its largest value at the lower left corner of the subrectangle and its smallest value at the upper right corner, and any other value will lie between these two. So for this partition (and in fact for any partition) $U < V < L$.

11. To calculate the estimates using a programmable calculator, we can use an algorithm similar to that of Exercise 4.2.19 (5.2.19 in the Early Transcendentals version). In Maple, we can define the function $f(x,y) = e^{-x^2-y^2}$ (calling it f), load the student package, and then use the command $\texttt{middlesum(middlesum(f,x=0..1,m),y=0..1,m)}$; to get the estimate with $n = m^2$ squares of equal size. Mathematica has no special Riemann sum command, but we can define f and then use nested Sum commands to calculate the estimates.

n	estimate
1	0.6065
4	0.5694
16	0.5606
64	0.5585
256	0.5579
1024	0.5578

13. $z = f(x,y) = 4 - 2y \geq 0$ for $0 \leq y \leq 1$.

Thus the integral represents the volume of that part of the rectangular solid $[0,1] \times [0,1] \times [0,4]$ which lies below the plane $z = 4 - 2y$. So $\iint_R (4-y)\,dA = (1)(1)(2) + \frac{1}{2}(1)(1)(2) = 3$.

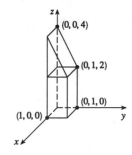

15. For any partition, $\iint_R k\,dA \approx \sum_i \sum_j f(x_{ij}^*, y_{ij}^*)\Delta A_{ij}$ but $f(x_{ij}^*, y_{ij}^*) = k$ always and $\sum_i \sum_j \Delta A_{ij} = $ area of $R = (b-a)(c-d)$. Thus for every partition $\sum_i \sum_j f(x_{ij}^*, y_{ij}^*)\Delta A_{ij} = k\sum_i \sum_j \Delta A_{ij} = k(b-a)(c-d)$ and so as $\|P\| \to 0$ the limit is $k(b-a)(c-d)$.

EXERCISES 13.2

1. $\int_0^2 x^2 y^3\,dy = x^2 \left[\frac{1}{4}y^4\right]_0^2 = 4x^2, \quad \int_0^1 x^2 y^3\,dx = y^3\left[\frac{1}{3}x^3\right]_0^1 = \frac{1}{3}y^3$

3. $\int_0^2 xe^{x+y}\,dy = xe^x[e^y]_0^2 = x(e^{x+2} - e^x) = xe^x(e^2-1), \quad \int_0^1 xe^{x+y}\,dx = e^y\int_0^1 xe^x\,dx = e^y[xe^x - e^x]_0^1 = e^y$

5. $\int_0^4 \int_0^2 x\sqrt{y}\,dx\,dy = \int_0^4 \sqrt{y}\left[\frac{1}{2}x^2\right]_0^2 dy = \int_0^4 2\sqrt{y}\,dy = \left[\frac{4}{3}y^{3/2}\right]_0^4 = \frac{32}{3}$

7. $\int_{-1}^1 \int_0^1 (x^3y^2 + 3xy^2)\,dy\,dx = \int_{-1}^1 \left[\frac{1}{3}x^3y^3 + xy^3\right]_{y=0}^{y=1} dx = \int_{-1}^1 \left[\frac{1}{3}x^3 + x\right]dx = \left[\frac{1}{16}x^4 + \frac{1}{2}x^2\right]_{-1}^1 = 0$

Alternate Solution: Applying Fubini's Theorem, the integral equals
$\int_0^1 \int_{-1}^1 (x^3y^2 + 3xy^2)\,dx\,dy = \int_0^1 \left[\frac{1}{4}y^2x^4 + \frac{3}{2}y^2x^2\right]_{x=-1}^{x=1} dy = \int_0^1 0\,dy = 0.$

9. $\int_0^3 \int_0^1 \sqrt{x+y}\,dx\,dy = \int_0^3 \left[\frac{2}{3}(x+y)^{3/2}\right]_{x=0}^{x=1} dy = \frac{2}{3}\int_0^3 \left[(1+y)^{3/2} - y^{3/2}\right]dy$

$= \frac{2}{3}\left[\frac{2}{5}(1+y)^{5/2} - \frac{2}{5}y^{5/2}\right]_0^3 = \frac{4}{15}[32 - 3^{5/2} - 1] = \frac{4}{15}\left(31 - 9\sqrt{3}\right)$

11. $\int_0^{\pi/4}\int_0^3 \sin x \, dy \, dx = 3\int_0^{\pi/4} \sin x \, dx = 3[-\cos x]_0^{\pi/4} = 3\left(1 - \frac{1}{\sqrt{2}}\right)$

13. $\int_0^{\ln 2}\int_0^{\ln 5} e^{2x-y} \, dx \, dy = \left(\int_0^{\ln 5} e^{2x} \, dx\right)\left(\int_0^{\ln 2} e^{-y} \, dy\right) = \left[\frac{1}{2}e^{2x}\right]_0^{\ln 5}\left[-e^{-y}\right]_0^{\ln 2} = \left(\frac{25}{2} - \frac{1}{2}\right)\left(-\frac{1}{2} + 1\right) = 6$

15. $\int_1^2\int_0^3 (2y^2 - 3xy^3)dy \, dx = \int_1^2 \left[\frac{2}{3}y^3 - \frac{3}{4}xy^4\right]_{y=0}^{y=3} dx = \int_1^2 \left(18 - \frac{243}{4}x\right)dx = \left[18x - \frac{243}{8}x^2\right]_1^2 = -\frac{585}{8}$

17. $\int_0^{\pi/6}\int_1^4 x \sin y \, dx \, dy = \left(\int_0^{\pi/6} \sin y \, dy\right)\left(\int_1^4 x \, dx\right) = \left(1 - \frac{\sqrt{3}}{2}\right)\frac{15}{2} = \frac{15(2-\sqrt{3})}{4}$

19. $\int_0^{\pi/6}\int_0^{\pi/3} x \sin(x+y)dy \, dx = \int_0^{\pi/6}[-x\cos(x+y)]_0^{\pi/3} \, dx = \int_0^{\pi/6}\left[x\cos x - x\cos\left(x + \frac{\pi}{3}\right)\right]dx$

$= x\left[\sin x - \sin\left(x + \frac{\pi}{3}\right)\right]_0^{\pi/6} - \int_0^{\pi/6}\left[\sin x - \sin\left(x + \frac{\pi}{3}\right)\right]dx$

$= \frac{\pi}{6}\left[\frac{1}{2} - 1\right] - \left[-\cos x + \cos\left(x + \frac{\pi}{3}\right)\right]_0^{\pi/6} = -\frac{\pi}{12} - \left[-\frac{\sqrt{3}}{2} + 0 - \left(-1 + \frac{1}{2}\right)\right] = \frac{\sqrt{3}-1}{2} - \frac{\pi}{12}$

21. $\int_0^1\int_1^2 \frac{1}{x+y} \, dx \, dy = \int_0^1 [\ln(x+y)]_1^2 \, dy = \int_0^1 [\ln(2+y) - \ln(1+y)]dy$

$= \left[[(2+y)\ln(2+y) - (2+y)] - [(1+y)\ln(1+y) - (1+y)]\right]_0^1$

$= (3\ln 3) - 3 - (2\ln 2) + 2 - [(2\ln 2 - 2) - (0-1)] = 3\ln 3 - 4\ln 2 = \ln\frac{27}{16}$

23. $z = f(x,y) = 4 - x - 2y \geq 0$ for $0 \leq x \leq 1$
and $0 \leq y \leq 1$. So the solid is the region
in the first octant which lies below the plane
$z = 4 - x - 2y$ and above $[0,1] \times [0,1]$.

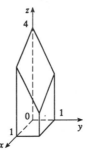

25. $V = \int_1^4\int_{-1}^0 (2x + 5y + 1)dx \, dy = \int_1^4 [x^2 + 5xy + x]_{-1}^0 \, dy = \int_1^4 5y \, dy = \frac{5}{2}y^2\big|_1^4 = \frac{75}{2}$

27. $V = \int_{-2}^2\int_{-1}^1 \left(1 - \frac{1}{4}x^2 - \frac{1}{9}y^2\right)dx \, dy = 4\int_0^2\int_0^1 \left(1 - \frac{1}{4}x^2 - \frac{1}{9}y^2\right)dx \, dy$

$= 4\int_0^2 \left[x - \frac{1}{12}x^3 - \frac{1}{9}y^2x\right]_0^1 dy = 4\int_0^2 \left(\frac{11}{12} - \frac{1}{9}y^2\right)dy = 4\left[\frac{11}{12}y - \frac{1}{27}y^3\right]_0^2 = 4 \cdot \frac{83}{54} = \frac{166}{27}$

29. Here we need the volume of the solid lying under the surface $z = x\sqrt{x^2 + y}$ and above the square
$R = [0,1] \times [0,1]$ in the xy-plane.
$V = \int_0^1\int_0^1 x\sqrt{x^2 + y} \, dx \, dy = \int_0^1 \frac{1}{3}\left[(x^2 + y)^{3/2}\right]_0^1 dy = \frac{1}{3}\int_0^1 \left[(1+y)^{3/2} - y^{3/2}\right]dy$

$= \frac{1}{3} \cdot \frac{2}{5}\left[(1+y)^{5/2} - y^{5/2}\right]_0^1 = \frac{4}{15}\left(2\sqrt{2} - 1\right)$

31. In the first octant, $z \geq 0 \implies y \leq 3$, so
$V = \int_0^3\int_0^2 (9 - y^2)dx \, dy = \int_0^3 [9x - y^2x]_0^2 \, dy = \int_0^3 (18 - 2y^2)dy = \left[18y - \frac{2}{3}y^3\right]_0^3 = 36$

33. In Maple, we can calculate the integral by defining

the integrand as f and then using the command

`int(int(f,x=0..1),y=0..1);`.

In Mathematica, we can use the command

`Integrate[Integrate[f,{x,0,1}],{y,0,1}]`.

We find that $\iint_R x^5 y^3 e^{xy}\, dA = 21e - 57 \approx 0.0839$.

We can use `plot3d` (in Maple) or `Plot3d`

(in Mathematica) to graph the function.

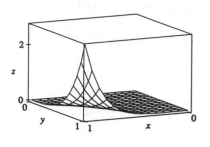

35. $A(R) = 2 \cdot 5 = 10$, so

$$f_{\text{ave}} = \frac{1}{A(R)} \iint\limits_R f(x,y)\,dA = \frac{1}{10}\int_0^5\int_{-1}^1 x^2 y\, dx\, dy = \frac{1}{10}\int_0^5 \left[\frac{x^3}{3}y\right]_{-1}^1 dy = \frac{1}{10}\int_0^5 \frac{2y}{3}\, dy = \frac{1}{10}\left[\frac{y^2}{3}\right]_0^5 = \frac{5}{6}.$$

37. Let $f(x,y) = \dfrac{x-y}{(x+y)^3}$. Then a CAS gives $\int_0^1\int_0^1 f(x,y)\,dy\, dx = \frac{1}{2}$ and $\int_0^1\int_0^1 f(x,y)\,dx\, dy = -\frac{1}{2}$.

To explain the seeming violation of Fubini's Theorem, note that f has an infinite discontinuity at $(0,0)$ and thus does not satisfy the conditions of Fubini's Theorem. In fact, both iterated integrals involve improper integrals which diverge at their lower limits of integration.

EXERCISES 13.3

1. $\int_0^1\int_0^y x\, dx\, dy = \int_0^1 \left[\frac{1}{2}x^2\right]_0^y dy = \int_0^1 \left[\frac{1}{2}y^2\right]dy = \frac{1}{6}$

3. $\int_0^2\int_{\sqrt{x}}^3 (x^2+y)\,dy\, dx = \int_0^2 \left[x^2 y + \frac{1}{2}y^2\right]_{\sqrt{x}}^3 dx = \int_0^2 \left[3x^2 + \frac{9}{2} - x^{5/2} - \frac{1}{2}x\right]dx$

$$= \left[x^3 + \frac{9}{2}x - \frac{2}{7}x^{7/2} - \frac{1}{4}x^2\right]_0^2 = 16\left(1 - \frac{\sqrt{2}}{7}\right)$$

5. $\int_0^1\int_0^x \sin(x^2)\,dy\, dx = \int_0^1 x\sin(x^2)\,dx = \frac{1}{2}[-\cos(x^2)]_0^1 = \frac{1}{2}(1 - \cos 1)$

7. $\int_0^1\int_{x^2}^{\sqrt{x}} xy\, dy\, dx = \int_0^1 \left[\frac{1}{2}xy^2\right]_{x^2}^{\sqrt{x}} dx = \frac{1}{2}\int_0^1 (x^2 - x^5)\,dx = \frac{1}{2}\left[\frac{1}{3}x^3 - \frac{1}{6}x^6\right]_0^1 = \frac{1}{12}$

9. $\int_0^1\int_{\sqrt{x}}^{2-x} (x^2 - 2xy)\,dy\, dx = \int_0^1 [x^2 y - xy^2]_{\sqrt{x}}^{2-x} dx = \int_0^1 \left(-2x^3 + 7x^2 - 4x - x^{5/2}\right)dx$

$$= \left[-\frac{1}{2}x^4 + \frac{7}{3}x^3 - 2x^2 - \frac{2}{7}x^{7/2}\right]_0^1 = -\frac{19}{42}$$

11. $\int_1^2\int_y^{y^3} e^{x/y}\,dx\, dy = \int_1^2 \left[ye^{x/y}\right]_y^{y^3} dy = \int_1^2 \left(ye^{y^2} - ey\right)dy = \left[\frac{1}{2}e^{y^2} - \frac{1}{2}ey^2\right]_1^2 = \frac{1}{2}\left(e^4 - 4e\right)$

13. $\int_0^1\int_0^{x^2} x\cos y\, dy\, dx = \int_0^1 [x\sin y]_0^{x^2} dx = \int_0^1 x\sin x^2\, dx = -\frac{1}{2}\cos x^2\big|_0^1 = \frac{1}{2}(1 - \cos 1)$

15.

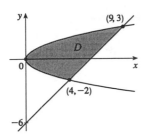

$$\int_{-2}^{3}\int_{y^2}^{y+6} 4y^3\, dx\, dy = \int_{-2}^{3}\left(4y^4 + 24y^3 - 4y^5\right) dy$$

$$= \left[\frac{4y^5}{5} + 6y^4 - \frac{2y^6}{3}\right]_{-2}^{3}$$

$$= 3^4\left(\frac{12}{5}\right) - 16\left(\frac{26}{15}\right)$$

$$= \frac{500}{3}$$

17.

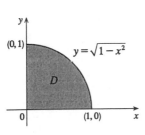

$$\int_{0}^{1}\int_{0}^{\sqrt{1-x^2}} xy\, dy\, dx = \int_{0}^{1}\left[\tfrac{1}{2}xy^2\right]_{0}^{\sqrt{1-x^2}} dx$$

$$= \int_{0}^{1}\frac{x - x^3}{2}\, dx$$

$$= \frac{1}{2}\left[\frac{x^2}{2} - \frac{x^4}{4}\right]_{0}^{1}$$

$$= \frac{1}{8}$$

19.

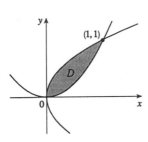

$$V = \int_{0}^{1}\int_{x^2}^{\sqrt{x}}\left(x^2 + y^2\right) dy\, dx$$

$$= \int_{0}^{1}\left(x^{5/2} - x^4 + \tfrac{1}{3}x^{3/2} - \tfrac{1}{3}x^6\right) dx$$

$$= \left[\tfrac{2}{7}x^{7/2} - \tfrac{1}{5}x^5 + \tfrac{2}{15}x^{5/2} - \tfrac{1}{21}x^7\right]_{0}^{1}$$

$$= \frac{18}{105} = \frac{6}{35}$$

21.

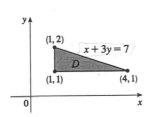

$$V = \int_{1}^{2}\int_{1}^{7-3y} xy\, dx\, dy = \int_{1}^{2}\left[\frac{yx^2}{2}\right]_{1}^{7-3y} dy$$

$$= \frac{1}{2}\int_{1}^{2}\left(48y - 42y^2 + 9y^3\right) dy$$

$$= \tfrac{1}{2}\left(24y^2 - 14y^3 + \tfrac{9}{4}y^4\right)_{1}^{2}$$

$$= \frac{31}{8}$$

23.

$$V = \int_{0}^{2}\int_{0}^{1-x/2}\sqrt{9 - x^2}\, dy\, dx = \int_{0}^{2}\left(\sqrt{9 - x^2} - \tfrac{1}{2}x\sqrt{9 - x^2}\right) dx$$

$$= \int_{0}^{2}\sqrt{9 - x^2}\, dx + \tfrac{1}{4}\int_{0}^{2}\left(-2x\sqrt{9 - x^2}\right) dx$$

$$= \left[\tfrac{1}{2}x\sqrt{9 - x^2} + \tfrac{9}{2}\sin^{-1}(x/3) + \tfrac{1}{6}(9 - x^2)^{3/2}\right]_{0}^{2}$$

$$= \sqrt{5} + \tfrac{9}{2}\sin^{-1}\tfrac{2}{3} + \tfrac{5}{6}\sqrt{5} - \tfrac{1}{6}(27)$$

$$= \tfrac{1}{6}\left(11\sqrt{5} - 27\right) + \tfrac{9}{2}\sin^{-1}\tfrac{2}{3}$$

25.

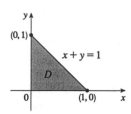

$$V = \int_0^1 \int_0^{1-x} (1 - x - y)\,dy\,dx$$

$$= \int_0^1 \left[(1 - x)^2 - \tfrac{1}{2}(1 - x)^2\right]dx$$

$$= \int_0^1 \tfrac{1}{2}(1 - x)^2\,dx = \left[-\tfrac{1}{6}(1 - x)^3\right]_0^1$$

$$= \frac{1}{6}$$

27.

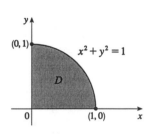

$$V = \int_0^1 \int_0^{\sqrt{1-x^2}} y\,dy\,dx$$

$$= \int_0^1 \frac{1 - x^2}{2}\,dx$$

$$= \tfrac{1}{2}\left[x - \tfrac{1}{3}x^3\right]_0^1$$

$$= \frac{1}{3}$$

29.

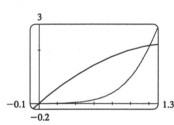

From the graph, it appears that the two

curves intersect at $x = 0$ and at $x \approx 1.213$.

Thus the desired integral is

$$\iint_D x\,dA \approx \int_0^{1.213}\int_{x^4}^{3x-x^2} x\,dy\,dx = \int_0^{1.213}[xy]_{x^4}^{3x-x^2}\,dx$$

$$= \left[(x^3 - \tfrac{1}{4}x^4) - \tfrac{1}{6}x^6\right]_0^{1.213} \approx 0.713$$

31. The two bounding curves $y = x^3 - x$ and $y = x^2 + x$ intersect at the origin and at $x = 2$, with $x^2 + x > x^3 - x$

on $(0, 2)$. Using a CAS, we find that the volume is

$$V = \int_0^2\int_{x^3-x}^{x^2+x} z\,dy\,dx = \int_0^2\int_{x^3-x}^{x^2+x}(x^3y^4 + xy^2)\,dy\,dx = \frac{13{,}984{,}735{,}616}{14{,}549{,}535}.$$

33.

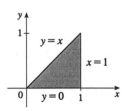

Because the region of integration is

$$D = \{(x, y) \mid 0 \le y \le x, 0 \le x \le 1\}$$

$$= \{(x, y) \mid y \le x \le 1, 0 \le y \le 1\}$$

we have

$$\int_0^1\int_0^x f(x,y)\,dy\,dx = \iint_D f(x,y)\,dA = \int_0^1\int_y^1 f(x,y)\,dx\,dy$$

35.

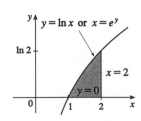

Because the region of integration is

$$D = \{(x, y) \mid 0 \le y \le \ln x, 1 \le x \le 2\}$$

$$= \{(x, y) \mid e^y \le x \le 2, 0 \le y \le \ln 2\}$$

we have

$$\int_1^2\int_0^{\ln x} f(x,y)\,dy\,dx = \iint_D f(x,y)\,dA = \int_0^{\ln 2}\int_{e^y}^2 f(x,y)\,dx\,dy$$

37.

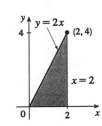

Because the region of integration is

$$D = \{(x,y) \mid y/2 \le x \le 2, 0 \le y \le 4\}$$
$$= \{(x,y) \mid 0 \le y \le 2x, 0 \le x \le 2\}$$

we have

$$\int_0^4 \int_{y/2}^2 f(x,y)dx\,dy = \iint_D f(x,y)dA = \int_0^2 \int_0^{2x} f(x,y)dy\,dx$$

39.

$$\int_0^1 \int_{3y}^3 e^{x^2}\,dx\,dy = \int_0^3 \int_0^{x/3} e^{x^2}\,dy\,dx$$

$$= \int_0^3 \left(\frac{x}{3}\right)e^{x^2}\,dx$$

$$= \tfrac{1}{6}e^{x^2}\Big|_0^3 = \frac{e^0 - 1}{6}$$

41.

$$\int_0^3 \int_{y^2}^9 y\cos x^2\,dx\,dy = \int_0^9 \int_0^{\sqrt{x}} y\cos x^2\,dy\,dx$$

$$= \int_0^9 \cos x^2 \left[\frac{y^2}{2}\right]_0^{\sqrt{x}}\,dx$$

$$= \int_0^9 \tfrac{1}{2}x\cos x^2\,dx = \tfrac{1}{4}\sin x^2\Big|_0^9$$

$$= \tfrac{1}{4}\sin 81$$

43.

$$\int_0^1 \int_{\arcsin y}^{\pi/2} \cos x\sqrt{1+\cos^2 x}\,dx\,dy = \int_0^{\pi/2} \int_0^{\sin x} \cos x\sqrt{1+\cos^2 x}\,dy\,dx$$

$$= \int_0^{\pi/2} \cos x\sqrt{1+\cos^2 x}\,[y]_0^{\sin x}\,dx$$

$$= \int_0^{\pi/2} \cos x\sqrt{1+\cos^2 x}\,\sin x\,dx \quad \left[\begin{matrix}\text{Let } u = \cos x,\ du = -\sin x\,dx, \\ dx = du/(-\sin x)\end{matrix}\right]$$

$$= \int_1^0 -u\sqrt{1+u^2}\,du = -\tfrac{1}{3}(1+u^2)^{3/2}\Big|_1^0 = \tfrac{1}{3}\left(\sqrt{8}-1\right) = \tfrac{1}{3}\left(2\sqrt{2}-1\right)$$

45. $D = \{(x,y) \mid 0 \le x \le 1, -x+1 \le y \le 1\} \cup \{(x,y) \mid -1 \le x \le 0, x+1 \le y \le 1\}$

$\cup\, \{(x,y) \mid 0 \le x \le 1, -1 \le y \le x-1\} \cup \{(x,y) \mid -1 \le x \le 0, -1 \le y \le -x-1\}$, all type I.

$\iint_D x^2\,dA = \int_0^1 \int_{1-x}^1 x^2\,dy\,dx + \int_{-1}^0 \int_{x+1}^1 x^2\,dy\,dx + \int_0^1 \int_{-1}^{x-1} x^2\,dy\,dx + \int_{-1}^0 \int_{-1}^{-x-1} x^2\,dy\,dx$

$= 4\int_0^1 \int_{1-x}^1 x^2\,dy\,dx$ (by symmetry of the regions, and because $f(x,y) = x^2 \ge 0$) $= 4\int_0^1 x^3\,dx = 1$

47. For $D = [0,1] \times [0,1], 0 \le \sqrt{x^3+y^3} \le \sqrt{2}$ and $A(D) = 1$, so $0 \le \iint_D \sqrt{x^3+y^3}\,dA \le \sqrt{2}$.

49. Since $m \le f(x,y) \le M$, $\iint_D m\,dA \le \iint_D f(x,y)dA \le \iint_D M\,dA$ by (8) $\Rightarrow$

$m\iint_D 1\,dA \le \iint_D f(x,y)\,dA \le M\iint_D 1\,dA$ by (7) $\Rightarrow$ $mA(D) \le \iint_D f(x,y)dA \le MA(D)$ by (10).

51. $\iint_D (x^2\tan x + y^3 + 4)dA = \iint_D x^2\tan x\,dA + \iint_D y^3\,dA + \iint_D 4\,dA$. But $x^2\tan x$ is an odd function and D is symmetric with respect to the y-axis, so $\iint_D x^2\tan x\,dA = 0$. Similarly, y^3 is an odd function and D is symmetric with respect to the x-axis, so $\iint_D y^3\,dA = 0$. Thus

$\iint_D (x^2\tan x + y^3 + 4)dA = 4\iint_D dA = 4(\text{area of } D) = 4 \cdot \pi\left(\sqrt{2}\right)^2 = 8\pi.$

EXERCISES 13.4

1. $\iint_R x\, dA = \int_0^{2\pi}\int_0^5 r^2\cos\theta\, dr\, d\theta = \left(\int_0^{2\pi}\cos\theta\, d\theta\right)\left(\int_0^5 r^2\, dr\right) = 0$

3. $\iint_R xy\, dA = \int_0^{\pi/2}\int_2^5 r^3\cos\theta\sin\theta\, dr\, d\theta = \left(\int_0^{\pi/2}\frac{\sin 2\theta}{2}\, d\theta\right)\left(\int_2^5 r^3\, dr\right) = \frac{1}{2}\cdot\frac{5^4-2^4}{4} = \frac{609}{8}$

5. The circle $r=1$ intersects the cardioid $r = 1+\sin\theta$ when $1 = 1+\sin\theta \;\;\Rightarrow\;\; \theta = 0$ or $\theta = \pi$, so

$\displaystyle\iint_D \frac{1}{\sqrt{x^2+y^2}}\, dA = \int_0^{\pi}\int_1^{1+\sin\theta}\left(\frac{1}{r}\right)r\, dr\, d\theta = \int_0^{\pi}[r]_1^{1+\sin\theta}\, d\theta = \int_0^{\pi}\sin\theta\, d\theta = [-\cos\theta]_0^{\pi} = 2.$

7. $\int_0^{2\pi}\int_0^{2\theta} r^3\, dr\, d\theta = \int_0^{2\pi}\frac{15}{4}\theta^4\, d\theta = \left[\frac{3}{4}\theta^5\right]_0^{2\pi} = 24\pi^5$

9. $A = \int_{-\pi/6}^{\pi/6}\int_0^{\cos 3\theta} r\, dr\, d\theta = \int_{-\pi/6}^{\pi/6}\frac{1}{2}r^2\Big|_0^{\cos 3\theta}\, d\theta = \frac{1}{2}\int_{-\pi/6}^{\pi/6}\cos^2 3\theta\, d\theta$

$\qquad = \int_0^{\pi/6}\frac{1}{2}(1+\cos 6\theta)d\theta = \frac{1}{2}\left[\theta+\frac{1}{6}\sin 6\theta\right]_0^{\pi/6} = \frac{\pi}{12}$

11. By symmetry, the two loops of the lemniscate are equal in area, so

$A = 2\int_{-\pi/4}^{\pi/4}\int_0^{2\sqrt{\cos 2\theta}} r\, dr\, d\theta = \int_{-\pi/4}^{\pi/4} 4\cos 2\theta\, d\theta = 8\int_0^{\pi/4}\cos 2\theta\, d\theta = 4\sin 2\theta\big|_0^{\pi/4} = 4.$

13. $3\cos\theta = 1+\cos\theta$ implies $\cos\theta = \frac{1}{2}$, so $\theta = \pm\frac{\pi}{3}$. Then by symmetry

$A = 2\int_0^{\pi/3}\int_{1+\cos\theta}^{3\cos\theta} r\, dr\, d\theta = 2\int_0^{\pi/3}\left[\frac{1}{2}r^2\right]_{1+\cos\theta}^{3\cos\theta}\, d\theta = \int_0^{\pi/3}(9\cos^2\theta - 1 - 2\cos\theta - \cos^2\theta)d\theta$

$\qquad = \int_0^{\pi/3}\left[8\cdot\frac{1}{2}(1+\cos 2\theta) - 2\cos\theta - 1\right]d\theta = [4\theta + 2\sin 2\theta - 2\sin\theta - \theta]_0^{\pi/3} = \pi.$

15. $V = \displaystyle\iint_{x^2+y^2\le 9}(x^2+y^2)dA = \int_0^{2\pi}\int_0^3 r^3\, dr\, d\theta = 2\pi\left(\frac{81}{4}\right) = \frac{81\pi}{2}$

17.

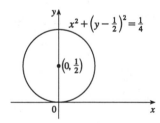

$x^2 + \left(y - \frac{1}{2}\right)^2 = \frac{1}{4}$

$\left(0, \frac{1}{2}\right)$

$V = \displaystyle\iint_{x^2+(y-\frac{1}{2})^2\le\frac{1}{4}}(12 - 6x - 4y)dA$

$\qquad = \int_0^{\pi}\int_0^{\sin\theta}(12r - 6r^2\cos\theta - 4r^2\sin\theta)dr\, d\theta$

$\qquad = \int_0^{\pi}\left(6\sin^2\theta - 2\sin^3\theta\cos\theta - \frac{4}{3}\sin^4\theta\right)d\theta$

$\qquad = \left[\frac{5}{2}(\theta - \sin\theta\cos\theta) - \frac{1}{2}\sin^4\theta + \frac{1}{3}\sin^2\theta\cos\theta\right]_0^{\pi} = \frac{5\pi}{2}$

This can also be done without calculus:

the volume of this cylinder is $\pi\left(\frac{1}{4}\right)\left(\frac{12+8}{2}\right) = \frac{5\pi}{2}$.

19. The cone $z = \sqrt{x^2+y^2}$ intersects the sphere $x^2+y^2+z^2 = 1$ when $x^2+y^2+\left(\sqrt{x^2+y^2}\right)^2 = 1$ or $x^2+y^2 = \frac{1}{2}$. So

$V = \displaystyle\iint_{x^2+y^2\le\frac{1}{2}}\left(\sqrt{1-x^2-y^2} - \sqrt{x^2+y^2}\right)dA = \int_0^{2\pi}\int_0^{1/\sqrt{2}}\left(\sqrt{1-r^2} - r\right)r\, dr\, d\theta$

$\qquad = \frac{2\pi}{3}\left[-(1-r^2)^{3/2} - r^3\right]_0^{1/\sqrt{2}} = \frac{2\pi}{3}\left(-\frac{1}{\sqrt{2}} + 1\right) = \frac{\pi}{3}\left(2 - \sqrt{2}\right)$

21. The given solid is the region inside the cylinder $x^2 + y^2 = 4$ between the surfaces $z = \sqrt{64 - 4x^2 - 4y^2}$ and $z = -\sqrt{64 - 4x^2 - 4y^2}$. So

$$V = \iint_{x^2+y^2 \le 4} \left[\sqrt{64 - 4x^2 - 4y^2} - \left(-\sqrt{64 - 4x^2 - 4y^2} \right) \right] dA = \iint_{x^2+y^2 \le 4} 2\sqrt{64 - 4x^2 - 4y^2} \, dA$$

$$= 4 \int_0^{2\pi} \int_0^2 \sqrt{16 - r^2} \, r \, dr \, d\theta = 8\pi \left[-\tfrac{1}{3}(16 - r^2)^{3/2} \right]_0^2 = \tfrac{8\pi}{3}\left(64 - 12^{3/2}\right) = \tfrac{8\pi}{3}\left(64 - 24\sqrt{3}\right)$$

23. $V = 2 \iint_{x^2+y^2 \le a^2} \sqrt{a^2 - x^2 - y^2} \, dA = 2\int_0^{2\pi}\int_0^a \sqrt{a^2 - r^2} \, r \, dr \, d\theta = \tfrac{4\pi}{3}\left[-(a^2 - r^2)^{3/2} \right]_0^a = \tfrac{4\pi}{3}a^3$

25.

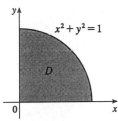

$x^2 + y^2 = 1$

D

$\int_0^{\pi/2}\int_0^1 re^{r^2} \, dr \, d\theta = \tfrac{\pi}{2}\left[\tfrac{1}{2}e^{r^2} \right]_0^1 = \tfrac{1}{4}\pi(e - 1)$

27.

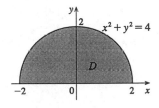

$x^2 + y^2 = 4$

D

$\int_0^\pi \int_0^2 (r^4 \cos^2\theta \sin^2\theta) r \, dr \, d\theta = \int_0^\pi \int_0^2 \left(\tfrac{1}{4}r^5 \sin^2 2\theta \right) dr \, d\theta$

$\qquad = \tfrac{8}{3}\int_0^\pi \sin^2 2\theta \, d\theta = \tfrac{8}{12}[2\theta - \sin 2\theta \cos 2\theta]_0^\pi = \tfrac{4\pi}{3}$

29. $\displaystyle \int_0^1 \int_0^{\sqrt{1-x^2}} e^{-(x^2+y^2)^2} \, dy \, dx = \int_0^{\pi/2} \int_0^1 re^{-(r^2)^2} \, dr \, d\theta = \tfrac{\pi}{2}\int_0^1 re^{-r^4} \, dr \approx 0.587$

31. $\displaystyle \int_{1/\sqrt{2}}^1 \int_{\sqrt{1-x^2}}^x xy \, dy \, dx + \int_1^{\sqrt{2}} \int_0^x xy \, dy \, dx + \int_{\sqrt{2}}^2 \int_0^{\sqrt{4-x^2}} xy \, dy \, dx = \int_0^{\pi/4} \int_1^2 r^3 \cos\theta \sin\theta \, dr \, d\theta$

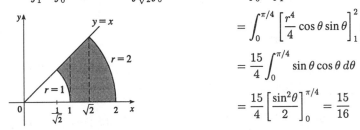

$y = x$

$r = 2$

$r = 1$

$\qquad = \int_0^{\pi/4} \left[\frac{r^4}{4} \cos\theta \sin\theta \right]_1^2 d\theta$

$\qquad = \frac{15}{4} \int_0^{\pi/4} \sin\theta \cos\theta \, d\theta$

$\qquad = \frac{15}{4} \left[\frac{\sin^2\theta}{2} \right]_0^{\pi/4} = \frac{15}{16}$

33. (a) We integrate by parts with $u = x$ and $dv = xe^{-x^2} dx$. Then $du = dx$ and $v = -\tfrac{1}{2}e^{-x^2}$, so

$$\int_0^\infty x^2 e^{-x^2} \, dx = \lim_{t\to\infty} \int_0^t x^2 e^{-x^2} \, dx = \lim_{t\to\infty} \left(-\tfrac{1}{2}xe^{-x^2} \big|_0^t + \int_0^t \tfrac{1}{2}e^{-x^2} \, dx \right)$$

$$= \lim_{t\to\infty} \left(-\tfrac{1}{2}te^{-t^2} \right) + \tfrac{1}{2}\int_0^\infty e^{-x^2} \, dx = 0 + \tfrac{1}{2}\int_0^\infty e^{-x^2} \, dx \quad \text{(by l'Hospital's Rule)}$$

$$= \tfrac{1}{4}\int_{-\infty}^\infty e^{-x^2} \, dx \quad \text{(since } e^{-x^2} \text{ is an even function)} \quad = \tfrac{1}{4}\sqrt{\pi} \quad \text{[by Exercise 32(c)]}.$$

(b) Let $u = \sqrt{x}$. Then $u^2 = x \;\Rightarrow\; dx = 2u \, du \;\Rightarrow\;$

$$\int_0^\infty \sqrt{x}e^{-x} \, dx = \lim_{t\to\infty} \int_0^t \sqrt{x}e^{-x} \, dx = \lim_{t\to\infty} \int_0^{\sqrt{t}} ue^{-u^2} 2u \, du$$

$$= 2\int_0^\infty u^2 e^{-u^2} \, du = 2\left(\tfrac{1}{4}\sqrt{\pi} \right) = \tfrac{1}{2}\sqrt{\pi} \quad \text{[by part (a)]}$$

EXERCISES 13.5

1. $Q = \iint_D (x^2 + 3y^2) dA = \int_0^2 \int_1^2 (x^2 + 3y^2) dy\, dx = \int_0^2 (x^2 + 7) dx = 14 + \frac{8}{3} = \frac{50}{3}$ C

3. $m = \int_{-1}^1 \int_0^1 x^2\, dy\, dx = \int_{-1}^1 x^2\, dx = \frac{2}{3}, \bar{x} = \frac{3}{2} \int_{-1}^1 \int_0^1 x^3\, dy\, dx = 0,$

$\bar{y} = \frac{3}{2} \int_{-1}^1 \int_0^1 x^2 y\, dy\, dx = \frac{3}{2} \int_{-1}^1 \frac{1}{2} x^2\, dx = \frac{3}{2} \left(\frac{2}{6} \right) = \frac{1}{2}.$ Hence $(\bar{x}, \bar{y}) = \left(0, \frac{1}{2} \right).$

5. $m = \int_0^2 \int_{x/2}^{3-x} (x+y) dy\, dx = \int_0^2 \left[x\left(3 - \frac{3}{2}x \right) + \frac{1}{2}(3-x)^2 - \frac{1}{8}x^2 \right] dx = \int_0^2 \left[-\frac{9}{8}x^2 + \frac{9}{2} \right] dx = 6,$

$M_y = \int_0^2 \int_{x/2}^{3-x} (x^2 + xy) dy\, dx = \int_0^2 \left[x^2 y + \frac{1}{2}xy^2 \right]_{x/2}^{3-x} dx = \int_0^2 \left(\frac{9}{2}x - \frac{9}{8}x^3 \right) dx = \frac{9}{2},$ and

$M_x = \int_0^2 \int_{x/2}^{3-y} (xy + y^2) dy\, dx = \int_0^2 \left(9 - \frac{9}{2}x \right) dx = 9.$ Hence $m = 6, (\bar{x}, \bar{y}) = \left(\frac{3}{4}, \frac{3}{2} \right).$

7. $m = \int_0^1 \int_{x^2}^x xy\, dy\, dx = \int_0^1 \left(\frac{1}{2}x - \frac{1}{2}x^5 \right) dx = \frac{1}{4} - \frac{1}{12} = \frac{1}{6},$

$M_y = \int_0^1 \int_{x^2}^x x^2 y\, dy\, dx = \int_0^1 \left(\frac{1}{2}x^2 - \frac{1}{2}x^6 \right) dx = \frac{1}{6} - \frac{1}{14} = \frac{2}{21}$ and

$M_x = \int_0^1 \int_{x^2}^x xy^2\, dy\, dx = \int_0^1 \left(\frac{1}{3}x - \frac{1}{3}x^7 \right) dx = \frac{1}{6} - \frac{1}{24} = \frac{1}{8}.$ Hence $m = \frac{1}{6}, (\bar{x}, \bar{y}) = \left(\frac{4}{7}, \frac{3}{4} \right).$

9.

$m = \int_{-1}^2 \int_{y^2}^{y+2} 3\, dx\, dy = \int_{-1}^2 (3y + 6 - 3y^2) dy = \frac{27}{2},$

$M_y = \int_{-1}^2 \int_{y^2}^{y+2} 3x\, dx\, dy = \int_{-1}^2 \frac{3}{2} \left[(y+2)^2 - y^4 \right] dy$

$= \left[\frac{1}{2}(y+2)^3 - \frac{3}{10}y^5 \right]_{-1}^2 = \frac{108}{5}$ and

$M_x = \int_{-1}^2 \int_{y^2}^{y+2} 3y\, dx\, dy = \int_{-1}^2 (3y^2 + 6y - 3y^3) dy$

$= \left[y^3 + 3y^2 - \frac{3}{4}y^4 \right]_{-1}^2 = \frac{27}{4}.$

Hence $m = \frac{27}{2}, (\bar{x}, \bar{y}) = \left(\frac{8}{5}, \frac{1}{2} \right).$

11. $m = \int_0^\pi \int_0^{\sin x} y\, dy\, dx = \int_0^\pi \frac{1}{2} \sin^2 x\, dx = \left[\frac{1}{4}x - \frac{1}{8} \sin 2x \right]_0^\pi = \frac{1}{4}\pi,$

$M_y = \int_0^\pi \int_0^{\sin x} xy\, dy\, dx = \int_0^\pi \frac{1}{2}x \sin^2 x\, dx = \left[\frac{1}{8}x^2 - \frac{1}{8}x \sin 2x - \frac{1}{16} \cos 2x \right]_0^\pi = \frac{1}{8}\pi^2,$ and

$M_x = \int_0^\pi \int_0^{\sin x} y^2\, dy\, dx = \int_0^\pi \frac{1}{3} \sin^3 x\, dx = \frac{1}{3} \left[-\cos x + \frac{1}{3} \cos^3 x \right]_0^\pi = \frac{4}{9}.$ Hence $m = \frac{\pi}{4}, (\bar{x}, \bar{y}) = \left(\frac{\pi}{2}, \frac{16}{9\pi} \right).$

13. $\rho(x, y) = ky = kr \sin \theta, m = \int_0^{\pi/2} \int_0^1 kr^2 \sin \theta\, dr\, d\theta = \frac{1}{3}k \int_0^{\pi/2} \sin \theta\, d\theta = \frac{1}{3}k[-\cos \theta]_0^{\pi/2} = \frac{1}{3}k,$

$M_y = \int_0^{\pi/2} \int_0^1 kr^3 \sin \theta \cos \theta\, dr\, d\theta = \frac{1}{4}k \int_0^{\pi/2} \sin \theta \cos \theta\, d\theta = \frac{1}{8}k[-\cos 2\theta]_0^{\pi/2} = \frac{1}{8}k,$

$M_x = \int_0^{\pi/2} \int_0^1 kr^3 \sin^2\theta\, dr\, d\theta = \frac{1}{4}k \int_0^{\pi/2} \sin^2\theta\, d\theta = \frac{1}{8}k[\theta + \sin 2\theta]_0^{\pi/2} = \frac{\pi}{16}k.$ Hence $(\bar{x}, \bar{y}) = \left(\frac{3}{8}, \frac{3\pi}{16} \right).$

15. Placing the vertex opposite the hypotenuse at $(0, 0), \rho(x, y) = k(x^2 + y^2).$ Then

$m = \int_0^a \int_0^{a-x} k(x^2 + y^2) dy\, dx = k \int_0^a \left[ax^2 - x^3 + \frac{1}{3}(a-x)^3 \right] dx = k \left[\frac{1}{3}ax^3 - \frac{1}{4}x^4 - \frac{1}{12}(a-x)^4 \right]_0^a = \frac{1}{6}ka^4.$

By symmetry,

$M_y = M_x = \int_0^a \int_0^{a-x} ky(x^2 + y^2) dy\, dx = k \int_0^a \left[\frac{1}{2}(a-x)^2 x^2 + \frac{1}{4}(a-x)^4 \right] dx$

$= k \left[\frac{1}{6}a^2 x^3 - \frac{1}{4}ax^4 + \frac{1}{10}x^5 - \frac{1}{20}(a-x)^5 \right]_0^a = \frac{1}{15}ka^5.$

Hence $(\bar{x}, \bar{y}) = \left(\frac{2}{5}a, \frac{2}{5}a \right).$

17. $I_x = \int_0^1 \int_{x^2}^1 y^2 (xy) dy \, dx = \int_0^1 \frac{1}{4} (x - x^9) dx = \frac{1}{8} - \frac{1}{40} - \frac{1}{10}$,

$I_y = \int_0^1 \int_{x^2}^1 x^3 y \, dy \, dx = \int_0^1 \frac{1}{2} (x^3 - x^7) dx = \frac{1}{8} - \frac{1}{16} = \frac{1}{16}$, $I_0 = I_x + I_y = \frac{13}{80}$.

19. $I_x = \int_{-1}^2 \int_{y^2}^{y+2} 3y^2 \, dx \, dy = \int_{-1}^2 (3y^3 + 6y^2 - 3y^4) dy = \left[\frac{3}{4}y^4 + 2y^3 - \frac{3}{5}y^5 \right]_{-1}^2 = \frac{189}{20}$,

$I_y = \int_{-1}^2 \int_{y^2}^{y+2} 3x^2 \, dx \, dy = \int_{-1}^2 [(y+2)^3 - y^6] dy = \left[\frac{1}{4}(y+2)^4 - \frac{1}{7}y^7 \right]_{-1}^2 = \frac{1269}{28}$, and $I_0 = I_x + I_y = \frac{1917}{35}$.

21. $I_x = \int_0^a \int_0^a \rho y^2 \, dx \, dy = \rho a \left(\frac{1}{3}a^3 \right) = \frac{1}{3}\rho a^4 = I_y$ by symmetry, and $m = \rho a^2$ since the lamina is homogeneous.

Hence $\overline{x} = \overline{y} = \left[\left(\frac{1}{3}\rho a^4 \right) / (\rho a^2) \right]^{1/2} = \frac{1}{\sqrt{3}}a$.

23. Since $m = \iint_D \rho(x,y) dA = \rho \iint_D dA = \rho(\text{area of } D) = \rho A(D)$,

$$\overline{x} = \frac{1}{\rho A(D)} \iint_D x\rho \, dA = \frac{1}{A} \int_a^b \int_{g(x)}^{f(x)} x \, dy \, dx = \frac{1}{A} \int_a^b x[f(x) - g(x)] dx \text{ and}$$

$$\overline{y} = \frac{1}{\rho A(D)} \iint_D y\rho \, dA = \frac{1}{A} \int_a^b \int_{g(x)}^{f(x)} y \, dy \, dx = \frac{1}{A} \int_a^b \left[\frac{1}{2}y^2 \right]_{g(x)}^{f(x)} dx = \frac{1}{A} \int_a^b \frac{1}{2} [f(x)^2 - g(x)^2] dx.$$

EXERCISES 13.6

1. Here $z = f(x,y) = 4 - x - 2y$ with $0 \le x^2 + y^2 \le 4$. Thus by (2)

$$A(S) = \iint_D \sqrt{(-1)^2 + (-2)^2 + 1} \, dA = \sqrt{6} \iint_{x^2 + y^2 \le 4} dA = \sqrt{6}\pi(2)^2 = 4\sqrt{6}\pi.$$

3. $y^2 + z^2 = 9 \quad \Rightarrow \quad z = \sqrt{9 - y^2}$. $f_x = 0$, $f_y = -y(9-y^2)^{-1/2} \quad \Rightarrow$

$$A(S) = \int_0^4 \int_0^2 \sqrt{\left[-y(9-y^2)^{-1/2} \right]^2 + 1} \, dy \, dx = \int_0^4 \int_0^2 \sqrt{\frac{y^2}{9-y^2} + 1} \, dy \, dx$$

$$= \int_0^4 \int_0^2 \frac{3 \, dy}{\sqrt{9-y^2}} dx = 3 \int_0^4 \left[\sin^{-1}\frac{y}{3} \right]_0^2 dx = 3 \left[(\sin^{-1}\frac{2}{3})x \right]_0^4 = 12 \sin^{-1}\frac{2}{3}$$

5. $z = f(x,y) = y^2 - x^2$ with $1 \le x^2 + y^2 \le 4$. $f_x = -2x$, $f_y = 2y \quad \Rightarrow$

$A(S) = \iint_D \sqrt{1 + 4x^2 + 4y^2} \, dA = \int_0^{2\pi} \int_1^2 \sqrt{1 + 4r^2} \, r \, dr \, d\theta$

$$= \frac{4\pi}{24} \left[(1 + 4r^2)^{3/2} \right]_1^2 = \frac{\pi}{6} \left(17\sqrt{17} - 5\sqrt{5} \right)$$

7. $z = f(x,y) = xy$ with $0 \le x^2 + y^2 \le 1$, so $f_x = y$, $f_y = x \quad \Rightarrow$

$A(S) = \iint_D \sqrt{y^2 + x^2 + 1} \, dA = \int_0^{2\pi} \int_0^1 \sqrt{r^2 + 1} \, r \, dr \, d\theta = \int_0^{2\pi} \left[\frac{1}{3}(r^2 + 1)^{3/2} \right]_0^1 d\theta$

$$= \int_0^{2\pi} \frac{1}{3} \left(2\sqrt{2} - 1 \right) d\theta = \frac{2\pi}{3} \left(2\sqrt{2} - 1 \right).$$

9. $z = \sqrt{a^2 - x^2 - y^2}$, $z_x = -x(a^2 - x^2 - y^2)^{-1/2}$, $z_y = -y(a^2 - x^2 - y^2)^{-1/2}$,

$$A(S) = \iint_D \sqrt{\frac{x^2 + y^2}{a^2 - x^2 - y^2} + 1}\, dA$$

$$= \int_{-\pi/2}^{\pi/2} \int_0^{a\cos\theta} \sqrt{\frac{r^2}{a^2 - r^2} + 1}\, r\, dr\, d\theta$$

$$= \int_{-\pi/2}^{\pi/2} \int_0^{a\cos\theta} \frac{ar}{\sqrt{a^2 - r^2}}\, dr\, d\theta$$

$$= \int_{-\pi/2}^{\pi/2} \left[-a\sqrt{a^2 - r^2}\right]_0^{a\cos\theta} d\theta$$

$$= \int_{-\pi/2}^{\pi/2} -a\left(\sqrt{a^2 - a^2\cos^2\theta} - a\right) d\theta = 2a^2 \int_0^{\pi/2} \left(1 - \sqrt{1 - \cos^2\theta}\right) d\theta$$

$$= 2a^2 \int_0^{\pi/2} d\theta - 2a^2 \int_0^{\pi/2} \sqrt{\sin^2\theta}\, d\theta = a^2\pi - 2a^2 \int_0^{\pi/2} \sin\theta\, d\theta = a^2(\pi - 2)$$

(figure: circle $r = a\cos\theta$ passing through origin 0 and point a on horizontal axis)

11.   The midpoints of the four squares are $\left(\frac{1}{4}, \frac{1}{4}\right)$, $\left(\frac{1}{4}, \frac{3}{4}\right)$, $\left(\frac{3}{4}, \frac{1}{4}\right)$, and $\left(\frac{3}{4}, \frac{3}{4}\right)$, the derivatives of the function

$f(x, y) = x^2 + y^2$ are $f_x(x, y) = 2x$, $f_y(x, y) = 2y$, so the Midpoint Rule gives

$$A(S) = \int_0^1 \int_0^1 \sqrt{[f_x(x, y)]^2 + [f_y(x, y)]^2 + 1}\, dy\, dx$$

$$\approx \frac{1}{4}\left(\sqrt{\left[2\left(\frac{1}{4}\right)\right]^2 + \left[2\left(\frac{1}{4}\right)\right]^2 + 1} + \sqrt{\left[2\left(\frac{1}{4}\right)\right]^2 + \left[2\left(\frac{3}{4}\right)\right]^2 + 1}\right.$$

$$\left. + \sqrt{\left[2\left(\frac{3}{4}\right)\right]^2 + \left[2\left(\frac{1}{4}\right)\right]^2 + 1} + \sqrt{\left[2\left(\frac{3}{4}\right)\right]^2 + \left[2\left(\frac{3}{4}\right)\right]^2 + 1}\right)$$

$$= \frac{1}{4}\left(\sqrt{\frac{3}{2}} + 2\sqrt{\frac{7}{2}} + \sqrt{\frac{11}{2}}\right) \approx 1.8279.$$

13.   Since $f(x, y) = x^2 + 2y$, we have $f_x = 2x$, $f_y = 2$. We use a CAS to calculate the integral

$A(S) = \int_0^1 \int_0^1 \sqrt{f_x^2 + f_y^2 + 1}\, dy\, dx = \int_0^1 \int_0^1 \sqrt{4x^2 + 5}\, dy\, dx = \int_0^1 \sqrt{4x^2 + 5}\, dx$, and find that

$A(S) = \frac{3}{2} + \frac{5}{8}\ln 5$.

15. $f(x, y) = 1 + x^2 y^2$ $\Rightarrow$ $f_x = 2xy^2$, $f_y = 2x^2 y$. We use a CAS (with precision reduced to five significant

digits, to speed up the calculation) to estimate the integral

$$A(S) = \int_{-1}^1 \int_{-\sqrt{1-x^2}}^{\sqrt{1-x^2}} \sqrt{f_x^2 + f_y^2 + 1}\, dy\, dx = \int_{-1}^1 \int_{-\sqrt{1-x^2}}^{\sqrt{1-x^2}} \sqrt{4x^2 y^4 + 4x^4 y^2 + 1}\, dy\, dx, \text{ and find that}$$

$A(S) \approx 3.3213.$

17.   Here $z = f(x, y) = ax + by + c$, $f_x(x, y) = a$, $f_y(x, y) = b$, so by (2),

$$A(S) = \iint_D \sqrt{a^2 + b^2 + 1}\, dA = \sqrt{a^2 + b^2 + 1} \iint_D dA = \sqrt{a^2 + b^2 + 1}\, A(D).$$

EXERCISES 13.7

1. $\int_0^1\int_{-1}^2\int_0^3 xyz^2\,dz\,dx\,dy = \int_0^1\int_{-1}^2 xy(9)dx\,dy = 9\int_0^1 x(2-\tfrac12)dx = 9(\tfrac34\cdot1^2) = \tfrac{27}{4}$

3. $\int_0^1\int_0^z\int_0^y xyz\,dx\,dy\,dz = \int_0^1\int_0^z (\tfrac12 y^3 z)dy\,dz = \int_0^1 \tfrac18 z^5\,dz = \tfrac1{48}z^6\big|_0^1 = \tfrac1{48}$

5. $\int_0^\pi\int_0^2\int_0^{\sqrt{4-z^2}} z\sin y\,dx\,dz\,dy = \int_0^\pi\int_0^2 z\sqrt{4-z^2}\sin y\,dz\,dy$

$$= \int_0^\pi\left[-\tfrac13(4-z^2)^{3/2}\right]_0^2\sin y\,dy = \int_0^\pi \tfrac83\sin y\,dy = -\tfrac83\cos y\big|_0^\pi = \tfrac{16}{3}$$

7. $\int_0^1\int_0^{2z}\int_0^{z+2} yz\,dx\,dy\,dz = \int_0^1\int_0^{2z} yz(z+2)dy\,dz = \int_0^1 (2z^4+4z^3)dz = \tfrac75$

9. $\int_0^1\int_0^{x^2}\int_0^{x+2y} y\,dz\,dy\,dx = \int_0^1\int_0^{x^2}(yx+2y^2)dy\,dx = \int_0^1\left[\tfrac12 xy^2+\tfrac23 y^3\right]_0^{x^2} dx$

$$= \int_0^1\left(\tfrac12 x^5+\tfrac23 x^6\right)dx = \left[\tfrac1{12}x^6+\tfrac2{21}x^7\right]_0^1 = \tfrac5{28}$$

11. Here E is the region that lies below the plane with x-, y-, and z-intercepts $1, 2,$ and 3 respectively, that is, below the plane $2z+6x+3y=6$ and above the region in the xy-plane bounded by the lines $x=0, y=0$ and $6x+3y=6$. So

$\iiint_E xy\,dV = \int_0^1\int_0^{2-2x}\int_0^{3-3x-3y/2} xy\,dz\,dy\,dx = \int_0^1\int_0^{2-2x}\left(3xy-3x^2y-\tfrac32 xy^2\right)dy\,dx$

$= \int_0^1\left[\tfrac32 xy^2-\tfrac32 x^2y^2-\tfrac12 xy^3\right]_0^{2-2x} dx = \int_0^1\left(2x-6x^2+6x^3-2x^4\right)dx = \left[x^2-2x^3+\tfrac32 x^4-\tfrac25 x^5\right]_0^1 = \tfrac1{10}.$

13.

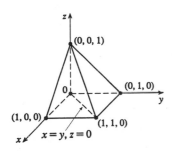

By symmetry $\iiint_E z\,dV = 2\iiint_{E'} z\,dV$ where E' is the part of E to the left [as viewed from $(10,10,0)$] of the plane $x=y$. So

$$\iiint_E z\,dV = \int_0^1\int_y^1\int_0^{1-x} 2z\,dz\,dx\,dy = \int_0^1\int_y^1 (1-x)^2\,dx\,dy$$

$$= \int_0^1\left[-\tfrac13(1-x)^3\right]_y^1 dy = \int_0^1 \tfrac13(1-y)^3\,dy$$

$$= \tfrac1{12}(1-y)^4\Big|_0^1 = \tfrac1{12}.$$

15.

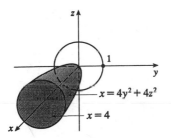

The projection E on the yz-plane is the disk $y^2+z^2\le1$. Using polar coordinates $y=r\cos\theta$ and $z=r\sin\theta$, we get

$$\iiint_E x\,dV = \iint_D\left[\int_{4y^2+4z^2}^4 x\,dx\right]dA$$

$$= \frac12\iint_D\left[4^2-(4y^2+4z^2)^2\right]dA = 8\int_0^{2\pi}\int_0^1 (1-r^4)r\,dr\,d\theta$$

$$= 8\int_0^{2\pi} d\theta\int_0^1(r-r^5)d\theta = 8(2\pi)\left[\tfrac12 r^2-\tfrac16 r^6\right]_0^1 = \tfrac{16\pi}{3}.$$

17. The plane $2x + 3y + 6z = 12$ intersects the xy-plane when $2x + 3y + 6(0) = 12$ $\Rightarrow$ $y = 4 - \frac{2}{3}x$. So

$E = \left\{(x, y, z) \mid 0 \le x \le 6, 0 \le y \le 4 - \frac{2}{3}x, 0 \le z \le \frac{1}{6}(12 - 2x - 3y)\right\}$ and

$V = \int_0^6 \int_0^{4 - \frac{2}{3}x} \int_0^{(12 - 2x - 3y)/6} dz\, dy\, dx = \frac{1}{6} \int_0^6 \int_0^{4 - \frac{2}{3}x} (12 - 2x - 3y)\, dy\, dx$

$= \frac{1}{6} \int_0^6 \left[\frac{(12 - 2x)^2}{3} - \frac{3}{2} \frac{12 - 2x}{9} \right] dx = \frac{1}{36} \int_0^6 (12 - 2x)^2\, dx = \left[\frac{1}{36}\left(-\frac{1}{6}\right)(12 - 2x)^3 \right]_0^6 = 8$

19.

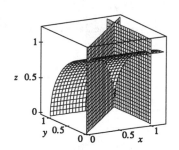

$V = \int_0^1 \int_{-\sqrt{x}}^{\sqrt{x}} \int_0^{1-x} dz\, dy\, dx = \int_0^1 \int_{-\sqrt{x}}^{\sqrt{x}} (1 - x)\, dy\, dx$

$= \int_0^1 2\sqrt{x}(1 - x)\, dx = \int_0^1 2\left(\sqrt{x} - x^{3/2}\right) dx$

$= 2\left(\frac{2}{3} - \frac{2}{5}\right) = \frac{8}{15}$

$z = 1 - x$

$x = y^2$

21. (a) The wedge can be described as the region

$D = \left\{(x, y, z) \mid y^2 + z^2 \le 1, 0 \le x \le 1, 0 \le y \le x\right\}$

$= \left\{(x, y, z) \mid 0 \le x \le 1, 0 \le y \le x, 0 \le z \le \sqrt{1 - y^2}\right\}.$

So the integral expressing the volume of the wedge is

$\iiint_D dV = \int_0^1 \int_0^x \int_0^{\sqrt{1-y^2}} dz\, dy\, dx.$

(b) A CAS gives $\int_0^1 \int_0^x \int_0^{\sqrt{1-y^2}} dz\, dy\, dx = \frac{\pi}{4} - \frac{1}{3}.$

(Or use Formulas 30 and 87.)

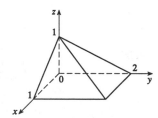

23. Note that $\Delta V_{ijk} = \left(\frac{1}{2}\right)^3 = \frac{1}{8}$, so the Midpoint Rule gives

$\iiint_B f(x, y, z)\, dV \approx \frac{1}{8}\left[f\left(\frac{1}{4}, \frac{1}{4}, \frac{1}{4}\right) + f\left(\frac{1}{4}, \frac{1}{4}, \frac{3}{4}\right) + f\left(\frac{1}{4}, \frac{3}{4}, \frac{1}{4}\right) + f\left(\frac{3}{4}, \frac{1}{4}, \frac{1}{4}\right) \right.$

$\left. + f\left(\frac{1}{4}, \frac{3}{4}, \frac{3}{4}\right) + f\left(\frac{3}{4}, \frac{1}{4}, \frac{3}{4}\right) + f\left(\frac{3}{4}, \frac{3}{4}, \frac{1}{4}\right) + f\left(\frac{3}{4}, \frac{3}{4}, \frac{3}{4}\right) \right]$

$= \frac{1}{8}\left[e^{-3\left(\frac{1}{4}\right)^2} + 3e^{-2\left(\frac{1}{4}\right)^2 - \left(\frac{3}{4}\right)^2} + 3e^{-\left(\frac{1}{4}\right)^2 - 2\left(\frac{3}{4}\right)^2} + e^{-3\left(\frac{3}{4}\right)^2} \right] \approx 0.42968.$

The norm of this partition is the longest diagonal of the eight sub-boxes. Of course, all diagonals have the same

length: $\|P\| = \sqrt{\left(\frac{1}{2}\right)^2 + \left(\frac{1}{2}\right)^2 + \left(\frac{1}{2}\right)^2} = \frac{\sqrt{3}}{2}.$

25. $E = \{(x, y, z) \mid 0 \le x \le 1,$

$0 \le z \le 1 - x, 0 \le y \le 2 - 2z\}$

27.

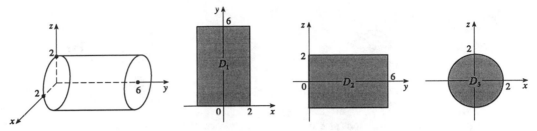

If D_1, D_2, D_3 are the projections of E on the xy-, yz-, and xz-planes, then

$D_1 = \{(x, y) \mid -2 \leq x \leq 2, 0 \leq y \leq 6\}$, $D_2 = \{(y, z) \mid -2 \leq z \leq 2, 0 \leq y \leq 6\}$,

$D_3 = \{(x, z) \mid x^2 + z^2 \leq 4\}$. Therefore

$E = \{(x, y, z) \mid -\sqrt{4 - x^2} \leq z \leq \sqrt{4 - x^2}, -2 \leq x \leq 2, 0 \leq y \leq 6\}$

$\quad = \{(x, y, z) \mid -\sqrt{4 - z^2} \leq x \leq \sqrt{4 - z^2}, -2 \leq z \leq 2, 0 \leq y \leq 6\}$.

$\iiint_E f(x, y, z) dV = \int_{-2}^{2} \int_{0}^{6} \int_{-\sqrt{4-x^2}}^{\sqrt{4-x^2}} f(x, y, z) dz\, dy\, dx = \int_{0}^{6} \int_{-2}^{2} \int_{-\sqrt{4-x^2}}^{\sqrt{4-x^2}} f(x, y, z) dz\, dx\, dy$

$\quad = \int_{0}^{6} \int_{-2}^{2} \int_{-\sqrt{4-z^2}}^{\sqrt{4-z^2}} f(x, y, z) dx\, dz\, dy = \int_{-2}^{2} \int_{0}^{6} \int_{-\sqrt{4-z^2}}^{\sqrt{4-z^2}} f(x, y, z) dx\, dy\, dz$

$\quad = \int_{-2}^{2} \int_{-\sqrt{4-x^2}}^{\sqrt{4-x^2}} \int_{0}^{6} f(x, y, z) dy\, dz\, dx = \int_{-2}^{2} \int_{-\sqrt{4-z^2}}^{\sqrt{4-z^2}} \int_{0}^{6} f(x, y, z) dy\, dx\, dz$.

29.

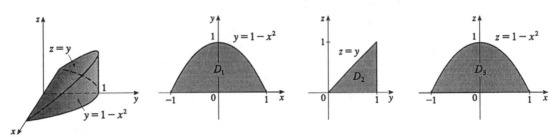

If D_1, D_2, and D_3 are the projections of E on the xy-, yz-, and xz-planes, then

$D_1 = \{(x, y) \mid -1 \leq x \leq 1, 0 \leq y \leq 1 - x^2\} = \{(x, y) \mid 0 \leq y \leq 1, -\sqrt{1-y} \leq x \leq \sqrt{1-y}\}$,

$D_2 = \{(y, z) \mid 0 \leq y \leq 1, 0 \leq z \leq y\} = \{(y, z) \mid 0 \leq z \leq 1, z \leq y \leq 1\}$,

$D_3 = \{(x, z) \mid -1 \leq x \leq 1, 0 \leq z \leq 1 - x^2\} = \{(x, z) \mid 0 \leq z \leq 1, -\sqrt{1-z} \leq x \leq \sqrt{1-z}\}$. Therefore

$E = \{(x, y, z) \mid -1 \leq x \leq 1, 0 \leq y \leq 1 - x^2, 0 \leq z \leq y\}$

$\quad = \{(x, y, z) \mid 0 \leq y \leq 1, -\sqrt{1-y} \leq x \leq \sqrt{1-y}, 0 \leq z \leq y\}$

$\quad = \{(x, y, z) \mid 0 \leq y \leq 1, 0 \leq z \leq y, -\sqrt{1-y} \leq x \leq \sqrt{1-y}\}$

$\quad = \{(x, y, z) \mid 0 \leq z \leq 1, z \leq y \leq 1, -\sqrt{1-y} \leq x \leq \sqrt{1-y}\}$

$\quad = \{(x, y, z) \mid -1 \leq x \leq 1, 0 \leq z \leq 1 - x^2, z \leq y \leq 1 - x^2\}$

$\quad = \{(x, y, z) \mid 0 \leq z \leq 1, -\sqrt{1-z} \leq x \leq \sqrt{1-z}, z \leq y \leq 1 - x^2\}$. Then

$\iiint_E f(x, y, z) dV = \int_{-1}^{1} \int_{0}^{1-x^2} \int_{0}^{y} f(x, y, z) dz\, dy\, dx = \int_{0}^{1} \int_{-\sqrt{1-y}}^{\sqrt{1-y}} \int_{0}^{y} f(x, y, z) dz\, dx\, dy$

$\quad = \int_{0}^{1} \int_{0}^{y} \int_{-\sqrt{1-y}}^{\sqrt{1-y}} f(x, y, z) dx\, dz\, dy = \int_{0}^{1} \int_{z}^{1} \int_{-\sqrt{1-y}}^{\sqrt{1-y}} f(x, y, z) dx\, dy\, dz$

$\quad = \int_{-1}^{1} \int_{0}^{1-x^2} \int_{z}^{1-x^2} f(x, y, z) dy\, dz\, dx = \int_{0}^{1} \int_{-\sqrt{1-z}}^{\sqrt{1-z}} \int_{z}^{1-x^2} f(x, y, z) dy\, dx\, dz$.

31.

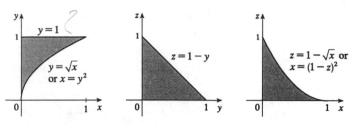

The diagrams show the projections of E on the xy-, yz-, and xz-planes. Therefore

$$\int_0^1 \int_{\sqrt{x}}^1 \int_0^{1-y} f(x,y,z)dz\,dy\,dx = \int_0^1 \int_0^{y^2} \int_0^{1-y} f(x,y,z)dz\,dx\,dy$$

$$= \int_0^1 \int_0^{1-z} \int_0^{y^2} f(x,y,z)dx\,dy\,dz = \int_0^1 \int_0^{1-y} \int_0^{y^2} f(x,y,z)dx\,dz\,dy$$

$$= \int_0^1 \int_0^{1-\sqrt{x}} \int_{\sqrt{x}}^{1-z} f(x,y,z)dy\,dz\,dx = \int_0^1 \int_0^{(1-z)^2} \int_{\sqrt{x}}^{1-z} f(x,y,z)dy\,dx\,dz.$$

33.

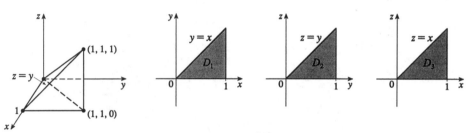

$\int_0^1 \int_y^1 \int_0^y f(x,y,z)dz\,dx\,dy = \iiint_E f(x,y,z)dV$ where $E = \{(x,y,z) \mid 0 \le z \le y, y \le x \le 1, 0 \le y \le 1\}$. If

D_1, D_2, and D_3 are the projections of E on the xy-, yz- and xz-planes then

$D_1 = \{(x,y) \mid 0 \le y \le 1, y \le x \le 1\} = \{(x,y) \mid 0 \le x \le 1, 0 \le y \le x\}$,

$D_2 = \{(y,z) \mid 0 \le y \le 1, 0 \le z \le y\} = \{(y,z) \mid 0 \le z \le 1, z \le y \le 1\}$, and

$D_3 = \{(x,z) \mid 0 \le x \le 1, 0 \le z \le x\} = \{(x,z) \mid 0 \le z \le 1, z \le x \le 1\}$. Thus we also have

$E = \{(x,y,z) \mid 0 \le x \le 1, 0 \le y \le x, 0 \le z \le y\} = \{(x,y,z) \mid 0 \le y \le 1, 0 \le z \le y, y \le x \le 1\}$

$\quad = \{(x,y,z) \mid 0 \le z \le 1, z \le y \le 1, y \le x \le 1\} = \{(x,y,z) \mid 0 \le x \le 1, 0 \le z \le x, z \le y \le x\}$

$\quad = \{(x,y,z) \mid 0 \le z \le 1, z \le x \le 1, z \le y \le x\}$.

Then

$$\int_0^1 \int_y^1 \int_0^y f(x,y,z)dz\,dx\,dy = \int_0^1 \int_0^x \int_0^y f(x,y,z)dz\,dy\,dx = \int_0^1 \int_0^y \int_y^1 f(x,y,z)dx\,dz\,dy$$

$$= \int_0^1 \int_z^1 \int_y^1 f(x,y,z)dx\,dy\,dz = \int_0^1 \int_0^x \int_z^x f(x,y,z)dy\,dz\,dx = \int_0^1 \int_z^1 \int_z^x f(x,y,z)dy\,dx\,dz.$$

35. $m = \int_0^1 \int_0^{x^2} \int_0^{x+2y} 2\,dz\,dy\,dx = 2\int_0^1 \int_0^{x^2} (x+2y)dy\,dx = 2\int_0^1 (x^3 + x^4)dx = \frac{9}{10}$,

$M_{yz} = \int_0^1 \int_0^{x^2} \int_0^{x+2y} 2x\,dz\,dy\,dx = \int_0^1 \int_0^{x^2} 2(x^2 + 2xy)dy\,dx = \int_0^1 2(x^4 + x^5)dx = \frac{11}{15}$, $M_{xz} = \frac{5}{14}$, and

$M_{xy} = \int_0^1 \int_0^{x^2} \int_0^{x+2y} 2z\,dz\,dy\,dx = \int_0^1 \int_0^{x^2} 2(x+2y)^2/2\,dy\,dx = \int_0^1 2\left[\frac{1}{2}x^2 y + xy^2 + \frac{2}{3}y^3\right]_0^{x^2} dx$

$\quad = \int_0^1 2\left(\frac{1}{2}x^4 + x^5 + \frac{2}{3}x^6\right)dx = \frac{76}{105}$. Hence $(\bar{x}, \bar{y}, \bar{z}) = \left(\frac{22}{27}, \frac{25}{63}, \frac{152}{189}\right)$.

37. $m = \int_0^a\int_0^a\int_0^a (x^2 + y^2 + z^2)dx\,dy\,dz = \int_0^a\int_0^a[\frac{1}{3}a^3 + a(y^2 + z^2)]dy\,dz = \int_0^a[\frac{2}{3}a^4 + a^2z^2]dz = \frac{2}{3}a^5 + \frac{1}{3}a^5 = a^5$,

$M_{yz} = \int_0^a\int_0^a\int_0^a[x^3 + x(y^2 + z^2)]dx\,dy\,dz = \int_0^a\int_0^a\left[\frac{1}{4}a^4 + \frac{1}{2}a^2(y^2 + z^2)\right]dy\,dz$

$= \int_0^a\left[\frac{1}{4}a^5 + \frac{1}{6}a^5 + \frac{1}{2}a^3z^2\right]dz = \frac{1}{4}a^6 + \frac{1}{3}a^6 = \frac{7}{12}a^6 = M_{xz} = M_{xy}$ by symmetry of E and $\rho(x, y, z)$.

Hence $(\bar{x}, \bar{y}, \bar{z}) = \left(\frac{7}{12}a, \frac{7}{12}a, \frac{7}{12}a\right)$.

39. (a) $m = \int_0^1\int_0^{\sqrt{1-x^2}}\int_0^y(1 + x + y + z)dz\,dy\,dx$

(b) $(\bar{x}, \bar{y}, \bar{z}) = \left(m^{-1}\int_0^1\int_0^{\sqrt{1-x^2}}\int_0^y x(1 + x + y + z)dz\,dy\,dx, m^{-1}\int_0^1\int_0^{\sqrt{1-x^2}}\int_0^y y(1 + x + y + z)dz\,dy\,dx,\right.$

$\left. m^{-1}\int_0^1\int_0^{\sqrt{1-x^2}}\int_0^y z(1 + x + y + z)dz\,dy\,dx\right)$

(c) $I_z = \int_0^1\int_0^{\sqrt{1-x^2}}\int_0^y(x^2 + y^2)(1 + x + y + z)dz\,dy\,dx$

41. (a) $m = \int_{-1}^1\int_{-\sqrt{1-y^2}}^{\sqrt{1-y^2}}\int_{4y^2+4z^2}^4(x^2 + y^2 + z^2)dx\,dz\,dy$

(b) $(\bar{x}, \bar{y}, \bar{z})$ where $\bar{x} = m^{-1}\int_{-1}^1\int_{-\sqrt{1-y^2}}^{\sqrt{1-y^2}}\int_{4y^2+4z^2}^4 x(x^2 + y^2 + z^2)dx\,dz\,dy$,

$\bar{y} = m^{-1}\int_{-1}^1\int_{-\sqrt{1-y^2}}^{\sqrt{1-y^2}}\int_{4y^2+4z^2}^4 y(x^2 + y^2 + z^2)dx\,dz\,dy$, and

$\bar{z} = m^{-1}\int_{-1}^1\int_{-\sqrt{1-y^2}}^{\sqrt{1-y^2}}\int_{4y^2+4z^2}^4 z(x^2 + y^2 + z^2)dx\,dz\,dy$

(c) $I_z = \int_{-1}^1\int_{-\sqrt{1-y^2}}^{\sqrt{1-y^2}}\int_{4y^2+4z^2}^4(x^2 + y^2)(x^2 + y^2 + z^2)dx\,dz\,dy$

43. Using the formulas in the solution to Exercise 39, a CAS gives

(a) $m = \dfrac{3\pi}{32} + \dfrac{11}{24}$ **(b)** $(\bar{x}, \bar{y}, \bar{z}) = \left(\dfrac{28}{9\pi + 44}, \dfrac{30\pi + 128}{45\pi + 220}, \dfrac{45\pi + 208}{135\pi + 660}\right)$ **(c)** $I_z = \dfrac{68 + 15\pi}{240}$

45. $I_x = \int_0^L\int_0^L\int_0^L k(y^2 + z^2)dz\,dy\,dx = kL\int_0^L[y^2L + \frac{1}{3}L^3]dy = kL[\frac{1}{3}L^4 + \frac{1}{3}L^4] = \frac{2}{3}kL^5$.

By symmetry, $I_x = I_y = I_z = \frac{2}{3}kL^5$.

47. $V(E) = L^3$, $f_{\text{ave}} = \dfrac{1}{L^3}\int_0^L\int_0^L\int_0^L xyz\,dx\,dy\,dz = \dfrac{1}{L^3}\int_0^L x^2\,dx\int_0^L y^2\,dy\int_0^L z^2\,dz = \dfrac{1}{L^3}\dfrac{L^2}{2}\dfrac{L^2}{2}\dfrac{L^2}{2} = \dfrac{L^3}{8}$

49. The triple integral will attain its maximum when the integrand $1 - x^2 - 2y^2 - 3z^2$ is positive in the region E and negative everywhere else. For if E contains some region F where the integrand is negative, the integral could be increased by excluding F from E, and if E fails to contain some part G of the region where the integrand is positive, the integral could be increased by including G in E. So we require that $x^2 + 2y^2 + 3z^2 \leq 1$. This describes the region bounded by the ellipsoid $x^2 + 2y^2 + 3z^2 = 1$.

EXERCISES 13.8

1.

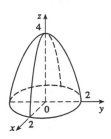

$$\int_0^{2\pi}\int_0^2\int_0^{4-r^2} r\,dz\,dr\,d\theta = \int_0^{2\pi}\int_0^2 (4r - r^3)\,dr\,d\theta$$

$$= \int_0^{2\pi} \left[2r^2 - \tfrac{1}{4}r^4\right]_0^2 d\theta$$

$$= \int_0^{2\pi} (8 - 4)d\theta = 8\pi$$

3.

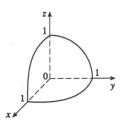

$$\int_0^{\pi/2}\int_0^{\pi/2}\int_0^1 \rho^2 \sin\phi\,d\rho\,d\theta\,d\phi = \int_0^{\pi/2}\int_0^{\pi/2} \tfrac{1}{3}\sin\phi\,d\theta\,d\phi$$

$$= \frac{1}{3}\int_0^{\pi/2} \tfrac{\pi}{2}\sin\phi\,d\phi$$

$$= \tfrac{\pi}{6}[-\cos\phi]_0^{\pi/2} = \tfrac{\pi}{6}$$

5. $\iiint_E (x^2 + y^2)dV = \int_{-1}^2\int_0^{2\pi}\int_0^2 (r^2)r\,dr\,d\theta\,dz = (3)(2\pi)\left[\tfrac{1}{4}r^4\right]_0^2 = 24\pi$

7. $\iiint_E y\,dV = \int_0^{2\pi}\int_1^2\int_0^{2+r\cos\theta} r^2 \sin\theta\,dz\,dr\,d\theta = \int_0^{2\pi}\int_1^2 [2r^2 \sin\theta + r^3 \cos\theta]dr\,d\theta$

$$= \int_0^{2\pi}\left[\tfrac{1}{3}(2r^3 \sin\theta) + \tfrac{1}{4}(r^4 \cos\theta)\right]_1^2 d\theta = \int_0^{2\pi}\left[\tfrac{14}{3}\sin\theta + \tfrac{15}{4}\cos\theta\right]d\theta = 0$$

9. $\iiint_E x^2\,dV = \int_0^{2\pi}\int_0^1\int_0^{2r} r^2 \cos^2\theta\,r\,dz\,dr\,d\theta = \int_0^{2\pi}\int_0^1 [r^3 \cos^2\theta\,z]_0^{2r}\,dr\,d\theta$

$$= \int_0^{2\pi}\int_0^1 2r^4 \cos^2\theta\,dr\,d\theta = \int_0^{2\pi}\left[\tfrac{2}{5}r^5 \cos^2\theta\right]_0^1 d\theta = \tfrac{2}{5}\int_0^{2\pi}\cos^2\theta\,d\theta$$

$$= \frac{2}{5}\int_0^{2\pi}\frac{1 + \cos 2\theta}{2}\,d\theta = \tfrac{1}{5}\left[\theta + \tfrac{1}{2}\sin 2\theta\right]_0^{2\pi} = \frac{2\pi}{5}$$

11. The paraboloids intersect when $x^2 + y^2 = 36 - 3x^2 - 3y^2 \;\;\Rightarrow\;\; D = \{(x,y) \mid x^2 + y^2 \le 9\}$. So, using cylindrical coordinates, $E = \{(r, \theta, z) \mid r^2 \le z \le 36 - r^2, 0 \le r \le 3, 0 \le \theta \le 2\pi\}$ and $V = \int_0^{2\pi}\int_0^3\int_{r^2}^{36-3r^2} r\,dz\,dr\,d\theta = 2\pi\int_0^3 (36r - 4r^3)dr = 2\pi\left[18r^2 - r^4\right]_0^3 = 162\pi$.

13. The paraboloid $z = 4x^2 + 4y^2$ intersects the plane $z = a$ when $a = 4x^2 + 4y^2$ or $x^2 + y^2 = \tfrac{1}{4}a$. So, using cylindrical coordinates, $E = \left\{(r, \theta, z) \mid 0 \le r \le \tfrac{1}{2}\sqrt{a}, 0 \le \theta \le 2\pi, 4r^2 \le z \le a\right\}$. Thus $m = \int_0^{2\pi}\int_0^{\sqrt{a}/2}\int_{4r^2}^a Kr\,dz\,dr\,d\theta = 2\pi K\int_0^{\sqrt{a}/2}(ar - 4r^3)dr = 2\pi K\left[\tfrac{1}{2}ar^2 - r^4\right]_0^{\sqrt{a}/2} = \tfrac{1}{8}a^2\pi K$. Since the region is homogeneous and symmetric, $M_{yz} = M_{xz} = 0$ and
$M_{xy} = \int_0^{2\pi}\int_0^{\sqrt{a}/2}\int_{4r^2}^a Krz\,dz\,dr\,d\theta = 2\pi K\int_0^{\sqrt{a}/2}(\tfrac{1}{2}a^2 r - 8r^5)dr = 2\pi K\left[\tfrac{1}{4}a^2 r^2 - \tfrac{4}{3}r^6\right]_0^{\sqrt{a}/2} = \tfrac{1}{12}a^3\pi K$.
Hence $(\overline{x}, \overline{y}, \overline{z}) = \left(0, 0, \tfrac{2}{3}a\right)$.

15. $\iiint_B (x^2 + y^2 + z^2)dV = \int_0^\pi\int_0^{2\pi}\int_0^1 \rho^4 \sin\phi\,d\rho\,d\theta\,d\phi = 2\pi\int_0^\pi \tfrac{1}{5}\sin\phi\,d\phi = \tfrac{2\pi}{5}[-\cos\phi]_0^\pi = \tfrac{4\pi}{5}$

17. $\iiint_E y^2 \, dV = \int_0^{\pi/2}\int_0^{\pi/2}\int_0^1 (\rho^2 \sin^2\phi \sin^2\theta)(\rho^2 \sin\phi) d\rho \, d\phi \, d\theta$

$\qquad = \int_0^{\pi/2}\int_0^{\pi/2}\int_0^1 \rho^4 \sin^3\phi \sin^2\theta \, d\rho \, d\phi \, d\theta = \int_0^{\pi/2}\int_0^{\pi/2} \frac{1}{5} \sin^3\phi \sin^2\theta \, d\phi \, d\theta$

$\qquad = \int_0^{\pi/2} \frac{2}{15} \sin^2\theta \, d\theta = \frac{2}{15}\left[\frac{1}{2}\theta - \frac{1}{4}\sin 2\theta\right]_0^{\pi/2} = \frac{\pi}{30}$

19. $\iiint_E \sqrt{x^2+y^2+z^2} \, dV = \int_0^{2\pi}\int_0^{\pi/6}\int_0^2 \rho^3 \sin\phi \, d\rho \, d\phi \, d\theta$

$\qquad = 8\pi\int_0^{\pi/6} \sin\phi \, d\phi = 8\pi[-\cos\phi]_0^{\pi/6} = 8\pi\left(1 - \frac{\sqrt{3}}{2}\right) = 4\pi\left(2 - \sqrt{3}\right)$

21. Since $\rho = 4\cos\phi$ implies $\rho^2 = 4\rho\cos\phi$, the equation is that of a sphere of radius 2 with center at $(0,0,2)$. Thus

$V = \int_0^{2\pi}\int_0^{\pi/3}\int_0^{4\cos\phi} \rho^2 \sin\phi \, d\rho \, d\phi \, d\theta = 2\pi\int_0^{\pi/3} \sin\phi\left(\frac{64}{3}\cos^3\phi\right)d\phi = \frac{32}{3}\pi\left[-\cos^4\phi\right]_0^{\pi/3} = 10\pi.$

23. Placing the center of the base at $(0,0,0)$, $\rho(x,y,z) = K\sqrt{x^2+y^2+z^2}$ is the density function. So

$m = \int_0^{2\pi}\int_0^{\pi/2}\int_0^a K\rho^3 \sin\phi \, d\rho \, d\phi \, d\theta = 2\pi K\int_0^{\pi/2} \frac{1}{4}a^4 \sin\phi \, d\phi = \frac{1}{2}\pi Ka^4[-\cos\phi]_0^{\pi/2} = \frac{1}{2}\pi Ka^4.$

25. $I_z = \int_0^{2\pi}\int_0^{\pi/2}\int_0^a (K\rho^3 \sin\phi)(\rho^2 \sin^2\phi) d\rho \, d\phi \, d\theta = 2\pi K\int_0^{\pi/2} \frac{1}{6}a^6 \sin^3\phi \, d\phi$

$\qquad = \frac{1}{3}\pi a^6 K\left[-\cos\phi + \frac{1}{3}\cos^3\phi\right]_0^{\pi/2} = \frac{2}{9}\pi Ka^6$

27. Place the center of the base at $(0,0,0)$; the density function is $\rho(x,y,z) = K$. By symmetry, the moments of inertia about any two such diameters will be equal, so we just need to find I_x:

$I_x = \int_0^{2\pi}\int_0^{\pi/2}\int_0^a (K\rho^2 \sin\phi)\rho^2(\sin^2\phi \sin^2\theta + \cos^2\phi) d\rho \, d\phi \, d\theta$

$\qquad = K\int_0^{2\pi}\int_0^{\pi/2} (\sin^3\phi \sin^2\theta + \sin\phi\cos^2\phi)\left(\frac{1}{5}a^5\right)d\phi \, d\theta$

$\qquad = \frac{1}{5}Ka^5\int_0^{2\pi}\left[\sin^2\theta\left(-\cos\phi + \frac{1}{3}\cos^3\phi\right) + \left(-\frac{1}{3}\cos^3\phi\right)\right]_0^{\pi/2} d\theta = \frac{1}{5}Ka^5\int_0^{2\pi}\left[\frac{2}{3}\sin^2\theta + \frac{1}{3}\right]d\theta$

$\qquad = \frac{1}{5}Ka^5\left[\frac{2}{3}\left(\frac{1}{2}\theta - \frac{1}{4}\sin 2\theta\right) + \frac{1}{3}\theta\right]_0^{2\pi} = \frac{1}{5}Ka^5\left[\frac{2}{3}(\pi - 0) + \frac{1}{3}(2\pi - 0)\right] = \frac{4}{15}Ka^5\pi$

29. In spherical coordinates $z = \sqrt{x^2+y^2}$ becomes $\cos\phi = \sin\phi$ or $\phi = \pi/4$. Then

$V = \int_0^{2\pi}\int_0^{\pi/4}\int_0^1 \rho^2 \sin\phi \, d\rho \, d\phi \, d\theta = 2\pi\int_0^{\pi/4} \sin\phi \, d\phi\left(\int_0^1 \rho^2 \, d\rho\right) = \frac{1}{3}\pi\left(2 - \sqrt{2}\right),$

$M_{xy} = \int_0^{2\pi}\int_0^{\pi/4}\int_0^1 \rho^3 \sin\phi\cos\phi \, d\rho \, d\phi \, d\theta = 2\pi\left[-\frac{1}{4}\cos 2\phi\right]_0^{\pi/4}\left(\frac{1}{4}\right) = \frac{\pi}{8}$ and by symmetry $M_{yz} = M_{xz} = 0.$

Hence $(\bar{x}, \bar{y}, \bar{z}) = \left(0, 0, \frac{3}{8(2-\sqrt{2})}\right).$

31. In cylindrical coordinates the paraboloid is given by $z = r^2$ and the plane by $z = 2r\sin\theta$ and they intersect in the

circle $r = 2\sin\theta$. Then $\iiint_E z \, dV = \int_0^\pi\int_0^{2\sin\theta}\int_{r^2}^{2r\sin\theta} rz \, dz \, dr \, d\theta = \frac{5\pi}{6}$ (using a CAS).

33. $\int_{-1}^1\int_{-\sqrt{1-x^2}}^{\sqrt{1-x^2}}\int_{x^2+y^2}^{2-x^2-y^2} (x^2+y^2)^{3/2} dz \, dy \, dx = \int_0^{2\pi}\int_0^1\int_{r^2}^{2-r^2} (r^2)^{3/2} r \, dz \, dr \, d\theta$

$\qquad = 2\pi\int_0^1 \left[r^4 z\right]_{r^2}^{2-r^2} dr = 2\pi\int_0^1 \left(2r^4 - r^6 - r^6\right)dr = 4\pi\left[\frac{1}{5}r^5 - \frac{1}{7}r^7\right]_0^1 = \frac{8\pi}{35}$

35. The region of integration E is the top half of the sphere $x^2 + y^2 + z^2 = 9$. So

$\int_{-3}^3\int_{-\sqrt{9-x^2}}^{\sqrt{9-x^2}}\int_0^{\sqrt{9-x^2-y^2}} z\sqrt{x^2+y^2+z^2} \, dz \, dy \, dx = \iiint_E z\sqrt{x^2+y^2+z^2} \, dV$

$\qquad = \int_0^{2\pi}\int_0^{\pi/2}\int_0^3 (\rho^2 \cos\phi)(\rho^2 \sin\phi) d\rho \, d\phi \, d\theta = 2\pi\int_0^{\pi/2} \frac{243}{5}\cos\phi \sin\phi \, d\phi = \frac{486}{5}\pi\left[-\frac{1}{4}\cos 2\phi\right]_0^{\pi/2} = \frac{243}{5}\pi$

37. (a) From the diagram, $z = r \cot \phi_0$ to $z = \sqrt{a^2 - r^2}$, $r = 0$ to
$r = a \sin \phi_0$ (or use $a^2 - r^2 = r^2 \cot^2 \phi_0$). Thus

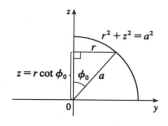

$$V = \int_0^{2\pi} \int_0^{a \sin \phi_0} \int_{r \cot \phi_0}^{\sqrt{a^2 - r^2}} r \, dz \, dr \, d\theta$$

$$= 2\pi \int_0^{a \sin \phi_0} \left(r\sqrt{a^2 - r^2} - r^2 \cot \phi_0 \right) dr$$

$$= \tfrac{2\pi}{3} \left[-(a^2 - r^2)^{3/2} - r^3 \cot \phi_0 \right]_0^{a \sin \phi_0}$$

$$= \tfrac{2\pi}{3} \left[-(a^2 - a^2 \sin^2 \phi_0)^{3/2} - a^3 \sin^3 \phi_0 \cot \phi_0 + a^3 \right]$$

$$= \tfrac{2}{3}\pi a^3 [1 - (\cos^3 \phi_0 + \sin^2 \phi_0 \cos \phi_0)] = \tfrac{2}{3}\pi a^3 (1 - \cos \phi_0)$$

(b) The wedge in question is the shaded area rotated from $\theta = \theta_1$ to $\theta = \theta_2$.

Letting

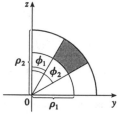

V_{ij} = volume of the region bounded by the sphere of radius ρ_i
    and the cone with angle ϕ_j ($\theta = \theta_1$ to θ_2)

and letting V be the volume of the wedge, we have

$$V = (V_{22} - V_{21}) - (V_{12} - V_{11})$$

$$= \tfrac{1}{3}(\theta_2 - \theta_1)[\rho_2^3(1 - \cos \phi_2) - \rho_2^3(1 - \cos \phi_1) - \rho_1^3(1 - \cos \phi_2) + \rho_1^3(1 - \cos \phi_1)]$$

$$= \tfrac{1}{3}(\theta_2 - \theta_1)[(\rho_2^3 - \rho_1^3)(1 - \cos \phi_2) - (\rho_2^3 - \rho_1^3)(1 - \cos \phi_1)]$$

$$= \tfrac{1}{3}(\theta_2 - \theta_1)[(\rho_2^3 - \rho_1^3)(\cos \phi_1 - \cos \phi_2)]$$

Or: Show that $V = \displaystyle\int_{\theta_1}^{\theta_2} \int_{\rho_1 \sin \phi_1}^{\rho_2 \sin \phi_2} \int_{r \cot \phi_2}^{r \cot \phi_1} r \, dz \, dr \, d\theta$.

(c) By the Mean Value Theorem with $f(\rho) = \rho^3$ there exists some $\tilde{\rho}$ with $\rho_1 \le \tilde{\rho} \le \rho_2$ such that
$f(\rho_2) - f(\rho_1) = f'(\tilde{\rho})(\rho_2 - \rho_1)$ or $\rho_1^3 - \rho_2^3 = 3\tilde{\rho}^2 \Delta \rho$. Similarly there exists ϕ with $\phi_1 \le \tilde{\phi} \le \phi_2$ such
that $\cos \phi_2 - \cos \phi_1 = (-\sin \tilde{\phi}) \Delta \phi$. Substituting into the result from (b) gives
$\Delta V = (\tilde{\rho}^2 \Delta \rho)(\theta_2 - \theta_1)(\sin \tilde{\phi}) \Delta \phi = \tilde{\rho}^2 \sin \tilde{\phi} \, \Delta \rho \, \Delta \phi \, \Delta \theta$.

EXERCISES 13.9

1. $\dfrac{\partial(x,y)}{\partial(u,v)} = \begin{vmatrix} \partial x/\partial u & \partial x/\partial v \\ \partial y/\partial u & \partial y/\partial v \end{vmatrix} = \begin{vmatrix} 1 & -2 \\ 2 & -1 \end{vmatrix} = 1(-1) - 2(-2) = 3$

3. $\dfrac{\partial(x,y)}{\partial(u,v)} = \begin{vmatrix} \partial x/\partial u & \partial x/\partial v \\ \partial y/\partial u & \partial y/\partial v \end{vmatrix} = \begin{vmatrix} 2e^{2u}\cos v & -e^{2u}\sin v \\ 2e^{2u}\sin v & e^{2u}\cos v \end{vmatrix} = 2e^{4u}(\cos^2 v + \sin^2 v) = 2e^{4u}$

5. $\dfrac{\partial(x,y,z)}{\partial(u,v,w)} = \begin{vmatrix} 1 & 1 & 1 \\ 1 & 1 & -1 \\ 1 & -1 & 1 \end{vmatrix} = 1(1-1) - 1(1+1) + 1(-1-1) = -4$

7. S_1: $v = 0, 0 \le u \le 2$, so $x = u$, $y = 2u$ and $y = 2x$. S_2: $u = 2, 0 \le v \le 1$, so $x = 2 - 2v$, $y = 4 - v$ and $x = 2y - 6$. S_3: $v = 1, 0 \le u \le 2$, so $x = u - 2$, $y = 2u - 1$ and $y = 2x + 3$. S_4: $u = 0, 0 \le v \le 1$, so $x = -2v$, $y = -v$ and $2y = x$.

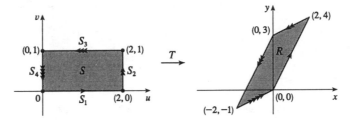

9. S_1: $0 \le u \le 1, v = 0, x = 4u, y = 0$ so $0 \le x \le 4, y = 0$ is the image of the first side.

S_2: $u + v = 1, x = 4(1 - v) + 3v = 4 - v \Leftrightarrow v = 4 - x, y = 4v = 4(4 - x) = 16 - 4x, 3 \le x \le 4$, so $y = 16 - 4x, 3 \le x \le 4$ is the image of the second side.

S_3: $u = 0, 0 \le v \le 1, x = 3v, y = 4v = \frac{4}{3}x, 0 \le x \le 3$, so $y = \frac{4}{3}x, 0 \le x \le 3$ is the image of the third side.

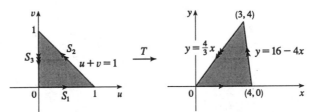

11. $\dfrac{\partial(x,y)}{\partial(u,v)} = \begin{vmatrix} 1/3 & 1/3 \\ -2/3 & 1/3 \end{vmatrix} = \dfrac{1}{3}$ and $3x + 4y = (u + v) + \frac{4}{3}(v - 2u) = \frac{1}{3}(7v - 5u)$. Then S is the region bounded by the lines $u = 0, \frac{1}{3}(v - 2) = \frac{1}{3}(u + v) - 2$ or $u = 2, \frac{1}{3}(v - 2u) = -\frac{2}{3}(u + v)$ or $v = 0$, and $\frac{1}{3}(v - 2u) = 3 - \frac{2}{3}(u + v)$ or $v = 3$. Thus

$\iint_R (3x + 4y)\,dA = \int_0^3 \int_0^2 \frac{1}{3}(7v - 5u)\left(\frac{1}{3}du\,dv\right) = \frac{1}{9}\int_0^3 (14v - 10)\,dv = \frac{1}{9}(33) = \frac{11}{3}$.

13. $\dfrac{\partial(x,y)}{\partial(u,v)} = \begin{vmatrix} 2 & 0 \\ 0 & 3 \end{vmatrix} = 6$, $x^2 = 4u^2$ and the planar ellipse $9x^2 + 4y^2 \le 36$ is the image of the disk $u^2 + v^2 \le 1$.

Thus $\iint_R x^2\,dA = \iint_{u^2+v^2\le 1} (4u^2)(6)\,du\,dv = \int_0^{2\pi}\int_0^1 (24r^2\cos^2\theta)r\,dr\,d\theta = 6\pi$.

15. $\dfrac{\partial(x,y)}{\partial(u,v)} = \begin{vmatrix} 1/v & -u/v^2 \\ 0 & 1 \end{vmatrix} = \dfrac{1}{v}$, $xy = u$, $y = x$ is the image of the parabola $v^2 = u$, $y = 3x$ is the image of the

parabola $v^2 = 3u$, and the hyperbolas $xy = 1$, $xy = 3$ are the images of the lines $u = 1$ and $u = 3$ respectively.

Thus $\displaystyle\iint_R xy\,dA = \int_1^3 \int_{\sqrt{u}}^{\sqrt{3u}} u\left(\frac{1}{v}\right)dv\,du = \int_1^3 u\left(\ln\sqrt{3u} - \ln\sqrt{u}\right)du = \int_1^3 u\ln\sqrt{3}\,du = 4\ln\sqrt{3} = 2\ln 3$.

17. $\dfrac{\partial(x,y,z)}{\partial(u,v,w)} = \begin{vmatrix} a & 0 & 0 \\ 0 & b & 0 \\ 0 & 0 & c \end{vmatrix} = abc$ and the solid enclosed by the ellipsoid is the image of the ball

$u^2 + v^2 + w^2 \le 1$. So $\displaystyle\iiint_E dV = \iiint_{u^2+v^2+w^2\le 1} abc\,du\,dv\,dw = (abc)(\text{volume of the ball}) = \frac{4}{3}\pi abc$.

19. Letting $u = 2x - y$ and $v = 3x + y$, we have $x = \frac{1}{5}(u + v)$, $y = \frac{1}{5}(2v - 3u)$. Then

$\dfrac{\partial(x,y)}{\partial(u,v)} = \begin{vmatrix} 1/5 & 1/5 \\ -3/5 & 2/5 \end{vmatrix} = \dfrac{1}{5}$ and

$\displaystyle\iint_R xy\,dA = \int_{-2}^1 \int_{-3}^1 \frac{(u+v)(2v-3u)}{25}\left(\frac{1}{5}\right)du\,dv = \frac{1}{125}\int_{-2}^1\int_{-3}^1 (2v^2 - uv - 3u^2)\,du\,dv$

$= \dfrac{1}{125}\int_{-2}^1 (8v^2 + 4v - 28)\,dv = -\dfrac{66}{125}$.

21. Letting $u = y - x$, $v = y + x$, we have $y = \frac{1}{2}(u + v)$, $x = \frac{1}{2}(v - u)$. Then $\dfrac{\partial(x,y)}{\partial(u,v)} = \begin{vmatrix} -1/2 & 1/2 \\ 1/2 & 1/2 \end{vmatrix} = -\dfrac{1}{2}$ and

R is the image of the trapezoidal region with vertices $(-1,1)$, $(-2,2)$, $(2,2)$, and $(1,1)$. Thus

$\displaystyle\iint_R \cos\frac{y-x}{y+x}\,dA = \int_1^2\int_{-v}^v \left|-\tfrac{1}{2}\right|\cos\frac{u}{v}\,du\,dv = \frac{1}{2}\int_1^2\left[v\sin\frac{u}{v}\right]_{-v}^v dv = \frac{1}{2}\int_1^2 2v\sin(1)\,dv = \frac{3}{2}\sin 1$.

23.

Let $u = x + y$ and $v = -x + y$. Then $u + v = 2y$ $\Rightarrow$

$y = \frac{1}{2}(u + v)$ and $u - v = 2x$ $\Rightarrow$ $x = \frac{1}{2}(u - v)$.

$\dfrac{\partial(x,y)}{\partial(u,v)} = \begin{vmatrix} 1/2 & -1/2 \\ 1/2 & 1/2 \end{vmatrix} = \dfrac{1}{2}$. Now

$|u| = |x + y| \le |x| + |y| \le 1$ $\Rightarrow$ $-1 \le u \le 1$, and

$|v| = |-x + y| \le |x| + |y| \le 1$ $\Rightarrow$ $-1 \le v \le 1$.

R is the image of the square region with vertices $(1,1)$, $(1,-1)$, $(-1,-1)$, and $(-1,1)$. So

$\displaystyle\iint_R e^{x+y}\,dA = \frac{1}{2}\int_{-1}^1\int_{-1}^1 e^u\,du\,dv = \frac{1}{2}\cdot 2\cdot e^u\big|_{-1}^1 = e - e^{-1}$.

REVIEW EXERCISES FOR CHAPTER 13

1. This is true by Fubini's Theorem.

3. $\iint_D \sqrt{4 - x^2 - y^2}\, dA$ = the volume under the surface $x^2 + y^2 + z^2 = 4$ and above the
xy-plane = $\frac{1}{2}$(the volume of the sphere $x^2 + y^2 + z^2 = 4$) = $\frac{1}{2} \cdot \frac{4}{3}\pi(2)^3 = \frac{16}{3}\pi$.

5. The volume enclosed by the cone $z = \sqrt{x^2 + y^2}$ and the plane $z = 2$ is, using cylindrical coordinates,
$V = \int_0^{2\pi}\int_0^2\int_r^2 r\, dz\, dr\, d\theta \neq \int_0^{2\pi}\int_0^2\int_r^2 dz\, dr\, d\theta$, so the assertion is false.

7. $\int_{-2}^2\int_0^4 (4x^3 + 3xy^2)dx\, dy = \int_{-2}^2 (256 + 24y^2)dy = 2[256y + 8y^3]_0^2 = 1152$

9. $\int_1^2\int_0^{x^2} \frac{1}{x+y}\, dy\, dx = \int_1^2 [\ln|x+y|]_0^{x^2}\, dx = \int_1^2 [\ln(1+x)]dx = [(1+x)\ln(1+x) - (1+x)]_1^2 = \ln\left(\frac{27}{4}\right) - 1$

11. $\int_0^1\int_0^{x^2}\int_0^y y^2 z\, dz\, dy\, dx = \int_0^1\int_0^{x^2} \frac{1}{2}y^4\, dy\, dx = \int_0^1 \frac{1}{10}x^{10}\, dx = \frac{1}{110}$

13.

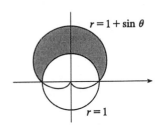

The region whose area is given by $\int_0^\pi\int_1^{1+\sin\theta} r\, dr\, d\theta$ is
$\{(r,\theta)\mid 0 \leq \theta \leq \pi, 1 \leq r \leq 1+\sin\theta\}$, which is the region
outside the circle $r = 1$ and inside the cardioid $r = 1 + \sin\theta$.

15.

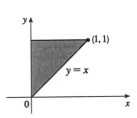

$\int_0^1\int_x^1 e^{x/y}\, dy\, dx = \int_0^1\int_0^y e^{x/y}\, dx\, dy$
$= \int_0^1 (ey - y)dy = \frac{1}{2}(e - 1)$

17. $\int_2^4\int_0^1 \frac{1}{(x-y)^2}\, dx\, dy = \int_2^4\left(-\frac{1}{y} - \frac{1}{1-y}\right)dy = [-\ln y + \ln|1 - y|]_2^4$
$= -\ln 4 + \ln 3 + \ln 2 = \ln\frac{3}{2}$

19. The curves $y^2 = x^3$ and $y = x$ intersect when $x^3 = x$, that is when $x = 0$ and $x = 1$ (note that $x \neq -1$ since
$x^0 = y^0 \Rightarrow x \geq 0$.) So $\int_0^1\int_{x^{3/2}}^x xy\, dy\, dx - \int_0^1 [\frac{1}{2}x^7 - \frac{1}{2}x^4]dx - [\frac{1}{8}x^4 - \frac{1}{10}x^5]_0^1 - \frac{1}{40}$.

21. $\int_0^1\int_0^{1-y^2} (xy + 2x + 3y)dx\, dy = \int_0^1 \left[(y+2)\frac{1}{2}(1-y^2)^2 + 3y(1-y^2)\right]dy$
$= \frac{1}{2}y^5 + y^4 - 4y^3 - 2y^2 + \frac{7}{2}y + 1 = \frac{1}{5} + \frac{1}{12} - \frac{2}{3} + \frac{7}{4} = \frac{41}{30}$

23.

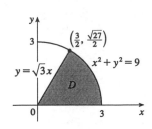

$$\iint_D (x^2 + y^2)^{3/2} \, dA = \int_0^{\pi/3} \int_0^3 (r^2)^{3/2} \, r \, dr \, d\theta$$

$$= \frac{\pi}{3} \frac{3^5}{5} = \frac{81\pi}{5}$$

25. $\iiint_E x^2 z \, dV = \int_0^2 \int_0^{2x} \int_0^x x^2 z \, dz \, dy \, dx = \int_0^2 \int_0^{2x} \frac{1}{2} x^4 \, dy \, dx = \int_0^2 x^5 \, dx = \frac{1}{6} \cdot 2^6 = \frac{32}{3}$

27. $\iiint_E y^2 z^2 \, dV = \int_{-1}^1 \int_{-\sqrt{1-y^2}}^{\sqrt{1-y^2}} \int_0^{1-y^2-z^2} y^2 z^2 \, dx \, dz \, dy = \int_{-1}^1 \int_{-\sqrt{1-y^2}}^{\sqrt{1-y^2}} y^2 z^2 (1 - y^2 - z^2) dz \, dy$

$$= \int_0^{2\pi} \int_0^1 (r^2 \cos^2\theta)(r^2 \sin^2\theta)(1 - r^2) r \, dr \, d\theta = \int_0^{2\pi} \int_0^1 \frac{1}{4} \sin^2 2\theta (r^5 - r^7) \, dr \, d\theta$$

$$= \int_0^{2\pi} \frac{1}{8}(1 - \cos 4\theta)\left[\frac{1}{6} r^6 - \frac{1}{8} r^8\right]_0^1 d\theta = \frac{1}{192}\left[\theta - \frac{1}{4}\sin 4\theta\right]_0^{2\pi} = \frac{2\pi}{192} = \frac{\pi}{96}$$

29. $\iiint_E yz \, dV = \int_{-2}^2 \int_0^{\sqrt{4-x^2}} \int_0^y yz \, dz \, dy \, dx = \int_{-2}^2 \int_0^{\sqrt{4-x^2}} \frac{1}{2} y^3 \, dy \, dx = \int_0^\pi \int_0^2 \frac{1}{2} r^3 \sin^3\theta \, r \, dr \, d\theta$

$$= \frac{16}{5} \int_0^\pi \sin^3\theta \, d\theta = \frac{16}{5}\left[-\cos\theta + \frac{1}{3}\cos^3\theta\right]_0^\pi = \frac{64}{15}$$

31. $V = \int_0^2 \int_1^4 (x^2 + 4y^2) dy \, dx = \int_0^2 (3x^2 + 84) dx = 176$

33.

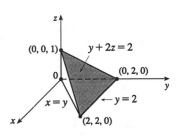

$$V = \int_0^2 \int_0^y \int_0^{(2-y)/2} dz \, dx \, dy = \int_0^2 \int_0^y \left(1 - \frac{1}{2}y\right) dx \, dy$$

$$= \int_0^2 \left(y - \frac{1}{2}y^2\right) dy = \frac{2}{3}$$

35. Using the wedge above the plane $z = 0$ and below the plane $z = mx$ and noting that we have the same volume for $m < 0$ as for $m > 0$ (so use $m > 0$), we have

$V = 2\int_0^{a/3} \int_0^{\sqrt{a^2-9y^2}} mx \, dx \, dy = 2\int_0^{a/3} \frac{1}{2} m(a^2 - 9y^2) dy = m[a^2 y - 3y^3]_0^{a/3} = m\left(\frac{1}{3}a^3 - \frac{1}{9}a^3\right) = \frac{2}{9}ma^3$.

37. $m = \int_0^1 \int_0^{1-y^2} y \, dx \, dy = \int_0^1 (y - y^3) dy = \frac{1}{2} - \frac{1}{4} = \frac{1}{4}$,

$M_y = \int_0^1 \int_0^{1-y^2} xy \, dx \, dy = \int_0^1 \frac{1}{2} y(1 - y^2)^2 \, dy = -\frac{1}{12}(1 - y^2)^3 \Big|_0^1 = \frac{1}{12}$,

$M_x = \int_0^1 \int_0^{1-y^2} y^2 \, dx \, dy = \int_0^1 (y^2 - y^4) dy = \frac{2}{15}$. Hence $(\overline{x}, \overline{y}) = \left(\frac{1}{3}, \frac{8}{15}\right)$.

39. $m = \frac{1}{4}\pi K a^2$ where K is constant,

$M_y = \iint\limits_{x^2+y^2 \le a^2} Kx \, dA = K\int_0^{\pi/2} \int_0^a r^2 \cos\theta \, dr \, d\theta = \frac{1}{3} Ka^3 \int_0^{\pi/2} \cos\theta \, d\theta = \frac{1}{3}a^3 K$, and

$M_x = K\int_0^{\pi/2} \int_0^a r^2 \sin\theta \, dr \, d\theta = \frac{1}{3}a^3 K$ (by symmetry $M_y = M_x$). Hence the centroid is $(\overline{x}, \overline{y}) = \left(\frac{4}{3\pi}a, \frac{4}{3\pi}a\right)$.

41. The equation of the cone with the suggested orientation is $(h - z) = \dfrac{h}{a}\sqrt{x^2 + y^2},\, 0 \le z \le h$. Then $V - \frac{1}{3}\pi a^2 h$

is the volume of one frustum of a cone; by symmetry $M_{yz} = M_{xz} = 0$; and

$$M_{xy} = \iint_{x^2+y^2 \le a^2} \int_0^{h-(h/a)\sqrt{x^2+y^2}} z\, dz\, dA = \int_0^{2\pi} \int_0^a \int_0^{(h/a)(a-r)} rz\, dz\, dr\, d\theta = \pi \int_0^a r\frac{h^2}{a^2}(a - r)^2\, dr$$

$$= \frac{\pi h^2}{a^2}\int_0^a (a^2 r - 2ar^2 + r^3)\, dr = \frac{\pi h^2}{a^2}\left(\frac{a^4}{2} - \frac{2a^4}{3} + \frac{a^4}{4}\right) = \frac{\pi h^2 a^2}{12}.$$

Hence the centroid is $(\overline{x}, \overline{y}, \overline{z}) = \left(0, 0, \frac{1}{4}h\right)$.

43. $z = 1 - \frac{1}{2}x - \frac{1}{3}y \;\Rightarrow\; \partial z/\partial x = -\frac{1}{2}$ and $\partial z/\partial y = -\frac{1}{3}$. The projection onto the xy-plane of the part of the

given plane in the first octant is the triangular region D bounded by the x- and y-axes and the line $3x + 2y = 6$,

which has intercepts 2 and 3. So $A(D) = \frac{1}{2}\cdot 2 \cdot 3 = 3$. Thus the surface area is

$$A(S) = \iint_D \sqrt{1 + \left(-\frac{1}{2}\right)^2 + \left(-\frac{1}{3}\right)^2}\, dA = \frac{7}{6}A(D) = \frac{7}{6}\cdot 3 = \frac{7}{2}.$$

45.

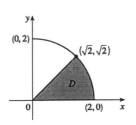

$$\int_0^{\sqrt{2}} \int_y^{\sqrt{4-y^2}} \frac{1}{1 + x^2 + y^2}\, dx\, dy = \int_0^{\pi/4} \int_0^2 \frac{1}{1 + r^2}\, r\, dr\, d\theta$$

$$= \frac{1}{2}\int_0^{\pi/4} \ln\left|1 + r^2\right|\Big|_0^2\, d\theta$$

$$= \frac{1}{2}\int_0^{\pi/4} \ln 5\, d\theta = \frac{\pi}{8}\ln 5$$

47. From the graph, it appears that $1 - x^2 = e^x$ at $x \approx -0.71$ and

at $x = 0$, with $1 - x^2 > e^x$ on $(-0.71, 0)$. So the desired integral is

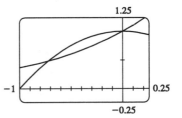

$$\iint_D y^2\, dA \approx \int_{-0.71}^0 \int_{e^x}^{1-x^2} y^2\, dy\, dx = \frac{1}{3}\int_{-0.71}^0 \left[(1 - x^2)^3 - e^{3x}\right] dx$$

$$= \frac{1}{3}\left[x - x^3 + \frac{3}{5}x^5 - \frac{1}{7}x^7 - \frac{1}{3}e^{3x}\right]_{-0.71}^0 \approx 0.0512.$$

49. Let the tetrahedron be called T. The front face of T is given by the plane $x + \frac{1}{2}y + \frac{1}{3}z = 1$, or $z = 3 - 3x - \frac{3}{2}y$,

which intersects the xy-plane in the line $y = 2 - 2x$. So the total mass is

$$m = \iiint_T \rho(x, y, z)\, dV = \int_0^1 \int_0^{2-2x} \int_0^{3-3x-\frac{3}{2}y} (x^2 + y^2 + z^2)\, dz\, dy\, dx = \frac{7}{5}.\text{ The center of mass is}$$

$$(\overline{x}, \overline{y}, \overline{z}) = \left(m^{-1}\iiint_T x\rho(x, y, z)\, dV, m^{-1}\iiint_T y\rho(x, y, z)\, dV, m^{-1}\iiint_T z\rho(x, y, z)\, dV\right) = \left(\tfrac{4}{21}, \tfrac{11}{21}, \tfrac{8}{7}\right).$$

51.

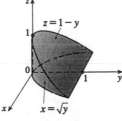

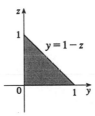

$$\int_{-1}^1 \int_{x^2}^1 \int_0^{1-y} f(x, y, z)\, dz\, dy\, dx = \int_0^1 \int_0^{1-z} \int_{-\sqrt{y}}^{\sqrt{y}} f(x, y, z)\, dx\, dy\, dz$$

53. Since $u = x - y$, $v = x + y$, $x = \frac{1}{2}(u + v)$ and $y = \frac{1}{2}(v - u)$. Thus $\dfrac{\partial(x, y)}{\partial(u, v)} = \begin{vmatrix} 1/2 & 1/2 \\ -1/2 & 1/2 \end{vmatrix} = \dfrac{1}{2}$ and

$$\iint_R \frac{x - y}{x + y}\, dA = \int_2^4 \int_{-2}^0 \frac{u}{v}\left(\tfrac{1}{2}\, du\, dv\right) = -\int_2^4 \frac{dv}{v} = -\ln 2.$$

55. Let $u = y - x$ and $v = y + x$ so $x = y - u = (v - x) - u \quad \Rightarrow \quad x = \frac{1}{2}(v - u)$ and

$y = v - \frac{1}{2}(v - u) = \frac{1}{2}(v + u)$. $\left|\dfrac{\partial(x, y)}{\partial(u, v)}\right| = \left|\dfrac{\partial x}{\partial u}\dfrac{\partial y}{\partial v} - \dfrac{\partial x}{\partial v}\dfrac{\partial y}{\partial u}\right| = \left|-\dfrac{1}{2}\left(\dfrac{1}{2}\right) - \dfrac{1}{2}\left(\dfrac{1}{2}\right)\right| = \left|-\dfrac{1}{2}\right| = \dfrac{1}{2}$. R is the

image under this transformation of the square with vertices $(u, v) = (0, 0)$, $(-2, 0)$, $(0, 2)$, and $(-2, 2)$. So

$$\iint_R xy\, dA = \int_0^2 \int_{-2}^0 \frac{v^2 - u^2}{4}\left(\frac{1}{2}\right) du\, dv = \frac{1}{8}\int_0^2 \left[v^2 u - \tfrac{1}{3}u^3\right]_{-2}^0 dv = \frac{1}{8}\int_0^2 \left(2v^2 - \tfrac{8}{3}\right) dv = \tfrac{1}{8}\left[\tfrac{2}{3}v^3 - \tfrac{8}{3}v\right]_0^2 = 0.$$

This result could have been anticipated by symmetry, since the integrand is an odd function of y and R is

symmetric about the x-axis.

57. For each r such that D_r lies within the domain D, $A(D_r) = \pi r^2$, and by the Mean Value Theorem for Double

Integrals there exists (x_r, y_r) in D_r such that $f(x_r, y_r) = \dfrac{1}{\pi r^2}\iint_{D_r} f(x, y)dA$. But $\lim\limits_{r \to 0^+}(x_r, y_r) = (a, b)$, so

$\lim\limits_{r \to 0^+} \dfrac{1}{\pi r^2}\iint_{D_r} f(x, y)dA = \lim\limits_{r \to 0^+} f(x_r, y_r) = f(a, b)$ by the continuity of f.

APPLICATIONS PLUS (after Chapter 13)

1. **(a)** The area of a trapezoid is $\frac{1}{2}h(b_1 + b_2)$, where h is the height (the distance between the two parallel sides) and b_1, b_2 are the lengths of the bases (the parallel sides). From the figure in the text, we see that $h = x \sin \theta$, $b_1 = w - 2x$, and $b_2 = w - 2x + 2x \cos \theta$. Therefore the cross-sectional area of the rain gutter is

$$A(x, \theta) = \tfrac{1}{2}x \sin \theta[(w - 2x) + (w - 2x + 2x \cos \theta)] = (x \sin \theta)(w - 2x + x \cos \theta)$$
$$= wx \sin \theta - 2x^2 \sin \theta + x^2 \sin \theta \cos \theta, \ 0 < x \le \tfrac{1}{2}w, 0 < \theta \le \tfrac{\pi}{2}$$

We look for the critical points of A: $\partial A/\partial x = w \sin \theta - 4x \sin \theta + 2x \sin \theta \cos \theta$ and

$\partial A/\partial \theta = wx \cos \theta - 2x^2 \cos \theta + x^2(\cos^2 \theta - \sin^2 \theta)$, so $\partial A/\partial x = 0 \ \Leftrightarrow$

$$\sin \theta(w - 4x + 2x \cos \theta) = 0 \quad \Leftrightarrow \quad \cos \theta = \frac{4x - w}{2x} = 2 - \frac{w}{2x} \ (0 < \theta \le \tfrac{\pi}{2} \quad \Rightarrow \quad \sin \theta > 0). \text{ If, in}$$

addition, $\partial A/\partial \theta = 0$, then

$$0 = wx \cos \theta - 2x^2 \cos \theta + x^2(2 \cos^2 \theta - 1)$$
$$= wx\left(2 - \frac{w}{2x}\right) - 2x^2\left(2 - \frac{w}{2x}\right) + x^2\left[2\left(2 - \frac{w}{2x}\right)^2 - 1\right]$$
$$= 2wx - \tfrac{1}{2}w^2 - 4x^2 + wx + x^2\left[8 - \frac{4w}{x} + \frac{w^2}{2x^2} - 1\right] = -wx + 3x^2 = x(3x - w)$$

Since $x > 0$, we must have $x = \frac{1}{3}w$, in which case $\cos \theta = \frac{1}{2}$, so $\theta = \frac{\pi}{3}$, $\sin \theta = \frac{\sqrt{3}}{2}$, $k = \frac{\sqrt{3}}{6}w$, $b_1 = \frac{1}{3}w$, $b_2 = \frac{2}{3}w$, and $A = \frac{\sqrt{3}}{12}w^2$. As in Example 12.7.6, we can argue from the physical nature of this problem that we have found a relative maximum of A. Now checking the boundary of A, let

$g(\theta) = A(w/2, \theta) = \frac{1}{2}w^2 \sin \theta - \frac{1}{2}w^2 \sin \theta + \frac{1}{4}w^2 \sin \theta \cos \theta = \frac{1}{8}w^2 \sin 2\theta, 0 < \theta \le \frac{\pi}{2}$. Clearly g is maximized when $\sin 2\theta = 1$ in which case $A = \frac{1}{8}w^2$. Also along the line $\theta = \frac{\pi}{2}$, let

$h(x) = A\left(x, \frac{\pi}{2}\right) = wx - 2x^2, 0 < x < \frac{1}{2}w \quad \Rightarrow \quad h'(x) = w - 4x = 0 \quad \Leftrightarrow \quad x = \frac{1}{4}w$, and

$h\left(\frac{1}{4}w\right) = w\left(\frac{1}{4}w\right) - 2\left(\frac{1}{4}w\right)^2 = \frac{1}{8}w^2$. Since $\frac{1}{8}w^2 < \frac{\sqrt{3}}{12}w^2$, we conclude that the relative maximum found earlier was an absolute maximum.

(b) If the metal were bent into a semi-circular gutter of radius r, we would have $w = \pi r$ and

$A = \frac{1}{2}\pi r^2 = \frac{1}{2}\pi\left(\frac{w}{\pi}\right)^2 = \frac{w^2}{2\pi}$. Since $\frac{w^2}{2\pi} > \frac{\sqrt{3}w^2}{12}$, it *would* be better to bend the metal into a gutter with a semi-circular cross-section.

3. **(a)** If $f(P, A)$ is the probability that an individual at A will be infected by an individual at P, and $k\, dA$ is the number of infected individuals in an element of area dA, then $f(P, A)k\, dA$ is the number of infections that should result from exposure of the individual at A to infected people in the element of area dA. Integration over D gives the number of infections of the person at A due to all the infected people in D. In rectangular coordinates (with the origin at the city's center), the exposure of a person at A is

$$E = \iint_D kf(P, A)dA = k \iint_D \frac{20 - d(P, A)}{20}\, dA = k \iint_D \left[1 - \frac{\sqrt{(x - x_0)^2 + (y - y_0)^2}}{20} \right] dx\, dy.$$

(b) If $A = (0, 0)$, then

$$E = k \iint_D \left[1 - \tfrac{1}{20}\sqrt{x^2 + y^2} \right] dx\, dy = k \int_0^{2\pi} \int_0^{10} \left(1 - \frac{r}{20} \right) r\, dr\, d\theta$$

$$= 2\pi k \left[\frac{r^2}{2} - \frac{r^3}{60} \right]_0^{10} = 2\pi k \left(50 - \tfrac{50}{3} \right) = \tfrac{200}{3}\pi k \approx 209k.$$

For A at the edge of the city, it is convenient to use a polar coordinate system centered at A. Then the polar equation for the circular boundary of the city becomes $r = 20 \cos\theta$ instead of $r = 10$, and the distance from A to a point P in the city is again r (see the figure.) So

$$E = k \int_{-\pi/2}^{\pi/2} \int_0^{20\cos\theta} \left(1 - \frac{r}{20} \right) r\, dr\, d\theta$$

$$= k \int_{-\pi/2}^{\pi/2} \left[\frac{r^2}{2} - \frac{r^3}{60} \right]_0^{20\cos\theta} d\theta$$

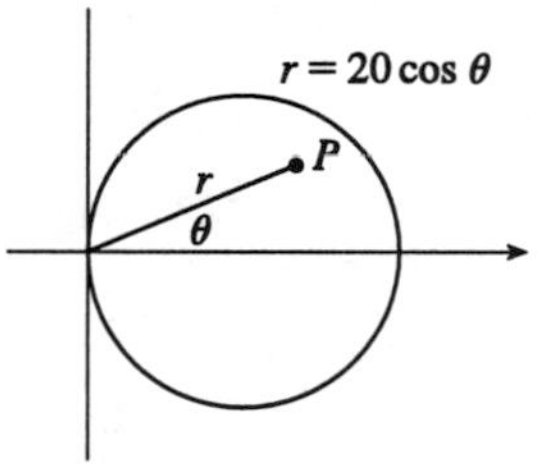

$$= k \int_{-\pi/2}^{\pi/2} \left(200 \cos^2\theta - \tfrac{400}{3}\cos^3\theta \right) d\theta$$

$r = 20 \cos\theta$

$$= 200k \int_{-\pi/2}^{\pi/2} \left[\tfrac{1}{2} + \tfrac{1}{2}\cos 2\theta - \tfrac{2}{3}(1 - \sin^2\theta)\cos\theta \right] d\theta$$

$$= 200k \left[\tfrac{1}{2}\theta + \tfrac{1}{4}\sin 2\theta - \tfrac{2}{3}\sin\theta + \tfrac{2}{3}\cdot\tfrac{1}{3}\sin^3\theta \right]_{-\pi/2}^{\pi/2}$$

$$= 200k \left[\frac{\pi}{4} + 0 - \frac{2}{3} + \frac{2}{9} + \frac{\pi}{4} + 0 - \frac{2}{3} + \frac{2}{9} \right] = 200k \left(\frac{\pi}{2} - \frac{8}{9} \right) \approx 136k. \text{ Therefore the risk of}$$

infection is much lower at the edge of the city than in the middle, so it is better to live at the edge.

APPLICATIONS PLUS

5. **(a)** The mountain comprises a solid conical region C. The work done in lifting a small volume of material ΔV with density $g(P)$ to a height $h(P)$ above sea level is $h(P)g(P)\Delta V$. Summing over the whole mountain we get $W = \iiint_C h(P)g(P)dV$.

(b) Here C is a solid right circular cone with radius $R = 62,000$ ft, height $H = 12,400$ ft, and density $g(P) = 200\,\text{lb}/\text{ft}^3$ at all points P in C. We use cylindrical coordinates:

$$W = \int_0^{2\pi} \int_0^H \int_0^{R(1-z/H)} z \cdot 200\, r\, dr\, dz\, d\theta$$

$$= 2\pi \int_0^H 200z\left[\tfrac{1}{2}r^2\right]_0^{R(1-z/H)} dz$$

$$= 400\pi \int_0^H z\frac{R^2}{2}\left(1 - \frac{z}{H}\right)^2 dz$$

$$= 200\pi R^2 \int_0^H \left(z - \frac{2z^2}{H} + \frac{z^3}{H^2}\right) dz$$

$$= 200\pi R^2 \left[\frac{z^2}{2} - \frac{2z^3}{3H} + \frac{z^4}{4H^2}\right]_0^H$$

$$= 200\pi R^2 \left(\frac{H^2}{2} - \frac{2H^2}{3} + \frac{H^2}{4}\right) = \tfrac{50}{3}\pi R^2 H^2 = \tfrac{50}{3}\pi(62,000)^2(12,400)^2 \approx 3.1 \times 10^{19} \text{ ft-lb.}$$

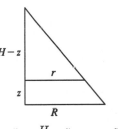

$$\frac{r}{R} = \frac{H-z}{H} = 1 - \frac{z}{H}$$

7. **(a)** $mgh = \tfrac{1}{2}mv^2 + \tfrac{1}{2}I\omega^2 = \tfrac{1}{2}(m + I/r^2)v^2$, so $v^2 = \dfrac{2mgh}{m + I/r^2} = \dfrac{2gh}{1 + I^*}$.

(b)

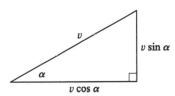

The vertical component of the speed is $v\sin\alpha$, so $\dfrac{dy}{dt} = \sqrt{\dfrac{2gy}{1+I^*}}\sin\alpha = \sqrt{\dfrac{2g}{1+I^*}}\sin\alpha\,\sqrt{y}$.

(c) Solving the separable differential equation, we get $\dfrac{dy}{\sqrt{y}} = \sqrt{\dfrac{2g}{1+I^*}}\sin\alpha\,dt \Rightarrow$

$2\sqrt{y} = \sqrt{\dfrac{2g}{1+I^*}}(\sin\alpha)t + C$. But $y = 0$ when $t = 0$, so $C = 0$ and we have $2\sqrt{y} = \sqrt{\dfrac{2g}{1+I^*}}(\sin\alpha)t$.

Solving for t when $y = h$ gives $T = \dfrac{2\sqrt{h}}{\sin\alpha}\sqrt{\dfrac{1+I^*}{2g}} = \sqrt{\dfrac{2h(1+I^*)}{g\sin^2\alpha}}$.

(d) Assume that the length of each cylinder is ℓ. Then the density of the solid cylinder is $\dfrac{m}{\pi r^2 \ell}$, and from

(13.7.16), its moment of inertia (using cylindrical coordinates) is

$$I_z = \iiint \frac{m}{\pi r^2 \ell}(x^2 + y^2)\,dV = \int_0^\ell \int_0^{2\pi} \int_0^r \frac{m}{\pi r^2 \ell} R^2 R\,dR\,d\theta\,dz = \frac{m}{\pi r^2 \ell} 2\pi \ell \left[\tfrac{1}{4}R^4\right]_0^r = \frac{mr^2}{2}, \text{ and so}$$

$$I^* = \frac{I_z}{mr^2} = \frac{1}{2}.$$

For the hollow cylinder, we consider its entire mass to lie a distance r from the axis of rotation, so

$x^2 + y^2 = r^2$ is a constant. We express the density in terms of mass per unit area as $\rho = \dfrac{m}{2\pi r \ell}$, and then

the moment of inertia is calculated as a double integral:

$$I_z = \iint (x^2 + y^2)\frac{m}{2\pi r \ell}\,dA = \frac{mr^2}{2\pi r \ell}\iint dA = mr^2, \text{ so } I^* = \frac{I_z}{mr^2} = 1.$$

(e), (f) Before considering the specific cases, we calculate I^* for a partly hollow ball with inner radius a and outer

radius r. The volume of such a ball is $\frac{4}{3}\pi(r^3 - a^3) = \frac{4}{3}\pi r^3(1 - b^3)$, and so its density is $\dfrac{m}{\frac{4}{3}\pi r^3(1 - b^3)}$.

Using Formula 13.8.4, we get

$$I_z = \iiint (x^2 + y^2)\frac{m}{\frac{4}{3}\pi r^3(1 - b^3)}\,dV = \frac{m}{\frac{4}{3}\pi r^3(1 - b^3)}\int_a^r \int_0^{2\pi} \int_0^\pi (\rho^2 \sin^2\phi)(\rho^2 \sin\phi)\,d\phi\,d\theta\,d\rho$$

$$= \frac{m}{\frac{4}{3}\pi r^3(1 - b^3)} \cdot 2\pi \left[-\frac{(2 + \sin^2\phi)\cos\phi}{3}\right]_0^\pi \left[\frac{\rho^5}{5}\right]_a^r \quad \text{(from the Table of Integrals)}$$

$$= \frac{m}{\frac{4}{3}\pi r^3(1 - b^3)} \cdot 2\pi \cdot \frac{4}{3} \cdot \frac{r^5 - a^5}{5} = \frac{2mr^5(1 - b^5)}{5r^3(1 - b^3)} = \frac{2(1 - b^5)mr^2}{5(1 - b^3)}$$

Therefore $I^* = \dfrac{2(1 - b^5)}{5(1 - b^3)}$. Now for the solid ball, we let $a \to 0$, so $b \to 0$ and $I^* \to \dfrac{2}{5}$. For the hollow

ball, we let $a \to r$, so $b \to 1$ and we use l'Hospital's Rule: $\lim_{b \to 1} I^* = \frac{2}{5}\lim_{b \to 1}\dfrac{-5b^4}{-3b^2} = \dfrac{2}{3}$.

Note: Instead of using l'Hospital's Rule, we could have found this limit by factoring our expression for I^*:

$$\lim_{b \to 1} I^* = \lim_{b \to 1}\frac{2(1 - b)(1 + b + b^2 + b^3 + b^4)}{5(1 - b)(1 + b + b^2)} = \frac{2 \cdot 5}{5 \cdot 3} = \frac{2}{3}.$$

CHAPTER FOURTEEN

EXERCISES 14.1

1. $\mathbf{F}(x, y) = x\mathbf{i} + y\mathbf{j}$

   The length of the vector $x\mathbf{i} + y\mathbf{j}$ is the distance from $(0, 0)$ to (x, y). Flow lines are rays emanating from the origin.

   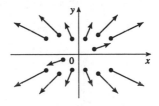

3. $\mathbf{F}(x, y) = y\mathbf{i} + \mathbf{j}$

   The length of the vector $y\mathbf{i} + \mathbf{j}$ is $\sqrt{y^2 + 1}$. Flow lines are parabolas opening about the x-axis.

   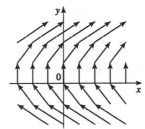

5. $\mathbf{F}(x, y) = \dfrac{y\mathbf{i} + x\mathbf{j}}{\sqrt{x^2 + y^2}}$

   The length of the vector $\dfrac{y\mathbf{i} + x\mathbf{j}}{\sqrt{x^2 + y^2}}$ is 1.

   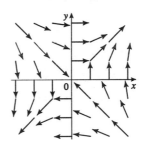

7. $\mathbf{F}(x, y, z) = \mathbf{j}$

   All vectors in this field are parallel to the y-axis.

   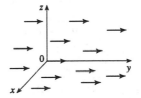

9. $\mathbf{F}(x, y, z) = y\mathbf{j}$

   The length of $\mathbf{F}(x, y, z)$ is $|y|$. No vectors emanate from the xz-plane since $y = 0$ there. In each plane $y = b$, all the vectors are identical.

   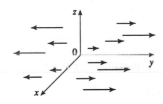

11. $\mathbf{F}(x, y) = \langle y, x \rangle$ corresponds to graph III, since in the first quadrant all the vectors have positive x- and y-components, in the second quadrant all vectors have positive x-components and negative y-components, in the third quadrant all vectors have negative x- and y-components, and in the fourth quadrant all vectors have negative x-components and positive y-components.

13. $\mathbf{F}(x, y) = \langle \sin x, \sin y \rangle$ corresponds to graph II, since the vector field is the same on each square of the form $[2n\pi, 2(n+1)\pi] \times [2m\pi, 2(m+1)\pi]$, m, n any integers.

15.

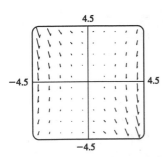

The vector field seems to have very short vectors near the line $y = 2x$. For $\mathbf{F}(x, y) = \langle 0, 0 \rangle$ we must have $y^2 - 2xy = 0$ and $3xy - 6x^2 = 0$. The first equation holds if $y = 0$ or $y = 2x$, and the second holds if $x = 0$ or $y = 2x$.

So both equations hold $\left[\text{and thus } \mathbf{F}(x, y) = \mathbf{0}\right]$ along the line $y = 2x$.

17. $\nabla f(x, y) = f_x(x, y)\mathbf{i} + f_y(x, y)\mathbf{j} = \left(5x^4 - 8xy^3\right)\mathbf{i} - \left(12x^2y^2\right)\mathbf{j}$

19. $\nabla f(x, y) = \langle f_x, f_y \rangle = \langle 3e^{3x} \cos 4y, -4e^{3x} \sin 4y \rangle$

21. $\nabla f(x, y, z) = \langle f_x, f_y, f_z \rangle = \langle y^2, 2xy - z^3, -3yz^2 \rangle$

23. $f(x, y) = x^2 - \frac{1}{2}y^2$, $\nabla f(x, y) = 2x\mathbf{i} - y\mathbf{j}$

The length of $\nabla f(x, y)$ is $\sqrt{4x^2 + y^2}$, and $\nabla f(x, y)$ terminates on the x-axis at the point $(3x, 0)$.

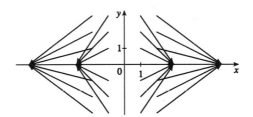

25. (a) The flow lines appear to be hyperbolas with equations $y = C/x$.

(b) $dx/dt = x \quad \Rightarrow \quad dx/x = dt \quad \Rightarrow$
$\ln|x| = t + C \quad \Rightarrow \quad x = \pm e^{t+C} = Ae^t$
for some constant A.
$dy/dt = -y \quad \Rightarrow \quad dy/y = -dt \quad \Rightarrow$
$\ln|y| = -t + K \quad \Rightarrow \quad y = \pm e^{-t+K} = Be^{-t}$

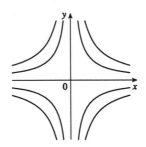

for some constant B. Therefore $xy = Ae^t Be^{-t} = AB = $ constant. If the flow line passes through $(1, 1)$ then $(1)(1) = $ constant $= 1 \quad \Rightarrow \quad xy = 1 \quad \Rightarrow \quad y = \frac{1}{x}, x > 0$.

EXERCISES 14.2

1. $\int_C x\,ds = \int_0^1 (t^3)\sqrt{9t^4+1}\,dt = \frac{1}{54}(9t^4+1)^{3/2}\big|_0^1 = \frac{1}{54}(10^{3/2}-1)$

3. $x = 4\cos t,\ y = 4\sin t,\ -\frac{\pi}{2} \le t \le \frac{\pi}{2}.$

$$\int_C xy^4\,ds = \int_{-\pi/2}^{\pi/2} \left[(4)^5\cos t\sin^4 t\right](4)dt = (4)^6\left[\tfrac{1}{5}\sin^5 t\right]_{-\pi/2}^{\pi/2} = \frac{2\cdot 4^6}{5} = 1638.4$$

5. $x = x,\ y = x^2,\ -2 \le x \le 1.$ Then
$$\int_C (x-2y^2)dy = \int_{-2}^1 (x-2x^4)2x\,dx = \int_{-2}^1 (2x^2-4x^5)dx = \tfrac{2}{3}[x^3-x^6]_{-2}^1 = 48.$$

7. $C = C_1 + C_2$

On C_1: $x = x,\ y = 0,\ 0 \le x \le 2.$

On C_2: $x = x,\ y = 2x - 4,\ 2 \le x \le 3.$ Then

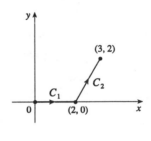

$$\int_C xy\,dx + (x-y)dy$$
$$= \int_{C_1} xy\,dx + (x-y)dy + \int_{C_2} xy\,dx + (x-y)dy$$
$$= \int_0^2 0\,dx + \int_2^3 [(2x^2 - 4x) + (-x+4)(2)]dx$$
$$= \int_2^3 (2x^2 - 6x + 8)dx - \tfrac{17}{3}.$$

9. $\int_C xyz\,ds = \int_0^{\pi/2}(18t\sin t\cos t)\sqrt{4+9}\,dt = 18\sqrt{13}\int_0^{\pi/2}(t\sin t\cos t)dt$
$$= 18\sqrt{13}\int_0^{\pi/2}\tfrac{1}{2}t\sin 2t\,dt = 9\sqrt{13}\left[-\tfrac{1}{2}t\cos 2t + \tfrac{1}{4}\sin 2t\right]_0^{\pi/2} = \tfrac{9\sqrt{13}}{4}\pi$$

11. $x = -t+1,\ y = 3t,\ z = 5t+1,\ 0 \le t \le 1.$
$$\int_C xy^2 z\,ds = \int_0^1 (1-t)(9t^2)(5t+1)\sqrt{1+9+25}\,dt = 9\sqrt{35}\int_0^1 (t^2 + 4t^3 - 5t^4)dt = 3\sqrt{35}$$

13. $\int_C x^3 y^2\,dz = \int_0^1 (8t^3)(t^4)(t^2)(2t)dt = \int_0^1 16t^{10}\,dt = \tfrac{16}{11}$

15. On C_1: $x = 0,\ y = t,\ z = t,\ 0 \le t \le 1,$

C_2: $x = t,\ y = t+1,\ z = 2t+1,\ 0 \le t \le 1,$

C_3: $x = 1,\ y = 2,\ z = t+3,\ 0 \le t \le 1.$ Then

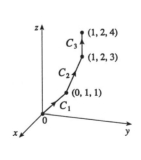

$$\int_C z^2\,dx - z\,dy + 2y\,dz$$
$$= \int_0^1 (0 - t + 2t)dt + \int_0^1 \left[(2t+1)^2 - (2t+1) + 2(t+1)(2)\right]dt$$
$$\quad + \int_0^1 (0 + 0 + 4)dt$$
$$= \tfrac{1}{2} + \left[\tfrac{4}{3}t^3 + 3t^2 + 4t\right]_0^1 + 4 = \tfrac{77}{6}$$

17. $\mathbf{F}(\mathbf{r}(t)) = t^{10}\mathbf{i} - t^7\mathbf{j},\ \mathbf{r}'(t) = 3t^2\mathbf{i} + 4t^3\mathbf{j}.\ \int_C \mathbf{F}\cdot d\mathbf{r} = \int_0^1 (3t^{12} - 4t^{10})dt = \tfrac{3}{13} - \tfrac{4}{11} = -\tfrac{19}{143}$

19. $\int_C \mathbf{F}\cdot d\mathbf{r} = \int_0^1 \langle \sin t^3, \cos(-t^2), t^4\rangle \cdot \langle 3t^2, -2t, 1\rangle dt$
$$= \int_0^1 \left(3t^2\sin t^3 - 2t\cos t^2 + t^4\right)dt = \left[-\cos t^3 - \sin t^2 + \tfrac{1}{5}t^5\right]_0^1 = \tfrac{6}{5} - \cos 1 - \sin 1$$

21. (a) $\displaystyle \int_C \mathbf{F} \cdot d\mathbf{r} = \int_0^1 \left\langle e^{t^2-1}, t^5 \right\rangle \cdot \left\langle 2t, 3t^2 \right\rangle dt = \int_0^1 \left(2te^{t^2-1} + 3t^7 \right) dt$

$\displaystyle = \left[e^{t^2-1} + \tfrac{3}{8}t^8 \right]_0^1 = \tfrac{11}{8} - 1/e$

(b) $\mathbf{r}(0) = \mathbf{0}$, $\mathbf{F}(\mathbf{r}(0)) = \left\langle e^{-1}, 0 \right\rangle$;

$\mathbf{r}\!\left(\tfrac{1}{\sqrt{2}}\right) = \left\langle \tfrac{1}{2}, \tfrac{1}{2\sqrt{2}} \right\rangle$, $\mathbf{F}\!\left(\mathbf{r}\!\left(\tfrac{1}{\sqrt{2}}\right)\right) = \left\langle e^{-1/2}, \tfrac{1}{4\sqrt{2}} \right\rangle$;

$\mathbf{r}(1) = \langle 1, 1 \rangle$, $\mathbf{F}(\mathbf{r}(1)) = \langle 1, 1 \rangle$.

In order to generate the graph with Maple, we use the PLOT command (not to be confused with the plot command) to define each of the vectors. For example,

`v1:=PLOT(CURVES([[0,0],[evalf(1/E),0]]));` generates the vector from the vector field at the point $(0, 0)$ (but without an arrowhead) and gives it the name v1. To show everything on the same screen, we use the display command.

In Mathematica, we use ListPlot (with the PlotJoined -> True option) to generate the vectors, and then Show to show everything on the same screen.

23. A calculator or CAS gives $\int_C x \sin y \, ds = \int_1^2 \ln t \sin(e^{-t}) \sqrt{(1/t)^2 + (-e^{-t})^2} \, dt \approx 0.052$.

25. The part of the astroid that lies in the quadrant is parametrized by $x = \cos^3 t$, $y = \sin^3 t$, $0 \le t \le \tfrac{\pi}{2}$. Now

$dx/dt = 3\cos^2 t\,(-\sin t)$ and $dy/dt = 3\sin^2 t\cos t$, so

$\sqrt{(dx/dt)^2 + (dy/dt)^2} = \sqrt{9\cos^4 t \sin^2 t + 9\sin^4 t \cos^2 t} = 3\cos t \sin t \sqrt{\cos^2 t + \sin^2 t} = 3\cos t \sin t$. Therefore $\int_C x^3 y^5 \, ds = \int_0^{\pi/2} \cos^9 t \sin^{15} t \, (3\cos t \sin t) dt = \tfrac{945}{16,777,216}\pi$.

27. (a) Along the line $x = -3$, the vectors of $\mathbf{F}$ have positive y-components, so since the path goes upward, the integrand $\mathbf{F} \cdot \mathbf{T}$ is always positive. Therefore $\int_{C_1} \mathbf{F} \cdot d\mathbf{r}$ is positive.

(b) All of the (nonzero) field vectors along the circle with radius 3 are pointed in the clockwise direction, that is, opposite the direction to the path. So $\mathbf{F} \cdot \mathbf{T}$ is negative, and therefore $\int_{C_2} \mathbf{F} \cdot d\mathbf{r}$ is negative.

29. $x = 4\cos t$, $y = 4\sin t$, $-\tfrac{\pi}{2} \le t \le \tfrac{\pi}{2}$, $m = \int_C k\,ds = k\int_{-\pi/2}^{\pi/2} \sqrt{4\cos^2 t + 4\sin^2 t}\,dt = 2k(\pi)$,

$\overline{x} = \dfrac{1}{2\pi k}\int_C xk\,dS = \dfrac{1}{2\pi}\int_{-\pi/2}^{\pi/2} 4\cos t\,dt = \dfrac{1}{2\pi}[4\sin t]_{-\pi/2}^{\pi/2} = \dfrac{4}{\pi}$, $\overline{y} = \dfrac{1}{2\pi k}\int_C yk\,dS = \dfrac{1}{2\pi}\int_{-\pi/2}^{\pi/2} 4\sin t\,dt = 0$.

Hence $(\overline{x}, \overline{y}) = \left(\tfrac{4}{\pi}, 0\right)$.

31. (a) $\overline{x} = \dfrac{1}{m}\int_C x\rho(x, y, z)ds$, $\overline{y} = \dfrac{1}{m}\int_C y\rho(x, y, z)ds$, $\overline{z} = \dfrac{1}{m}\int_C z\rho(x, y, z)ds$ where $m = \int_C \rho(x,y,z)ds$.

(b) $m = \displaystyle\int_C k\,ds = k\int_0^{2\pi} \sqrt{4\sin^2 t + 4\cos^2 t + 9}\,dt = k\sqrt{13}\int_0^{2\pi} dt = 2\pi k\sqrt{13}$,

$\overline{x} = \dfrac{1}{2\pi k\sqrt{13}}\int_0^{2\pi} k2\sqrt{13}\sin t\,dt = 0$, $\overline{y} = \dfrac{1}{2\pi k\sqrt{13}}\int_0^{2\pi} k2\sqrt{13}\cos t\,dt = 0$,

$\overline{z} = \dfrac{1}{2\pi k\sqrt{13}}\int_0^{2\pi} \left(k\sqrt{13}\right)(3t)dt = \dfrac{3}{2\pi}(2\pi^2) = 3\pi$. Hence $(\overline{x}, \overline{y}, \overline{z}) = (0, 0, 3\pi)$.

33. From Example 3, $\rho(x, y) = k(1 - y)$, $x = \cos t$, $y = \sin t$, and $ds = dt$, $0 \le t \le \pi$ $\Rightarrow$

$$I_x = \int_C y^2 \rho(x, y) ds = \int_0^\pi \sin^2 t[k(1 - \sin t)] dt = k \int_0^\pi (\sin^2 t - \sin^3 t) dt$$

$$= \frac{k}{2} \int_0^\pi (1 - \cos 2t) dt - k \int_0^\pi (1 - \cos^2 t) \sin t \, dt \quad \text{(let } u = \cos t, \, du = -\sin t \text{ in the second integral)}$$

$$= k \left[\frac{\pi}{2} + \int_1^{-1} (1 - u^2) du \right] = k \left(\frac{\pi}{2} - \frac{4}{3} \right).$$

$$I_y = \int_C x^2 \rho(x, y) ds = k \int_0^\pi \cos^2 t(1 - \sin t) dt = \frac{k}{2} \int_0^\pi (1 + \cos 2t) dt - k \int_0^\pi \cos^2 t \sin t \, dt = k \left[\frac{\pi}{2} - \frac{2}{3} \right], \text{ using}$$

the same substitution as above.

35. $W = \int_C \mathbf{F} \cdot d\mathbf{r} = \int_0^{2\pi} \langle t - \sin t, 3 - \cos t \rangle \cdot \langle 1 - \cos t, \sin t \rangle dt$

$$= \int_0^{2\pi} (t - t \cos t - \sin t + \sin t \cos t + 3 \sin t - \sin t \cos t) dt = 2\pi^2$$

37. $W = \int_0^1 \langle t^6, -t^5, -t^7 \rangle \cdot \langle 2t, -3t^2, 4t^3 \rangle dt = \int_0^1 (5t^7 - 4t^{10}) dt = \frac{5}{8} - \frac{4}{11} = \frac{23}{88}$

39. Let $\mathbf{F} = 185\mathbf{k}$. To parametrize the staircase, let $x = 20 \cos t$, $y = 20 \sin t$, $z = \frac{90}{6\pi} t = \frac{15}{\pi} t$, $0 \le t \le 6\pi$ $\Rightarrow$

$$W = \int_C \mathbf{F} \cdot d\mathbf{r} = \int_0^{6\pi} \langle 0, 0, 185 \rangle \cdot \langle -20 \sin t, 20 \cos t, \frac{15}{\pi} \rangle dt = (185) \frac{15}{\pi} \int_0^{6\pi} dt = (185)(90) \approx 1.67 \times 10^4 \text{ ft-lb.}$$

41. Use the orientation pictured in the figure. Then since $\mathbf{B}$ is tangent to any circle that

lies in the plane perpendicular to the wire, $\mathbf{B} = |\mathbf{B}|\mathbf{T}$ where $\mathbf{T}$ is the unit tangent to the circle C:

$x = r \cos \theta$, $y = r \sin \theta$. Thus $\mathbf{B} = |\mathbf{B}|\langle -\sin \theta, \cos \theta \rangle$. Then

$\int_C \mathbf{B} \cdot d\mathbf{r} = \int_0^{2\pi} |\mathbf{B}|\langle -\sin \theta, \cos \theta \rangle \cdot \langle -r \sin \theta, r \cos \theta \rangle d\theta = \int_0^{2\pi} |\mathbf{B}| r \, d\theta = 2\pi r |\mathbf{B}|$. (Note that $|\mathbf{B}|$ here is the

magnitude of the field at a distance r from the wire's center.) But by Ampere's Law $\int_C \mathbf{B} \cdot d\mathbf{r} = \mu_0 I$. Hence

$|\mathbf{B}| = \mu_0 I / (2\pi r)$.

EXERCISES 14.3

1. $\partial(2x - 3y)/\partial y = -3 = \partial(2y - 3x)/\partial x$ and the domain of $\mathbf{F}$ is $\mathbb{R}^2$ which is open and simply-connected, so $\mathbf{F}$ is conservative. Thus there exists f such that $\nabla f = \mathbf{F}$, that is, $f_x(x, y) = 2x - 3y$ and $f_y(x, y) = 2y - 3x$. But $f_x(x, y) = 2x - 3y$ implies $f(x, y) = x^2 - 3yx + g(y)$ and differentiating both sides of this equation with respect to y gives $f_y(x, y) = -3x + g'(y)$. Thus $2y - 3x = -3x + g'(y)$ so $g'(y) = 2y$ and $g(y) = y^2 + K$ where K is a constant. Hence $f(x, y) = x^2 - 3xy + y^2 + K$ is a potential for $\mathbf{F}$.

3. $\partial(x^2 + y)/\partial y = 1$, $\partial(x^2)/\partial x = 2x$ and these are not equal, so $\mathbf{F}$ is not conservative.

5. $\partial(1 + 4x^3y^3)/\partial y = 12x^3y^2 = \partial(3x^4y^2)/\partial x$ and the domain of $\mathbf{F}$ is $\mathbb{R}^2$ which is open and simply-connected. Thus $\mathbf{F}$ is conservative so there exists f such that $\nabla f = \mathbf{F}$. Then $f_x(x, y) = 1 + 4x^3y^3$ implies $f(x, y) = x + x^4y^3 + g(y)$ and $f_y(x, y) = 3x^4y^3 + g'(y)$. But $f_y(x, y) = 3x^4y^2$ implies $g(y) = K$. Hence a potential for $\mathbf{F}$ is $f(x, y) = x + x^4y^3 + K$.

7. $\partial(e^{2x} + x \sin y)/\partial y = x \cos y$, $\partial(x^2 \cos y)/\partial x = 2x \cos y$, so $\mathbf{F}$ is not conservative.

9. $\partial(ye^x + \sin y)/\partial y = e^x + \cos y = \partial(e^x + x \cos y)/\partial x$ and the domain of $\mathbf{F}$ is $\mathbb{R}^2$. Hence $\mathbf{F}$ is conservative so there exists f such that $\nabla f = \mathbf{F}$. Then $f_x(x, y) = ye^x + \sin y$ implies $f(x, y) = ye^x + x \sin y + g(y)$ and $f_y(x, y) = e^x + x \cos y + g'(y)$. But $f_y(x, y) = e^x + x \cos y$ so $g(y) = K$ and $f(x, y) = ye^x + x \sin y + K$ is a potential for $\mathbf{F}$.

11. **(a)** $f_x(x, y) = x$ implies $f(x, y) = \frac{1}{2}x^2 + g(y)$ and $f_y(x, y) = g'(y)$. But $f_y(x, y) = y$ so $g(y) = \frac{1}{2}y^2 + K$ and $f(x, y) = \frac{1}{2}x^2 + \frac{1}{2}y^2 + K$ (or set $K = 0$.)

    (b) $\int_C \mathbf{F} \cdot d\mathbf{r} = f(3, 9) - f(-1, 1) = 44$

13. **(a)** $f_x(x, y) = 2xy^3$ implies $f(x, y) = x^2y^3 + g(y)$ and $f_y(x, y) = 3x^2y^2 + g'(y)$. But $f_y(x, y) = 3x^2y^2$ so $f(x, y) = x^2y^3$ (setting $K = 0$).

    (b) Since $\mathbf{r}(0) = \langle 0, 1 \rangle$ and $\mathbf{r}\left(\frac{\pi}{2}\right) = \langle 1, \frac{1}{4}(\pi^2 + 4) \rangle$, $\int_C \mathbf{F} \cdot d\mathbf{r} = f\left(1, \frac{1}{4}(\pi^2 + 4)\right) - f(0, 1) = \frac{1}{64}(\pi^2 + 4)^3$.

15. **(a)** $f_x(x, y, z) = y$ implies $f(x, y, z) = xy + g(y, z)$ and $f_y(x, y, z) = x + \partial g/\partial y$. But $f_y(x, y, z) = x + z$ so $\partial g/\partial y = z$ and $g(y, z) = yz + h(z)$. Thus $f(x, y, z) = xy + yz + h(z)$ and $f_z(x, y, z) = y + h'(z)$. But $f_z(x, y, z) = y$ so $h'(z) = 0$ or $h(z) = K$. Hence $f(x, y, z) = xy + yz$ (setting $K = 0$).

    (b) $\int_C \mathbf{F} \cdot d\mathbf{r} = f(8, 3, -1) - f(2, 1, 4) = 21 - 6 = 15$

17. **(a)** $f_x(x, y, z) = 2xz + \sin y$ implies $f(x, y, z) = x^2z + x \sin y + g(y, z)$ and $f_y(x, y, z) = x \cos y + g_y(y, z)$. But $f_y(x, y, z) = x \cos y$ so $g_y(y, z) = 0$ and $f(x, y, z) = x^2z + x \sin y + h(z)$. Thus $f_z(x, y, z) = x^2 + h'(z)$. But $f_z(x, y, z) = x^2$ so $h'(z) = 0$ and $f(x, y, z) = x^2z + x \sin y$ (setting $K = 0$).

    (b) $\mathbf{r}(0) = \langle 1, 0, 0 \rangle$, $\mathbf{r}(2\pi) = \langle 1, 0, 2\pi \rangle$. Thus $\int_C \mathbf{F} \cdot d\mathbf{r} = f(1, 0, 2\pi) - f(1, 0, 0) = 2\pi$.

19. Here $\mathbf{F}(x, y) = (2x \sin y)\mathbf{I} + (x^2 \cos y - 3y^2)\mathbf{j}$. Then $f(x, y) = x^2 \sin y - y^3$ is a potential for $\mathbf{F}$, that is,

$\nabla f = \mathbf{F}$ so $\mathbf{F}$ is conservative and thus its line integral is independent of path. Hence

$\int_C 2x \sin y \, dx + (x^2 \cos y - 3y^2) dy = \int_C \mathbf{F} \cdot d\mathbf{r} = f(5, 1) - f(-1, 0) = 25 \sin 1 - 1.$

21. Here $\mathbf{F}(x, y) = x^2 y^3 \mathbf{i} + x^3 y^2 \mathbf{j}$. $W = \displaystyle\int_C \mathbf{F} \cdot d\mathbf{r}$. Since $\dfrac{\partial}{\partial y}(x^2 y^3) = 3x^2 y^2 = \dfrac{\partial}{\partial x}(x^3 y^2)$, there exists f such that

$\nabla f = \mathbf{F}$. In fact, $f_x = x^2 y^3 \quad \Rightarrow \quad f(x, y) = \frac{1}{3}x^3 y^3 + g(y) \quad \Rightarrow \quad f_y = x^3 y^2 + g'(y) \quad \Rightarrow \quad g'(y) = 0$, so

we can take $f(x, y) = \frac{1}{3}x^3 y^3$. Thus $W = \int_C \mathbf{F} \cdot d\mathbf{r} = f(2, 1) - f(0, 0) = \frac{1}{3}(2^3)(1^3) - 0 = \frac{8}{3}$.

23. We know that if the vector field (call it $\mathbf{F}$) is conservative, then around any closed path C, $\int_C \mathbf{F} \cdot d\mathbf{r} = 0$. But

take C to be some circle centered at the origin, oriented counterclockwise. All of the field vectors along C point

"against" the direction of C (that is, within $90°$ of $-C$) so the integral around C will be negative. Therefore the

field is not conservative.

25.

From the graph, it appears that $\mathbf{F}$ is not conservative. For example, any

closed curve containing the point $(2, 1)$ seems to have many field vectors

pointing counterclockwise along it, and none pointing clockwise. So along

this path the integral $\int \mathbf{F} \cdot d\mathbf{r} \neq 0$. To confirm our guess, we calculate

$$\frac{\partial}{\partial y}\left(\frac{x - 2y}{\sqrt{1 + x^2 + y^2}}\right) = (x - 2y)\left[\frac{-y}{(1 + x^2 + y^2)^{3/2}}\right] - \frac{2}{\sqrt{1 + x^2 + y^2}}$$

$$= \frac{-2 - 2x^2 - xy}{(1 + x^2 + y^2)^{3/2}}, \quad \frac{\partial}{\partial x}\left(\frac{x - 2}{\sqrt{1 + x^2 + y^2}}\right) = (x - 2)\left[\frac{-x}{(1 + x^2 + y^2)^{3/2}}\right] + \frac{1}{\sqrt{1 + x^2 + y^2}}$$

$$= \frac{1 + y^2 + 2x}{(1 + x^2 + y^2)^{3/2}}.$$ These are not equal, so the field is not conservative, by Theorem 5.

27. Since $\mathbf{F}$ is conservative, there exists a function f such that $\mathbf{F} = \nabla f$, that is, $P = f_x$, $Q = f_y$, and $R = f_z$. Since

P, Q and R have continuous first order partial derivatives, Clairaut's Theorem says

$\partial P / \partial y = f_{xy} = f_{yx} = \partial Q / \partial x$, $\partial P / \partial z = f_{xz} = f_{zx} = \partial R / \partial x$, and $\partial Q / \partial z = f_{yz} = f_{zy} = \partial R / \partial y$.

29. $D = \{(x, y) \mid x > 0, y > 0\} = $ the first quadrant (excluding the axes).

(a) D is open because around every point in D we can put a disk that lies in D.

(b) D is connected because the straight line segment joining any two points in D lies in D.

(c) D is simply-connected because it's connected and has no holes.

31. $D = \{(x, y) \mid 1 < x^2 + y^2 < 4\} = $ the annular region between the circles with center $(0, 0)$ and radii 1 and 2.

(a) D is open.

(b) D is connected.

(c) D is not simply-connected. For example, $x^2 + y^2 = (1.5)^2$ is simple and closed and lies within D but

encloses points that are not in D. (Or, D has a hole, so is not simply-connected.)

33. (a) $P = -\dfrac{y}{x^2 + y^2}$, $\dfrac{\partial P}{\partial y} = \dfrac{y^2 - x^2}{(x^2 + y^2)^2}$ and $Q = \dfrac{x}{x^2 + y^2}$, $\dfrac{\partial Q}{\partial x} = \dfrac{y^2 - x^2}{(x^2 + y^2)^2}$. Thus $\dfrac{\partial P}{\partial y} = \dfrac{\partial Q}{\partial x}$.

(b) C_1: $x = \cos t$, $y = \sin t$, $0 \le t \le \pi$, C_2: $x = \cos t$, $y = \sin t$, $t = 2\pi$ to $t = \pi$. Then

$$\int_{C_1} \mathbf{F} \cdot d\mathbf{r} = \int_0^\pi \frac{(-\sin t)(-\sin t) + (\cos t)(\cos t)}{\cos^2 t + \sin^2 t} \, dt = \int_0^\pi dt = \pi \text{ and } \int_{C_2} \mathbf{F} \cdot d\mathbf{r} = \int_{2\pi}^\pi dt = -\pi. \text{ Since}$$

these aren't equal, the line integral of $\mathbf{F}$ isn't independent of path. $\left(\text{Or notice that } \int_{C_3} \mathbf{F} \cdot d\mathbf{r} = \int_0^{2\pi} dt = 2\pi \right.$

where C_3 is the circle $x^2 + y^2 = 1$, and apply the contrapositive of Theorem 3.$\left.\right)$ This doesn't contradict

Theorem 6, since the domain of $\mathbf{F}$, which is $\mathbb{R}^2$ except the origin, isn't simply-connected.

EXERCISES 14.4

1. (a)

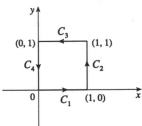

$\oint_C x^2 y \, dx + xy^3 \, dy$

$= \oint_{C_1 + C_2 + C_3 + C_4} x^2 y \, dx + xy^3 \, dy$

$= \int_0^1 0 \, dx + \int_0^1 y^3 \, dy + \int_1^0 x^2 \, dx + \int_1^0 0 \, dy$

$= \frac{1}{4} - \frac{1}{3} = -\frac{1}{12}$

(b) $\oint_C x^2 y \, dx + xy^3 \, dy = \int_0^1 \int_0^1 (y^3 - x^2) \, dx \, dy = \int_0^1 \left(y^3 - \frac{1}{3}\right) dy = \frac{1}{4} - \frac{1}{3} = -\frac{1}{12}$

3. (a)

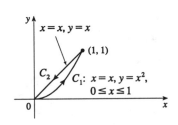

$\oint_C (x + 2y) \, dx + (x - 2y) \, dy$

$= \oint_{C_1 + C_2} (x + 2y) \, dx + (x - 2y) \, dy$

$= \int_0^1 [x + 2x^2 + (x - 2x^2)(2x)] \, dx$

$\qquad + \int_1^0 [3x + (-x)] \, dx$

$= \int_0^1 (x + 4x^2 - 4x^3) \, dx + \int_1^0 2x \, dx$

$= \left(\frac{1}{2} + \frac{4}{3} - 1\right) - 1 = -\frac{1}{6}$

(b) $\oint_C (x + 2y) \, dx + (x - 2y) \, dy = \int_0^1 \int_{x^2}^x (1 - 2) \, dy \, dx = \int_0^1 (x^2 - x) \, dx = \frac{1}{3} - \frac{1}{2} = -\frac{1}{6}$

5. We can parametrize C as $x = \cos\theta$, $y = \sin\theta$, $0 \le \theta \le 2\pi$. Then the line integral is

$\oint_C P \, dx + Q \, dy = \int_0^{2\pi} \cos^4\theta \sin^5\theta(-\sin\theta) d\theta + \int_0^{2\pi} (-\cos^7\theta \sin^6\theta)\cos\theta \, d\theta = -\frac{29\pi}{1024}$, according to a CAS. The

double integral is $\displaystyle\iint_D \left(\frac{\partial Q}{\partial x} - \frac{\partial P}{\partial y}\right) dA = \int_{-1}^1 \int_{-\sqrt{1-x^2}}^{\sqrt{1-x^2}} (-7x^6 y^6 - 5x^4 y^4) \, dy \, dx = -\frac{29\pi}{1024}$, verifying Green's

Theorem in this case.

7. $\oint_C xy \, dx + y^5 \, dy = \int_0^2 \int_0^{x/2} (0 - x) \, dy \, dx = \int_0^2 \left(-\frac{1}{2}x^2\right) dx = -\frac{4}{3}$

9. $\int_0^1 \int_{y^2}^{\sqrt{y}} (2-1)dx\,dy - \int_0^1 (y^{1/2} - y^2)\,dy = \frac{1}{3}$

11. $\iint_D (0-0)dA = 0$

13. $\iint\limits_{\substack{0 \le x^2 + y^2 \le 4 \\ y \ge 0}} (4x - x)dA = 3\int_0^\pi \int_0^2 r^2 \cos\theta\,dr\,d\theta = 0$ since $\int_0^\pi \cos\theta\,d\theta = 0$ or $\iint_D 3x\,dA = 3M_y = 0$.

15. $\iint_D (2x - x)dA = \int_0^\pi \int_0^{\sin x} x\,dy\,dx = \int_0^\pi x\sin x\,dx = [-x\cos x + \sin x]_0^\pi = \pi$

17. $\int_C \mathbf{F} \cdot d\mathbf{r} = \int_C (y^2 - x^2 y)dx + xy^2\,dy = \iint\limits_{\substack{x^2 + y^2 \le 4 \\ 0 \le y \le x}} (y^2 - 2y + x^2)dA = \int_0^{\pi/4} \int_0^2 (r^2 - 2r\sin\theta)r\,dr\,d\theta$

$$= \int_0^{\pi/4} \left[4 - \frac{16}{3}\sin\theta\right]d\theta = \left[4\theta + \frac{16}{3}\cos\theta\right]_0^{\pi/4} = \pi + \frac{8}{3}\left(\sqrt{2} - 2\right)$$

19. By Green's Theorem, $W = \int_C \mathbf{F} \cdot d\mathbf{r} = \int_C x(x+y)dx + xy^2\,dy = \iint_D (y^2 - x)dy\,dx$ where C is the path

described in the question and D is the triangle bounded by C. So

$W = \int_0^1 \int_0^{1-x} (y^2 - x)dy\,dx = \int_0^1 \left[\frac{1}{3}y^3 - xy\right]_0^{1-x}dx = \int_0^1 \left(\frac{1}{3}(1-x)^3 - x(1-x)\right)dx$

$= \left[-\frac{1}{12}(1-x)^4 - \frac{1}{2}x^2 + \frac{1}{3}x^3\right]_0^1 = \left(-\frac{1}{2} + \frac{1}{3}\right) - \left(-\frac{1}{12}\right) = -\frac{1}{12}$.

21. $A = \oint_C x\,dy = \int_0^{2\pi} (\cos^3 t)(3\sin^2 t\cos t)dt = 3\int_0^{2\pi} (\cos^4 t\sin^2 t)dt$

$= 3\left[-\frac{1}{6}(\sin t\cos^5 t) + \frac{1}{6}\left[\frac{1}{4}(\sin t\cos^3 t) + \frac{3}{8}(\cos t\sin t) + \frac{3}{8}t\right]\right]_0^{2\pi} = 3\left(\frac{1}{6}\right)\left(\frac{6}{8}\pi\right) = \frac{3}{8}\pi$

Or: $\int_0^{2\pi} (\cos^4 t\sin^2 t)dt = \int_0^{2\pi} \frac{1}{8}\left[\frac{1}{2}(1 - \cos 4t) + \sin^2 2t\cos 2t\right]dt = \frac{\pi}{8}$

23. **(a)** Using Equation 14.2.8, we write parametric equations of the line segment as $x = (1-t)x_1 + tx_2$,

$y = (1-t)y_1 + ty_2$, $0 \le t \le 1$. Then $dx = (x_2 - x_1)dt$ and $dy = (y_2 - y_1)dt$, so

$\int_C x\,dy - y\,dx = \int_0^1 [(1-t)x_1 + tx_2](y_2 - y_1)dt + [(1-t)y_1 + ty_2](x_2 - x_1)dt$

$= \int_0^1 (x_1(y_2 - y_1) - y_1(x_2 - x_1) + t[(y_2 - y_1)(x_2 - x_1) - (x_2 - x_1)(y_2 - y_1)])dt$

$= \int_0^1 (x_1 y_2 - x_2 y_1)dt = x_1 y_2 - x_2 y_1$.

(b) We apply Green's Theorem to the path $C = C_1 \cup C_2 \cup \cdots \cup C_n$, where C_i is the line segment that joins

(x_i, y_i) to (x_{i+1}, y_{i+1}) for $i = 1, 2, \ldots, n-1$, and C_n is the line segment that joins (x_n, y_n) to (x_1, y_1).

From (6), $\frac{1}{2}\int_C x\,dy - y\,dx = \iint_D dA$, where D is the polygon bounded by C. Therefore

area of polygon $= A(D) = \iint_D dA = \frac{1}{2}\int_C x\,dy - y\,dx$

$= \frac{1}{2}\left(\int_{C_1} x\,dy - y\,dx + \int_{C_2} x\,dy - y\,dx + \cdots + \int_{C_{n-1}} x\,dy - y\,dx + \int_{C_n} x\,dy - y\,dx\right)$.

To evaluate these integrals we use the formula from (a) to get

$A(D) = \frac{1}{2}[(x_1 y_2 - x_2 y_1) + (x_2 y_3 - x_3 y_2) + \cdots + (x_{n-1} y_n - x_n y_{n-1}) + (x_n y_1 - x_1 y_n)]$.

(c) $A = \frac{1}{2}[(0 \cdot 1 - 2 \cdot 0) + (2 \cdot 3 - 1 \cdot 1) + (1 \cdot 2 - 0 \cdot 3)$

$+ (0 \cdot 1 - (-1) \cdot 2) + (-1 \cdot 0 - 0 \cdot 1)] = \frac{1}{2}(0 + 5 + 2 + 2) = \frac{9}{2}$

25. Here $A = \frac{1}{2}(1)(1) = \frac{1}{2}$ and $C = C_1 + C_2 + C_3$, where $C_1: x = x, y = 0, 0 \le x \le 1$; $C_2: x = x, y = 1 - x$,

$x = 1$ to $x = 0$; and $C_3: x = 0, y = 1$ to $y = 0$. Then

$$\bar{x} = \frac{1}{2A} \int_C x^2 \, dy = \int_{C_1} x^2 \, dy + \int_{C_2} x^2 \, dy + \int_{C_3} x^2 \, dy = 0 + \int_1^0 (x^2)(-dx) + 0 = \frac{1}{3}. \text{ Similarly,}$$

$$\bar{y} = -\frac{1}{2A} \int_C y^2 \, dx = \int_{C_1} y^2 \, dx + \int_{C_2} y^2 \, dx + \int_{C_3} y^2 \, dx = 0 + \int_1^0 (1-x)^2(-dx) + 0 = \frac{1}{3}. \text{ Therefore}$$

$(\bar{x}, \bar{y}) = \left(\frac{1}{3}, \frac{1}{3}\right)$.

27. By Green's Theorem, $-\frac{1}{3}\rho \oint_C y^3 \, dx = -\frac{1}{3}\rho \iint_D (-3y^2) dA = \iint_D y^2 \rho \, dA = I_x$ and

$\frac{1}{3}\rho \oint_C x^3 \, dy = \frac{1}{3}\rho \iint_D (3x^2) dA = \iint_D x^2 \rho \, dA = I_y$.

29. Since C is a simple closed path which doesn't pass through or enclose the origin, there exists an open region that

doesn't contain the origin but does contain D. Thus $P = -\dfrac{y}{x^2 + y^2}$ and $Q = \dfrac{x}{x^2 + y^2}$ have continuous partials

on this open region containing D and we can apply Green's Theorem. But by Exercise 14.3.33(a), $\dfrac{\partial P}{\partial y} = \dfrac{\partial Q}{\partial x}$,

so $\oint_C \mathbf{F} \cdot d\mathbf{r} = \iint_D 0 \, dA = 0$.

31. Using the first part of (6) we have that $\iint_R dx \, dy = A(R) = \int_{\partial R} x \, dy$. But $x = g(u, v)$, and

$dy = \dfrac{\partial h}{\partial u} \, du + \dfrac{\partial h}{\partial v} \, dv$, and we orient ∂S by taking the positive direction to be that which corresponds, under the

mapping, to the positive direction along ∂R, so

$$\int_{\partial R} x \, dy = \int_{\partial S} g(u, v) \left(\frac{\partial h}{\partial u} \, du + \frac{\partial h}{\partial v} \, dv \right) = \int_{\partial S} g(u, v) \frac{\partial h}{\partial u} \, du + g(u, v) \frac{\partial h}{\partial v} \, dv$$

$$= \pm \iint_S \left[\frac{\partial}{\partial u} \left(g(u, v) \frac{\partial h}{\partial v} \right) - \frac{\partial}{\partial v} \left(g(u, v) \frac{\partial h}{\partial u} \right) \right] dA \quad \text{(using Green's Theorem in the } uv\text{-plane)}$$

$$= \pm \iint_S \left(\frac{\partial g}{\partial u} \frac{\partial h}{\partial v} + g(u, v) \frac{\partial^2 h}{\partial u \partial v} - \frac{\partial g}{\partial v} \frac{\partial h}{\partial u} - g(u, v) \frac{\partial^2 h}{\partial v \partial u} \right) dA \quad \text{(using the Chain Rule)}$$

$$= \pm \iint_S \left(\frac{\partial x}{\partial u} \frac{\partial y}{\partial v} - \frac{\partial x}{\partial v} \frac{\partial y}{\partial u} \right) dA \quad \text{(by the equality of mixed partials)} \quad = \pm \iint_S \frac{\partial(x, y)}{\partial(u, v)} \, du \, dv.$$

The sign is chosen to be positive if the orientation that we gave to ∂S corresponds to the usual positive

orientation, and it is negative otherwise. In either case, since $A(R)$ is positive, the sign chosen must be the same

as the sign of $\dfrac{\partial(x, y)}{\partial(u, v)}$. Therefore $A(R) = \iint_R dx \, dy = \iint_S \left| \dfrac{\partial(x, y)}{\partial(u, v)} \right| du \, dv$.

EXERCISES 14.5

1. **(a)** $\operatorname{curl}\mathbf{F}=\nabla\times\mathbf{F}=\begin{vmatrix} \mathbf{i} & \mathbf{j} & \mathbf{k} \\ \partial/\partial x & \partial/\partial y & \partial/\partial z \\ x & y & z \end{vmatrix}=(0-0)\mathbf{i}+(0-0)\mathbf{j}+(0-0)\mathbf{k}=\mathbf{0}$

   (b) $\operatorname{div}\mathbf{F}=\nabla\cdot\mathbf{F}=\dfrac{\partial}{\partial x}(x)+\dfrac{\partial}{\partial y}(y)+\dfrac{\partial}{\partial z}(z)=1+1+1=3$

3. **(a)** $\operatorname{curl}\mathbf{F}=\nabla\times\mathbf{F}=\begin{vmatrix} \mathbf{i} & \mathbf{j} & \mathbf{k} \\ \partial/\partial x & \partial/\partial y & \partial/\partial z \\ yz & xz & xy \end{vmatrix}=(x-x)\mathbf{i}+(y-y)\mathbf{j}+(z-z)\mathbf{k}=\mathbf{0}$

   (b) $\operatorname{div}\mathbf{F}=\nabla\cdot\mathbf{F}=\dfrac{\partial}{\partial x}(yz)+\dfrac{\partial}{\partial y}(xz)+\dfrac{\partial}{\partial z}(xy)=0+0+0=0$

5. **(a)** $\operatorname{curl}\mathbf{F}=\nabla\times\mathbf{F}=\begin{vmatrix} \mathbf{i} & \mathbf{j} & \mathbf{k} \\ \partial/\partial x & \partial/\partial y & \partial/\partial z \\ 0 & xy & xyz \end{vmatrix}=xz\mathbf{i}-yz\mathbf{j}+y\mathbf{k}$

   (b) $\operatorname{div}\mathbf{F}=\nabla\cdot\mathbf{F}=\dfrac{\partial}{\partial x}(0)+\dfrac{\partial}{\partial y}(xy)+\dfrac{\partial}{\partial z}(xyz)=0+x+xy=x(1+y)$

7. **(a)** $\nabla\times\mathbf{F}=\begin{vmatrix} \mathbf{i} & \mathbf{j} & \mathbf{k} \\ \partial/\partial x & \partial/\partial y & \partial/\partial z \\ e^{xz} & -2e^{yz} & 3xe^{y} \end{vmatrix}=(3xe^{y}+2ye^{yz})\mathbf{i}+(xe^{xz}-3e^{y})\mathbf{j}$

   (b) $\nabla\cdot\mathbf{F}=\dfrac{\partial}{\partial x}(e^{xz})+\dfrac{\partial}{\partial y}(-2e^{yz})+\dfrac{\partial}{\partial z}(3xe^{y})=ze^{xz}-2ze^{yz}$

9. **(a)** $\operatorname{curl}\mathbf{F}=\begin{vmatrix} \mathbf{i} & \mathbf{j} & \mathbf{k} \\ \partial/\partial x & \partial/\partial y & \partial/\partial z \\ xe^{y} & -ze^{-y} & y\ln z \end{vmatrix}=(e^{-y}+\ln z)\mathbf{i}-xe^{y}\mathbf{k}$

   (b) $\operatorname{div}\mathbf{F}=\dfrac{\partial}{\partial x}(xe^{y})+\dfrac{\partial}{\partial y}(-ze^{-y})+\dfrac{\partial}{\partial z}(y\ln z)=e^{y}+ze^{-y}+\dfrac{y}{z}$

11. $\operatorname{curl}\mathbf{F}=\begin{vmatrix} \mathbf{i} & \mathbf{j} & \mathbf{k} \\ \partial/\partial x & \partial/\partial y & \partial/\partial z \\ y & x & 1 \end{vmatrix}=\mathbf{0}$ and $\mathbf{F}$ is defined on all of $\mathbb{R}^3$ with component functions which have

continuous partial derivatives, so by (4), $\mathbf{F}$ is conservative. Thus there exists f such that $\mathbf{F}=\nabla f$. Then

$f_x(x,y,z)=y$ implies $f(x,y,z)=xy+g(y,z)$ and $f_y(x,y,z)=x+g_y(y,z)$. But $f_y(x,y,z)=x$, so

$g(y,z)=h(z)$ and $f(x,y,z)=xy+h(z)$. Thus $f_z(x,y,z)=h'(z)$ but $f_z(x,y,z)=1$ so $h(z)=z+k$.

Hence a potential for $\mathbf{F}$ is $f(x,y,z)=xy+z+k$.

13. $\operatorname{curl}\mathbf{F}=\begin{vmatrix} \mathbf{i} & \mathbf{j} & \mathbf{k} \\ \partial/\partial x & \partial/\partial y & \partial/\partial z \\ yz & -z^{2} & x^{2} \end{vmatrix}=2z\mathbf{i}+(y-2x)\mathbf{j}-z\mathbf{k}\neq\mathbf{0}$. Hence $\mathbf{F}$ isn't conservative.

15. $\operatorname{curl}\mathbf{F}=\begin{vmatrix} \mathbf{i} & \mathbf{j} & \mathbf{k} \\ \partial/\partial x & \partial/\partial y & \partial/\partial z \\ \cos y & \sin x & \tan z \end{vmatrix}=(\cos x-\sin y)\mathbf{k}\neq\mathbf{0}$. Hence $\mathbf{F}$ isn't conservative.

17. Since $\text{curl}\,\mathbf{F} = \begin{vmatrix} \mathbf{i} & \mathbf{j} & \mathbf{k} \\ \partial/\partial x & \partial/\partial y & \partial/\partial z \\ yz & y^2+xz & xy \end{vmatrix} = (x-x)\mathbf{i} + (y-y)\mathbf{j} + (z-z)\mathbf{k} = \mathbf{0}$, $\mathbf{F}$ is defined on $\mathbb{R}^3$, and since

the partial derivatives of the components of $\mathbf{F}$ are continuous, $\mathbf{F}$ is conservative. Thus there exists f such that

$\nabla f = \mathbf{F}$. Then $f_x(x,y,z) = yz$ implies $f(x,y,z) = xyz + g(y,z)$ and $f_y(x,y,z) = xz + g_y(y,z)$. But

$f_y(x,y,z) = xz + y^2$ so $g(y,z) = \frac{1}{3}y^3 + h(z)$ and $f(x,y,z) = xyz + \frac{1}{3}y^3 + h(z)$. Then

$f_z(x,y,z) = xy + h'(z)$. But $f_z(x,y,z) = xy$ so $h(z) = k$. Hence $f(x,y,z) = xyz + \frac{1}{3}y^3 + k$ is a potential

for $\mathbf{F}$.

19. No. Assume there is such a $\mathbf{G}$. Then $\text{div}(\text{curl}\,\mathbf{G}) = y^2 + z^2 + x^2 \neq 0$, which contradicts Theorem 11.

21. $\text{curl}\,\mathbf{F} = \begin{vmatrix} \mathbf{i} & \mathbf{j} & \mathbf{k} \\ \partial/\partial x & \partial/\partial y & \partial/\partial z \\ f(x) & g(y) & h(z) \end{vmatrix} = (0-0)\mathbf{i} + (0-0)\mathbf{j} + (0-0)\mathbf{k} = \mathbf{0}$. Hence $\mathbf{F} = f(x)\mathbf{i} + g(y)\mathbf{j} + h(z)\mathbf{k}$ is

irrotational.

Note: **For Exercises 23-29, let $\mathbf{F}(x,y,z) = P_1\mathbf{i} + Q_1\mathbf{j} + R_1\mathbf{k}$ and $\mathbf{G}(x,y,z) = P_2\mathbf{i} + Q_2\mathbf{j} + R_2\mathbf{k}$.**

23. $\text{div}(\mathbf{F}+\mathbf{G}) = \dfrac{\partial(P_1+P_2)}{\partial x} + \dfrac{\partial(Q_1+Q_2)}{\partial y} + \dfrac{\partial(R_1+R_2)}{\partial z}$

$= \left(\dfrac{\partial P_1}{\partial x} + \dfrac{\partial Q_1}{\partial y} + \dfrac{\partial R_1}{\partial z}\right) + \left(\dfrac{\partial P_2}{\partial x} + \dfrac{\partial Q_2}{\partial y} + \dfrac{\partial R_3}{\partial z}\right) = \text{div}\,\mathbf{F} + \text{div}\,\mathbf{G}$

25. $\text{div}(f\mathbf{F}) = \dfrac{\partial(fP_1)}{\partial x} + \dfrac{\partial(fQ_1)}{\partial y} + \dfrac{\partial(fR_1)}{\partial z} = \left[f\dfrac{\partial P_1}{\partial x} + P_1\dfrac{\partial f}{\partial x}\right] + \left[f\dfrac{\partial Q_1}{\partial y} + Q_1\dfrac{\partial f}{\partial y}\right] + \left[f\dfrac{\partial R_1}{\partial z} + R_1\dfrac{\partial f}{\partial z}\right]$

$= f\left(\dfrac{\partial P_1}{\partial x} + \dfrac{\partial Q_1}{\partial y} + \dfrac{\partial R_1}{\partial z}\right) + \langle P_1, Q_1, R_1\rangle \cdot \left\langle \dfrac{\partial f}{\partial x}, \dfrac{\partial f}{\partial y}, \dfrac{\partial f}{\partial z}\right\rangle = f\,\text{div}\,\mathbf{F} + \mathbf{F}\cdot\nabla f$

27. $\text{div}(\mathbf{F}\times\mathbf{G}) = \nabla\cdot(\mathbf{F}\times\mathbf{G}) = \begin{vmatrix} \partial/\partial x & \partial/\partial y & \partial/\partial z \\ P_1 & Q_1 & R_1 \\ P_2 & Q_2 & R_2 \end{vmatrix} = \dfrac{\partial}{\partial x}\begin{vmatrix} Q_1 & R_1 \\ Q_2 & R_2 \end{vmatrix} - \dfrac{\partial}{\partial y}\begin{vmatrix} P_1 & R_1 \\ P_2 & R_2 \end{vmatrix} + \dfrac{\partial}{\partial z}\begin{vmatrix} P_1 & Q_1 \\ P_2 & Q_2 \end{vmatrix}$

$= \left[Q_1\dfrac{\partial R_2}{\partial x} + R_2\dfrac{\partial Q_1}{\partial x} - Q_2\dfrac{\partial R_1}{\partial x} - R_1\dfrac{\partial Q_2}{\partial x}\right] - \left[P_1\dfrac{\partial R_2}{\partial y} + R_2\dfrac{\partial P_1}{\partial y} - P_2\dfrac{\partial R_1}{\partial y} - R_1\dfrac{\partial P_2}{\partial y}\right]$

$\qquad + \left[P_1\dfrac{\partial Q_2}{\partial z} + Q_2\dfrac{\partial P_1}{\partial z} - P_2\dfrac{\partial Q_1}{\partial z} - Q_1\dfrac{\partial P_2}{\partial z}\right]$

$= \left[P_2\left(\dfrac{\partial R_1}{\partial y} - \dfrac{\partial Q_1}{\partial z}\right) + Q_2\left(\dfrac{\partial P_1}{\partial z} - \dfrac{\partial R_1}{\partial x}\right) + R_2\left(\dfrac{\partial Q_1}{\partial x} - \dfrac{\partial P_1}{\partial y}\right)\right]$

$\qquad - \left[P_1\left(\dfrac{\partial R_2}{\partial y} - \dfrac{\partial Q_2}{\partial z}\right) + Q_1\left(\dfrac{\partial P_2}{\partial z} - \dfrac{\partial R_2}{\partial x}\right) + R_1\left(\dfrac{\partial Q_2}{\partial x} - \dfrac{\partial P_2}{\partial y}\right)\right]$

$= \mathbf{G}\cdot\text{curl}\,\mathbf{F} - \mathbf{F}\cdot\text{curl}\,\mathbf{G}$

29. $\text{curl curl } \mathbf{F} = \nabla \times (\nabla \times \mathbf{F}) = \begin{vmatrix} \mathbf{i} & \mathbf{j} & \mathbf{k} \\ \partial/\partial x & \partial/\partial y & \partial/\partial z \\ \partial R_1/\partial y - \partial Q_1/\partial z & \partial P_1/\partial z - \partial R_1/\partial x & \partial Q_1/\partial x - \partial P_1/\partial y \end{vmatrix}$

$$= \left(\frac{\partial^2 Q_1}{\partial y \partial x} - \frac{\partial^2 P_1}{\partial y^2} - \frac{\partial^2 P_1}{\partial z^2} + \frac{\partial^2 R_1}{\partial z \partial x} \right) \mathbf{i} + \left(\frac{\partial^2 R_1}{\partial z \partial y} - \frac{\partial^2 Q_1}{\partial z^2} - \frac{\partial^2 Q_1}{\partial x^2} + \frac{\partial^2 P_1}{\partial x \partial y} \right) \mathbf{j}$$

$$+ \left(\frac{\partial^2 P_1}{\partial x \partial z} - \frac{\partial^2 R_1}{\partial x^2} - \frac{\partial^2 R_1}{\partial y^2} + \frac{\partial^2 Q_1}{\partial y \partial z} \right) \mathbf{k}.$$

Now let's consider grad div $\mathbf{F} - \nabla^2 \mathbf{F}$ and compare with the above.

grad div $\mathbf{F} - \nabla^2 \mathbf{F}$

$$= \left[\left(\frac{\partial^2 P_1}{\partial x^2} + \frac{\partial^2 Q_1}{\partial x \partial y} + \frac{\partial^2 R_1}{\partial x \partial z} \right) \mathbf{i} + \left(\frac{\partial^2 P_1}{\partial y \partial x} + \frac{\partial^2 Q_1}{\partial y^2} + \frac{\partial^2 R_1}{\partial y \partial z} \right) \mathbf{j} + \left(\frac{\partial^2 P_1}{\partial z \partial x} + \frac{\partial^2 Q_1}{\partial z \partial y} + \frac{\partial^2 R_1}{\partial z^2} \right) \mathbf{k} \right]$$

$$- \left[\left(\frac{\partial^2 P_1}{\partial x^2} + \frac{\partial^2 P_1}{\partial y^2} + \frac{\partial^2 P_1}{\partial z^2} \right) \mathbf{i} + \left(\frac{\partial^2 Q_1}{\partial x^2} + \frac{\partial^2 Q_1}{\partial y^2} + \frac{\partial^2 Q_1}{\partial z^2} \right) \mathbf{j} + \left(\frac{\partial^2 R_1}{\partial x^2} + \frac{\partial^2 R_1}{\partial y^2} + \frac{\partial^2 R_1}{\partial z^2} \right) \mathbf{k} \right]$$

$$= \left(\frac{\partial^2 Q_1}{\partial x \partial y} + \frac{\partial^2 R_1}{\partial x \partial z} - \frac{\partial^2 P_1}{\partial y^2} - \frac{\partial^2 P_1}{\partial z^2} \right) \mathbf{i} + \left(\frac{\partial^2 P_1}{\partial y \partial x} + \frac{\partial^2 R_1}{\partial y \partial z} - \frac{\partial^2 Q_1}{\partial x^2} - \frac{\partial^2 Q_1}{\partial z^2} \right) \mathbf{j}$$

$$+ \left(\frac{\partial^2 P_1}{\partial z \partial x} + \frac{\partial^2 Q_1}{\partial z \partial y} - \frac{\partial^2 R_1}{\partial x^2} - \frac{\partial^2 R_2}{\partial y^2} \right) \mathbf{k}.$$

Then applying Clairaut's Theorem to reverse the order of differentiation in the second partial derivatives as needed and comparing, we have curl curl $\mathbf{F} = $ grad div $\mathbf{F} - \nabla^2 \mathbf{F}$ as desired.

31. (a) curl $f = \nabla \times \mathbf{f}$ is meaningless because f is a scalar field. **(b)** grad f is a vector field.

(c) div $\mathbf{F}$ is a scalar field. **(d)** curl(grad f) is a vector field.

(e) grad $\mathbf{F}$ is meaningless. **(f)** grad(div $\mathbf{F}$) is a vector field.

(g) div(grad f) is a scalar field. **(h)** grad(div f) is meaningless.

(i) curl(curl $\mathbf{F}$) is a vector field. **(j)** div(div $\mathbf{F}$) is meaningless.

(k) (grad f) $\times$ (div $\mathbf{F}$) is meaningless because **(l)** div(curl(grad f)) is a scalar field.
div $\mathbf{F}$ is a scalar field.

33. $\nabla \cdot \mathbf{r} = \left(\frac{\partial}{\partial x} \mathbf{i} + \frac{\partial}{\partial y} \mathbf{j} + \frac{\partial}{\partial z} \mathbf{k} \right) \cdot (x\mathbf{i} + y\mathbf{j} + z\mathbf{k}) = 1 + 1 + 1 = 3$

35. $\nabla \left(\frac{1}{r} \right) = \nabla \left(\frac{1}{\sqrt{x^2 + y^2 + z^2}} \right)$

$$= \frac{-\frac{1}{2\sqrt{x^2 + y^2 + z^2}}(2x)}{x^2 + y^2 + z^2} \mathbf{i} - \frac{\frac{1}{2\sqrt{x^2 + y^2 + z^2}}(2y)}{x^2 + y^2 + z^2} \mathbf{j} - \frac{\frac{1}{2\sqrt{x^2 + y^2 + z^2}}(2z)}{x^2 + y^2 + z^2} \mathbf{k}$$

$$= -\frac{x\mathbf{i} + y\mathbf{j} + z\mathbf{k}}{(x^2 + y^2 + z^2)^{3/2}} = -\frac{\mathbf{r}}{r^3}$$

37. $\nabla \ln r = \nabla \ln\left(x^2 + y^2 + z^2\right)^{1/2} = \frac{1}{2}\nabla \ln(x^2 + y^2 + z^2)$

$$= \frac{x}{x^2 + y^2 + z^2}\mathbf{i} + \frac{y}{x^2 + y^2 + z^2}\mathbf{j} + \frac{z}{x^2 + y^2 + z^2}\mathbf{k} = \frac{x\mathbf{i} + y\mathbf{j} + z\mathbf{k}}{x^2 + y^2 + z^2} = \frac{\mathbf{r}}{r^2}$$

39. If the vector field is $\mathbf{F} = P\mathbf{i} + Q\mathbf{j} + R\mathbf{k}$, then we are assuming here that $R = 0$, that is, the vector field does not vary in the z-direction, so $\dfrac{\partial R}{\partial z} = 0$. Since the x-component of each vector of $\mathbf{F}$ is 0, $P = 0$, so $\dfrac{\partial P}{\partial x} = 0$. But Q is decreasing as y increases, so $\dfrac{\partial Q}{\partial y} < 0$. Hence $\operatorname{div}\mathbf{F} = \dfrac{\partial P}{\partial x} + \dfrac{\partial Q}{\partial y} + \dfrac{\partial R}{\partial z}$ is negative at every point.

41. By (13), $\oint_C f(\nabla g) \cdot \mathbf{n}\, ds = \iint_D \operatorname{div}(f\nabla g)\,dA = \iint_D[f \operatorname{div}(\nabla g) + \nabla g \cdot \nabla f]\,dA$ by Exercise 25. But $\operatorname{div}(\nabla g) = \nabla^2 g$. Hence $\iint_D f\nabla^2 g\, dA = \oint_C f(\nabla g) \cdot \mathbf{n}\, ds - \iint_D \nabla g \cdot \nabla f\, dA$.

43. **(a)** We know that $\omega = \dfrac{v}{d}$, and from the diagram $\sin\theta = \dfrac{d}{r} \ \Rightarrow\ v = d\omega = (\sin\theta)r\omega = |\omega \times \mathbf{r}|$. But $\mathbf{v}$ is perpendicular to both $\mathbf{w}$ and $\mathbf{r}$, so that $\mathbf{v} = \mathbf{w} \times \mathbf{r}$.

(b) From (a), $\mathbf{v} = \mathbf{w} \times \mathbf{r} = \begin{vmatrix} \mathbf{i} & \mathbf{j} & \mathbf{k} \\ 0 & 0 & \omega \\ x & y & z \end{vmatrix} = (0 \cdot z - \omega y)\mathbf{i} + (\omega x - 0 \cdot z)\mathbf{j} + (0 \cdot y - x \cdot 0)\mathbf{k} = -\omega y\mathbf{i} + \omega x\mathbf{j}$

(c) $\operatorname{curl}\mathbf{v} = \nabla \times \mathbf{v} = \begin{vmatrix} \mathbf{i} & \mathbf{j} & \mathbf{k} \\ \partial/\partial x & \partial/\partial y & \partial/\partial z \\ -\omega y & \omega x & 0 \end{vmatrix}$

$$= \left[\frac{\partial}{\partial y}(0) - \frac{\partial}{\partial z}(\omega x)\right]\mathbf{i} + \left[\frac{\partial}{\partial z}(-\omega y) - \frac{\partial}{\partial x}(0)\right]\mathbf{j} + \left[\frac{\partial}{\partial x}(\omega x) - \frac{\partial}{\partial y}(-\omega y)\right]\mathbf{k}$$

$$= [\omega - (-\omega)]\mathbf{k} = 2\omega\mathbf{k} = 2\mathbf{w}$$

EXERCISES 14.6

1. Letting x and y be the parameters, the parametric equations are $x = x$, $y = y$, $z = \sqrt{1 - 3x^2 - 2y^2}$ where
$-\frac{1}{\sqrt{3}} \leq x \leq \frac{1}{\sqrt{3}}$ and $-\frac{1}{\sqrt{2}} \leq y \leq \frac{1}{\sqrt{2}}$. Then the vector equation of the surface is
$\mathbf{r}(x, y) = x\mathbf{i} + y\mathbf{j} + \sqrt{1 - 3x^2 - 2y^2}\mathbf{k}$.
Alternate Solution: Letting ϕ and θ be the parameters, the parametric equations are $x = \frac{1}{\sqrt{3}} \sin \phi \cos \theta$,
$y = \frac{1}{\sqrt{2}} \sin \phi \sin \theta$, $z = \cos \phi$ where $0 \leq \phi \leq \frac{\pi}{2}$ and $0 \leq \theta \leq 2\pi$.

Note: There are many parametric representations of a given surface.

3. $x = x$, $y = 6 - 3x^2 - 2z^2$, $z = z$ where $3x^2 + 2z^2 \leq 6$ since $y \geq 0$. Then the associated vector equation is
$\mathbf{r}(x, y) = x\mathbf{i} + (6 - 3x^2 - 2z^2)\mathbf{j} + z\mathbf{k}$.

5. Since the cone intersects the sphere in the circle $x^2 + y^2 = 2$, $z = 2$ and we want the portion of the sphere above
this, we can parametrize the surface as $x = x$, $y = y$, $z = \sqrt{4 - x^2 - y^2}$ where $2 \leq x^2 + y^2 \leq 4$.
Or: Using spherical coordinates, $x = 2 \sin \phi \cos \theta$, $y = 2 \sin \phi \sin \theta$, $z = 2 \cos \phi$ where $0 \leq \phi \leq \frac{\pi}{4}$ and
$0 \leq \theta \leq 2\pi$.

7. The surface is a disc of radius 4 and center $(0, 0, 5)$. Thus $x = r \cos \theta$, $y = r \sin \theta$, $z = 5$ where $0 \leq r \leq 4$,
$0 \leq \theta \leq 2\pi$ is a parametric representation of the surface.
Or: In rectangular coordinates we could represent the surface as $x = x$, $y = y$, $z = 5$ where $0 \leq x^2 + y^2 \leq 16$.

9. $\mathbf{r}(u, v) = u \cos v\mathbf{i} + u \sin v\mathbf{j} + v\mathbf{k}$. This equation must correspond to graph I, since for fixed v, $\mathbf{r}$ parametrizes a
straight line in the plane $z = v$. As v increases, the line rotates and moves upward, generating a spiral ramp.

11. $x = (u - \sin u)\cos v$, $y = (1 - \cos u)\sin v$, $z = u$. This corresponds to graph II: when $u = 0$, $x = y = z = 0$,
so $(0, 0, 0)$ is on the surface; and when $u = \pi$, $z = \pi$ and $y = 0$ while x ranges between $-\pi$ and π, giving the
upper "seam" on the surface.

13. Using Equations 3, we have the parametrization
$x = x$, $y = e^{-x} \cos \theta$, $z = e^{-x} \sin \theta$,
$0 \leq x \leq 3$, $0 \leq \theta \leq 2\pi$.

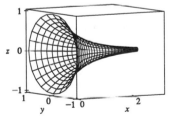

15. $\mathbf{r}(u, v) = (u + v)\mathbf{i} + 3u^2\mathbf{j} + (u - v)\mathbf{k}$. $\mathbf{r}_u = \mathbf{i} + 6u\mathbf{j} + \mathbf{k}$
and $\mathbf{r}_v = \mathbf{i} - \mathbf{k}$, so $\mathbf{r}_u \times \mathbf{r}_v = -6u\mathbf{i} + 2\mathbf{j} - 6u\mathbf{k}$.
Since the point $(2, 3, 0)$ corresponds to $u = 1$, $v = 1$,
a normal vector to the surface at $(2, 3, 0)$ is
$-6\mathbf{i} + 2\mathbf{j} - 6\mathbf{k}$, and an equation of the tangent plane is
$-6x + 2y - 6z = -6$ or $3x - y + 3z = 3$.

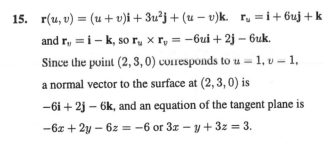

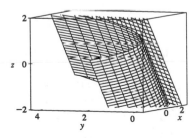

17. $r(u, v) = uv\mathbf{i} + ue^v\mathbf{j} + ve^u\mathbf{k}.$

$r_u = \langle v, e^v, ve^u \rangle, r_v = \langle u, ue^v, e^u \rangle,$ and

$r_u \times r_v = e^{u+v}(1 - uv)\mathbf{i} + e^u(uv - v)\mathbf{j} + e^v(uv - u)\mathbf{k}.$

The point $(0, 0, 0)$ corresponds to $u = 0, v = 0.$

Thus a normal vector to the surface at $(0, 0, 0)$ is $\mathbf{i}$, and

an equation of the tangent plane is $x = 0.$

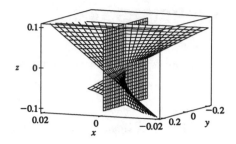

19. Here $z = f(x, y) = 4 - x - 2y$ with $0 \le x^2 + y^2 \le 4.$ Thus, by (9),

$$A(S) = \iint_D \sqrt{1 + (-1)^2 + (-2)^2}\, dA = \sqrt{6} \iint_{x^2 + y^2 \le 4} dA = 4\sqrt{6}\pi.$$

21. $z = f(x, y) = y^2 - x^2$ with $1 \le x^2 + y^2 \le 4.$ Then

$$A(S) = \iint_D \sqrt{1 + 4x^2 + 4y^2}\, dA = \int_0^{2\pi}\int_1^2 \sqrt{1 + 4r^2}\, r\, dr\, d\theta = 4\pi\left(\tfrac{1}{24}\right)(1 + 4r^2)^{3/2}\Big|_1^2 = \tfrac{\pi}{6}\left(17\sqrt{17} - 5\sqrt{5}\right).$$

23. A parametric representation of the surface is $x = x, y = 4x + z^2, z = z$

with $0 \le x \le 1, 0 \le z \le 1.$ Hence $r_x \times r_z = 4\mathbf{i} - \mathbf{j} + 2z\mathbf{k}.$

Note: In general, if $y = f(x, z)$ then $r_z \times r_x = -\dfrac{\partial f}{\partial x}\mathbf{i} + \mathbf{j} - \dfrac{\partial f}{\partial x}\mathbf{k}$ and

$$A(S) = \iint_D \sqrt{1 + \left(\frac{\partial f}{\partial x}\right)^2 + \left(\frac{\partial f}{\partial z}\right)^2}\, dA. \text{ Then}$$

$$A(S) = \int_0^1\int_0^1 \sqrt{17 + 4z^2}\, dx\, dz = \int_0^1 \sqrt{17 + 4z^2}\, dz = \tfrac{1}{2}\left(z\sqrt{17 + 4z^2} + \tfrac{17}{2}\ln\left|2z + \sqrt{4z^2 + 17}\right|\right)\Big|_0^1$$

$$= \tfrac{\sqrt{21}}{2} + \tfrac{17}{4}\left[\ln\left(2 + \sqrt{21}\right) - \ln\sqrt{17}\right].$$

25. Let $A(S_1)$ be the surface area of that portion of the surface which lies above the plane $z = 0$, then

$A(S) = 2A(S_1).$ Following Example 6, a parametric representation of S_1 is $x = a\sin\phi\cos\theta, \ y = a\sin\phi\sin\theta,$

$z = a\cos\phi$ and $|r_\phi \times r_\theta| = a^2 \sin\phi.$ For D, $0 \le \phi \le \tfrac{\pi}{2}$ and for each fixed ϕ, $\left(x - \tfrac{1}{2}a\right)^2 + y^2 \le \left(\tfrac{1}{2}a\right)^2$ or

$\left[a\sin\phi\cos\theta - \tfrac{1}{2}a\right]^2 + a^2\sin^2\phi\sin^2\theta \le (a/2)^2$ implies $a^2\sin^2\phi - a^2\sin\phi\cos\theta \le 0$ or $\sin\phi(\sin\phi - \cos\theta) \le 0.$

But $0 \le \phi \le \tfrac{\pi}{2}$, so $\cos\theta \ge \sin\phi$ or $\sin\left(\tfrac{\pi}{2} + \theta\right) \ge \sin\phi$ or $\phi - \tfrac{\pi}{2} \le \theta \le \tfrac{\pi}{2} - \phi.$ Hence

$D = \left\{(\phi, \theta) \mid 0 \le \phi \le \tfrac{\pi}{2}, \ \phi - \tfrac{\pi}{2} \le \theta \le \tfrac{\pi}{2} - \phi\right\}.$ Then

$A(S_1) = \int_0^{\pi/2}\int_{\phi - (\pi/2)}^{(\pi/2)-\phi} a^2\sin\phi\, d\theta\, d\phi = a^2\int_0^{\pi/2}(\pi - 2\phi)\sin\phi\, d\phi = a^2[(-\pi\cos\phi) - 2(-\phi\cos\phi + \sin\phi)]_0^{\pi/2}$

$= a^2(\pi - 2).$ Thus $A(S) = 2a^2(\pi - 2).$

Alternate Solution: Working on S_1 we could parametrize the portion of the sphere by $x = x, y = y,$

$z = \sqrt{a^2 - x^2 - y^2}.$ Then $|r_x \times r_y| = \sqrt{1 + \dfrac{x^2}{a^2 - x^2 - y^2} + \dfrac{y^2}{a^2 - x^2 - y^2}} = \dfrac{a}{\sqrt{a^2 - x^2 - y^2}}$ and

$$A(S_1) = \iint\limits_{0 \le (x-\frac{1}{2}a)^2 + y^2 \le (\frac{1}{2}a)^2} \frac{a}{\sqrt{a^2 - x^2 - y^2}} \, dA = \int_{-\pi/2}^{\pi/2} \int_0^{a\cos\theta} \frac{a}{\sqrt{a^2 - r^2}} \, r \, dr \, d\theta$$

$$= \int_{-\pi/2}^{\pi/2} -a(a^2 - r^2)^{1/2}\Big|_0^{a\cos\theta} \, d\theta = \int_{-\pi/2}^{\pi/2} a^2\Big[1 - (1 - \cos^2\theta)^{1/2}\Big] d\theta$$

$$= \int_{-\pi/2}^{\pi/2} a^2(1 - |\sin\theta|) d\theta = 2a^2 \int_0^{\pi/2} (1 - \sin\theta) d\theta = 2a^2\left(\frac{\pi}{2} - 1\right).$$

Thus $A(S) = 4a^2\left(\frac{\pi}{2} - 1\right) = 2a^2(\pi - 2)$.

Notes: (1) Perhaps working in spherical coordinates is the most obvious approach here. However, you must be careful in setting up D.

(2) In the alternate solution, you can avoid having to use $|\sin\theta|$ by working in the first octant and then multiplying by 8. However, if you set up S_1 as above and arrived at $A(S_1) = a^2\pi$, you now see your error.

27. $\mathbf{r}_u = \langle v, 1, 1 \rangle$, $\mathbf{r}_v = \langle u, 1, -1 \rangle$ and $\mathbf{r}_u \times \mathbf{r}_v = \langle -2, u+v, v-u \rangle$. Then

$$A(S) = \iint\limits_{u^2+v^2 \le 1} \sqrt{4 + 2u^2 + 2v^2} \, dA = \int_0^{2\pi}\int_0^1 r\sqrt{4 + 2r^2} \, dr \, d\theta = 2\pi\left(\tfrac{1}{6}\right)(4 + 2r^2)^{3/2}\Big|_0^1$$

$$= \tfrac{\pi}{3}\left(6\sqrt{6} - 8\right) = \pi\left(2\sqrt{6} - \tfrac{8}{3}\right).$$

29. (a) $x = a \sin u \cos v$, $y = b \sin u \sin v$, $z = c \cos u$ $\Rightarrow$ **(b)**

$$\frac{x^2}{a^2} + \frac{y^2}{b^2} + \frac{z^2}{c^2} = (\sin u \cos v)^2 + (\sin u \sin v)^2 + (\cos u)^2$$

$$= \sin^2 u + \cos^2 u = 1$$

and since the ranges of u and v are sufficient to generate the entire graph, the parametric equations represent an ellipsoid.

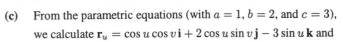

(c) From the parametric equations (with $a = 1$, $b = 2$, and $c = 3$), we calculate $\mathbf{r}_u = \cos u \cos v\,\mathbf{i} + 2\cos u \sin v\,\mathbf{j} - 3\sin u\,\mathbf{k}$ and

$\mathbf{r}_v = -\sin u \sin v\,\mathbf{i} + 2\sin u \cos v\,\mathbf{j}$. So $\mathbf{r}_u \times \mathbf{r}_v = 6\sin^2 u \cos v\,\mathbf{i} + 3\sin^2 u \sin v\,\mathbf{j} + 2\sin u \cos u\,\mathbf{k}$, and the surface area is given by

$$A(S) = \int_0^{2\pi}\int_0^\pi |\mathbf{r}_u \times \mathbf{r}_v| \, du \, dv = \int_0^{2\pi}\int_0^\pi \sqrt{36\sin^4 u \cos^2 v + 9\sin^4 u \sin^2 v + 4\cos^2 u \sin^2 u} \, du \, dv.$$

31.

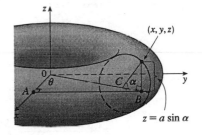

Here $z = a\sin\alpha$, $y = |AB|$, and $x = |OA|$. But $|OB| = |OC| + |CB| = b + a\cos\alpha$ and $\sin\theta = |AB|/|OB|$ so that $y = |OB|\sin\theta = (b + a\cos\alpha)\sin\theta$. Similarly $\cos\theta = |OA|/|OB|$ so $x = (b + a\cos\alpha)\cos\theta$. Hence a parametric representation for the torus is

$x = b\cos\theta + a\cos\alpha\cos\theta$, $y = b\sin\theta + a\cos\alpha\sin\theta$, $z = a\sin\alpha$,

where $0 \le \alpha \le 2\pi$, $0 \le \theta \le 2\pi$.

EXERCISES 14.7

1. Here $f(x, y, z) = \sqrt{x^2 + 2y^2 + 3z^2}$ and by Definition 1,

    $\iint_S f(x, y, z)dS \approx [f(1, 0, 0)](4) + [f(-1, 0, 0)](4) + [f(0, 1, 0)](4) + [f(0, -1, 0)](4)$

    $\qquad\qquad\qquad + [f(0, 0, 1)](4) + [f(0, 0, -1)](4)$

    $\qquad\qquad = 4\left(1 + 1 + 2\sqrt{2} + 2\sqrt{3}\right) = 8\left(1 + \sqrt{2} + \sqrt{3}\right) \approx 33.170.$

3. $\mathbf{r}(x, y) = x\mathbf{i} + y\mathbf{j} + (6 - 3x - 2y)\mathbf{k}$, $\mathbf{r}_x \times \mathbf{r}_y = 3\mathbf{i} + 2\mathbf{j} + \mathbf{k}$ (the normal to the plane) and $|\mathbf{r}_x \times \mathbf{r}_y| = \sqrt{14}$. The given plane meets the first octant in the line $3x + 2y = 6$, $z = 0$, $x \geq 0$, $y \geq 0$, so
    $D = \left\{(x, y) \mid 0 \leq x \leq \frac{1}{3}(6 - 2y), 0 \leq y \leq 3\right\}$. Then
    $\iint_S y \, dS = \int_0^3 \int_0^{(6-2y)/3} y\sqrt{14} \, dx \, dy = \sqrt{14}\int_0^3 \left(2y - \frac{2}{3}y^2\right)dy = 3\sqrt{14}.$

5. $\mathbf{r}(x, z) = x\mathbf{i} + (x^2 + 4z)\mathbf{j} + z\mathbf{k}$, $0 \leq x \leq 2$, $0 \leq z \leq 2$, $|\mathbf{r}_x \times \mathbf{r}_z| = \sqrt{4x^2 + 17}$ (see Exercise 14.6.23.) Then
    $\iint_S x \, dS = \int_0^2 \int_0^2 x\sqrt{4x^2 + 17} \, dx \, dz = 2\left[\frac{1}{12}(4x^2 + 17)^{3/2}\right]_0^2 = \frac{33\sqrt{33} - 17\sqrt{17}}{6}.$

7. Since $z = y + 3$, $|\mathbf{r}_x \times \mathbf{r}_y| = \sqrt{2}$ and $\iint_S yz \, dS = \iint\limits_{x^2 + y^2 \leq 1} \sqrt{2}y(y + 3)dA$
    $= \sqrt{2}\int_0^{2\pi}\int_0^1 (r^2 \sin^2\theta + 3r \sin\theta)r \, dr \, d\theta = \sqrt{2}\int_0^{2\pi}\left[\frac{1}{4}\sin^2\theta + \sin\theta\right]d\theta = \frac{\pi}{2\sqrt{2}}.$

9. Using spherical coordinates and Example 6 in Section 14.6 we have
    $\mathbf{r}(\phi, \theta) = 2\sin\phi\cos\theta\,\mathbf{i} + 2\sin\phi\sin\theta\,\mathbf{j} + 2\cos\phi\,\mathbf{k}$ and $|\mathbf{r}_\phi \times \mathbf{r}_\theta| = 4\sin\phi$. Then
    $\iint_S (x^2z + y^2z)dS = \int_0^{2\pi}\int_0^{\pi/2}(4\sin^2\phi)(2\cos\phi)(4\sin\phi)d\phi \, d\theta = 16\pi \sin^4\phi\big|_0^{\pi/2} = 16\pi.$

11. Using cylindrical coordinates, $\mathbf{r}(\theta, z) = 3\cos\theta\,\mathbf{i} + 3\sin\theta\,\mathbf{j} + z\mathbf{k}$, $0 \leq \theta \leq 2\pi$, $0 \leq z \leq 2$, and $|\mathbf{r}_\theta \times \mathbf{r}_z| = 3$.
    $\iint_S (x^2y + z^2)dS = \int_0^{2\pi}\int_0^2 (27\cos^2\theta\sin\theta + z^2)3 \, dz \, d\theta = \int_0^{2\pi}(162\cos^2\theta\sin\theta + 8)d\theta = 16\pi$

13. $\mathbf{r}(u, v) = uv\mathbf{i} + (u + v)\mathbf{j} + (u - v)\mathbf{k}$, $u^2 + v^2 \leq 1$ and $|\mathbf{r}_u \times \mathbf{r}_v| = \sqrt{4 + 2u^2 + 2v^2}$ (see Exercise 14.6.27).
    Then $\iint\limits_S yx \, dS = \iint\limits_{u^2 + v^2 \leq 1} (u^2 - v^2)\sqrt{4 + 2u^2 + 2v^2} \, dA = \int_0^{2\pi}\int_0^1 r^2(\cos^2\theta - \sin^2\theta)\sqrt{4 + 2r^2}\, r \, dr \, d\theta$

    $\qquad = \left[\int_0^{2\pi}(\cos^2\theta - \sin^2\theta)d\theta\right]\left[\int_0^1 r^3\sqrt{4 + 2r^2} \, dr\right] = 0$ since the first integral is 0.

15. $\mathbf{F}(\mathbf{r}(x, y)) = e^y\mathbf{i} + ye^x\mathbf{j} + x^2y\mathbf{k}$ and $\mathbf{r}_x \times \mathbf{r}_y = -2x\mathbf{i} - 2y\mathbf{j} + \mathbf{k}$. Then
    $\mathbf{F}(\mathbf{r}(x, y)) \cdot (\mathbf{r}_x \times \mathbf{r}_y) = -2xe^y - 2y^2e^x + x^2y$ and
    $\iint_S \mathbf{F} \cdot d\mathbf{S} = \int_0^1\int_0^1 (-2xe^y - 2y^2e^x + x^2y)dx \, dy = \int_0^1 \left(-e^y - 2ey^2 + \frac{1}{3}y + 2y^2\right)dy = \frac{1}{6}(11 - 10e).$

17. As in Exercise 3, $D = \left\{(x, y) \mid 0 \leq x \leq 2, 0 \leq y \leq \frac{1}{2}(6 - 3x)\right\}$.
    $\iint_S \mathbf{F} \cdot d\mathbf{S} = \int_0^2\int_0^{(6-3x)/2}[x\mathbf{i} + xy\mathbf{j} + x(6 - 3x - 2y)\mathbf{k}] \cdot (3\mathbf{i} + 2\mathbf{j} + \mathbf{k})dy \, dx$
    $= \int_0^2\int_0^{(6-3x)/2}(9x - 3x^2)dy \, dx = \int_0^2 \left[27x - \frac{45}{2}x^2 + \frac{9}{2}x^3\right]dx = 12.$

19. $\mathbf{F}(\mathbf{r}(\phi,\theta)) = 3\sin\phi\cos\theta\,\mathbf{i} + 3\sin\phi\sin\theta\,\mathbf{j} + 3\cos\phi\,\mathbf{k}$ and

$\mathbf{r}_\phi \times \mathbf{r}_\theta = 9\sin^2\phi\cos\theta\,\mathbf{i} + 9\sin^2\phi\sin\theta\,\mathbf{j} + 9\sin\phi\cos\phi\,\mathbf{k}$. Then

$\mathbf{F}(\mathbf{r}(\phi,\theta))\cdot(\mathbf{r}_\phi\times\mathbf{r}_\theta) = 27\sin^3\phi\cos^2\theta + 27\sin^3\phi\sin^2\theta + 27\sin\phi\cos^2\phi = 27\sin\phi$ and

$\iint_S \mathbf{F}\cdot d\mathbf{S} = \int_0^{2\pi}\int_0^\pi 27\sin\phi\,d\phi\,d\theta = (2\pi)(54) = 108\pi.$

21. Let S_1 be the paraboloid $y = x^2 + z^2, 0 \le y \le 1$ and S_2 the disc $x^2 + z^2 \le 1, y = 1$. Since S is a closed

surface, we use the outward orientation. On S_1: $\mathbf{F}(\mathbf{r}(x,z)) = (x^2 + z^2)\mathbf{j} - z\mathbf{k}$ and $\mathbf{r}_x \times \mathbf{r}_z = 2x\mathbf{i} - \mathbf{j} + 2z\mathbf{k}$

(since the $\mathbf{j}$-component must be negative on S_1). Then

$\iint\limits_{S_1} \mathbf{F}\cdot d\mathbf{S} = \iint\limits_{x^2 + z^2 \le 1} [-(x^2 + z^2) - 2z^2]dA = -\int_0^{2\pi}\int_0^1 (r^2 + 2r^2\cos^2\theta)r\,dr\,d\theta$

$= -\int_0^{2\pi} \frac{1}{4}(1 + 2\cos^2\theta)d\theta = -\left(\frac{\pi}{2} + \frac{\pi}{2}\right) = -\pi.$

On S_2: $\mathbf{F}(\mathbf{r}(x,z)) = \mathbf{j} - z\mathbf{k}$ and $\mathbf{r}_z \times \mathbf{r}_x = \mathbf{j}$. Then $\iint\limits_{S_2} \mathbf{F}\cdot d\mathbf{S} = \iint\limits_{x^2 + z^2 \le 1} (1)dA = \pi.$ Hence

$\iint\limits_S \mathbf{F}\cdot d\mathbf{S} = -\pi + \pi = 0.$

23. Here S consists of the six faces of the cube as labeled in the figure.

On S_1: $\mathbf{F} = \mathbf{i} + 2y\mathbf{j} + 3z\mathbf{k}, \mathbf{r}_y \times \mathbf{r}_z = \mathbf{i}$ and $\iint\limits_{S_1} \mathbf{F}\cdot d\mathbf{S} = \int_{-1}^1\int_{-1}^1 dy\,dz = 4;$

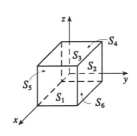

S_2: $\mathbf{F} = x\mathbf{i} + 2\mathbf{j} + 3z\mathbf{k}, \mathbf{r}_z \times \mathbf{r}_x = \mathbf{j}$ and $\iint\limits_{S_2} \mathbf{F}\cdot d\mathbf{S} = \int_{-1}^1\int_{-1}^1 2\,dx\,dz = 8;$

S_3: $\mathbf{F} = x\mathbf{i} + 2y\mathbf{j} + 3\mathbf{k}, \mathbf{r}_x \times \mathbf{r}_y = \mathbf{k}$ and $\iint\limits_{S_3} \mathbf{F}\cdot d\mathbf{S} = \int_{-1}^1\int_{-1}^1 3\,dx\,dy = 12;$

S_4: $\mathbf{F} = -\mathbf{i} + 2y\mathbf{j} + 3z\mathbf{k}, \mathbf{r}_z \times \mathbf{r}_y = -\mathbf{i}$ and $\iint\limits_{S_4} \mathbf{F}\cdot d\mathbf{S} = 4;$

S_5: $\mathbf{F} = x\mathbf{i} - 2\mathbf{j} + 3z\mathbf{k}, \mathbf{r}_x \times \mathbf{r}_z = -\mathbf{j}$ and $\iint\limits_{S_5} \mathbf{F}\cdot d\mathbf{S} = 8;$

S_6: $\mathbf{F} = x\mathbf{i} + 2y\mathbf{j} - 3\mathbf{k}, \mathbf{r}_y \times \mathbf{r}_x = -\mathbf{k}$ and $\iint\limits_{S_6} \mathbf{F}\cdot d\mathbf{S} = \int_{-1}^1\int_{-1}^1 3\,dx\,dy = 12.$

Hence $\iint_S \mathbf{F}\cdot d\mathbf{S} = \sum_{i=1}^6 \iint_{S_i} \mathbf{F}\cdot d\mathbf{S} = 48.$

25. $z = xy \quad\Rightarrow\quad \partial z/\partial x = y, \partial z/\partial y = x$, so by Formula 2, a CAS gives

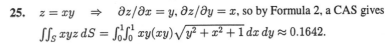

$\iint_S xyz\,dS = \int_0^1\int_0^1 xy(xy)\sqrt{y^2 + x^2 + 1}\,dx\,dy \approx 0.1642.$

27. We use Formula 2 with $z = 3 - 2x^2 - y^2 \quad\Rightarrow\quad \partial z/\partial x = -4x, \partial z/\partial y = -2y$. The boundaries of the region

$3 - 2x^2 - y^2 \ge 0$ are $-\sqrt{\frac{3}{2}} \le x \le \sqrt{\frac{3}{2}}$ and $-\sqrt{3 - 2x^2} \le y \le \sqrt{3 - 2x^2}$, so we use a CAS (with precision

reduced to seven or fewer digits; otherwise the calculation takes a very long time) to calculate

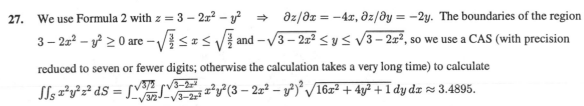

$\iint_S x^2 y^2 z^2\,dS = \int_{-\sqrt{3/2}}^{\sqrt{3/2}}\int_{-\sqrt{3-2x^2}}^{\sqrt{3-2x^2}} x^2 y^2 (3 - 2x^2 - y^2)^2 \sqrt{16x^2 + 4y^2 + 1}\,dy\,dx \approx 3.4895.$

29. If S is given by $y = h(x, z)$, then S is also the level surface $f(x, y, z) = y - h(x, z) = 0$.

$\mathbf{n} = \dfrac{\nabla f(x, y, z)}{|\nabla f(x, y, z)|} = \dfrac{-h_x \mathbf{i} + \mathbf{j} - h_z \mathbf{k}}{\sqrt{h_x^2 + 1 + h_z^2}}$, and $-\mathbf{n}$ is the unit normal that points to the left. Now we proceed as in

the derivation of (8), using Formula 2 to evaluate

$$\iint_S \mathbf{F} \cdot d\mathbf{S} = \iint_S \mathbf{F} \cdot \mathbf{n}\, dS = \iint_D (P\mathbf{i} + Q\mathbf{j} + R\mathbf{k}) \dfrac{\dfrac{\partial h}{\partial x}\mathbf{i} - \mathbf{j} + \dfrac{\partial h}{\partial z}\mathbf{k}}{\sqrt{\left(\dfrac{\partial h}{\partial x}\right)^2 + 1 + \left(\dfrac{\partial h}{\partial z}\right)^2}} \sqrt{\left(\dfrac{\partial h}{\partial x}\right)^2 + 1 + \left(\dfrac{\partial h}{\partial z}\right)^2}\, dA$$

where D is the projection of $f(x, y, z)$ onto the xz-plane. Therefore

$$\iint_S \mathbf{F} \cdot d\mathbf{S} = \iint_D \left(P\dfrac{\partial h}{\partial x} - Q + R\dfrac{\partial h}{\partial z}\right) dA.$$

31. $m = \iint_S K\, dS = K \cdot 4\pi\left(\frac{1}{2}a^2\right) = 2\pi a^2 K$; by symmetry $M_{xz} = M_{yz} = 0$, and

$M_{xy} = \iint_S zK\, dS = K \int_0^{2\pi} \int_0^{\pi/2} (a \cos\phi)(a^2 \sin\phi)d\phi\, d\theta = 2\pi K a^3 \left[-\frac{1}{4} \cos 2\phi\right]_0^{\pi/2} = \pi K a^3$. Hence

$(\overline{x}, \overline{y}, \overline{z}) = \left(0, 0, \frac{1}{2}a\right)$.

33. **(a)** $I_z = \iint_S (x^2 + y^2)\rho(x, y, z)dS$

(b) $I_z = \iint_S (x^2 + y^2)\left(10 - \sqrt{x^2 + y^2}\right)dS = \iint_{1 \le x^2 + y^2 \le 16} (x^2 + y^2)\left(10 - \sqrt{x^2 + y^2}\right)\sqrt{2}\, dA$

$\qquad = \int_0^{2\pi} \int_1^4 \sqrt{2}\left(10r^3 - r^4\right)dr\, d\theta = 2\sqrt{2}\pi\left(\frac{4329}{10}\right) = \frac{4329}{5}\sqrt{2}\pi$

35. $\rho(x, y, z) = 1200,\ \mathbf{V} = y\mathbf{i} + \mathbf{j} + z\mathbf{k},\ \mathbf{F} = \rho\mathbf{V} = (1200)(y\mathbf{i} + \mathbf{j} + z\mathbf{k})$. S is given by

$\mathbf{r}(x, y) = x\mathbf{i} + y\mathbf{j} + \left[9 - \frac{1}{4}(x^2 + y^2)\right]\mathbf{k},\ 0 \le x^2 + y^2 \le 36$ and $\mathbf{r}_x \times \mathbf{r}_y = \frac{1}{2}x\mathbf{i} + \frac{1}{2}y\mathbf{j} + \mathbf{k}$. Thus the rate of flow

is given by

$\iint_S \mathbf{F} \cdot d\mathbf{S} = \iint_{0 \le x^2 + y^2 \le 36} (1200)\left(\frac{1}{2}xy + \frac{1}{2}y + \left[9 - \frac{1}{4}(x^2 + y^2)\right]\right)dA$

$\qquad = 1200 \int_0^6 \int_0^{2\pi} \left[\frac{1}{2}r^2 \sin\theta \cos\theta + \frac{1}{2}r \sin\theta + 9 - \frac{1}{4}r^2\right]r\, d\theta\, dr = 1200 \int_0^6 2\pi\left(9r - \frac{1}{4}r^3\right)dr$

$\qquad = (1200)(2\pi)(81) = 194{,}400\pi$.

37. S consists of the hemisphere S_1 given by $z = \sqrt{a^2 - x^2 - y^2}$ and the disk S_2 given by $0 \le x^2 + y^2 \le a^2,\ z = 0$.

On S_1: $\mathbf{E} = a \sin\phi \cos\theta\, \mathbf{i} + a \sin\phi \sin\theta\, \mathbf{j} + 2a \cos\phi\, \mathbf{k}$,

$\mathbf{T}_\phi \times \mathbf{T}_\theta = a^2 \sin^2\phi \cos\theta\, \mathbf{i} + a^2 \sin^2\phi \sin\theta\, \mathbf{j} + a^2 \sin\phi \cos\phi\, \mathbf{k}$. Thus

$\iint_{S_1} \mathbf{E} \cdot d\mathbf{S} = \int_0^{2\pi} \int_0^{\pi/2} (a^3 \sin^3\phi + 2a^3 \sin\phi \cos^2\phi)d\phi\, d\theta$

$\qquad = \int_0^{2\pi} \int_0^{\pi/2} (a^3 \sin\phi + a^3 \sin\phi \cos^2\phi)d\phi\, d\theta = (2\pi)a^3\left(1 + \frac{1}{3}\right) = \frac{8}{3}\pi a^3$.

On S_2: $\mathbf{E} = x\mathbf{i} + y\mathbf{j}$, and $\mathbf{r}_y \times \mathbf{r}_x = -\mathbf{k}$ so $\iint_{S_2} \mathbf{E} \cdot d\mathbf{S} = 0$. Hence the total charge is $q = \epsilon_0 \iint_S \mathbf{E} \cdot d\mathbf{S} = \frac{8}{3}\pi a^3 \epsilon_0$.

39. $K\nabla u = 6.5(4y\mathbf{j} + 4z\mathbf{k})$. S is given by $\mathbf{r}(x, \theta) = x\mathbf{i} + \sqrt{6} \cos\theta\, \mathbf{j} + \sqrt{6} \sin\theta\, \mathbf{k}$ and since we want the inward

heat flow, we use $\mathbf{r}_x \times \mathbf{r}_\theta = -\sqrt{6} \cos\theta\, \mathbf{j} - \sqrt{6} \sin\theta\, \mathbf{k}$. Then the rate of heat flow inward is given by

$\iint_S (-K\nabla u) \cdot d\mathbf{S} = \int_0^{2\pi} \int_0^4 -(6.5)(-24)dx\, d\theta = (2\pi)(156)(4) = 1248\pi$.

EXERCISES 14.8

1. The boundary curve is C: $x^2 + y^2 = 1$, $z = 0$ oriented in the counterclockwise direction. The vector equation of C is $\mathbf{r}(t) = \cos t\,\mathbf{i} + \sin t\,\mathbf{j}$, $0 \le t \le 2\pi$. Then $\mathbf{F}(\mathbf{r}(t)) = \cos t\,\mathbf{j} + e^{\cos t \sin t}\,\mathbf{k}$ and $\mathbf{F}(\mathbf{r}(t)) \cdot \mathbf{r}'(t) = \cos^2 t$. Hence $\iint_S \operatorname{curl} \mathbf{F} \cdot d\mathbf{S} = \oint_C \mathbf{F} \cdot d\mathbf{r} = \int_0^{2\pi} \cos^2 t\, dt = \int_0^{2\pi} \frac{1}{2}(1 + \cos 2t)dt = \pi$.

3. C is the circle $x^2 + z^2 = 1$, $y = 0$ and the vector equation is $\mathbf{r}(t) = \cos t\,\mathbf{i} + \sin t\,\mathbf{k}$, $0 \le t \le 2\pi$ since the surface is oriented toward the xy-plane. Then $\mathbf{F}(\mathbf{r}(t)) = \cos^3 t\,\mathbf{k}$ and $\mathbf{F}(\mathbf{r}(t)) \cdot \mathbf{r}'(t) = \cos^4 t$. Hence $\iint_S \operatorname{curl} \mathbf{F} \cdot d\mathbf{S} = \oint_C \mathbf{F} \cdot d\mathbf{r} = \int_0^{2\pi} \cos^4 t\, dt = \int_0^{2\pi} \left[\frac{3}{8} + \frac{1}{2}\cos 2t + \frac{1}{8}\cos 4t\right]dt = \frac{3\pi}{4}$.

5. C is the square in the plane $z = -1$. By (4), $\iint_{S_1} \operatorname{curl} \mathbf{F} \cdot d\mathbf{S} = \oint_C \mathbf{F} \cdot d\mathbf{r} = \iint_{S_2} \operatorname{curl} \mathbf{F} \cdot d\mathbf{S}$ where S_1 is the original cube without the bottom and S_2 is the bottom face of the cube.
    $\operatorname{curl} \mathbf{F} = x^2 z\,\mathbf{i} + (xy - 2xyz)\mathbf{j} + (y - xz)\mathbf{k}$. For S_2, $\mathbf{n} = -\mathbf{k}$ and $\operatorname{curl} \mathbf{F} \cdot \mathbf{n} = xz - y = -x - y$ on S_2, where $z = -1$. Then $\iint_{S_2} \operatorname{curl} \mathbf{F} \cdot d\mathbf{S} = -\int_{-1}^{1}\int_{-1}^{1}(x + y)dx\,dy = 0$ so $\iint_{S_1} \operatorname{curl} \mathbf{F} \cdot d\mathbf{S} = 0$.

7. $\operatorname{curl} \mathbf{F} = 3x\,\mathbf{i} + (x - 3y)\mathbf{j} + 2y\,\mathbf{k}$, $\mathbf{n} = \frac{1}{\sqrt{11}}(3\mathbf{i} + \mathbf{j} + \mathbf{k})$ and
    $$\oint_C \mathbf{F} \cdot d\mathbf{r} = \iint_S \operatorname{curl} \mathbf{F} \cdot \mathbf{n}\, dS = \int_0^1\int_0^{3-3x} \frac{1}{\sqrt{11}}[9x + (x - 3y) + 2y]\left(\sqrt{11}\right)dy\,dx$$
    $$= \int_0^1\int_0^{3-3x}(10x - y)dy\,dx = \int_0^1\left[10(3x - 3x^2) - \frac{1}{2}(3 - 3x)^2\right]dx$$
    $$= \left[15x^2 - 10x^3 + \frac{3}{2}(1 - x^3)\right]_0^1 = \frac{7}{2}.$$

9. The curve of intersection is an ellipse in the plane $z = x + 4$ with unit normal $\mathbf{n} = \frac{1}{\sqrt{2}}(-\mathbf{i} + \mathbf{k})$ and $\operatorname{curl} \mathbf{F} = 5\mathbf{i} + 2\mathbf{j} + 4\mathbf{k}$ so $\operatorname{curl} \mathbf{F} \cdot \mathbf{n} = -\frac{1}{\sqrt{2}}$. Then
    $$\oint_C \mathbf{F} \cdot d\mathbf{r} = -\iint_S \frac{1}{\sqrt{2}}dS = -\frac{1}{\sqrt{2}}(\text{surface area of planar ellipse}) = -\frac{1}{\sqrt{2}}\pi(2)\left(2\sqrt{2}\right) = -4\pi.$$

11. (a) The curve of intersection is an ellipse in the plane $x + y + z = 1$ with unit normal $\mathbf{n} = \frac{1}{\sqrt{3}}(\mathbf{i} + \mathbf{j} + \mathbf{k})$, $\operatorname{curl} \mathbf{F} = x^2\mathbf{j} + y^2\mathbf{k}$ and $\operatorname{curl} \mathbf{F} \cdot \mathbf{n} = \frac{1}{\sqrt{3}}(x^2 + y^2)$. Then
    $$\oint_C \mathbf{F} \cdot d\mathbf{r} = \iint_S \frac{1}{\sqrt{3}}(x^2 + y^2)dS = \iint_{x^2+y^2 \le 9}(x^2 + y^2)dx\,dy = \int_0^{2\pi}\int_0^3 r^3\,dr\,d\theta = 2\pi\left(\frac{81}{4}\right) = \frac{81\pi}{2}.$$

    (b)
    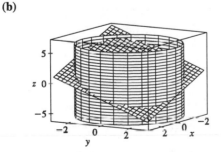

    (c) One possible parametrization is $x = 3\cos t$, $y = 3\sin t$, $z = 1 - 3\cos t - 3\sin t$, $0 \le t \le 2\pi$.

    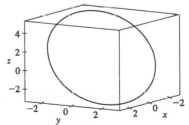

13. The boundary curve C is the circle $x^2 + y^2 = 9$, $z = 0$ oriented in the counterclockwise direction as viewed from

$(0, 0, 1)$. Then $\mathbf{r}(t) = 3\cos t\,\mathbf{i} + 3\sin t\,\mathbf{j}$, $0 \le t \le 2\pi$, so $\mathbf{F}(\mathbf{r}(t)) = 9\sin t\,\mathbf{i} - 18\cos t\,\mathbf{k}$ and

$\mathbf{F} \cdot \mathbf{r}'(t) = -27\sin^2 t$. Thus $\oint_C \mathbf{F} \cdot d\mathbf{r} = \int_0^{2\pi}(-27\sin^2 t)dt = -27\pi$. Now curl $\mathbf{F} = -4\mathbf{i} + 6\mathbf{j} - 3\mathbf{k}$,

$\mathbf{r}_x \times \mathbf{r}_y = 2x\mathbf{i} + 2y\mathbf{j} + \mathbf{k}$, so

$$\iint\limits_{S} \text{curl }\mathbf{F} \cdot d\mathbf{S} = \iint\limits_{x^2+y^2 \le 9}(-8x + 12y - 3)dA = \int_0^{2\pi}\int_0^3(-8r\cos\theta + 12r\sin\theta - 3)r\,dr\,d\theta$$

$$= \int_0^3(-3r)(2\pi)dr = -27\pi.$$

15. The x-, y-, and z-intercepts of the plane are all 1, so C consists of the three line segments

C_1: $\mathbf{r}_1(t) = (1 - t)\mathbf{i} + t\mathbf{j}$, $0 \le t \le 1$, C_2: $\mathbf{r}_2(t) = (1 - t)\mathbf{j} + t\mathbf{k}$, $0 \le t \le 1$, and

C_3: $\mathbf{r}_3(t) = t\mathbf{i} + (1 - t)\mathbf{k}$, $0 \le t \le 1$. Then

$\oint_C \mathbf{F} \cdot d\mathbf{r} = \int_0^1[t\mathbf{i} + (1 - t)\mathbf{k}] \cdot (-\mathbf{i} + \mathbf{j})dt + \int_0^1[(1 - t)\mathbf{i} + t\mathbf{j}] \cdot (-\mathbf{j} + \mathbf{k})dt + \int_0^1[(1 - t)\mathbf{j} + t\mathbf{k}] \cdot (\mathbf{i} - \mathbf{k})dt$

$\quad = \int_0^1(-3t)dt = -\frac{3}{2}$.

Now curl $\mathbf{F} = -\mathbf{i} - \mathbf{j} - \mathbf{k}$ and $\mathbf{r}_x \times \mathbf{r}_y = \mathbf{i} + \mathbf{j} + \mathbf{k}$. Hence $\iint_S \text{curl }\mathbf{F} \cdot d\mathbf{S} = \int_0^1\int_0^{1-x}(-3)dy\,dx = -\frac{3}{2}$.

17. $\text{curl }\mathbf{F} = \begin{vmatrix} \mathbf{i} & \mathbf{j} & \mathbf{k} \\ \partial/\partial x & \partial/\partial y & \partial/\partial z \\ x^x + z^2 & y^y + x^2 & z^z + y^2 \end{vmatrix} = 2y\mathbf{i} + 2z\mathbf{j} + 2x\mathbf{k}$ and $W = \int_C \mathbf{F} \cdot d\mathbf{r} = \iint_S \text{curl }\mathbf{F} \cdot d\mathbf{S}$.

To parametrize the surface, let $x = 2\cos\theta\sin\phi$, $y = 2\sin\theta\sin\phi$, $z = 2\cos\phi$, so that

$\mathbf{r}(\phi, \theta) = 2\sin\phi\cos\theta\,\mathbf{i} + 2\sin\phi\sin\theta\,\mathbf{j} + 2\cos\phi\,\mathbf{k}$, $0 \le \phi \le \frac{\pi}{2}$, $0 \le \theta \le \frac{\pi}{2}$, and

$\mathbf{r}_\phi \times \mathbf{r}_\theta = 4\sin^2\phi\cos\theta\,\mathbf{i} + 4\sin^2\phi\sin\theta\,\mathbf{j} + 4\sin\phi\cos\phi\,\mathbf{k}$. Then

$\text{curl }\mathbf{F}(\mathbf{r}(\phi, \theta)) = 4\sin\phi\sin\theta\,\mathbf{i} + 4\cos\phi\,\mathbf{j} + 4\sin\phi\cos\theta\,\mathbf{k}$, and

$\text{curl }\mathbf{F} \cdot (\mathbf{r}_\phi \times \mathbf{r}_\theta) = 16\sin^3\phi\sin\theta\cos\theta + 16\cos\phi\sin^2\phi\sin\theta + 16\sin^2\phi\cos\phi\cos\theta$. Therefore

$\iint_S \text{curl }\mathbf{F} \cdot d\mathbf{S} = \iint_D \text{curl }\mathbf{F} \cdot (\mathbf{r}_\phi \times \mathbf{r}_\theta)dA$

$$= 16\left[\int_0^{\pi/2}\sin\theta\cos\theta\,d\theta\right]\left[\int_0^{\pi/2}\sin^3\phi\,d\phi\right] + 16\left[\int_0^{\pi/2}\sin\theta\,d\theta\right]\left[\int_0^{\pi/2}\sin^2\phi\cos\phi\,d\phi\right]$$

$$+ 16\left[\int_0^{\pi/2}\cos\theta\,d\theta\right]\left[\int_0^{\pi/2}\sin^2\phi\cos\phi\,d\phi\right]$$

$$= 8\left[-\cos\phi + \tfrac{1}{3}\cos^3\phi\right]_0^{\pi/2} + 16(1)\left[\tfrac{1}{3}\sin^3\phi\right]_0^{\pi/2} + 16(1)\left[\tfrac{1}{3}\sin^3\phi\right]_0^{\pi/2}$$

$$= 8\left[0 + 1 + 0 - \tfrac{1}{3}\right] + 16\left(\tfrac{1}{3}\right) + 16\left(\tfrac{1}{3}\right) = \tfrac{16}{3} + \tfrac{16}{3} + \tfrac{16}{3} = 16.$$

19. Assume S is centered at the origin with radius a and let H_1 and H_2 be the upper and lower hemispheres,

respectively, of S. Then $\iint_S \text{curl }\mathbf{F} \cdot d\mathbf{S} = \iint_{H_1} \text{curl }\mathbf{F} \cdot d\mathbf{S} + \iint_{H_2} \text{curl }\mathbf{F} \cdot d\mathbf{S} = \oint_{C_1} \mathbf{F} \cdot d\mathbf{r} + \oint_{C_2} \mathbf{F} \cdot d\mathbf{r}$ by Stokes'

Theorem. But C_1 is the circle $x^2 + y^2 = a^2$ oriented in the counterclockwise direction while C_2 is the same

circle but oriented in the clockwise direction. Hence $\oint_{C_2} \mathbf{F} \cdot d\mathbf{r} = -\oint_{C_1} \mathbf{F} \cdot d\mathbf{r}$ so $\iint_S \text{curl }\mathbf{F} \cdot d\mathbf{S} = 0$ as desired.

EXERCISES 14.9

1.

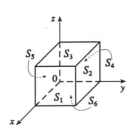

$\operatorname{div}\mathbf{F} = 3 + x + 2x = 3 + 3x$, so $\iiint_E \operatorname{div}\mathbf{F}\, dV = \int_0^1\int_0^1\int_0^1 (3x+3)\, dx\, dy\, dz = \frac{9}{2}$
(notice the triple integral is three times the volume of the cube plus three times $\overline{x}$).
To compute $\iint_S \mathbf{F}\cdot d\mathbf{S}$: on

S_1: $\mathbf{n} = \mathbf{i}$, $\mathbf{F} = 3\mathbf{i} + y\mathbf{j} + 2z\mathbf{k}$, and $\iint_{S_1}\mathbf{F}\cdot d\mathbf{S} = \iint_{S_1} 3\, dS = 3$;

S_2: $\mathbf{F} = 3x\mathbf{i} + x\mathbf{j} + 2xz\mathbf{k}$, $\mathbf{n} = \mathbf{j}$ and $\iint_{S_2}\mathbf{F}\cdot d\mathbf{S} = \iint_{S_2} x\, dS = \frac{1}{2}$;

S_3: $\mathbf{F} = 3x\mathbf{i} + xy\mathbf{j} + 2x\mathbf{k}$, $\mathbf{n} = \mathbf{k}$ and $\iint_{S_3}\mathbf{F}\cdot d\mathbf{S} = \iint_{S_3} 2x\, dS = 1$;

S_4: $\mathbf{F} = \mathbf{0}$, $\iint_{S_4}\mathbf{F}\cdot d\mathbf{S} = 0$; S_5: $\mathbf{F} = 3x\mathbf{i} + 2x\mathbf{k}$, $\mathbf{n} = -\mathbf{j}$ and $\iint_{S_5}\mathbf{F}\cdot d\mathbf{S} = \iint_{S_5} 0\, dS = 0$;

S_6: $\mathbf{F} = 3x\mathbf{i} + xy\mathbf{j}$, $\mathbf{n} = -\mathbf{k}$ and $\iint_{S_6}\mathbf{F}\cdot d\mathbf{S} = \iint_{S_6} 0\, dS = 0$. Thus $\iint_S \mathbf{F}\cdot d\mathbf{S} = \frac{9}{2}$.

3. $\operatorname{div}\mathbf{F} = \dfrac{\partial}{\partial x}(3y^2 z^3) + \dfrac{\partial}{\partial y}(9x^2 yz^2) + \dfrac{\partial}{\partial z}(4xy^2) = 9x^2 z^2$, so by the Divergence Theorem,

$\iint_S \mathbf{F}\cdot d\mathbf{S} = \iiint_E 9x^2 z^2\, dV = \int_{-1}^1\int_{-1}^1\int_{-1}^1 9x^2 z^2\, dx\, dy\, dz = 8$.

5. $\iint_S \mathbf{F}\cdot d\mathbf{S} = \iiint_E (-z - z + 2z)dV = 0$

7. $\iint_S \mathbf{F}\cdot d\mathbf{S} = \iiint_E x\, dV = \int_0^1\int_0^{2-2x}\int_0^{2-2x-y} x\, dz\, dy\, dx = \int_0^1\int_0^{2-2x}[x(2-2x) - xy]dy\, dx$

$\qquad = \int_0^1\left[x(2-2x)^2 - \frac{1}{2}x(2-2x)^2\right]dx = \frac{1}{6}$

9. $\iint_S \mathbf{F}\cdot d\mathbf{S} = \iiint_E 3(x^2 + y^2 + z^2)dV = \int_0^{2\pi}\int_0^\pi\int_0^1 3\rho^4 \sin\phi\, d\rho\, d\phi\, d\theta = 2\pi\int_0^\pi \frac{3}{5}\sin\phi\, d\phi = \frac{12}{5}\pi$

11. $\iint_S \mathbf{F}\cdot d\mathbf{S} = \iiint_E 2y\, dV = \iint_{x^2+y^2\le 9}\int_{y-3}^0 2y\, dz\, dA = \int_0^{2\pi}\int_0^3\int_{-3+r\sin\theta}^0 (2r^2 \sin\theta)dz\, dr\, d\theta$

$\qquad = \int_0^{2\pi}\int_0^3 (6r^2 \sin\theta - 2r^3 \sin^2\theta)dr\, d\theta = \int_0^{2\pi}\left[54\sin\theta - \frac{81}{2}\sin^2\theta\right]d\theta = -\frac{81}{2}\pi$

13. $\iint_S \mathbf{F}\cdot d\mathbf{S} = \iiint_E (x^2 + y^2 + z)dV = \int_0^{2\pi}\int_1^2\int_1^3 (r^2 + z)r\, dz\, dr\, d\theta = 2\pi\int_1^2 (2r^3 + 4r)dr = 27\pi$

15. $\iint_S \mathbf{F}\cdot d\mathbf{S} = \iiint_E \sqrt{3 - x^2}\, dV = \int_{-1}^1\int_{-1}^1\int_0^{2-x^4-y^4}\sqrt{3 - x^2}\, dz\, dy\, dx = \frac{341}{60}\sqrt{2} + \frac{81}{20}\sin^{-1}\left(\frac{\sqrt{3}}{3}\right)$

17. For S_1 we have $\mathbf{n} = -\mathbf{k}$, so $\mathbf{F}\cdot\mathbf{n} = \mathbf{F}\cdot(-\mathbf{k}) = -x^2 z - y^2 = -y^2$ (since $z = 0$ on S_1). So if D is the unit disk,

we get $\iint_{S_1}\mathbf{F}\cdot d\mathbf{S} = \iint_{S_1}\mathbf{F}\cdot\mathbf{n}\, dS = \iint_D (-y^2)dA = -\int_0^{2\pi}\int_0^1 r^2 \sin^2\theta\, r\, dr\, d\theta = -\frac{1}{4}\pi$. Now since S_2 is closed,

we can use the Divergence Theorem. Since

$\operatorname{div}\mathbf{F} = \dfrac{\partial}{\partial x}(z^2 x) + \dfrac{\partial}{\partial y}(\frac{1}{3}y^3 + \tan z) + \dfrac{\partial}{\partial z}(x^2 z + y^2) = z^2 + y^2 + x^2$, we use spherical coordinates to get

$\iint_{S_2}\mathbf{F}\cdot d\mathbf{S} = \iiint_E \operatorname{div}\mathbf{F}\, dV = \int_0^{2\pi}\int_0^{\pi/2}\int_0^1 \rho^2\cdot\rho^2 \sin\phi\, d\rho\, d\phi\, d\theta = \frac{2}{5}\pi$. Finally

$\iint_S \mathbf{F}\cdot d\mathbf{S} = \iint_{S_2}\mathbf{F}\cdot d\mathbf{S} - \iint_{S_1}\mathbf{F}\cdot d\mathbf{S} = \frac{2}{5}\pi - (-\frac{1}{4}\pi) = \frac{13}{20}\pi$.

19. Since $\dfrac{\mathbf{x}}{|\mathbf{x}|^3} = \dfrac{x\mathbf{i} + y\mathbf{j} + z\mathbf{k}}{(x^2 + y^2 + z^2)^{3/2}}$ and $\dfrac{\partial}{\partial x}\left[\dfrac{x}{(x^2 + y^2 + z^2)^{3/2}}\right] = \dfrac{(x^2 + y^2 + z^2) - 3x^2}{(x^2 + y^2 + z^2)^{5/2}}$ with similar expressions

for $\dfrac{\partial}{\partial y}\left[\dfrac{y}{(x^2 + y^2 + z^2)^{3/2}}\right]$ and $\dfrac{\partial}{\partial z}\left[\dfrac{z}{(x^2 + y^2 + z^2)^{3/2}}\right]$, we have

$\text{div}\left(\dfrac{\mathbf{x}}{|\mathbf{x}|^3}\right) = \dfrac{3(x^2 + y^2 + z^2) - 3(x^2 + y^2 + z^2)}{(x^2 + y^2 + z^2)^{5/2}} = 0$, except at $(0, 0, 0)$ where it is undefined.

21. $\iint_S \mathbf{a} \cdot \mathbf{n}\, dS = \iiint_E \text{div}\, \mathbf{a}\, dV = 0$ since $\text{div}\, \mathbf{a} = 0$.

23. $\iint_S \text{curl}\, \mathbf{F} \cdot d\mathbf{S} = \iiint_E \text{div}(\text{curl}\, \mathbf{F})dV = 0$ by Theorem 14.5.11.

25. $\iint_S (f\nabla g) \cdot \mathbf{n}\, dS = \iiint_E \text{div}(f\nabla g)dV = \iiint_E (f\nabla^2 g + \nabla g \cdot \nabla f)dV$ by Exercise 14.5.25.

REVIEW EXERCISES FOR CHAPTER 14

1. False; $\text{div}\, \mathbf{F}$ is a scalar field.

3. True, by (14.5.3) and the fact that $\text{div}\, \mathbf{0} = 0$.

5. False. See Exercise 14.3.33. [But the assertion is true if D is simply-connected; see (14.3.6).]

7. True. Apply the Divergence Theorem and use the fact that $\text{div}\, \mathbf{F} = 0$.

9. $x = \frac{1}{2}y^2, y = y, \int_C y\, dS = \int_0^2 y\sqrt{y^2 + 1}\, dy = \frac{1}{3}\left(5\sqrt{5} - 1\right)$

11. $\int_C x^3 z\, ds = \int_0^{\pi/2}(16\sin^3 t \cos t)\sqrt{5}\, dt = 4\sqrt{5}\sin^4 t\Big|_0^{\pi/2} = 4\sqrt{5}$

13. $x = \cos t, y = \sin t, 0 \le t \le 2\pi$ and $\int_C x^3 y\, dx - x\, dy = \int_0^{2\pi}(-\cos^3 t \sin^2 t - \cos^2 t)dt = -\pi$

Or: Since C is a simple closed curve, apply Green's Theorem giving

$\displaystyle\iint_{x^2 + y^2 \le 1} (-1 - x^3)dA = \int_0^1\int_0^{2\pi}(-r - r^4\cos^3\theta)\,d\theta = -\pi.$

15.

C_1: $x = t, y = t, z = 2t, 0 \le t \le 1$;

C_2: $x = 1 + 2t, y = 1, z = 2 + 2t, 0 \le t \le 1$. Then

$\int_C y\, dx + z\, dy + x\, dz = \int_0^1 5t\, dt + \int_0^1(4 + 4t)dt = \frac{17}{2}.$

17. $\mathbf{F}(\mathbf{r}(t)) = (2t + t^2)\mathbf{i} + t^4\mathbf{j} + 4t^4\mathbf{k}, \mathbf{F} \cdot \mathbf{r}'(t) = 4t + 2t^2 + 2t^5 + 16t^7$ and

$\int_C \mathbf{F} \cdot d\mathbf{r} = \int_0^1\left(4t + 2t^2 + 2t^5 + 16t^7\right)dt = 5.$

19. $\dfrac{\partial(\sin y)}{\partial y} = \cos y$ and $\dfrac{\partial(x \cos y + \sin y)}{\partial x} = \cos y$ and the domain of $\mathbf{F}$ is $\mathbb{R}^2$ so $\mathbf{F}$ is conservative. Hence there

exists f such that $\nabla f = \mathbf{F}$. Then $f_x(x,y) = \sin y$ implies $f(x,y) = x \sin y + g(y)$ and

$f_y(x,y) = x \cos y + g'(y)$. But $f_y(x,y) = x \cos y + \sin y$ so $g'(y) = \sin y$ and $f(x,y) = x \sin y - \cos y + K$ is

a potential for $\mathbf{F}$.

21. Since $\dfrac{\partial(2x + y^2 + 3x^2 y)}{\partial y} = 2y + 3x^2 = \dfrac{\partial(2xy + x^3 + 3y^2)}{\partial x}$ and the domain of $\mathbf{F}$ is $\mathbb{R}^2$, $\mathbf{F}$ is conservative.

Furthermore $f(x,y) = x^2 + xy^2 + x^3 y + y^3 + K$ is a potential for $\mathbf{F}$. Then $\int_C \mathbf{F} \cdot d\mathbf{r} = f(\pi, 0) - f(0,0) = \pi^2$.

23.

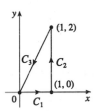

C_1: $0 \le x \le 1, y = 0$; C_2: $x = 1, 0 \le y \le 2$; C_3: $x = x, y = 2x, x = 1$ to $x = 0$. Then

$\oint_C xy\,dx + x^2\,dy = \int_0^1 0\,dx + \int_0^2 (0+1)dy + \int_1^0 (2x^2 + 2x^2)dx = \frac{2}{3}$. And

$\iint_D (2x - x)dA = \int_0^1 \int_0^{2x} x\,dy\,dx = \frac{2}{3}$.

25. $\int_C x^2 y\,dx - xy^2\,dy = \displaystyle\iint_{x^2+y^2 \le 4} (-y^2 - x^2)dA = -\int_0^{2\pi}\int_0^2 r^3\,dr\,d\theta = -8\pi$

27. If we assume there is such a vector field $\mathbf{G}$, then $\operatorname{div}(\operatorname{curl}\mathbf{G}) = 2 + 3z - 2xz$. But $\operatorname{div}(\operatorname{curl}\mathbf{F}) = 0$ for all vector

fields $\mathbf{F}$. Thus such a $\mathbf{G}$ cannot exist.

29. For any piecewise-smooth simple closed plane curve C, bounding a region D, we can apply Green's Theorem to

$\mathbf{F}(x,y) = f(x)\mathbf{i} + g(y)\mathbf{j}$ to get $\displaystyle\int_C f(x)dx + g(y)dy = \iint_D \left[\frac{\partial}{\partial x} g(y) - \frac{\partial}{\partial y} f(x) \right] dA = \iint_D 0\,dA = 0$.

31. $\nabla^2 f = 0$ means that $\dfrac{\partial^2 f}{\partial x^2} + \dfrac{\partial^2 f}{\partial y^2} = 0$. Now if $\mathbf{F} = f_y \mathbf{i} - f_x \mathbf{j}$ and C is any closed path in D, then applying

Green's Theorem, we get

$\displaystyle\int_C \mathbf{F} \cdot d\mathbf{r} = \int_C f_y\,dx - f_x\,dy = \iint_D \left[\frac{\partial}{\partial x}(-f_x) - \frac{\partial}{\partial y}(f_y) \right] dA = -\iint_D (f_{xx} + f_{yy})dA = -\iint_D 0\,dA = 0$

Therefore the line integral is independent of path by Theorem 14.3.3.

33. $z = f(x,y) = x^2 + 2y$ with $0 \le x \le 1, 0 \le y \le 2x$. Thus

$A(S) = \iint_D \sqrt{1 + 4x^2 + 4}\,dA = \int_0^1 \int_0^{2x} \sqrt{5 + x^2}\,dy\,dx = \int_0^1 2x\sqrt{5 + x^2}\,dx = \frac{2}{3}\left(6\sqrt{6} - 5\sqrt{5}\right)$.

35. $z = f(x,y) = x^2 + y^2$ with $0 \le x^2 + y^2 \le 4$ so $\mathbf{r}_x \times \mathbf{r}_y = -2x\mathbf{i} - 2y\mathbf{j} + \mathbf{k}$ (using upward orientation). Then

$\displaystyle\iint_S z\,dS = \iint_{x^2+y^2 \le 4} (x^2 + y^2)\sqrt{4x^2 + 4y^2 + 1}\,dA = \int_0^{2\pi}\int_0^2 r^3\sqrt{1 + 4r^2}\,dr\,d\theta = \frac{1}{60}\pi\left(391\sqrt{17} + 1\right)$.

(Substitute $u = 1 + 4r^2$ and use tables.)

37. Since the sphere bounds a simple solid region, the Divergence Theorem applies and

$\iint_S \mathbf{F} \cdot d\mathbf{S} = \iiint_E (z - 2)dV = \iiint_E z\, dV - 2\iiint_E dV = m\bar{z} - 2(\frac{4}{3}\pi 2^3) = -\frac{64}{3}\pi.$

Alternate Solution: $\mathbf{F}(\mathbf{r}(\phi,\theta)) = 4\sin\phi\cos\theta\cos\phi\,\mathbf{i} - 4\sin\phi\sin\theta\,\mathbf{j} + 6\sin\phi\cos\theta\,\mathbf{k},$

$\mathbf{r}_\phi \times \mathbf{r}_\theta = 4\sin^2\phi\cos\theta\,\mathbf{i} + 4\sin^2\phi\sin\theta\,\mathbf{j} + 4\sin\phi\cos\phi\,\mathbf{k},$ and

$\mathbf{F} \cdot (\mathbf{r}_\phi \times \mathbf{r}_\theta) = 16\sin^3\phi\cos^2\theta\cos\phi - 16\sin^3\phi\sin^2\theta + 24\sin^2\phi\cos\phi\cos\theta.$ Then

$\iint_S \mathbf{F} \cdot d\mathbf{S} = \int_0^{2\pi}\int_0^\pi (16\sin^3\phi\cos\phi\cos^2\theta - 16\sin^3\phi\sin^2\theta + 24\sin^2\phi\cos\phi\cos\theta)d\phi\, d\theta$

$\qquad = \int_0^{2\pi} \frac{4}{3}(-16\sin^2\theta)d\theta = -\frac{64}{3}\pi.$

39. Since curl $\mathbf{F} = \mathbf{0}$, $\iint_S (\text{curl }\mathbf{F}) \cdot d\mathbf{S} = 0$. And C: $\mathbf{r}(t) = \cos t\,\mathbf{i} + \sin t\,\mathbf{j}, 0 \le t \le 2\pi$ and

$\oint_C \mathbf{F} \cdot d\mathbf{r} = \int_0^{2\pi}(-\cos^2 t\sin t + \sin^2 t\cos t)dt = \frac{1}{3}\cos^3 t + \frac{1}{3}\sin^3 t\Big|_0^{2\pi} = 0.$

41. The surface is given by $x + y + z = 1$ or $z = 1 - x - y$, $0 \le x \le 1$, $0 \le y \le 1 - x$ and $\mathbf{r}_x \times \mathbf{r}_y = \mathbf{i} + \mathbf{j} + \mathbf{k}$.

Then $\oint_C \mathbf{F} \cdot d\mathbf{r} = \iint_S \text{curl }\mathbf{F} \cdot d\mathbf{S} = \iint_D (-y\mathbf{i} - z\mathbf{j} - x\mathbf{k}) \cdot (\mathbf{i} + \mathbf{j} + \mathbf{k})dA = \iint_D (-1)dA = -(\text{area of D}) = -\frac{1}{2}.$

43. $\iiint_E \text{div }\mathbf{F}\, dV = \iiint_{x^2+y^2+z^2 \le 1} 3\, dV = 3(\text{volume of sphere}) = 4\pi.$ Then

$\mathbf{F}(\mathbf{r}(\phi,\theta)) \cdot (\mathbf{r}_\phi \times \mathbf{r}_\theta) = \sin^3\phi\cos^2\theta + \sin^3\phi\sin^2\theta + \sin\phi\cos^2\phi = \sin\phi$ and

$\iint_S \mathbf{F} \cdot d\mathbf{S} = \int_0^{2\pi}\int_0^\pi \sin\phi\, d\phi\, d\theta = (2\pi)(2) = 4\pi.$

45. Because curl $\mathbf{F} = \mathbf{0}$, $\mathbf{F}$ is conservative, and if $f(x, y, z) = x^3yz - 3xy + z^2$, then $\nabla f = \mathbf{F}$. Hence

$\int_C \mathbf{F} \cdot d\mathbf{r} = \int_C \nabla f \cdot d\mathbf{r} = f(0, 3, 0) - f(0, 0, 2) = 0 - 4 = -4.$

47. By the Divergence Theorem, $\iint_S \mathbf{F} \cdot \mathbf{n}\, dS = \iiint_E \text{div }\mathbf{F}\, dV = 3(\text{volume of }E) = 3(8 - 1) = 21.$

PROBLEMS PLUS (after Chapter 14)

1. Since $|xy| < 1$, except at $(1, 1)$, the formula for the sum of a geometric series gives $\dfrac{1}{1 - xy} = \sum_{n=0}^{\infty}(xy)^n$, so

$$\int_0^1\int_0^1 \frac{1}{1 - xy}\,dx\,dy = \int_0^1\int_0^1\sum_{n=0}^{\infty}(xy)^n\,dx\,dy = \sum_{n=0}^{\infty}\int_0^1\int_0^1(xy)^n\,dx\,dy = \sum_{n=0}^{\infty}\left[\int_0^1 x^n\,dx\right]\left[\int_0^1 y^n\,dy\right]$$

$$= \sum_{n=0}^{\infty}\frac{1}{n+1}\cdot\frac{1}{n+1} = \sum_{n=0}^{\infty}\frac{1}{(n+1)^2} = \frac{1}{1^2} + \frac{1}{2^2} + \frac{1}{3^2} + \cdots = \sum_{n=1}^{\infty}\frac{1}{n^2}.$$

3. **(a)** Since $|xyz| < 1$ except at $(1, 1, 1)$, the formula for the sum of a geometric series gives

$$\frac{1}{1 - xyz} - \sum_{n=0}^{\infty}(xyz)^n, \text{ so}$$

$$\int_0^1\int_0^1\int_0^1 \frac{1}{1 - xyz}\,dx\,dy\,dz = \int_0^1\int_0^1\int_0^1\sum_{n=0}^{\infty}(xyz)^n\,dx\,dy\,dz = \sum_{n=0}^{\infty}\int_0^1\int_0^1\int_0^1(xyz)^n\,dx\,dy\,dz$$

$$= \sum_{n=0}^{\infty}\left[\int_0^1 x^n\,dx\right]\left[\int_0^1 y^n\,dy\right]\left[\int_0^1 z^n\,dz\right] = \sum_{n=0}^{\infty}\frac{1}{n+1}\cdot\frac{1}{n+1}\cdot\frac{1}{n+1}$$

$$= \sum_{n=0}^{\infty}\frac{1}{(n+1)^3} = \frac{1}{1^3} + \frac{1}{2^3} + \frac{1}{3^3} + \cdots = \sum_{n=1}^{\infty}\frac{1}{n^3}.$$

(b) Since $|-xyz| < 1$, except at $(1, 1, 1)$, the formula for the sum of a geometric series gives

$$\frac{1}{1 + xyz} = \sum_{n=0}^{\infty}(-xyz)^n, \text{ so}$$

$$\int_0^1\int_0^1\int_0^1 \frac{1}{1 + xyz}\,dx\,dy\,dz = \int_0^1\int_0^1\int_0^1\sum_{n=0}^{\infty}(-xyz)^n\,dx\,dy\,dz = \sum_{n=0}^{\infty}\int_0^1\int_0^1\int_0^1(-xyz)^n\,dx\,dy\,dz$$

$$= \sum_{n=0}^{\infty}(-1)^n\left[\int_0^1 x^n\,dx\right]\left[\int_0^1 y^n\,dy\right]\left[\int_0^1 z^n\,dz\right] = \sum_{n=0}^{\infty}(-1)^n\frac{1}{n+1}\cdot\frac{1}{n+1}\cdot\frac{1}{n+1}$$

$$= \sum_{n=0}^{\infty}\frac{(-1)^n}{(n+1)^3} = \frac{1}{1^3} - \frac{1}{2^3} + \frac{1}{3^3} - \cdots = \sum_{n=1}^{\infty}\frac{(-1)^{n-1}}{n^3}.$$

To evaluate this sum, we first write out a few terms: $s = 1 - \dfrac{1}{2^3} + \dfrac{1}{3^3} - \dfrac{1}{4^3} + \dfrac{1}{5^3} - \dfrac{1}{6^3} \approx 0.8998.$

Notice that $a_7 = \frac{1}{7^3} < 0.003$. By the Alternating Series Estimation Theorem (10.5.1), we have

$|s - s_6| \le a_7 < 0.003$. This error of 0.003 will not affect the second decimal place, so we have $s \approx 0.90$.

5.

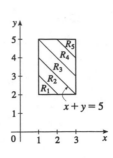

Let $R = \bigcup_{i=1}^{5} R_i$, where

$R_i = \{(x,y) \mid x+y \geq i+2, x+y < i+3, 1 \leq x \leq 3, 2 \leq y \leq 5\}$.

$\iint_R [\![x+y]\!] \, dA = \sum_{i=1}^{5} \iint_{R_i} [\![x+y]\!] \, dA = \sum_{i=1}^{5} [\![x+y]\!] \iint_{R_i} dA$,

since $[\![x+y]\!] = \text{constant} = i+2$ for $(x,y) \in R_i$. Therefore

$\iint_R [\![x+y]\!] \, dA = \sum_{i=1}^{5} (i+2)[A(R_i)]$

$= 3A(R_1) + 4A(R_2) + 5A(R_3) + 6A(R_4) + 7A(R_5)$

$= 3\left(\tfrac{1}{2}\right) + 4\left(\tfrac{3}{2}\right) + 5(2) + 6\left(\tfrac{3}{2}\right) + 7\left(\tfrac{1}{2}\right) = 30$.

7.
$f_{ave} = \dfrac{1}{b-a} \displaystyle\int_a^b f(x)\,dx = \dfrac{1}{1-0} \int_0^1 \left[\int_x^1 \cos(t^2)\,dt \right] dx = \int_0^1 \int_x^1 \cos(t^2)\,dt\,dx$

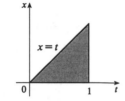

$= \displaystyle\int_0^1 \int_0^t \cos(t^2)\,dx\,dt$ (changing the order of integration)

$= \int_0^1 t\cos(t^2)\,dt = \tfrac{1}{2}\sin(t^2)\big|_0^1 = \tfrac{1}{2}\sin 1$

9.
$\int_0^x \int_0^y \int_0^z f(t)\,dt\,dz\,dy = \iiint_E f(t)\,dV$, where

$E = \{(t,z,y) \mid 0 \leq t \leq z, 0 \leq z \leq y, 0 \leq y \leq x\}$.

If we let D be the projection of E on the yt-plane

then $D = \{(y,t) \mid 0 \leq t \leq x, t \leq y \leq x\}$.   And

we see from the diagram that

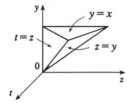

$E = \{(t,z,y) \mid t \leq z \leq y, t \leq y \leq x, 0 \leq t \leq x\}$. So

$\int_0^x \int_0^y \int_0^z f(t)\,dt\,dz\,dy = \int_0^x \int_t^x \int_t^y f(t)\,dz\,dy\,dt = \int_0^x \left[\int_t^x (y-t)f(t)\,dy \right] dt = \int_0^x \left[\left(\tfrac{1}{2}y^2 - ty\right) f(t) \right]_t^x dt$

$= \int_0^x \left[\tfrac{1}{2}x^2 - tx - \tfrac{1}{2}t^2 + t^2 \right] f(t)\,dt = \int_0^x \left[\tfrac{1}{2}x^2 - tx + \tfrac{1}{2}t^2 \right] f(t)\,dt$

$= \int_0^x \tfrac{1}{2}(x^2 - 2tx + t^2)f(t)\,dt = \tfrac{1}{2}\int_0^x (x-t)^2 f(t)\,dt$.

11. Let $u = \mathbf{a} \cdot \mathbf{r}$, $v = \mathbf{b} \cdot \mathbf{r}$, $w = \mathbf{c} \cdot \mathbf{r}$, where $\mathbf{a} = \langle a_1, a_2, a_3 \rangle$, $\mathbf{b} = \langle b_1, b_2, b_3 \rangle$, $\mathbf{c} = \langle c_1, c_2, c_3 \rangle$. Under this change of

variables, E corresponds to the rectangular box $0 \leq u \leq \alpha$, $0 \leq v \leq \beta$, $0 \leq w \leq \gamma$. So, by the Change of

Variables Theorem, $\displaystyle\int_0^\gamma \int_0^\beta \int_0^\alpha uvw\,du\,dv\,dw = \iiint_E (\mathbf{a}\cdot\mathbf{r})(\mathbf{b}\cdot\mathbf{r})(\mathbf{c}\cdot\mathbf{r}) \left| \dfrac{\partial(u,v,w)}{\partial(x,y,z)} \right| dV$. But

$\left| \dfrac{\partial(u,v,w)}{\partial(x,y,z)} \right| = \begin{Vmatrix} a_1 & a_2 & a_3 \\ b_1 & b_2 & b_3 \\ c_1 & c_2 & c_3 \end{Vmatrix} = |\mathbf{a}\cdot\mathbf{b}\times\mathbf{c}| \quad \Rightarrow$

$\iiint_E (\mathbf{a}\cdot\mathbf{r})(\mathbf{b}\cdot\mathbf{r})(\mathbf{c}\cdot\mathbf{r})\,dV = \dfrac{1}{|\mathbf{a}\cdot\mathbf{b}\times\mathbf{c}|} \int_0^\gamma \int_0^\beta \int_0^\alpha uvw\,du\,dv\,dw \quad \Rightarrow$

$\iiint_E (\mathbf{a}\cdot\mathbf{r})(\mathbf{b}\cdot\mathbf{r})(\mathbf{c}\cdot\mathbf{r})\,dV = \dfrac{1}{|\mathbf{a}\cdot\mathbf{b}\times\mathbf{c}|} \left(\dfrac{\alpha^2}{2}\right) \left(\dfrac{\beta^2}{2}\right) \left(\dfrac{\gamma^2}{2}\right) = \dfrac{(\alpha\beta\gamma)^2}{8|\mathbf{a}\cdot\mathbf{b}\times\mathbf{c}|}$.

13. Let S_1 be the portion of $\Omega(S)$ between $S(a)$ and S, and let ∂S_1 be its boundary. Also let S_L be the lateral surface of S_1 [that is, the surface of S_1 except S and $S(a)$]. Applying the Divergence Theorem we have

$$\iint_{\partial S_1} \frac{\mathbf{r} \cdot \mathbf{n}}{r^3} \, dS = \iiint_{S_1} \nabla \cdot \frac{\mathbf{r}}{r^3} \, dV. \text{ But}$$

$$\nabla \cdot \frac{\mathbf{r}}{r^3} = \left\langle \frac{\partial}{\partial x}, \frac{\partial}{\partial y}, \frac{\partial}{\partial z} \right\rangle \cdot \left\langle \frac{x}{(x^2 + y^2 + z^2)^{3/2}}, \frac{y}{(x^2 + y^2 + z^2)^{3/2}}, \frac{z}{(x^2 + y^2 + z^2)^{3/2}} \right\rangle$$

$$= \frac{(x^2 + y^2 + z^2 - 3x^2) + (x^2 + y^2 + z^2 - 3y^2) + (x^2 + y^2 + z^2 - 3z^2)}{(x^2 + y^2 + z^2)^{5/2}} = 0 \quad \Rightarrow$$

$$\iint_{\partial S_1} \frac{\mathbf{r} \cdot \mathbf{n}}{r^3} \, dS = \iiint_{S_1} 0 \, dV = 0. \text{ On the other hand, notice that for the surfaces of } \partial S_1 \text{ other than } S(a)$$

and S, $\mathbf{r} \cdot \mathbf{n} = 0 \quad \Rightarrow$

$$0 = \iint_{\partial S_1} \frac{\mathbf{r} \cdot \mathbf{n}}{r^3} \, dS = \iint_{S} \frac{\mathbf{r} \cdot \mathbf{n}}{r^3} \, dS + \iint_{S(a)} \frac{\mathbf{r} \cdot \mathbf{n}}{r^3} \, dS + \iint_{S_L} \frac{\mathbf{r} \cdot \mathbf{n}}{r^3} \, dS = \iint_{S} \frac{\mathbf{r} \cdot \mathbf{n}}{r^3} \, dS + \iint_{S(a)} \frac{\mathbf{r} \cdot \mathbf{n}}{r^3} \, dS \quad \Rightarrow$$

$$\iint_{S} \frac{\mathbf{r} \cdot \mathbf{n}}{r^3} \, dS = -\iint_{S(a)} \frac{\mathbf{r} \cdot \mathbf{n}}{r^3} \, dS. \text{ Notice that on } S(a), r = a \quad \Rightarrow \quad \mathbf{n} = -\frac{\mathbf{r}}{r} = -\frac{\mathbf{r}}{a} \text{ and } \mathbf{r} \cdot \mathbf{r} = r^2 = a^2, \text{ so}$$

that $-\iint_{S(a)} \frac{\mathbf{r} \cdot \mathbf{n}}{r^3} \, dS = \iint_{S(a)} \frac{\mathbf{r} \cdot \mathbf{r}}{a^4} \, dS = \iint_{S(a)} \frac{a^2}{a^4} \, dS = \frac{1}{a^2} \iint_{S(a)} dS = \frac{\text{area of } S(a)}{a^2} = |\Omega(S)|.$ Therefore

$$|\Omega(S)| = \iint_{S} \frac{\mathbf{r} \cdot \mathbf{n}}{r^3} \, dS.$$

CHAPTER FIFTEEN

EXERCISES 15.1

1. $x^2 y' + y = 0$ $\Rightarrow$ $\dfrac{dy}{dx} = -\dfrac{y}{x^2}$ $\Rightarrow$ $\displaystyle\int \dfrac{dy}{y} = \int -\dfrac{dx}{x^2}$ $(y \neq 0)$ $\Rightarrow$ $\ln|y| = \dfrac{1}{x} + K$ $\Rightarrow$

   $|y| = e^K e^{1/x}$ $\Rightarrow$ $y = C e^{1/x}$, where we allow C to be any constant $(C = \pm e^K$ or $C = 0)$.

3. $\dfrac{dy}{dx} = \dfrac{x\sqrt{x^2 + 1}}{y e^y}$ $\Rightarrow$ $\displaystyle\int y e^y \, dy = \int x\sqrt{x^2 + 1} \, dx$ $\Rightarrow$ $(y - 1)e^y = \frac{1}{3}(x^2 + 1)^{3/2} + C$

5. $\dfrac{dy}{dx} = y^2 + 1$, $y(1) = 0$. $\displaystyle\int \dfrac{dy}{y^2 + 1} = \int dx$ $\Rightarrow$ $\tan^{-1}y = x + C$. $y = 0$ when $x = 1$, so

   $1 + C = \tan^{-1}0 = 0$ and $C = -1$. Thus $\tan^{-1}y = x - 1$ and $y = \tan(x - 1)$.

7. $\dfrac{du}{dt} = \dfrac{2t + 1}{2(u - 1)}$, $u(0) = -1$. $\displaystyle\int 2(u - 1)du = \int(2t + 1)dt$ $\Rightarrow$ $u^2 - 2u = t^2 + t + C$. $u(0) = -1$ so

   $(-1)^2 - 2(-1) = 0^2 + 0 + C$ and $C = 3$. Thus $u^2 - 2u = t^2 + t + 3$; the quadratic formula gives

   $u = 1 - \sqrt{t^2 + t + 4}$.

9. $y' = y \sin x$, $y(0) = 1$. $\displaystyle\int \dfrac{dy}{y} = \int \sin x \, dx$

   $\Rightarrow$ $\ln|y| = -\cos x + C$ $\Rightarrow$ $|y| = e^{-\cos x + C}$

   $\Rightarrow$ $y(x) = A e^{-\cos x}$. $y(0) = A e^{-1}$ $\Rightarrow$ $A = e$, so

   $y = e^{1 - \cos x}$.

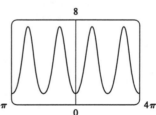

11. $y' = y - 1$. The slopes at each point are independent of x, so the slopes are the same along each line parallel to the x-axis. Thus IV is the graph of this equation.

13. $y' = y^2 - x^2 > 0$ when $y^2 > x^2$ $\Leftrightarrow$ $|y| > |x|$ $\Leftrightarrow$ $y > |x|$ or $y < -|x|$. Graph III has positive slopes in these regions and negative slopes elsewhere, so this equation corresponds to graph III.

15. (a)　　　　　　　　　　(b)　　　　　　　　　　(c)

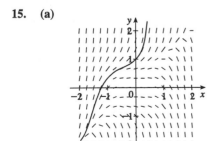
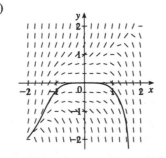
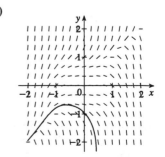

17.

x	y	$y' = y^2$
0	0	0
0	1	1
0	−1	1
1	0	0
−1	0	0
1	−1	1
1	1	1
1	2	4
1	−2	4
−1	2	4
−1	−2	4

The solution curve through $(0, 1)$

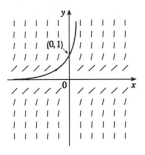

19.

x	y	$y' = x^2 + y^2$
0	0	0
0	1	1
1	0	1
1	1	2
−1	1	2
0	2	4
2	0	4
2	2	8
2	1	5
−2	−1	5
1	2	5

The solution curve through $(0, 0)$

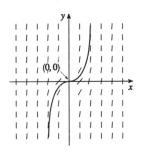

21. (a),(b)

x	y	$y' = 1/y$
0	0.5	2
0	−0.5	−2
0	1	1
0	−1	−1
0	2	0.5
0	−2	−0.5
0	4	0.25
0	3	$0.\overline{3}$
0	0.25	4
0	$0.\overline{3}$	3

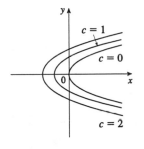

(d)

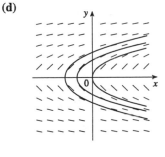

(c) $y\,dy = dx$ so $\frac{1}{2}y^2 = x + c$ or $y = \pm\sqrt{2(x + c)}$

23. In Maple, we can use either `directionfield` (in Maple's share library) or `plots[fieldplot]` to plot the direction field. To plot the solution, we can either use the initial-value option in `directionfield`, or actually solve the equation. In Mathematica, we use `PlotVectorField` for the direction field, and the `Plot[Evaluate[⋯]]` construction to plot the solution, which is $y = e^{(1-\cos 2x)/2}$.

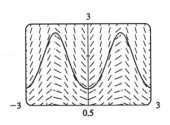

25. $h = 0.5$, $x_0 = 1$, $y_0 = 2$ and $F(x, y) = 1 + 3x - 2y$. So

$y_n = y_{n-1} + hF(x_{n-1}, y_{n-1}) = y_{n-1} + 0.5(1 + 3x_{n-1} - 2y_{n-1}) = 0.5 + 1.5x_{n-1}$. Thus $y_1 = 0.5 + 1.5 = 2$,

$y_2 = 0.5 + 1.5 \cdot 1.5 = 2.75$, $y_3 = 0.5 + 1.5 \cdot 2 = 3.5$, $y_4 = 0.5 + 1.5 \cdot 2.5 = 4.25$.

27. $h = 0.1$, $x_0 = 0$, $y_0 = 1$ and $F(x, y) = x^2 + y^2$. We need to find y_5, because $x_5 = 0.5$. So

$y_n = y_{n-1} + 0.1(x_{n-1}^2 + y_{n-1}^2)$. $y_1 = 1 + 0.1(0^2 + 1^2) = 1.1$, $y_2 = 1.1 + 0.1(0.1^2 + 1.1^2) = 1.222$,

$y_3 = 1.222 + 0.1(0.2^2 + 1.222^2) \approx 1.37533$, $y_4 = 1.37533 + 0.1(0.3^2 + 1.37533^2) \approx 1.57348$,

$y_5 = 1.57348 + 0.1(0.4^2 + 1.57348^2) \approx 1.8371 \approx y(0.5)$.

29. $x^2 + 1 + 2xyy' = 0$ isn't homogeneous since $y' = -\dfrac{x}{2y} - \dfrac{1}{2xy}$ can't be written as a function of $\dfrac{y}{x}$.

31. $y' = \ln y - \ln x \quad \Rightarrow \quad y' = \ln(y/x)$ and as a result the equation is homogeneous.

33. Since $y' = 1 - \dfrac{y}{x}$, setting $v = \dfrac{y}{x}$ gives $v' = \dfrac{1 - 2v}{x}$ or $\dfrac{dv}{1 - 2v} = \dfrac{dx}{x}$ or $-\frac{1}{2}\ln|1 - 2v| = c_1 + \ln|x|$ $(v \neq \frac{1}{2})$.

Then $\ln|1 - 2v| = c_2 + \ln x^{-2}$ or $v = \frac{1}{2}(1 - kx^{-2})$ and $y = xv = \frac{1}{2}(x - kx^{-1})$.

35. $y' = \dfrac{x}{y} + \dfrac{y}{x} = \left(\dfrac{y}{x}\right)^{-1} + \dfrac{y}{x}$. Setting $v = \dfrac{y}{x}$ gives $v' = \dfrac{1}{vx}$ or $v\,dv = \dfrac{dx}{x}$. Hence $v^2 = \ln x^2 + C$ or

$v = \pm(C + \ln x^2)^{1/2}$. Thus $y = xv = \pm x\sqrt{C + \ln x^2}$ is the solution.

37. $y' = \dfrac{y}{x} + e^{y/x}$, so setting $v = \dfrac{y}{x}$ gives $v' = \dfrac{e^v}{x}$ or $e^{-v}\,dv = \dfrac{dx}{x}$. Hence $-e^{-v} = \ln|x| + c_1$, or $-v = \ln[c - \ln|x|]$

or $v = -\ln(c - \ln|x|)$. Thus the solution is $y = -x\ln(c - \ln|x|)$.

39. The curves $y = kx^2$ form a family of parabolas with axis the y-axis.

Differentiating gives $y' = 2kx$, but $k = y/x^2$, so $y' = 2y/x$.

Thus the slope of the tangent line at any point (x, y) on one of

the parabolas is $y' = 2y/x$ so the orthogonal trajectories must

satisfy $y' = -x/(2y)$ or $y^2 = -x^2/2 + c_1$. This is a family

of ellipses: $x^2 + 2y^2 = c$.

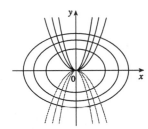

41. Differentiating gives $y' = -\dfrac{1}{(x + k)^2}$, but $k = \dfrac{1}{y} - x$, so

$y' = -\dfrac{1}{(1/y)^2} = -y^2$. So the orthogonal trajectories must

satisfy $y' = \dfrac{1}{y^2}$ or $\dfrac{y^3}{3} = x + c$ or $y = [3(x + c)]^{1/3}$.

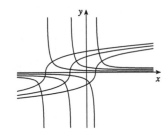

43. The curves $x^2 - 2y^2 = k$ form a family of hyperbolas.

Differentiating gives $2x - 4y(dy/dx) = 0$ or $y' = x/(2y)$.

Thus the orthogonal trajectories must satisfy $y' = -2y/x$ or

$\ln|y| = -2\ln|x| + C_1$ or $y = c/x^2$.

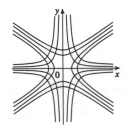

45. **(a)** We have $V(t) = \pi r^2 y(t)$ $\Rightarrow$ $\dfrac{dV}{dy} = \pi r^2 = 4\pi$ where $\dfrac{dV}{dt} = \dfrac{dV}{dy}\dfrac{dy}{dt}$ and $a - \pi\left(\frac{1}{12}\right)^2$. Thus

$\dfrac{dV}{dt} = -a\sqrt{2gy}$ $\Rightarrow$ $\dfrac{dV}{dy}\dfrac{dy}{dt} = -\pi\left(\frac{1}{12}\right)^2\sqrt{64y}$ $\Rightarrow$ $\dfrac{dy}{dt}4\pi = -\pi\frac{8}{144}\sqrt{y}$ $\Rightarrow$ $\dfrac{dy}{dt} = -\frac{1}{72}\sqrt{y}$.

(b) $\dfrac{dy}{dt} = -\frac{1}{72}\sqrt{y}$ $\Rightarrow$ $y^{-1/2}dy = -\frac{1}{72}dt$ $\Rightarrow$ $2\sqrt{y} = -\frac{1}{72}t + C$ where $y(0) = 6$. Thus

$2\sqrt{6} = 0 + C$ $\Rightarrow$ $C = 2\sqrt{6}$ $\Rightarrow$ $y = \left(-\frac{1}{144}t + \sqrt{6}\right)^2$.

(c) We want to find t when $y = 0$, so we set $y = 0 = \left(-\frac{1}{144}t + \sqrt{6}\right)^2$ $\Rightarrow$ $t = 144\sqrt{6} \approx 5\ \text{min}\ 53\ \text{s}$.

EXERCISES 15.2

1. $y' + x^2y = y^2$ is not linear since it cannot be put into the standard linear form (1).

3. $xy' = x - y$ $\Rightarrow$ $xy' + y = x$ $\Rightarrow$ $y' + \dfrac{1}{x}y = 1$, which is in the standard linear form (1), and thus this differential equation is linear.

5. $I(x) = e^{\int -3\,dx} = e^{-3x}$. Multiplying the differential equation by $I(x)$ gives $e^{-3x}y' - 3e^{-3x}y = e^{-2x}$ $\Rightarrow$ $\left(e^{-3x}y\right)' = e^{-2x}$ $\Rightarrow$ $y = e^{3x}[\int(e^{-2x})dx + C] = Ce^{3x} - \frac{1}{2}e^x$.

7. $I(x) = e^{\int -2x\,dx} = e^{-x^2}$. Multiplying the differential equation by $I(x)$ gives $e^{-x^2}(y' - 2xy) = xe^{-x^2}$ $\Rightarrow$ $\left(e^{-x^2}y\right)' = xe^{-x^2}$ $\Rightarrow$ $y = e^{x^2}[\int xe^{-x^2}\,dx + C] = Ce^{x^2} - \frac{1}{2}$.

9. $y' - y\tan x = \dfrac{\sin 2x}{\cos x} = 2\sin x$, so $I(x) = e^{\int -\tan x\,dx} = e^{\ln|\cos x|} = \cos x$ (since $-\frac{\pi}{2} < x < \frac{\pi}{2}$.) Multiplying the differential equation by $I(x)$ gives $(y' - y\tan x)\cos x = \dfrac{\sin 2x}{\cos x}\cos x$ $\Rightarrow$ $[y\cos x]' = \sin 2x$ $\Rightarrow$

$y\cos x = \int\sin 2x\,dx + C = -\frac{1}{2}\cos 2x + C = \frac{1}{2} - \cos^2 x + C$ $\Rightarrow$ $y = \dfrac{\sec x}{2} - \cos x + C\sec x$.

11. $I(x) = e^{\int 2x\,dx} = e^{x^2}$. Multiplying the differential equation by $I(x)$ gives $e^{x^2}y' + 2xe^{x^2}y = x^2e^{x^2}$ $\Rightarrow$ $\left(e^{x^2}y\right)' = x^2e^{x^2}$. Thus

$y = e^{-x^2}\left[\int x^2e^{x^2}\,dx + C\right] = e^{-x^2}\left[\frac{1}{2}xe^{x^2} - \int\frac{1}{2}e^{x^2}\,dx + C\right] = \frac{1}{2}x + Ce^{-x^2} - e^{-x^2}\int\frac{1}{2}e^{x^2}\,dx$.

13. $I(\theta) = e^{\int -\tan\theta\,d\theta} = e^{-\ln(\sec\theta)} = \cos\theta$. Multiplying the differential equation by $I(\theta)$ gives

$\cos\theta(dy/d\theta) - y\sin\theta = \cos\theta$ $\Rightarrow$ $(y\cos\theta)' = \cos\theta$ $\Rightarrow$ $y\cos\theta = \int\cos\theta\,d\theta$ $\Rightarrow$

$y\cos\theta = \sin\theta + C$ $\Rightarrow$ $y = \tan\theta + C\sec\theta$.

15. $I(x) = e^{\int dx} = e^x$. Multiplying the differential equation by $I(x)$ gives $e^xy' + e^xy = e^x(x + e^x)$ $\Rightarrow$

$(e^xy)' = e^x(x + e^x)$. Thus $y = e^{-x}\left[\int e^x(x + e^x)dx + C\right] = e^{-x}\left[xe^x - e^x + \dfrac{e^{2x}}{2} + C\right] = x - 1 + \dfrac{e^x}{2} + \dfrac{C}{e^x}$.

But $0 = y(0) = -1 + \frac{1}{2} + C$, so $C = \frac{1}{2}$, and the solution to the initial-value problem is

$y = x - 1 + \frac{1}{2}e^x + \frac{1}{2}e^{-x} = x - 1 + \cosh x$.

17. $I(x) = e^{\int -2x\,dx} = e^{-x^2}$. Multiplying the differential equation by $I(x)$ gives $e^{-x^2}y' - 2xe^{-x^2}y = 2x$ $\Rightarrow$

$(e^{-x^2}y)' = 2x$ $\Rightarrow$ $y = e^{x^2}[\int 2x\,dx + C] = x^2e^{x^2} + Ce^{x^2}$. But $3 = y(0) = C$, so the solution to the

initial-value problem is $y = (x^2 + 3)e^{x^2}$.

19. $y' + 2\dfrac{y}{x} = \dfrac{\cos x}{x^2}$ $(x \neq 0)$, so $I(x) = e^{\int (2/x)dx} = x^2$. Multiplying the differential equation by $I(x)$ gives

$x^2y' + 2xy = \cos x$ $\Rightarrow$ $(x^2y)' = \cos x$ $\Rightarrow$ $y = x^{-2}[\int \cos x\,dx + C] = x^{-2}(\sin x + C)$ $(x \neq 0)$. But

$0 = y(\pi) = C$, so the solution to the initial-value problem is $y = (\sin x)/x^2$.

21. $y' + \dfrac{1}{x}y = \cos x$ $(x \neq 0)$, so $I(x) = e^{\int (1/x)dx} = e^{\ln|x|} = x$

(for $x > 0$). Multiplying the differential equation by $I(x)$

gives $xy' + y = x\cos x$ $\Rightarrow$ $(xy)' = x\cos x$. Thus,

$y = \dfrac{1}{x}\left[\int x\cos x\,dx + C\right] = \dfrac{1}{x}[x\sin x + \cos x + C]$

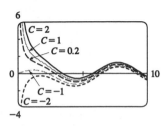

$= \sin x + \dfrac{\cos x}{x} + \dfrac{C}{x}$. The solutions are asymptotic to

the y-axis (except for $C = -1$). In fact, for $C > -1$, $y \to \infty$ as $x \to 0^+$, whereas for $C < -1$, $y \to -\infty$ as

$x \to 0^+$. As x gets larger, the solutions approximate $y = \sin x$ more closely. The graphs for larger C lie above

those for smaller C. The distance between the graphs lessens as x increases.

23. Setting $u = y^{1-n}$, $\dfrac{du}{dx} = (1-n)y^{-n}\dfrac{dy}{dx}$ or $\dfrac{dy}{dx} = \dfrac{y^n}{1-n}\dfrac{du}{dx} = \dfrac{u^{n/(1-n)}}{1-n}\dfrac{du}{dx}$. Then the Bernoulli differential

equation becomes $\dfrac{u^{n/(1-n)}}{1-n}\dfrac{du}{dx} + P(x)u^{1/(1-n)} = Q(x)u^{n/(1-n)}$ or $\dfrac{du}{dx} + (1-n)P(x)u = Q(x)(1-n)$.

25. Here $n = 3$, $P(x) = \dfrac{2}{x}$, $Q(x) = \dfrac{1}{x^2}$ and setting $u = y^{-2}$, u satisfies $u' - \dfrac{4u}{x} = -\dfrac{2}{x^2}$. Then

$I(x) = e^{\int -4/x\,dx} = x^{-4}$ and $u = x^4\left(\int -\dfrac{2}{x^6}\,dx + C\right) = x^4\left(\dfrac{2}{5x^5} + C\right) = Cx^4 + \dfrac{2}{5x}$. Thus

$y = \pm\left(Cx^4 + \dfrac{2}{5x}\right)^{-1/2}$.

27. (a) $2\dfrac{dI}{dt} + 10I = 40$ or $\dfrac{dI}{dt} + 5I = 20$. Then the integrating factor is $e^{\int 5\,dt} = e^{5t}$. Multiplying the

differential equation by the integrating factor gives $e^{5t}\dfrac{dI}{dt} + 5Ie^{5t} = 20e^{5t}$ $\Rightarrow$ $(e^{5t}I)' = 20e^{5t}$ $\Rightarrow$

$I(t) = e^{-5t}[\int 20e^{5t}\,dt + C] = 4 + Ce^{-5t}$. But $0 = I(0) = 4 + C$ so $I(t) = 4 - 4e^{-5t}$.

(b) $I(0.1) = 4 - 4e^{-0.5} \approx 1.57\,\text{A}$.

29. $5\dfrac{dQ}{dt} + 20Q = 60$ with $Q(0) = 0\,\text{C}$. Then the integrating factor is $e^{\int 4\,dt} = e^{4t}$ and multiplying the differential

equation by the integrating factor gives $e^{4t}\dfrac{dQ}{dt} + 4e^{4t}Q = 12e^{4t}$ $\Rightarrow$ $(e^{4t}Q)' = 12e^{4t}$ $\Rightarrow$

$Q(t) = e^{-4t}\left[\int 12e^{4t}\,dt + C\right] = 3 + Ce^{-4t}$. But $0 = Q(0) = 3 + C$ so $Q(t) = 3(1 - e^{-4t})$ is the charge at time

t and $I = dQ/dt = 12e^{-4t}$ is the current at time t.

31. $\dfrac{dP}{dt} + kP = kM$ so $I(t) = e^{\int k\,dt} = e^{kt}$. Multiplying the differential

equation by $I(t)$ gives $e^{kt}\dfrac{dP}{dt} + kPe^{kt} = kMe^{kt} \;\Rightarrow\; \left(e^{kt}P\right)' = kMe^{kt}$

$\Rightarrow\; P(t) = e^{-kt}\left[\int kMe^{kt}\,dt + C\right] = M + Ce^{-kt},\, k > 0.$ Furthermore

it is reasonable to assume $0 \le P(0) \le M$, so $-M \le C \le 0.$

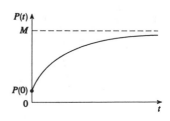

33. (a) $\dfrac{dv}{dt} + \dfrac{c}{m}v = g$ and $I(t) = e^{\int (c/m)dt} = e^{(c/m)t}$ and multiplying the differential equation by $I(t)$ gives

$e^{(c/m)t}\dfrac{dv}{dt} + \dfrac{vce^{(c/m)t}}{m} = ge^{(c/m)t} \;\Rightarrow\; \left[e^{(c/m)t}v\right]' = ge^{(c/m)t}.$ Hence

$v(t) = e^{-(c/m)t}\left[\int ge^{(c/m)t}\,dt + K\right] = \dfrac{mg}{c} + Ke^{-(c/m)t}.$ But the object is dropped from rest so $v(0) = 0$

and $K = -\dfrac{mg}{c}.$ Thus the velocity at time t is $v(t) = \dfrac{mg}{c}\left[1 - e^{-(c/m)t}\right].$

(b) $\lim\limits_{t\to\infty} v(t) = mg/c.$

(c) $s(t) = \displaystyle\int v(t)\,dt = \dfrac{mg}{c}\left[t + \dfrac{m}{c}e^{-(c/m)t}\right] + c_1$ where $c_1 = s(0) - \dfrac{m^2g}{c^2},\, s(0)$ is the initial position so

$s(0) = 0$ and $s(t) = \dfrac{mg}{c}\left[t + \dfrac{m}{c}e^{-(c/m)t}\right] - \dfrac{m^2g}{c^2}.$

EXERCISES 15.3

1. (a) $\dfrac{\partial(x\sin y)}{\partial y} = x\cos y$ and $\dfrac{\partial(y\cos x)}{\partial x} = -y\sin x,$ so the equation is not exact.

(b) $\dfrac{\partial(2xy^3)}{\partial y} = 6xy^2,\ \dfrac{\partial(3x^2y^2)}{\partial x} = 6xy^2,$ so the equation is exact.

(c) $\dfrac{\partial(x-y)}{\partial y} = -1,\ \dfrac{\partial(x+y)}{\partial x} = 1,$ so the equation is not exact.

3. $\dfrac{\partial(2x+y)}{\partial y} = 1 = \dfrac{\partial(x+2y)}{\partial x},$ so the equation is exact. Thus there exists f such that $f_x(x,y) = 2x + y$ and

$f_y(x,y) = x + 2y \;\Rightarrow\; f(x,y) = x^2 + xy + g(y) \;\Rightarrow\; f_y(x,y) = x + g'(y) \;\Rightarrow\; g(y) = y^2$ and

$f(x,y) = x^2 + xy + y^2.$ Thus the solution to the differential equation is given implicitly by $x^2 + xy + y^2 = C$

or $y = \frac{1}{2}\left(-x \pm \sqrt{4C - 3x^2}\right).$

5. $\dfrac{\partial(3xy-2)}{\partial y} = 3x,\ \dfrac{\partial(3y^2 - x^2)}{\partial x} = -2x,$ so the equation isn't exact.

7. $\dfrac{\partial(\sin y)}{\partial y} = \cos y = \dfrac{\partial(1 + x\cos y)}{\partial x},$ so the equation is exact and there exists f such that $f_x(x,y) = \sin y$ and

$f_y(x,y) = x\cos y + 1 \;\Rightarrow\; f(x,y) = x\sin y + g(y) \;\Rightarrow\; f_y(x,y) = x\cos y + g'(y),$ so $g'(y) = 1$ and

$f(x,y) = x\sin y + y.$ Hence the solution is $x\sin y + y = C.$

9. $\dfrac{\partial\left[(x+y)e^{x/y}\right]}{\partial y} = e^{x/y}\left[1 + (x+y)(-xy^{-2})\right] = e^{x/y}\left(1 - x^2y^{-2} - xy^{-1}\right)$ and

$\dfrac{\partial\left[(x - x^2y^{-1})e^{x/y}\right]}{\partial x} = e^{x/y}\left[1 - 2xy^{-1} + (x - x^2y^{-1})y^{-1}\right] = e^{x/y}\left(1 - x^2y^{-2} - xy^{-1}\right)$, so the equation is exact

and there exists f such that $f_x(x,y) = (x+y)e^{x/y}$ and $f_y(x,y) = (x - x^2y^{-1})e^{x/y}$ $\Rightarrow$

$f(x,y) = e^{x/y}\left[(x+y)y - y^2\right] + g(y) = xye^{x/y} + g(y)$ $\Rightarrow$

$f_y(x,y) = e^{x/y}\left[x + xy(-xy^{-2})\right] + g'(y) = e^{x/y}\left(x - x^2y^{-1}\right) + g'(y)$ $\Rightarrow$ $g'(y) = 0$ and $f(x,y) = xye^{x/y}$.

Thus the solution is given by $xye^{x/y} = C$.

11. $\dfrac{\partial(x\ln y)}{\partial y} = \dfrac{x}{y}$, $\dfrac{\partial(-x - y\ln x)}{\partial x} = -1 - \dfrac{y}{x}$, so the equation isn't exact.

13. $\dfrac{\partial(y^{-1} + 2yx^{-3})}{\partial y} = -y^{-2} + 2x^{-3} = \dfrac{\partial(-xy^{-2} - x^{-2})}{\partial x}$, so the equation is exact. Hence there exists f such that

$f_x(x,y) = y^{-1} + 2yx^{-3}$ and $f_y(x,y) = -xy^{-2} - x^{-2}$ $\Rightarrow$ $f(x,y) = xy^{-1} - yx^{-2} + g(y)$ $\Rightarrow$

$f_y(x,y) = -xy^{-2} - x^{-2} + g'(y)$ $\Rightarrow$ $g'(y) = 0$ and the solution is given by $xy^{-1} - yx^{-2} = C$. For an

explicit formula we could write $x^3 - y^2 = Cx^2y$ or $y^2 + Cx^2y - x^3 = 0$ and solve for y using the quadratic

formula. This gives $y = \tfrac{1}{2}\left(-Cx^2 \pm \sqrt{C^2x^4 + 4x^3}\right) = -Kx^2 \pm \sqrt{x^3 + K^2x^4}$, where $K = \tfrac{1}{2}C$.

15. Here the equation is exact. The function f such that $f_x = P$ and $f_y = Q$ is given by $f(x,y) = x^3 + x^2y + 3xy^2$.

Thus the general solution is given by $3xy^2 + x^2y + x^3 = C$. But when $x = 1$, $y = 2$ and $C = 15$, so the

solution is $3xy^2 + x^2y + x^3 = 15$. This is a quadratic in y, so

$y = \dfrac{-x^2 \pm \sqrt{x^4 - 4(3x)(x^3 - 15)}}{6x} = \dfrac{-x^2 \pm \sqrt{180x - 11x^4}}{6x}$. However $2 = y(1)$ so we must take the positive

sign. Therefore the solution is $y = \dfrac{-x^2 + \sqrt{180x - 11x^4}}{6x}$.

17. The equation is exact so there exists f such that $f_x(x,y) = 1 + y\cos xy$ and $f_y(x,y) = x\cos xy$ $\Rightarrow$

$f(x,y) = x + \sin xy + g(y)$ $\Rightarrow$ $f_y(x,y) = x\cos xy + g'(y)$. So the general solution is given by

$x + \sin xy = C$. But when $x = 1$, $y = 0$ so $C = 1$ and the solution to the initial-value problem is

$x + \sin xy = 1$ which we could rewrite as $y = \dfrac{1}{x}\sin^{-1}(1 - x)$.

19. $\dfrac{\partial y^2}{\partial y} = 2y$ while $\dfrac{\partial(1 + xy)}{\partial x} = y$; but $\dfrac{\partial(y^2e^{xy})}{\partial y} = e^{xy}\left(2y + y^2x\right)$ and

$\dfrac{\partial}{\partial x}\left[(1 + xy)e^{xy}\right] = e^{xy}\left[y + y(1 + xy)\right] = e^{xy}\left(2y + y^2x\right)$. Hence the new equation $y^2e^{xy} + (1 + xy)e^{xy}\dfrac{dy}{dx} = 0$

is exact, so there exists f such that $f_x(x,y) = y^2e^{xy}$ and $f_y(x,y) = e^{xy}(1 + xy)$ $\Rightarrow$ $f(x,y) = ye^{xy} + g(y)$

$\Rightarrow$ $f_y(x,y) = e^{xy}(1 + yx) + g'(y)$. So the solution is given by $ye^{xy} = C$.

21. $\dfrac{\partial}{\partial y}(y + y^3) = 1 + 3y^2$ while $\dfrac{\partial}{\partial x}(x + x^3) = 1 + 3x^2$; but

$$\frac{\partial}{\partial y}\left[\frac{y + y^3}{(1 + x^2 + y^2)^{3/2}}\right] = \frac{1 + x^2 + 3x^2 y^2 + y^2}{(1 + x^2 + y^2)^{5/2}} = \frac{\partial}{\partial x}\left[\frac{x + x^3}{(1 + x^2 + y^2)^{3/2}}\right]. \text{ So there exists } f \text{ such that}$$

$$f_x(x, y) = \frac{y + y^3}{(1 + x^2 + y^2)^{3/2}} \text{ and } f_y(x, y) = \frac{x + x^3}{(1 + x^2 + y^2)^{3/2}} \quad \Rightarrow$$

$$f(x, y) = (y + y^3)\left[\frac{x}{(1 + y^2)(1 + x^2 + y^2)^{1/2}}\right] + g(y) = \frac{xy}{(1 + x^2 + y^2)^{1/2}} + g(y) \quad \Rightarrow$$

$$f_y(x, y) = \frac{x + x^3}{(1 + x^2 + y^2)^{3/2}} + g'(y). \text{ So } g'(y) = 0 \text{ and the solution is } \frac{xy}{(1 + x^2 + y^2)^{1/2}} = C.$$

23. $P_y = 3x + 4y$ and $Q_x = 2x + 2y$ so the equation isn't exact, but $\dfrac{P_y - Q_x}{Q} = \dfrac{x + 2y}{x^2 + 2xy} = \dfrac{1}{x}$. So there exists an

integrating factor which satisfies $\dfrac{dI}{dx} = \dfrac{I}{x}$, so $I(x) = x$ and the new equation $3x^2 y + 2xy^2 + (x^3 + 2x^2 y)y' = 0$

is exact. So there exists f such that $f_x(x, y) = 3x^2 y + 2xy^2$ and $f_y(x, y) = x^3 + 2x^2 y$. Hence

$f(x, y) = x^3 y + x^2 y^2$ and the solution is $x^3 y + x^2 y^2 = C$.

25. $\dfrac{P_y - Q_x}{Q} = \dfrac{(2x + 3x^2 + 6y) - (2x)}{x^2 + 2y} = 3$. Thus there exists an integrating factor which satisfies $\dfrac{dI}{dx} = 3I$ or

$I(x) = e^{3x}$ and the new equation $e^{3x}(2xy + 3x^2 y + 3y^2) + e^{3x}(x^2 + 2y)y' = 0$ is exact. So there exists f such

that $f_x(x, y) = e^{3x}(2xy + 3x^2 y + 3y^2)$ and $f_y(x, y) = e^{3x}(x^2 + 2y) \quad \Rightarrow \quad f(x, y) = e^{3x}(x^2 y + y^2) + h(x)$

$\Rightarrow \quad f_x(x, y) = e^{3x}(3x^2 y + 3y^2 + 2xy) + h'(x)$ so $f(x, y) = e^{3x}(x^2 y + y^2)$ and the solution is given by

$e^{3x}(x^2 y + y^2) = C$.

27. The general form of a first order separable equation is $\dfrac{dy}{dx} = g(x)f(y)$ or $\dfrac{1}{f(y)}\dfrac{dy}{dx} = g(x)$ or

$-g(x) + \dfrac{1}{f(y)}\dfrac{dy}{dx} = 0$. Hence $\dfrac{\partial}{\partial y}[-g(x)] = 0 = \dfrac{\partial}{\partial x}\left[\dfrac{1}{f(y)}\right]$, so the equation is exact.

EXERCISES 15.4

1. **(a)** $yy' + x^2 = 0 \quad \Rightarrow \quad y\,dy = -x^2\,dx$ and so the equation is separable.

   (b) $y' = -x^2/y$ which cannot be written as a function of y/x. Thus, it is not homogeneous.

   (c) This equation is not linear since it cannot be placed into standard linear form (2).

   (d) $\partial(x^2)/\partial y = 0$, $\partial(y)/\partial x = 0$ and so the equation is exact.

3. **(a)** $xy' + y = 0 \quad \Rightarrow \quad (1/y)dy = -(1/x)dx$ and so the equation is separable.

   (b) $y' = -(y/x)$ and so it is homogeneous.

   (c) $xy' + y = 0 \quad \Rightarrow \quad y' + y/x = 0$ which is standard linear form and the equation is linear.

   (d) $\partial(y)/\partial y = 1$, $\partial(x)/\partial x = 1$ and so the equation is exact.

5. $y' - y = \sin x$ is linear with integrating factor $I(x) = e^{\int -1\,dx} = e^{-x}$. Multiplying the differential equation by
   $I(x)$ gives $e^{-x}y' - e^{-x}y = e^{-x}\sin x \quad \Rightarrow \quad (e^{-x}y)' = e^{-x}\sin x \quad \Rightarrow$
   $y = e^x\left(\int e^{-x}\sin x\,dx + C\right) = e^x\left[\frac{1}{2}e^{-x}(-\sin x - \cos x) + C\right] = Ce^x - \frac{1}{2}(\sin x + \cos x)$.

7. Since $y' = \dfrac{e^x - x}{4y^3}$ the equation is separable as well as exact. Then $4y^3\,dy = (e^x - x)dx$ or $y^4 = e^x - \frac{1}{2}x^2 + C$
   or $y = \pm\left(e^x - \frac{1}{2}x^2 + C\right)^{1/4}$.

9. $y' = \dfrac{x^2 + 2xy - y^2}{x^2 - 2xy - y^2} = \dfrac{1 + 2(y/x) - (y/x)^2}{1 - 2(y/x) - (y/x)^2}$ the equation is homogeneous (but not exact since $P_y = -Q_x$.)

   Setting $v = \dfrac{y}{x}$ gives $v' = \dfrac{1}{x}\left(\dfrac{1 + 2v - v^2}{1 - 2v - v^2} - v\right) = \dfrac{1 + v + v^2 + v^3}{(1 - 2v - v^2)x}$. Then $\dfrac{1 - 2v - v^2}{v^3 + v^2 + v + 1}dv = \dfrac{1}{x}dx \quad \Rightarrow$

   $\left(-\dfrac{2v}{v^2 + 1} + \dfrac{1}{v + 1}\right)dv = \dfrac{1}{x}dx \quad \Rightarrow \quad -\ln(v^2 + 1) + \ln|v + 1| = \ln|x| + C$. Hence $\dfrac{v + 1}{v^2 + 1} = kx$ and the

   solution is given by $\dfrac{x(y + x)}{x^2 + y^2} = kx$ or $x + y = k(x^2 + y^2)$.

11. Since $\partial(-3y\sec^2 x + y^2)/\partial y = -3\sec^2 x + 2y = \partial(2xy - 3\tan x)/\partial x$ the equation is exact, so there exists f
    such that $f_x(x, y) = y^2 - 3y\sec^2 x$ and $f_y(x, y) = 2xy - 3\tan x$. Thus $f(x, y) = xy^2 - 3y\tan x$ and the
    solution is given by $xy^2 - 3y\tan x = C$.

13. $y' = \dfrac{x}{x^2 y + y} = \dfrac{x}{x^2 + 1}\dfrac{1}{y}$, so the equation is separable and exact. Then $y\,dy = \dfrac{x}{x^2 + 1}dx$, so
    $\frac{1}{2}y^2 = \frac{1}{2}\ln(x^2 + 1) + C$ or $y^2 = \ln(x^2 + 1) + K$.

15. $y' - \dfrac{2}{x}y - \dfrac{\sqrt{1+x^2}}{x}$ so the equation is linear. $I(x) = e^{\int -2/x\,dx} = x^{-2}$ and multiplying the differential equation

by $I(x)$ gives $x^{-2}y' - 2x^{-3} = x^{-3}\sqrt{1+x^2}$ $\Rightarrow$ $\left(x^{-2}y\right)' = x^{-3}\sqrt{1+x^2}$ $\Rightarrow$

$y = x^2\left[\displaystyle\int \dfrac{\sqrt{1+x^2}}{x^3}\,dx + C\right] = x^2\left[-\dfrac{\sqrt{1+x^2}}{2x^2} - \dfrac{1}{2}\ln\left|\dfrac{\sqrt{1+x^2}+1}{x}\right| + C\right]$ by trig substitution. Hence the

solution is $y = Cx^2 - \dfrac{1}{2}\sqrt{1+x^2} - \dfrac{1}{2}x^2\ln\left|\dfrac{\sqrt{1+x^2}+1}{x}\right|$.

17. $y' = -y^2\sqrt{1+x^3}$ is separable and exact, so $\dfrac{dy}{y^2} = -\sqrt{1+x^3}\,dx$ and $-\dfrac{1}{y} = -\displaystyle\int\sqrt{1+x^3}\,dx + C$ $\Rightarrow$

$y = \left(K + \int\sqrt{1+x^3}\,dx\right)^{-1}$.

19. Rewriting the equation as $(2x + e^y) + (2y + xe^y)y' = 0$ we see the equation is exact. Thus there exists f such

that $f_x(x, y) = 2x + e^y$ and $f_y(x, y) = 2y + xe^y$. So $f(x, y) = x^2 + y^2 + xe^y$ and the solution is given by

$x^2 + y^2 + xe^y = C$.

21. $\dfrac{\partial[y^2\cos(xy) - 1]}{\partial y} = 2y\cos(xy) - y^2x\sin(xy) = \dfrac{\partial[\sin(xy) + xy\cos(xy)]}{\partial x}$ so the equation is exact and there

exists f such that $f_x(x, y) = y^2\cos(xy) - 1$ and $f_y(x, y) = \sin(xy) + xy\cos(xy)$. Thus

$f(x, y) = y\sin(xy) - x$ and the solution is given by $y\sin(xy) - x = C$.

23. $y' = \dfrac{2\sqrt{xy} - y}{x} = 2\sqrt{\dfrac{y}{x}} - \dfrac{y}{x}$ is homogeneous. Setting $v = \dfrac{y}{x}$ gives $v' = \dfrac{2\sqrt{v} - 2v}{x}$ or

$\left[\dfrac{1}{2}(\sqrt{v} - v)\right]dv = \dfrac{1}{x}\,dx$ $\Rightarrow$ $\dfrac{\frac{1}{2}v^{-1/2}}{1 - v^{1/2}}\,dv = \dfrac{1}{x}\,dx$ $\Rightarrow$ $-\ln|1 - v^{1/2}| = \ln|x| + C_1$ $\Rightarrow$

$\ln|1 - v^{1/2}| = K - \ln|x|$ $\Rightarrow$ $1 - \sqrt{v} = \dfrac{C}{x}$ $\Rightarrow$ $v = \left(1 - \dfrac{C}{x}\right)^2$ and the solution is $y = x\left(1 - \dfrac{C}{x}\right)^2$.

25. (a) $y' = 3x(y + x^n)$ is separable only if $n = 0$ since only then can we separate x and y factors to get an

equation of the form (1).

(b) $y' = 3x(y + x^n)$ $\Rightarrow$ $y' - 3xy = 3x^{n+1}$ which is linear [of the form (2)] for any integer n.

EXERCISES 15.5

1. The auxiliary equation is $r^2 - 3r + 2 = (r - 2)(r - 1) = 0$, so $y = c_1 e^x + c_2 e^{2x}$.

3. The auxiliary equation is $3r^2 - 8r - 3 = (3r + 1)(r - 3) = 0$, so $y = c_1 e^{-x/3} + c_2 e^{3x}$.

5. The auxiliary equation is $r^2 + 2r + 10 = 0 \Rightarrow r = -1 \pm 3i$, so $y = e^{-x}(c_1 \cos 3x + c_2 \sin 3x)$.

7. The auxiliary equation is $r^2 - 1 = (r - 1)(r + 1) = 0$, so $y = c_1 e^x + c_2 e^{-x}$.

9. The auxiliary equation is $r^2 + 25 = 0 \Rightarrow r = \pm 5i$, so $y = c_1 \cos 5x + c_2 \sin 5x$.

11. The auxiliary equation is $2r^2 + r = r(2r + 1) = 0$, so $y = c_1 + c_2 e^{-x/2}$.

13. The auxiliary equation is $r^2 - r + 2 = 0 \Rightarrow r = \frac{1}{2}\left(1 \pm \sqrt{7}i\right)$, so $y = e^{x/2}\left[c_1 \cos\left(\frac{\sqrt{7}}{2}x\right) + c_2 \sin\left(\frac{\sqrt{7}}{2}x\right)\right]$.

15. The auxiliary equation is $r^2 + 2r - 1 = 0 \Rightarrow r = -1 \pm \sqrt{2}$, so $y = c_1 e^{(-1+\sqrt{2})x} + c_2 e^{(-1-\sqrt{2})x}$.

17. The auxiliary equation is $2r^2 + r + 3 = 0 \Rightarrow r = \frac{1}{4}\left(-1 \pm \sqrt{23}i\right)$, so
$$y = e^{-x/4}\left[c_1 \cos\left(\frac{\sqrt{23}}{4}x\right) + c_2 \sin\left(\frac{\sqrt{23}}{4}x\right)\right].$$

19. $r^2 - 8r + 16 = (r - 4)^2 = 0$ so $y = c_1 e^{4x} + c_2 x e^{4x}$.

The graphs are all asymptotic to the x-axis as $x \to -\infty$,

and as $x \to \infty$ the solutions tend to $\pm\infty$.

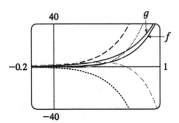

21. $r^2 + 3r - 4 = (r + 4)(r - 1) = 0$ so the general solution is $y = c_1 e^x + c_2 e^{-4x}$. Then $2 = y(0) = c_1 + c_2$ and $-3 = y'(0) = c_1 - 4c_2$ so $c_1 = 1$, $c_2 = 1$ and the solution to the initial-value problem is $y = e^x + e^{-4x}$.

23. $r^2 - 2r + 2 = 0 \Rightarrow r = 1 \pm i$ and the general solution is $y = e^x(c_1 \cos x + c_2 \sin x)$. But $1 = y(0) = c_1$ and $2 = y'(0) = c_1 + c_2$ so the solution to the initial-value problem is $y = e^x(\cos x + \sin x)$.

25. $r^2 - 2r - 3 = (r - 3)(r + 1) = 0$ so the general solution is $y = c_1 e^{-x} + c_2 e^{3x}$. However the conditions are given at $x = 1$ so rewrite the general solution as $y = k_1 e^{-(x-1)} + k_2 e^{3(x-1)}$. Then $3 = y(1) = k_1 + k_2$ and $1 = y'(1) = -k_1 + 3k_2$ so $k_1 = 2$, $k_2 = 1$ and the solution to the initial-value problem is $y = 2e^{-(x-1)} + e^{3(x-1)}$.

27. $r^2 + 9 = 0 \Rightarrow r = \pm 3i$ and the general solution is $y = c_1 \cos 3x + c_2 \sin 3x$. But $0 = y\left(\frac{\pi}{3}\right) = -c_1$ and $1 = y'\left(\frac{\pi}{3}\right) = -3c_2$, so the solution to the initial-value problem is $y = -\frac{1}{3}\sin 3x$.

29. $r^2 + 4r + 4 = (r+2)^2 = 0$ so the general solution is $y = c_1 e^{-2x} + c_2 x e^{-2x}$. Then $0 = y(0) = c_1$,

$3 = y(1) = c_2 e^{-2}$ so $c_2 = 3e^2$ and the solution of the boundary-value problem is $y = 3xe^{-2x+2}$.

31. $r^2 + 1 = 0 \quad \Rightarrow \quad r = \pm i$ and the general solution is $y = c_1 \cos x + c_2 \sin x$. But $1 = y(0) = c_1$ and

$0 = y(\pi) = -c_1$ so there is no solution.

33. $r^2 - r - 2 = (r-2)(r+1) = 0$ so the general solution is $y = c_1 e^{-x} + c_2 e^{2x}$. Then $1 = y(-1) = c_1 e + c_2 e^{-2}$

and $0 = y(1) = c_1 e^{-1} + c_2 e^2$ so $c_1 = \dfrac{e^5}{e^6 - 1}$ and $c_2 = \dfrac{e^2}{1 - e^6}$ so the solution to the boundary-value problem is

$y = \dfrac{e^5}{e^6 - 1} e^{-x} + \dfrac{e^2}{1 - e^6} e^{2x} = \dfrac{1}{e^6 - 1} \left[e^{5-x} - e^{2(1+x)} \right].$

35. $r^2 + 4r + 13 = 0 \quad \Rightarrow \quad r = -2 \pm 3i$ and the general solution is $y = e^{-2x}(c_1 \cos 3x + c_2 \sin 3x)$. But

$2 = y(0) = c_1$ and $1 = y\left(\frac{\pi}{2}\right) = e^{-\pi}(-c_2)$, so the solution to the boundary-value problem is

$y = e^{-2x}(2 \cos 3x - e^\pi \sin 3x)$.

37. (a) *Case 1* $(\lambda = 0)$: $y'' + \lambda y = 0 \quad \Rightarrow \quad y'' = 0$ which has an auxiliary equation $r^2 = 0 \quad \Rightarrow \quad r = 0 \quad \Rightarrow$

$y = c_1 + c_2 x$ where $y(0) = 0$ and $y(L) = 0$. Thus, $0 = y(0) = c_1$ and $0 = y(L) = c_2 L \quad \Rightarrow$

$c_1 = c_2 = 0$. Thus, $y = 0$.

Case 2 $(\lambda < 0)$: $y'' + \lambda y = 0$ has auxiliary equation $r^2 = -\lambda \quad \Rightarrow \quad r = \pm \sqrt{-\lambda}$ (distinct and real

since $\lambda < 0$) $\quad \Rightarrow \quad y = c_1 e^{\sqrt{-\lambda}x} + c_2 e^{-\sqrt{-\lambda}x}$ where $y(0) = 0$ and $y(L) = 0$. Thus,

$0 = y(0) = c_1 + c_2$ ($\bigstar$) and $0 = y(L) = c_1 e^{\sqrt{-\lambda}L} + c_2 e^{-\sqrt{-\lambda}L}$ (†).

Multiplying ($\bigstar$) by $e^{\sqrt{-\lambda}L}$ and subtracting (†) gives $c_2 \left(e^{\sqrt{-\lambda}L} - e^{-\sqrt{-\lambda}L} \right) = 0 \quad \Rightarrow \quad c_2 = 0$ and thus

$c_1 = 0$ from ($\bigstar$). Thus, $y = 0$ for the cases $\lambda = 0$ and $\lambda < 0$.

(b) $y'' + \lambda y = 0$ has an auxiliary equation $r^2 + \lambda = 0 \quad \Rightarrow \quad r = \pm i \sqrt{\lambda} \quad \Rightarrow$

$y = c_1 \cos \sqrt{\lambda}x + c_2 \sin \sqrt{\lambda}x$ where $y(0) = 0$ and $y(L) = 0$. Thus, $0 = y(0) = c_1$ and

$0 = y(L) = c_2 \sin \sqrt{\lambda}L$ since $c_1 = 0$. Since we cannot have a trivial solution, $c_2 \neq 0$ and thus

$\sin \sqrt{\lambda}L = 0 \quad \Rightarrow \quad \sqrt{\lambda}L = n\pi$ where n is an integer $\quad \Rightarrow \quad \lambda = n^2\pi^2/L^2$ and $y = c_2 \sin(n\pi x/L)$

where n is an integer.

EXERCISES 15.6

1. The auxiliary equation is $r^2 - r - 6 = (r-3)(r+2) = 0$, so the complementary solution is
$y_c(x) = c_1 e^{3x} + c_2 e^{-2x}$. Try the particular solution $y_p(x) = A\cos 3x + B\sin 3x$, so
$y_p' = 3B\cos 3x - 3A\sin 3x$ and $y_p'' = -9A\cos 3x - 9B\sin 3x$. Substitution gives
$(-9A\cos 3x - 9B\sin 3x) - (3B\cos 3x - 3A\sin 3x) - 6(A\cos 3x + B\sin 3x) = \cos 3x \quad\Rightarrow$
$\cos 3x = (-15A - 3B)\cos 3x + (3A - 15B)\sin 3x$. Hence $-15A - 3B = 1$ and $3A - 15B = 0 \quad\Rightarrow$
$A = -\frac{5}{78}, B = -\frac{1}{78}$ and the general solution is $y(x) = y_c + y_p = c_1 e^{3x} + c_2 e^{-2x} - \frac{5}{78}\cos 3x - \frac{1}{78}\sin 3x$.

3. The complementary solution is $y_c(x) = e^{2x}(c_1 x + c_2)$, so try a particular solution $y_p(x) = Ae^{-x}$. Then
$y_p'' = y_p' = y_p = Ae^{-x}$ and substitution into the differential equation gives $Ae^{-x} + 4Ae^{-x} + 4Ae^{-x} = e^{-x} \quad\Rightarrow$
$A = \frac{1}{9}$. Hence the general solution is $y(x) = e^{2x}(c_1 x + c_2) + \frac{1}{9}e^{-x}$.

5. The complementary solution is $y_c(x) = c_1 \cos 6x + c_2 \sin 6x$. Try $y_p(x) = Ax^2 + Bx + C$. Then
$y_p' = 2Ax + B$ and $y_p'' = 2A$; substitution gives $2A + 36(Ax^2 + Bx + C) = 2x^2 - x$. $A = \frac{1}{18}, B = -\frac{1}{36}$,
$C = -\frac{1}{324}$. The general solution is $y(x) = c_1 \cos 6x + c_2 \sin 6x + \frac{1}{18}x^2 - \frac{1}{36}x - \frac{1}{324}$.

7. Since the roots of $r^2 - 2r + 5 = 0$ are $1 \pm 2i$, $y_c(x) = e^x(c_1 \cos 2x + c_2 \sin 2x)$. For $y'' - 2y' + 5y = x$ try
$y_{p_1}(x) = Ax + B$. Then $0 - 2A + 5(Ax + B) = x$, so $y_{p_1}(x) = \frac{1}{5}x + \frac{2}{25}$. For $y'' - 2y' + 5y = \sin 3x$ try
$y_{p_2}(x) = A\cos 3x + B\sin 3x$. Then $y_{p_2}' = -3A\sin 3x + 3B\cos 3x$ and $y_{p_2}'' = -9A\cos 3x - 9B\sin 3x$.
Substituting into the differential equation gives
$-9A\cos 3x - 9B\sin 3x + 6A\sin 3x - 6B\cos 3x + 5A\cos 3x + 5B\sin 3x = \sin 3x$. Thus
$(-9A - 6B + 5A) = 0$ and $(-9B + 6A + 5B) = 1$, so $A = \frac{3}{26}$ and $B = -\frac{1}{13}$. Hence the general solution is
$y(x) = e^x(c_1 \cos 2x + c_2 \sin 2x) + \frac{1}{5}x + \frac{2}{25} + \frac{3}{26}\cos 3x - \frac{1}{13}\sin 3x$. But $1 = y(0) = c_1 + \frac{2}{25} + \frac{3}{26} \quad\Rightarrow$
$c_1 = \frac{523}{650}$, $2 = y'(0) = c_1 + 2c_2 + \frac{1}{5} - \frac{3}{13} \quad\Rightarrow \quad c_2 = \frac{797}{1300}$. Thus the solution to the initial-value problem is
$y(x) = e^x\left[\frac{523}{650}\cos 2x + \frac{797}{1300}\sin 2x\right] + \frac{1}{5}x + \frac{2}{25} + \frac{3}{26}\cos 3x - \frac{1}{13}\sin 3x$.

9. $y_c(x) = c_1 e^x + c_2 e^{-x}$. Try $y_p(x) = (Ax + B)e^{3x}$. Then $y_p' = e^{3x}(A + 3Ax + 3B)$ and
$y_p'' = e^{3x}(3A + 3A + 9Ax + 9B)$. Substitution into the differential equation gives
$e^{3x}[9Ax + 9B + 6A - (Ax + B)] = xe^{3x}$, so $A = \frac{1}{8}, B = -\frac{3}{32}$ and the general solution is
$y(x) = c_1 e^x + c_2 e^{-x} + \left(\frac{1}{8}x - \frac{3}{32}\right)e^{3x}$. But $0 = y(0) = c_1 + c_2 - \frac{3}{32}, 1 = y'(0) = c_1 - c_2 - \frac{9}{32} + \frac{1}{8}$ so the
solution to the initial-value problem is $y(x) = \frac{5}{8}e^x - \frac{17}{32}e^{-x} + e^{3x}\left(\frac{1}{8}x - \frac{3}{32}\right)$.

11. $y_c(x) = c_1 e^{-x/4} + c_2 e^{-x}$. Try $y_p(x) = Ae^x$. Then $10Ae^x = e^x$,
so $A = \frac{1}{10}$ and the general solution is $y(x) = c_1 e^{-x/4} + c_2 e^{-x} + \frac{1}{10}e^x$.
The solutions are all composed of exponential curves and with the
exception of the particular solution (which approaches 0 as $x \to -\infty$),
they all approach either ∞ or $-\infty$ as $x \to -\infty$. As $x \to \infty$, all solutions
are asymptotic to $y_p = \frac{1}{10}e^x$.

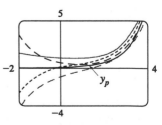

13. Since the roots of the auxiliary equation are complex, we try $y_p(x) = (Ax^4 + Bx^3 + Cx^2 + Dx + E)e^{2x}$.

15. Since $y_c(x) = e^x(c_1 \cos x + c_2 \sin x)$ we try $y_p(x) = xe^x(A\cos x + B\sin x)$.

Note: Solving Equations (7) and (9) in **The Method of Variation of Parameters** gives

$$u_1' = -\frac{Gy_2}{a(y_1y_2' - y_2y_1')} \quad \text{and} \quad u_2' = \frac{Gy_1}{a(y_1y_2' - y_2y_1')}$$

We will use these equations rather than resolving the system in each of the remaining exercises in this section.

17. (a) The complementary solution is $y_c(x) = c_1\cos 2x + c_2\sin 2x$. A particular solution is of the form $y_p(x) = Ax + B$. Thus, $4Ax + 4B = x \Rightarrow A = \frac{1}{4}$ and $B = 0 \Rightarrow y_p(x) = \frac{1}{4}x$. Thus, the general solution is $y = y_c + y_p = c_1\cos 2x + c_2\sin 2x + \frac{1}{4}x$.

(b) In (a), $y_c(x) = c_1\cos 2x + c_2\sin 2x$, so set $y_1 = \cos 2x$, $y_2 = \sin 2x$. Then

$y_1y_2' - y_2y_1' = 2\cos^2 2x + 2\sin^2 2x = 2$ so $u_1' = -\frac{1}{2}x\sin 2x \Rightarrow$

$u_1(x) = -\frac{1}{2}\int x\sin 2x\, dx = -\frac{1}{4}(-x\cos 2x + \frac{1}{2}\sin 2x)$ (by parts) and $u_2' = \frac{1}{2}x\cos 2x \Rightarrow$

$u_2(x) = \frac{1}{2}\int x\cos 2x\, dx = \frac{1}{4}(x\sin 2x + \frac{1}{2}\cos 2x)$ (by parts). Hence

$y_p(x) = -\frac{1}{4}(-x\cos 2x + \frac{1}{2}\sin 2x)\cos 2x + \frac{1}{4}(x\sin 2x + \frac{1}{2}\cos 2x)\sin 2x = \frac{1}{4}x$. Thus

$y(x) = y_c(x) + y_p(x) = c_1\cos 2x + c_2\sin 2x + \frac{1}{4}x$.

19. (a) $r^2 - r = r(r-1) = 0 \Rightarrow r = 0, 1$, so the complementary solution is $y_c(x) = c_1e^x + c_2xe^x$.

A particular solution is of the form $y_p(x) = Ae^{2x}$. Thus $4Ae^{2x} - 4Ae^{2x} + Ae^{2x} = e^{2x} \Rightarrow Ae^{2x} = e^{2x}$

$\Rightarrow A = 1 \Rightarrow y_p(x) = e^{2x}$. So a general solution is $y(x) = y_c(x) + y_p(x) = c_1e^x + c_2xe^x + e^{2x}$.

(b) From (a), $y_c(x) = c_1e^x + c_2xe^x$, so set $y_1 = e^x$, $y_2 = xe^x$. Then, $y_1y_2' - y_2y_1' = e^{2x}(1+x) - xe^{2x} = e^{2x}$

and so $u_1' = -xe^x \Rightarrow u_1(x) = -\int xe^x\, dx = -(x-1)e^x$ (by parts) and $u_2' = e^x \Rightarrow$

$u_2(x) = \int e^x\, dx = e^x$. Hence $y_p(x) = (1-x)e^{2x} + xe^{2x} = e^{2x}$ and the general solution is

$y(x) = y_c(x) + y_p(x) = c_1e^x + c_2xe^x + e^{2x}$.

21. As in Example 6, $y_c(x) = c_1\sin x + c_2\cos x$, so set $y_1 = \sin x$, $y_2 = \cos x$. Then

$y_1y_2' - y_2y_1' = -\sin^2 x - \cos^2 x = -1$, so $u_1' = -\frac{\sec x\cos x}{-1} = 1 \Rightarrow u_1(x) = x$ and $u_2' = \frac{\sec x\sin x}{-1} = -\tan x$

$\Rightarrow u_2(x) = -\int\tan x\, dx = \ln|\cos x| = \ln(\cos x)$ on $0 < x < \frac{\pi}{2}$. Hence $y_p(x) = x\sin x + \cos x\ln(\cos x)$ and the general solution is $y(x) = (c_1 + x)\sin x + [c_2 + \ln(\cos x)]\cos x$.

23. $y_1 = e^x$, $y_2 = e^{2x}$ and $y_1y_2' - y_2y_1' = e^{3x}$. So $u_1' = \frac{-e^{2x}}{(1+e^{-x})e^{3x}} = -\frac{e^{-x}}{1+e^{-x}}$ and

$u_1(x) = \int -\frac{e^{-x}}{1+e^{-x}}\, dx = \ln(1+e^{-x})$. $u_2' = \frac{e^x}{(1+e^{-x})e^{3x}} = \frac{e^x}{e^{3x}+e^{2x}}$ so

$u_2(x) = \int\frac{e^x}{e^{3x}+e^{2x}}\, dx = \ln\left(\frac{e^x+1}{e^x}\right) - e^{-x} = \ln(1+e^{-x}) - e^{-x}$. Hence

$y_p(x) = e^x\ln(1+e^{-x}) + e^{2x}[\ln(1+e^{-x}) - e^{-x}]$ and the general solution is

$y(x) = [c_1 + \ln(1+e^{-x})]e^x + [c_2 - e^{-x} + \ln(1+e^{-x})]e^{2x}$.

25. $y_1 = e^{-x}$, $y_2 = e^x$ and $y_1y_2' - y_2y_1' = 2$. So $u_1' = -\frac{e^x}{2x}$, $u_2' = \frac{e^{-x}}{2x}$ and $y_p(x) = -e^{-x}\int\frac{e^x}{2x}\, dx + e^x\int\frac{e^{-x}}{2x}\, dx$.

Hence the general solution is $y(x) = \left(c_1 - \int\frac{e^x}{2x}\, dx\right)e^{-x} + \left(c_2 + \int\frac{e^{-x}}{2x}\, dx\right)e^x$.

EXERCISES 15.7

1. By Hooke's Law $k(0.6) = 20$ so $k = \frac{100}{3}$ is the spring constant and the differential equation is $3x'' + \frac{100}{3}x = 0$.
    The general solution is $x(t) = c_1 \cos\left(\frac{10}{3}t\right) + c_2 \sin\left(\frac{10}{3}t\right)$. But $0 = x(0) = c_1$ and $1.2 = x'(0) = \frac{10}{3}c_2$, so the
    position of the mass after t seconds is $x(t) = 0.36 \sin\left(\frac{10}{3}t\right)$.

3. $k(0.5) = 6$ or $k = 12$ is the spring constant, so the initial-value problem is $2x'' + 14x' + 12x = 0$, $x(0) = 1$,
    $x'(0) = 0$. The general solution is $x(t) = c_1 e^{-6t} + c_2 e^{-t}$. But $1 = x(0) = c_1 + c_2$ and $0 = x'(0) = -6c_1 - c_2$.
    Thus the position is given by $x(t) = -\frac{1}{5}e^{-6t} + \frac{6}{5}e^{-t}$.

5. For critical damping we need $c^2 - 4mk = 0$ or $m = c^2/(4k) = 14^2/(4 \cdot 12) = \frac{49}{12}$ kg.

7. The differential equation is $mx'' + kx = F_0 \cos\omega_0 t$ and $\omega_0 \neq \omega = \sqrt{k/m}$. Here the auxiliary equation is
    $mr^2 + k = 0$ with roots $\pm\sqrt{k/m}i = \pm\omega i$ so $x_c(t) = c_1 \cos\omega t + c_2 \sin\omega t$. Since $\omega_0 \neq \omega$, try
    $x_p(t) = A \cos\omega_0 t + B \sin\omega_0 t$. Then we need
    $(m)(-\omega_0^2)(A\cos\omega_0 t + B\sin\omega_0 t) + k(A\cos\omega_0 t + B\sin\omega_0 t) = F_0\cos\omega_0 t$ or $A(k - m\omega_0^2) = F_0$ and
    $B(k - m\omega_0^2) = 0$. Hence $B = 0$ and $A = \dfrac{F_0}{k - m\omega_0^2} = \dfrac{F_0}{m(\omega^2 - \omega_0^2)}$ since $\omega^2 = \dfrac{k}{m}$. Thus the motion of the
    mass is given by $x(t) = c_1 \cos\omega t + c_2 \sin\omega t + \dfrac{F_0}{m(\omega^2 - \omega_0^2)}\cos\omega_0 t$.

9. Here the initial-value problem for the charge is $Q'' + 20Q' + 500Q = 12$, $Q(0) = Q'(0) = 0$. Then
    $Q_c(t) = e^{-10t}(c_1 \cos 20t + c_2 \sin 20t)$ and try $Q_p(t) = A$ $\Rightarrow$ $500A = 12$ or $A = \frac{3}{125}$. The general solution
    is $Q(t) = e^{-10t}(c_1 \cos 20t + c_2 \sin 20t) + \frac{3}{125}$. But $0 = Q(0) = c_1 + \frac{3}{125}$ and
    $Q'(t) = I(t) = e^{-10t}[(-10c_1 + 20c_2)\cos 20t + (-10c_2 - 20c_1)\sin 20t]$ but $0 = Q'(0) = -10c_1 + 20c_2$. Thus
    the charge is $Q(t) = -\frac{1}{250}e^{-10t}(6\cos 20t + 3\sin 20t) + \frac{3}{125}$ and the current is $I(t) = e^{-10t}\left(\frac{3}{5}\right)\sin 20t$.

11. As in Exercise 9, $Q_c(t) = e^{-10t}(c_1 \cos 20t + c_2 \sin 20t)$ but $E(t) = 12 \sin 10t$ so try
    $Q_p(t) = A \cos 10t + B \sin 10t$. Substituting into the differential equation gives
    $(-100A + 200B + 500A)\cos 10t + (-100B - 200A + 500B)\sin 10t = 12 \sin 10t$, $\Rightarrow$ $400A + 200B = 0$
    and $400B - 200A = 12$. Thus $A = -\frac{3}{250}$, $B = \frac{3}{125}$ and the general solution is
    $Q(t) = e^{-10t}(c_1 \cos 20t + c_2 \sin 20t) - \frac{3}{250}\cos 10t + \frac{3}{125}\sin 10t$. But $0 = Q(0) = c_1 - \frac{3}{250}$ so $c_1 = \frac{3}{250}$. Also
    $Q'(t) = \frac{3}{25}\sin 10t + \frac{6}{25}\cos 10t + e^{-10t}[(-10c_1 + 20c_2)\cos 20t + (-10c_2 - 20c_1)\sin 20t]$ and
    $0 = Q'(0) = \frac{6}{25} - 10c_1 + 20c_2$ so $c_2 = -\frac{3}{500}$. Hence the charge is given by
    $Q(t) = e^{-10t}\left[\frac{3}{250}\cos 20t - \frac{3}{500}\sin 20t\right] - \frac{3}{250}\cos 10t + \frac{3}{125}\sin 10t$.

13. $x(t) = A\cos(\omega t + \delta)$ $\Leftrightarrow$ $x(t) = A[\cos\omega t \cos\delta - \sin\omega t \sin\delta]$ $\Leftrightarrow$ $x(t) = A\left(\dfrac{c_1}{A}\cos\omega t + \dfrac{c_2}{A}\sin\omega t\right)$
    where $\cos\delta = c_1/A$ and $\sin\delta = -c_2/A$ $\Leftrightarrow$ $x(t) = c_1\cos\omega t + c_2\sin\omega t$. (Note that $\cos^2\delta + \sin^2\delta = 1$ $\Rightarrow$
    $c_1^2 + c_2^2 = A^2$.)

EXERCISES 15.8

1. Let $y(x) = \sum_{n=0}^{\infty} a_n x^n$. Then $\sum_{n=1}^{\infty} na_n x^{n-1} - 6\sum_{n=0}^{\infty} a_n x^n = 0 \Rightarrow \sum_{n=1}^{\infty} na_n x^{n-1} - 6\sum_{n=0}^{\infty} a_n x^n = 0$. Replacing

n by $n+1$ in the first sum gives $\sum_{n=0}^{\infty}[(n+1)a_{n+1} - 6a_n]x^n = 0$. Thus the recurrence relation is $a_{n+1} = \dfrac{6a_n}{n+1}$,

$n = 0, 1, 2, \ldots$. Then $a_1 = 6a_0$, $a_2 = \dfrac{6a_1}{2} = \dfrac{6^2 a_0}{2}$, $a_3 = \dfrac{6a_2}{3} = \dfrac{6^3 a_0}{2 \cdot 3}, \ldots, a_n = \dfrac{6a_{n-1}}{n} = \dfrac{6^n a_0}{n!}$. Thus the

solution is $y(x) = \sum_{n=0}^{\infty} a_0 \dfrac{6^n}{n!} x^n = \sum_{n=0}^{\infty} \left[a_0 \dfrac{(6x)^n}{n!} \right] = a_0 e^{6x}$.

3. Assuming $y(x) = \sum_{n=0}^{\infty} a_n x^n$, we have $y'(x) = \sum_{n=1}^{\infty} na_n x^{n-1} = \sum_{n=0}^{\infty}(n+1)a_{n+1}x^n$ and

$-x^2 y = -\sum_{n=0}^{\infty} a_n x^{n+2} = -\sum_{n=2}^{\infty} a_{n-2} x^n$. Hence the differential equation becomes

$\sum_{n=0}^{\infty}(n+1)a_{n+1}x^n - \sum_{n=2}^{\infty} a_{n-2}x^n = 0$ or $a_1 = 2a_2 x + \sum_{n=2}^{\infty}[(n+1)a_{n+1} - a_{n-2}]x^n = 0$. Equating

coefficients gives $a_1 = a_2 = 0$ and $a_{n+1} = \dfrac{a_{n-2}}{n+1}$ for $n = 2, 3, \ldots$. But $a_1 = 0$, so $a_4 = 0$ and $a_7 = 0$ and in

general $a_{3n+1} = 0$. Similarly $a_2 = 0$ so $a_{3n+2} = 0$. Finally $a_3 = \dfrac{a_0}{3}$, $a_6 = \dfrac{a_3}{6} = \dfrac{a_0}{6 \cdot 3} = \dfrac{a_0}{3^2 \cdot 2!}$,

$a_9 = \dfrac{a_6}{9} = \dfrac{a_0}{9 \cdot 6 \cdot 3} = \dfrac{a_0}{3^3 \cdot 3!}, \ldots$, and $a_{3n} = \dfrac{a_0}{3^n \cdot n!}$. Thus the solution is

$y(x) = \sum_{n=0}^{\infty} a_n x^n = \sum_{n=0}^{\infty} a_{3n} x^{3n} = a_0 \sum_{n=0}^{\infty} \dfrac{(x^3/3)^n}{n!} = a_0 e^{x^3/3}$.

5. Assuming $y(x) = \sum_{n=0}^{\infty} a_n x^n$, $y''(x) = \sum_{n=2}^{\infty} n(n-1)a_n x^{n-2} = \sum_{n=0}^{\infty}(n+2)(n+1)a_{n+2}x^n$,

$3xy'(x) = 3x\sum_{n=0}^{\infty} na_n x^{n-1} = \sum_{n=0}^{\infty} 3na_n x^n$ and the differential equation becomes

$\sum_{n=0}^{\infty}(n+2)(n+1)a_{n+2} + \sum_{n=0}^{\infty} 3na_n x^n + \sum_{n=0}^{\infty} 3a_n x^n = 0 \Rightarrow$

$\sum_{n=0}^{\infty}[(n+2)(n+1)a_{n+2} + 3(n+1)a_n]x^n = 0$. Thus the recurrence relation is $a_{n+2} = -\dfrac{3a_n}{n+2}$ for

$n = 0, 1, 2, \ldots$. Given a_0 and a_1, $a_2 = -\dfrac{3a_0}{2}$, $a_4 = -\dfrac{3a_2}{4} = (-1)^2 \dfrac{3^2 a_0}{2^2 \cdot 2!}$, $a_6 = -\dfrac{3a_4}{6} = (-1)^3 \dfrac{3^3 a_0}{2^3 \cdot 3!}, \ldots$,

$a_{2n} = (-1)^n \dfrac{3^n a_0}{2^n n!} = (-1)^n \left(\dfrac{3}{2}\right)^n \dfrac{a_0}{n!}$ and $a_3 = -\dfrac{3a_1}{3}$, $a_5 = -\dfrac{3a_3}{5} = (-1)^2 \dfrac{3^2 a_1}{5 \cdot 3}$, $a_7 = -\dfrac{3a_5}{7} = (-1)^3 \dfrac{3^3 a_1}{7 \cdot 5 \cdot 3}$,

$\ldots, a_{2n+1} = (-1)^n \dfrac{3^n a_1}{(2n+1)(2n-1)\cdot\cdots\cdot 5 \cdot 3} = (-1)^n \dfrac{3^n a_1 2^n n!}{(2n+1)!} = (-1)^n \dfrac{6^n n! a_1}{(2n+1)!}$ because

$(2n+1)(2n-1)\cdot\cdots\cdot 5 \cdot 3 = \dfrac{(2n+1)!}{2^n \cdot n!}$. Thus the solution is

$y(x) = \sum_{n=0}^{\infty} a_{2n} x^{2n} + \sum_{n=0}^{\infty} a_{2n+1} x^{2n+1} = a_0 \sum_{n=0}^{\infty}(-1)^n \dfrac{(3x^2/2)^n}{n!} + a_1 \sum_{n=0}^{\infty} \left[(-1)^n \dfrac{n! 6^n x^{2n+1}}{(2n+1)!}\right]$

$= a_0 e^{-3x^2/2} + a_1 \sum_{n=0}^{\infty} \left[(-1)^n \dfrac{6^n n! x^{2n+1}}{(2n+1)!}\right]$.

7. Let $y(x) = \sum_{n=0}^{\infty} a_n x^n$. Then $y'' = \sum_{n=0}^{\infty} n(n-1)a_n x^{n-2}$, $xy' = \sum_{n=0}^{\infty} na_n x^n$ and

$(x^2+1)y'' = \sum_{n=0}^{\infty} n(n-1)a_n x^n + \sum_{n=0}^{\infty}(n+2)(n+1)a_{n+2}x^n$. The differential equation becomes

$\sum_{n=0}^{\infty}[(n+2)(n+1)a_{n+2} + [n(n-1) + n - 1]a_n]x^n = 0$. The recurrence relation is $a_{n+2} = -\dfrac{(n-1)a_n}{n+2}$,

$n = 0, 1, 2, \ldots$. Given a_0 and a_1, $a_2 = \dfrac{a_0}{2}$, $a_4 = -\dfrac{a_2}{4} = -\dfrac{a_0}{2^2 \cdot 2!}$, $a_6 = -\dfrac{3a_4}{6} = (-1)^2 \dfrac{3a_0}{2^3 \cdot 3!}, \ldots$,

$a_{2n} = (-1)^{n-1} \dfrac{1 \cdot 3 \cdots (2n-3)a_0}{2^n n!} = (-1)^{n-1} \dfrac{(2n-3)! \, a_0}{2^n 2^{n-2} n! \, (n-2)!} = (-1)^{n-1} \dfrac{(2n-3)! \, a_0}{2^{2n-2} n! \, (n-2)!}$ for

$n = 2, 3, \ldots$. $a_3 = \dfrac{0 \cdot a_1}{3} = 0 \quad \Rightarrow \quad a_{2n+1} = 0$ for $n = 1, 2, \ldots$. Thus the solution is

$y(x) = a_0 + a_1 x + a_0 \dfrac{x^2}{2} + a_0 \sum_{n=2}^{\infty} (-1)^{n-1} \dfrac{(2n-3)! \, x^{2n}}{2^{2n-2} n! \, (n-2)!}$.

9. Let $y(x) = \sum_{n=0}^{\infty} a_n x^n$. Then $y''(x) = \sum_{n=0}^{\infty}(n+2)(n+1)a_{n+2}x^n$, $-xy'(x) = -\sum_{n=0}^{\infty} na_n x^n$ and the

differential equation becomes $\sum_{n=0}^{\infty}[(n+2)(n+1)a_{n+2} - (n+1)a_n]x^n = 0$. Thus the recurrence relation is

$a_{n+2} = \dfrac{a_n}{n+2}$ for $n = 0, 1, 2, \ldots$. But $a_0 = y(0) = 1$ so $a_2 = \dfrac{1}{2}$, $a_4 = \dfrac{a_2}{4} = \dfrac{1}{2 \cdot 4}$, $a_6 = \dfrac{a_4}{6} = \dfrac{1}{2 \cdot 4 \cdot 6}, \ldots$,

$a_{2n} = \dfrac{1}{2^n n!}$. Also $a_1 = y'(0) = 0$ and by the recurrence relation $a_{2n+1} = 0$ for $n = 0, 1, 2, \ldots$. Thus the

solution to the initial-value problem is $y(x) = \sum_{n=0}^{\infty} a_n x^n = \sum_{n=0}^{\infty} \dfrac{x^{2n}}{2^n n!} = \sum_{n=0}^{\infty} \dfrac{(x^2/2)^n}{n!} = e^{x^2/2}$.

11. Let $y(x) = \sum_{n=0}^{\infty} a_n x^n$. Then $y''(x) = \sum_{n=0}^{\infty} n(n-1)a_n x^{n-2} = \sum_{n=-1}^{\infty}(n+3)(n+2)a_{n+3}x^{n+1}$

$= 2a_2 + \sum_{n=0}^{\infty}(n+3)(n+2)a_{n+3}x^{n+1}$ and the differential equation becomes

$2a_2 + \sum_{n=0}^{\infty}[(n+3)(n+2)a_{n+3} + (n+1)a_n]x^{n+1} = 0$. Then $a_2 = 0$ and the recurrence relation is

$a_{n+3} = -\dfrac{(n+1)a_n}{(n+3)(n+2)}$, $n = 0, 1, 2, \ldots$. But $a_0 = y(0) = 0 = a_2$ and by the recurrence relation

$a_{3n} = a_{3n+2} = 0$ for $n = 0, 1, 2, \ldots$. Also $a_1 = y'(0) = 1$ so

$a_4 = -\dfrac{2}{4 \cdot 3}$, $a_7 = -\dfrac{5a_4}{7 \cdot 6} = (-1)^2 \dfrac{2 \cdot 5}{7 \cdot 6 \cdot 4 \cdot 3} = (-1)^2 \dfrac{2^2 5^2}{7!}, \ldots, a_{3n+1} = (-1)^n \dfrac{2^2 5^2 \cdots (3n-1)^2}{(3n+1)!}$. Thus the

solution is $y(x) = \sum_{n=0}^{\infty} a_n x^n = x + \sum_{n=0}^{\infty} \left[(-1)^n \dfrac{2^2 5^2 \cdots (3n-1)^2 x^{3n+1}}{(3n+1)!} \right]$.

REVIEW EXERCISES FOR CHAPTER 15

1. False. y' cannot be written as a function of y/x.

3. True. $\dfrac{\partial}{\partial y}(3x^2 + 6xy) = 6x$, $\dfrac{\partial}{\partial x}(3x^2 + 2y) = 6x$, so it is exact.

5. True. $y' + \dfrac{2}{x}y = \dfrac{\sin x}{x}$ $\Rightarrow$ $I(x) = e^{\int 2/x\,dx} = e^{2\ln|x|} = x^2$.

7. True. $\cosh x$ and $\sinh x$ are linearly independent solutions of this linear homogeneous equation.

9. The equation is exact, so there is an f such that $f_x(x,y) = 1 + 2xy^2$ and $f_y(x,y) = 2x^2y$. Hence
   $f(x,y) = x + x^2y^2 + g(y)$, $f_y(x,y) = 2x^2y + g'(y)$ and the solution is $x + x^2y^2 = c$.

11. Since it's linear, $I(x) = e^{\int -2/x\,dx} = x^{-2}$ and multiplying by $I(x)$ gives $x^{-2}y' - 2y^{-3} = 1$ $\Rightarrow$ $\left(x^{-2}y\right)' = 1$
    $\Rightarrow$ $y(x) = x^2(\int 1\,dx + C) = x^2(x + C) = Cx^2 + x^3$.

13. $(2y - 3y^2)dy = (x\cos x)dx$ $\Rightarrow$ $y^2 - y^3 = x\sin x + \cos x + C$.

15. $y' = \dfrac{x^2 + y^2}{x^2 + xy} = \dfrac{1 + (y/x)^2}{1 + (y/x)}$. Setting $v = \dfrac{y}{x}$ gives $v' = \left(\dfrac{1 + v^2}{1 + v} - v\right)\dfrac{1}{x}$ $\Rightarrow$ $\dfrac{1 + v}{1 - v}dv = \left(\dfrac{1}{x}\right)dx$ $\Rightarrow$
    $-v - 2\ln|1 - v| = \ln|x| + c_1$ $\Rightarrow$ $-\dfrac{y}{x} - 2\ln\left|\dfrac{x - y}{x}\right| = \ln|x| + c_1$ $\Rightarrow$ $\dfrac{y}{x} + \ln(x - y)^2 = c + \ln|x|$.

17. $y' = \dfrac{y^2 + 1}{\sqrt{1 - x^2}}$ $\Rightarrow$ $\dfrac{dy}{y^2 + 1} = \dfrac{dx}{\sqrt{1 - x^2}}$ $\Rightarrow$ $\tan^{-1}y = \sin^{-1}x + c$ $\Rightarrow$ $y = \tan\left(\sin^{-1}x + c\right)$.

19. The auxiliary equation is $r^2 - 6r + 34 = 0$ with roots $r = 3 \pm 5i$. Thus the solution is
    $y(x) = e^{3x}(c_1\cos 5x + c_2\sin 5x)$.

21. The auxiliary equation is $2r^2 + r - 1 = 0$ with roots $r_1 = -1$, $r_2 = \frac{1}{2}$. Thus the solution is
    $y(x) = c_1e^{-x} + c_2e^{x/2}$.

23. $y_c(x) = e^{-x}(c_1 + c_2x)$ and try $y_p(x) = A\cos 3x + B\sin 3x$. Then
    $-9A\cos 3x - 9B\sin 3x - 6A\sin 3x + 6B\cos 3x + A\cos 3x + B\sin 3x = \sin 3x$ $\Rightarrow$ $6B - 8A = 0$ and
    $-6A - 8B = 1$ $\Rightarrow$ $A = -\frac{3}{50}$, $B = -\frac{2}{25}$ and the general solution is
    $y(x) = e^{-x}(c_1 + c_2x) - \frac{3}{50}\cos 3x - \frac{2}{25}\sin 3x$.

25. $y_c(x) = c_1\cos\left(\frac{3}{2}x\right) + c_2\sin\left(\frac{3}{2}x\right)$ and try $y_p(x) = Ax^2 + Bx + C$. Then $8A + 9Ax^2 + 9Bx + 9C = 2x^2 - 3$
    $\Rightarrow$ $A = \frac{2}{9}$, $B = 0$, $C = -\frac{43}{81}$, and the general solution is $y(x) = c_1\cos\left(\frac{3}{2}x\right) + c_2\sin\left(\frac{3}{2}x\right) + \frac{2}{9}x^2 - \frac{43}{81}$.

27. $y_c(x) = c_1e^x + c_2e^{2x}$ so try $y_p(x) = Axe^{2x}$. Then $(4Ax + 4A - 6Ax - 3A)e^{2x} + 2Axe^{2x} = e^{2x}$ $\Rightarrow$
    $A = 1$. Thus the general solution is $y(x) = c_1e^x + c_2e^{2x} + xe^{2x}$.

29. Since the equation is linear, let $I(x) = e^{\int -1\,dx} = e^{-x}$.

Multiplying by $I(x)$ gives $e^{-x}y' - e^{-x}y = e^x \quad \Rightarrow$

$(e^{-x}y)' = e^x \quad \Rightarrow \quad y = e^x(\int e^x\,dx + C) = e^{2x} + Ce^x$.

All solutions approach 0 as $x \to -\infty$ and approach

∞ as $x \to \infty$. For $C \geq 0$, all solutions are increasing.

For $C < 0$, the solutions have a minimum point, which

moves downward and to the right as $C \to -\infty$.

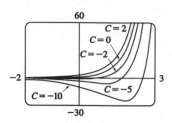

31. Since the equation is linear, let $I(x) = e^{\int dx} = e^x$. Then multiplying by $I(x)$ gives $e^x y' + e^x y = \sqrt{x} \quad \Rightarrow$

$(e^x y)' = \sqrt{x} \quad \Rightarrow \quad y(x) = e^{-x}(\int \sqrt{x}\,dx + c) = e^{-x}(\frac{2}{3}x^{3/2} + c)$. But $3 = y(0) = c$, so the solution to the

initial-value problem is $y(x) = e^{-x}(\frac{2}{3}x^{3/2} + 3)$.

33. The auxiliary equation is $r^2 + 6r = 0$ and the general solution is $y(x) = c_1 + c_2 e^{-6x} = k_1 + k_2 e^{-6(x-1)}$. But

$3 = y(1) = k_1 + k_2$ and $12 = y'(1) = -6k_2$. Thus $k_2 = -2$, $k_1 = 5$ and the solution is $y(x) = 5 - 2e^{-6(x-1)}$.

35. The auxiliary equation is $r^2 - 5r + 4 = 0$ and the general solution is $y(x) = c_1 e^x + c_2 e^{4x}$. But

$0 = y(0) = c_1 + c_2$ and $1 = y'(0) = c_1 + 4c_2$, so the solution is $y(x) = \frac{1}{3}(e^{4x} - e^x)$.

37. $h = 0.1$, $x_0 = 0$, $y_0 = 1$ and $F(x, y) = x^2 - y^2$. So $y_n = y_{n-1} + 0.1(x_{n-1}^2 - y_{n-1}^2)$. Thus

$y_1 = 1 + 0.1(0^2 - 1^2) = 0.9$, $y_2 = 0.9 + 0.1(0.1^2 - 0.9^2) = 0.82$,

$y_3 = 0.82 + 0.1(0.2^2 - 0.82^2) \approx 0.7568 \approx y(0.3)$

39. The curves $kx^2 + y^2 = 1$ form a family of ellipses for $k > 0$, a family of hyperbolas for $k < 0$, and two parallel

lines $y = \pm 1$ for $k = 0$. Differentiating gives $2kx + 2yy' = 0 \quad \Rightarrow \quad y' = -\dfrac{kx}{y} = -(1 - y^2)\dfrac{x}{yx^2} = \dfrac{y^2 - 1}{xy}$.

Thus the orthogonal trajectories must satisfy $y' = -\dfrac{xy}{y^2 - 1}$ (for $k \neq 0$) $\quad \Rightarrow \quad \dfrac{y^2}{2} - \ln|y| = \dfrac{-x^2}{2} + c$.

(For $k = 0$, the orthogonal trajectories are given by $x = c_1$ for c_1 an arbitrary constant.)

41. Let $y(x) = \displaystyle\sum_{n=0}^{\infty} a_n x^n$. Then $y''(x) = \displaystyle\sum_{n=2}^{\infty} n(n-1)a_n x^{n-2} = \sum_{n=0}^{\infty}(n+2)(n+1)a_{n+2}x^n$ and the differential

equation becomes $\displaystyle\sum_{n=0}^{\infty}[(n+2)(n+1)a_{n+2} + (n+1)a_n]x^n = 0$. Thus the recurrence relation is $a_{n+2} = -\dfrac{a_n}{n+2}$

for $n = 0, 1, 2, \ldots$. But $a_0 = y(0) = 0$, so $a_{2n} = 0$ for $n = 0, 1, 2, \ldots$. Also $a_1 = y'(0) = 1$, so

$a_3 = -\dfrac{1}{3}$, $a_5 = \dfrac{(-1)^2}{3 \cdot 5}$, $a_7 = \dfrac{(-1)^3}{3 \cdot 5 \cdot 7} = \dfrac{(-1)^3 2^3 3!}{7!}$, $\ldots$, $a_{2n+1} = \dfrac{(-1)^n 2^n n!}{(2n+1)!}$ for $n = 0, 1, 2, \ldots$. Thus the

solution to the initial-value problem is $y(x) = \displaystyle\sum_{n=0}^{\infty} a_n x^n = \sum_{n=0}^{\infty} \dfrac{(-1)^n 2^n n! \, x^{2n+1}}{(2n+1)!}$.

43. Here the initial-value problem is $2Q'' + 40Q' + 400Q = 12$, $Q(0) = 0.01$, $Q'(0) = 0$. Then

$Q_c(t) = e^{-10t}(c_1 \cos 10t + c_2 \sin 10t)$ and we try $Q_p(t) = A$. Thus the general solution is

$Q(t) = e^{-10t}(c_1 \cos 10t + c_2 \sin 10t) + \frac{3}{100}$. But $0.01 = Q(0) = c_1 + 0.03$ and $0 = Q'(0) = -10c_1 + 10c_2$, so

$c_1 = -0.02 = c_2$. Hence the charge is given by $Q(t) = -0.02e^{-10t}(\cos 10t + \sin 10t) + 0.03$.

45. **(a)** The differential equation is $\dfrac{dP}{dt} - kP = -m$, which is linear with integrating factor $I(t) = e^{-\int k\,dt} = e^{-kt}$.

Multiplying by $I(t)$ gives $e^{-kt}\dfrac{dP}{dt} - ke^{-kt}P = -me^{-kt}$ $\Rightarrow$ $\dfrac{d}{dt}\left(e^{-kt}P\right) = -me^{-kt}$ $\Rightarrow$

$e^{-kt}P(t) = \dfrac{m}{k}e^{-kt} + C$ $\Rightarrow$ $P(t) = \dfrac{m}{k} + Ce^{kt}$. But $P(0) = \dfrac{m}{k} + C = P_0$ $\Rightarrow$ $C = P_0 - \dfrac{m}{k}$ so

$P(t) = \dfrac{m}{k} + \left(P_0 - \dfrac{m}{k}\right)e^{kt}$.

(b) There will be an exponential expansion $\Leftrightarrow$ $P_0 - \dfrac{m}{k} > 0$ $\Leftrightarrow$ $m < kP_0$.

(c) The population will be constant if $P_0 - \dfrac{m}{k} = 0$ $\Leftrightarrow$ $m = kP_0$.

It will decline if $P_0 - \dfrac{m}{k} < 0$ $\Leftrightarrow$ $m > kP_0$.

APPLICATIONS PLUS (after Chapter 15)

1. **(a)** Since we are assuming that the earth is a solid sphere of uniform density, we can calculate the density ρ as

follows: $\rho = \dfrac{\text{mass of earth}}{\text{volume of earth}} = \dfrac{M}{\frac{4}{3}\pi R^3}$. If V_r is the volume of the portion of the earth which lies within a

distance r of the center, then $V_r = \frac{4}{3}\pi r^3$ and $M_r = \rho V_r = \dfrac{Mr^3}{R^3}$. Thus $F_r = -\dfrac{GM_r m}{r^2} = -\dfrac{GMm}{R^3}r$.

(b) The particle is acted upon by a varying gravitational force during its motion. By Newton's Second Law of

Motion, $m\dfrac{d^2y}{dt^2} = F_y = -\dfrac{GMm}{R^3}y$, so $y''(t) = -k^2y(t)$ where $k^2 = \dfrac{GM}{R^3}$. At the surface,

$-mg = F_R = -\dfrac{GMm}{R^2}$, so $g = \dfrac{GM}{R^2}$. Therefore $k^2 = \dfrac{g}{R}$.

(c) The differential equation $y'' + k^2y = 0$ has auxiliary equation $r^2 + k^2 = 0$. (This is the r of Section 15.5,

not the r measuring distance from the earth's center.) The roots of the auxiliary equation are $\pm ik$, so by

(15.5.11), the general solution of our differential equation for t is $y(t) = c_1\cos kt + c_2\sin kt$. It follows

that $y'(t) = -c_1 k\sin kt + c_2 k\cos kt$. Now $y(0) = R$ and $y'(0) = 0$, so $c_1 = R$ and $c_2 k = 0$. Thus

$y(t) = R\cos kt$ and $y'(t) = -kR\sin kt$. This is simple harmonic motion (see Section 15.7) with

amplitude R, frequency k, and phase angle 0. The period is $T = 2\pi/k$. $R \approx 3960$ mi $= 3960 \cdot 5280$ ft

and $g = 32$ ft/s^2, so $k = \sqrt{g/R} \approx 1.24 \times 10^{-3}\,s^{-1}$ and $T = 2\pi/k \approx 5079$ s ≈ 85 min.

(d) $y(t) = 0 \iff \cos kt = 0 \iff kt = \frac{\pi}{2} + \pi n$ for some integer $n \Rightarrow$

$y'(t) = -kR\sin\left(\frac{\pi}{2} + \pi n\right) = \pm kR$. Thus the particle passes through the center of the earth with speed

$kR \approx 4.899$ mi/s $\approx 17,600$ mi/h.

3.

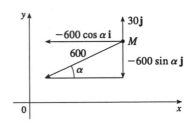

The figure shows the missile as viewed from above, with the x-axis pointing due east and the y-axis pointing due north. The angle α varies with time. When the missile is fired, it points due west, but the northerly wind pushes the missile in the positive y-direction. As the homing device turns it toward the target, the missile's velocity acquires a component in the negative y-direction. That component counteracts the wind and carries the missile back toward the x-axis and the target.

(a) We see that the effective velocity of the missile, taking the wind into consideration, is

$\mathbf{v} = -600\cos\alpha\,\mathbf{i} + (30 - 600\sin\alpha)\mathbf{j}$. Thus $dx/dt = -600\cos\alpha$ and $dy/dt = 30 - 600\sin\alpha$.

(b) $\dfrac{dy}{dx} = \dfrac{dy/dt}{dx/dt} = \dfrac{30 - 600\sin\alpha}{-600\cos\alpha}$. Since $x = r\cos\alpha = \sqrt{x^2 + y^2}\cos\alpha$, we have $\cos\alpha = \dfrac{x}{\sqrt{x^2 + y^2}}$

[for $(x, y) \neq (0, 0)$]. Similarly, $\sin\alpha = \dfrac{y}{\sqrt{x^2 + y^2}}$. Thus

$\dfrac{dy}{dx} = \dfrac{30 - 600y/\sqrt{x^2 + y^2}}{-600x/\sqrt{x^2 + y^2}} = \dfrac{30\sqrt{x^2 + y^2} - 600y}{-600x} = \dfrac{\sqrt{x^2 + y^2} - 20y}{-20x}$.

(c) From (b), $\dfrac{dy}{dx} = \dfrac{\sqrt{1 + (y/x)^2} - 20(y/x)}{-20}$, which is a homogeneous differential equation. Let $v = \dfrac{y}{x}$.

Then $y = vx \Rightarrow \dfrac{dy}{dx} = v + x\,\dfrac{dv}{dx}$. Then the differential equation becomes

$$v + x\,\frac{dv}{dx} = \frac{\sqrt{1 + v^2} - 20v}{-20} \quad\Rightarrow\quad -\frac{20\,dv}{\sqrt{1 + v^2}} = \frac{dx}{x}. \text{ Using Formula 25 (or substituting } v = \tan t)$$

gives $-20\ln\left(v + \sqrt{1 + v^2}\right) = \ln x + C$. When $x = 50$, $y = 0$ and $v = y/x = 0$, so $C = -\ln 50$ and

$-20\ln\left(v + \sqrt{1 + v^2}\right) = \ln x - \ln 50$. But $v + \sqrt{1 + v^2} = \dfrac{y}{x} + \sqrt{1 + (y/x)^2} = \dfrac{y + \sqrt{x^2 + y^2}}{x}$, so

$-20\left[\ln\left(y + \sqrt{x^2 + y^2}\right) - \ln x\right] = \ln x - \ln 50 \quad\Rightarrow\quad 19\ln x + \ln 50 = 20\ln\left(y + \sqrt{x^2 + y^2}\right) \quad\Rightarrow$

$\ln\left(50x^{19}\right) = \ln\left(y + \sqrt{x^2 + y^2}\right)^{20} \quad\Rightarrow\quad \left(y + \sqrt{x^2 + y^2}\right)^{20} = 50x^{19}$ and $y + \sqrt{x^2 + y^2} = 50^{1/20}x^{19/20}$

is the equation describing the path of the missile.

(d) In polar coordinates, the path is described by $r = 50\,\dfrac{\cos^{19}\theta}{(1 + \sin\theta)^{20}}$, $0 \le \theta \le \frac{\pi}{4}$.

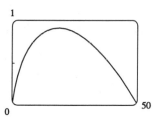

5. **(a)** We have $P' = -mP + kC$, where $C(t) = S - P(t)$, and $P(0) = S$. Thus

$P' = -mP + k(S - P) = -(m + k)P + kS$, or $P' + (m + k)P = kS$. Following the method of

Section 15.2, we multiply both sides by the integrating factor $e^{(m+k)t}$ to get $\dfrac{d}{dt}\left(e^{(m+k)t}P\right) = kSe^{(m+k)t}$

$\Rightarrow \quad e^{(m+k)t}P = \dfrac{kS}{m + k}e^{(m+k)t} + C$. When $t = 0$, this says $S = \dfrac{kS}{m + k} + C$, so

$C = S - \dfrac{kS}{m + k} = \dfrac{mS}{m + k}$ and so $e^{(m+k)t}P = \dfrac{kS}{m + k}e^{(m+k)t} + \dfrac{mS}{m + k}$. Therefore

$P(t) = \dfrac{kS}{m + k} + \dfrac{mS}{m + k}e^{-(m+k)t}$. $Q = \lim\limits_{t\to\infty}P(t) = \dfrac{kS}{m + k}$.

(b) $C(t) = S - P(t) = S - \dfrac{kS}{m + k} - \dfrac{mS}{m + k}e^{-(m+k)t} = \dfrac{mS}{m + k} - \dfrac{mS}{m + k}e^{-(m+k)t} = \dfrac{mS}{m + k}\left[1 - e^{-(m+k)t}\right]$,

so $\lim\limits_{t\to\infty}C(t) = \dfrac{mS}{m + k}$.

(c) $g(t) = \ln\left(\dfrac{P(t)}{Q} - 1\right) = \ln\left(1 + \dfrac{m}{k}e^{-(m+k)t} - 1\right) = \ln\left(\dfrac{m}{k}e^{-(m+k)t}\right) = -(m + k)t + \ln\dfrac{m}{k}$. If

$y(t) = at + b$, then $a - -(m + k)$ and $b = \ln(m/k)$. The second relation says $m/k = e^b$, so $m = ke^b$.

Now $a = -(m + k) = -ke^b - k = -k\left(e^b + 1\right)$. Therefore $k = -\dfrac{a}{e^b + 1}$ and $m = ke^b = -\dfrac{ae^b}{e^b + 1}$.

7. **(a)**

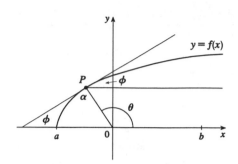

Consider a light ray emanating from the origin with angle of inclination θ. Suppose that the light ray intersects the graph of f at the point P with coordinates (x, y). Since P lies on the ray and also on the graph, its coordinates satisfy the relations $y = (\tan\theta)x$ (1) and $y = f(x)$. Let ϕ be the angle of inclination of the tangent line to the graph of f at P. Then ϕ is also the angle between the tangent line and the reflected ray (since the reflected ray is parallel to the x-axis.) We have $f'(x) = \tan\phi$ (2). Since the angle of incidence equals the angle of reflection at P, we have $\alpha = \phi$ (3). Also, we see from the figure that $\theta = \alpha + \phi$ (4). By (3) and (4), $\theta = 2\phi$, so the coordinates of P satisfy

$$\frac{f(x)}{x} = \tan\theta = \tan 2\phi = \frac{2\tan\phi}{1 - \tan^2\phi} = \frac{2f'(x)}{1 - [f'(x)]^2}.$$ Solving for $f'(x)$, we get

$$f(x)[f'(x)]^2 + 2xf'(x) - f(x) = 0 \text{ and } f'(x) = \frac{-2x \pm \sqrt{4x^2 + 4[f(x)]^2}}{2f(x)} = \frac{-x \pm \sqrt{x^2 + [f(x)]^2}}{f(x)}.$$

This condition holds at all points P on the graph of f, so solving for f will determine the nature of the graph. Writing $y = f(x)$, we have $\dfrac{dy}{dx} = \dfrac{-x \pm \sqrt{x^2 + y^2}}{y}$.

(b) Rewrite the equation in the form $\dfrac{x}{\sqrt{x^2 + y^2}} + \dfrac{y}{\sqrt{x^2 + y^2}}\dfrac{dy}{dx} = \pm 1$. This equation is exact by Theorem 5.

(The theorem applies even when the right-hand side of the equation is a nonzero constant). Integrating, we get $\sqrt{x^2 + y^2} = \pm x + K$. By hypothesis, f is non-negative, so $y(0) \geq 0$. Taking $x = 0$, we get $K = \sqrt{[y(0)]^2} = y(0) \geq 0$. Squaring, we get $x^2 + y^2 = x^2 \pm 2Kx + K^2$ or $y^2 = \pm 2Kx + K^2$. Set $C = \pm K$. Then our result becomes $y^2 = 2Cx + C^2$ where C is an arbitrary constant. When $C > 0$, this is the top half of a parabola with focus at the origin, opening to the right. When $C < 0$, the graph is the top half of a parabola with focus at the origin, opening to the left. In both cases, the vertex is at $\left(-\frac{1}{2}C, 0\right)$.

Another Method: The differential equation is also homogeneous, so substitute $v = y/x$.

EXERCISES H

1. $(3+2i)+(7-3i)=(3+7)+(2-3)i=10-i$

3. $(3-i)(4+i)=12+3i-4i-(-1)=13-i$

5. $\overline{12+7i}=12-7i$

7. $\dfrac{2+3i}{1-5i}=\dfrac{2+3i}{1-5i}\cdot\dfrac{1+5i}{1+5i}=\dfrac{2+10i+3i+15(-1)}{1-25(-1)}=\dfrac{-13+13i}{26}=-\tfrac{1}{2}+\tfrac{1}{2}i$

9. $\dfrac{1}{1+i}=\dfrac{1}{1+i}\cdot\dfrac{1-i}{1-i}=\dfrac{1-i}{1-(-1)}=\dfrac{1-i}{2}=\tfrac{1}{2}-\tfrac{1}{2}i$

11. $i^3=i^2\cdot i=(-1)i=-i$

13. $\sqrt{-25}=\sqrt{25}\,i=5i$

15. $\overline{3+4i}=3-4i,\ |3+4i|=\sqrt{3^2+4^2}=\sqrt{25}=5$

17. $\overline{-4i}=\overline{0-4i}=0+4i=4i,\ |-4i|=\sqrt{0^2+(-4)^2}=4$

19. $4x^2+9=0\ \Leftrightarrow\ 4x^2=-9\ \Leftrightarrow\ x^2=-\tfrac{9}{4}\ \Leftrightarrow\ x=\pm\sqrt{-\tfrac{9}{4}}=\pm\sqrt{\tfrac{9}{4}}i=\pm\tfrac{3}{2}i.$

21. By the quadratic formula, $x^2-8x+17=0\ \Leftrightarrow\ x=\dfrac{8\pm\sqrt{8^2-4(1)(17)}}{2(1)}=\dfrac{8\pm\sqrt{-4}}{2}=\dfrac{8\pm2i}{2}=4\pm i.$

23. By the quadratic formula, $z^2+z+2=0\ \Leftrightarrow\ z=\dfrac{-1\pm\sqrt{1-4(1)(2)}}{2(1)}=\dfrac{-1\pm\sqrt{-7}}{2}=-\tfrac{1}{2}\pm\tfrac{\sqrt{7}}{2}i.$

25. $r=\sqrt{(-3)^2+3^2}=3\sqrt{2},\ \tan\theta=\tfrac{3}{-3}=-1\ \Rightarrow\ \theta=\tfrac{3}{4}\pi$ (since the given number is in the second quadrant). Therefore $-3+3i=3\sqrt{2}\left(\cos\tfrac{3\pi}{4}+i\sin\tfrac{3\pi}{4}\right).$

27. $r=\sqrt{3^2+4^2}=5,\ \tan\theta=\tfrac{4}{3}\ \Rightarrow\ \theta=\tan^{-1}\tfrac{4}{3}$ (since the given number is in the second quadrant). Therefore $3+4i=5\left[\cos\left(\tan^{-1}\tfrac{4}{3}\right)+i\sin\left(\tan^{-1}\tfrac{4}{3}\right)\right].$

29. For $z=\sqrt{3}+i,r=\sqrt{\left(\sqrt{3}\right)^2+1^2}=2$, and $\tan\theta=\tfrac{1}{\sqrt{3}}\ \Rightarrow\ \theta=\tfrac{\pi}{6}$ so that $z=2\left(\cos\tfrac{\pi}{6}+i\sin\tfrac{\pi}{6}\right).$ For $w=1+\sqrt{3}i,r=2$, and $\tan\theta=\sqrt{3}\ \Rightarrow\ \theta=\tfrac{\pi}{3}$ so that $w=2\left(\cos\tfrac{\pi}{3}+i\sin\tfrac{\pi}{3}\right).$ Therefore
$zw=2\cdot2\left[\cos\left(\tfrac{\pi}{6}+\tfrac{\pi}{3}\right)+i\sin\left(\tfrac{\pi}{6}+\tfrac{\pi}{3}\right)\right]=4\left(\cos\tfrac{\pi}{2}+i\sin\tfrac{\pi}{2}\right),$
$z/w=\tfrac{2}{2}\left[\cos\left(\tfrac{\pi}{6}-\tfrac{\pi}{3}\right)+i\sin\left(\tfrac{\pi}{6}-\tfrac{\pi}{3}\right)\right]=\cos\left(-\tfrac{\pi}{6}\right)+i\sin\left(-\tfrac{\pi}{6}\right),$ and $1=1+0i=\cos0+i\sin0\ \Rightarrow$
$1/z=\tfrac{1}{2}\left[\cos\left(0-\tfrac{\pi}{6}\right)+i\sin\left(0-\tfrac{\pi}{6}\right)\right]=\tfrac{1}{2}\left[\cos\left(-\tfrac{\pi}{6}\right)+i\sin\left(-\tfrac{\pi}{6}\right)\right].$

31. For $z=2\sqrt{3}-2i,\ r=4,\ \tan\theta=\tfrac{-2}{2\sqrt{3}}=-\tfrac{1}{\sqrt{3}}\ \Rightarrow\ \theta=-\tfrac{\pi}{6}\ \Rightarrow\ z=4\left[\cos\left(-\tfrac{\pi}{6}\right)+i\sin\left(-\tfrac{\pi}{6}\right)\right].$ For $w=-1+i,r=\sqrt{2},\ \tan\theta=\tfrac{1}{-1}=-1\ \Rightarrow\ \theta=\tfrac{3\pi}{4}\ \Rightarrow\ z=\sqrt{2}\left(\cos\tfrac{3\pi}{4}+i\sin\tfrac{3\pi}{4}\right).$ Therefore
$zw=4\sqrt{2}\left[\cos\left(-\tfrac{\pi}{6}+\tfrac{3\pi}{4}\right)+i\sin\left(-\tfrac{\pi}{6}+\tfrac{3\pi}{4}\right)\right]=4\sqrt{2}\left(\cos\tfrac{7\pi}{12}+i\sin\tfrac{7\pi}{12}\right),$
$z/w=\tfrac{4}{\sqrt{2}}\left[\cos\left(-\tfrac{\pi}{6}-\tfrac{3\pi}{4}\right)+i\sin\left(-\tfrac{\pi}{6}-\tfrac{3\pi}{4}\right)\right]=\tfrac{4}{\sqrt{2}}\left[\cos\left(-\tfrac{11\pi}{12}\right)+i\sin\left(-\tfrac{11\pi}{12}\right)\right]=2\sqrt{2}\left(\cos\tfrac{13\pi}{12}+i\sin\tfrac{13\pi}{12}\right),$
and $1=1+0i=\cos0+i\sin0\ \Rightarrow\ 1/z=\tfrac{1}{4}\left[\cos\left(0-\left(-\tfrac{\pi}{6}\right)\right)+i\sin\left(0-\left(-\tfrac{\pi}{6}\right)\right)\right]=\tfrac{1}{4}\left(\cos\tfrac{\pi}{6}+i\sin\tfrac{\pi}{6}\right).$

33. For $z = 1 + i$, $r = \sqrt{2}$, $\tan\theta = \frac{1}{1} = 1$ $\Rightarrow$ $\theta = \frac{\pi}{4}$ $\Rightarrow$ $1 + i = \sqrt{2}\left(\cos\frac{\pi}{4} + i\sin\frac{\pi}{4}\right)$. So by De Moivre's Theorem, $(1 + i)^{20} = \left[\sqrt{2}\left(\cos\frac{\pi}{4} + i\sin\frac{\pi}{4}\right)\right]^{20} = \left(2^{1/2}\right)^{20}\left(\cos\frac{20\pi}{4} + i\sin\frac{20\pi}{4}\right) = 2^{10}(\cos 5\pi + i\sin 5\pi)$

$= 2^{10}[-1 + i(0)] = -2^{10} = -1024.$

35. For $z = 2\sqrt{3} + 2i$, $r = 4$, $\tan\theta = \frac{2}{2\sqrt{3}} = \frac{1}{\sqrt{3}}$ $\Rightarrow$ $\theta = \frac{\pi}{6}$ $\Rightarrow$ $2\sqrt{3} + 2i = 4\left(\cos\frac{\pi}{6} + i\sin\frac{\pi}{6}\right)$. So by

De Moivre's Theorem,

$\left(2\sqrt{3} + 2i\right)^5 = \left[4\left(\cos\frac{\pi}{6} + i\sin\frac{\pi}{6}\right)\right]^5 = 4^5\left(\cos\frac{5\pi}{6} + i\sin\frac{5\pi}{6}\right) = 4^5\left[-\frac{\sqrt{3}}{2} + i(0.5)\right] = -512\sqrt{3} + 512i.$

37. $1 = 1 + 0i = \cos 0 + i\sin 0$. Using Equation 3 with $r = 1$, $n = 8$, and $\theta = 0$ we have

$w_k = 1^{1/8}\left[\cos\left(\frac{0 + 2k\pi}{8}\right) + i\sin\left(\frac{0 + 2k\pi}{8}\right)\right] = \cos\frac{k\pi}{4} + i\sin\frac{k\pi}{4}$, where $k = 0, 1, 2, \ldots, 7$.

$w_0 = (\cos 0 + i\sin 0) = 1$ $\qquad$ $w_4 = (\cos\pi + i\sin\pi) = -1$

$w_1 = \left(\cos\frac{\pi}{4} + i\sin\frac{\pi}{4}\right)$ $\qquad$ $w_5 = \left(\cos\frac{5\pi}{4} + i\sin\frac{5\pi}{4}\right)$

$\quad = \frac{1}{\sqrt{2}} + \frac{1}{\sqrt{2}}i$ $\qquad\qquad\quad = -\frac{1}{\sqrt{2}} - \frac{1}{\sqrt{2}}i$

$w_2 = \left(\cos\frac{\pi}{2} + i\sin\frac{\pi}{2}\right) = i$ $\qquad$ $w_6 = \left(\cos\frac{3\pi}{2} + i\sin\frac{3\pi}{2}\right) = -i$

$w_3 = \left(\cos\frac{3\pi}{4} + i\sin\frac{3\pi}{4}\right)$ $\qquad$ $w_7 = \left(\cos\frac{7\pi}{4} + i\sin\frac{7\pi}{4}\right)$

$\quad = -\frac{1}{\sqrt{2}} + \frac{1}{\sqrt{2}}i$ $\qquad\qquad\quad = \frac{1}{\sqrt{2}} - \frac{1}{\sqrt{2}}i$

39. $0 = 0 + i = \cos\frac{\pi}{2} + i\sin\frac{\pi}{2}$. Using Equation 3 with $r = 1$, $n = 3$, and $\theta = \frac{\pi}{2}$,

we have

$w_k = 1^{1/3}\left[\cos\left(\frac{\pi/2 + 2k\pi}{3}\right) + i\sin\left(\frac{\pi/2 + 2k\pi}{3}\right)\right]$, where $k = 0, 1, 2$.

$w_0 = \left(\cos\frac{\pi}{6} + i\sin\frac{\pi}{6}\right) = \frac{\sqrt{3}}{2} + \frac{1}{2}i$, $w_1 = \left(\cos\frac{5\pi}{6} + i\sin\frac{5\pi}{6}\right) = -\frac{\sqrt{3}}{2} + \frac{1}{2}i$

$w_2 = \left(\cos\frac{9\pi}{6} + i\sin\frac{9\pi}{6}\right) = -i$

41. Using Euler's formula (6) with $y = \frac{\pi}{2}$, $e^{i\pi/2} = \cos\frac{\pi}{2} + i\sin\frac{\pi}{2} = i$.

43. Using Euler's formula with $y = \frac{3\pi}{4}$, $e^{i3\pi/4} = \cos\frac{3\pi}{4} + i\sin\frac{3\pi}{4} = -\frac{1}{\sqrt{2}} + \frac{1}{\sqrt{2}}i$.

45. Using Equation 7 with $x = 2$ and $y = \pi$, $e^{2+i\pi} = e^2 e^{i\pi} = e^2(\cos\pi + i\sin\pi) = e^2(-1 + 0) = -e^2$.

47. Take $r = 1$ and $n = 3$ in De Moivre's Theorem to get

$[1(\cos\theta + i\sin\theta)]^3 = 1^3(\cos 3\theta + i\sin 3\theta)$ $\Rightarrow$ $(\cos\theta + i\sin\theta)^3 = \cos 3\theta + i\sin 3\theta$ $\Rightarrow$

$\cos^3\theta + 3(\cos^2\theta)(i\sin\theta) + 3(\cos\theta)(i\sin\theta)^2 + (i\sin\theta)^3 = \cos 3\theta + i\sin 3\theta$ $\Rightarrow$

$(\cos^3\theta - 3\sin^2\theta\cos\theta) + (3\sin\theta\cos^2\theta - \sin^3\theta)i = \cos 3\theta + i\sin 3\theta$. Equating real and imaginary parts gives

$\cos 3\theta = \cos^3\theta - 3\sin^2\theta\cos\theta$ and $\sin 3\theta = 3\sin\theta\cos^2\theta - \sin^3\theta$.

49. $F(x) = e^{rx} = e^{(a+bi)x} = e^{ax+bxi} = e^{ax}(\cos bx + i\sin bx) = e^{ax}\cos bx + i(e^{ax}\sin bx)$ $\Rightarrow$

$F'(x) = (e^{ax}\cos bx)' + i(e^{ax}\sin bx)' = (ae^{ax}\cos bx - be^{ax}\sin bx) + i(ae^{ax}\sin bx + be^{ax}\cos bx)$

$\quad = a[e^{ax}(\cos bx + i\sin bx)] + b[e^{ax}(-\sin bx + i\cos bx)] = ae^{rx} + b\left[e^{ax}\left(i^2\sin bx + i\cos bx\right)\right]$

$\quad = ae^{rx} + bi[e^{ax}(\cos bx + i\sin bx)] = ae^{rx} + bie^{rx} = (a + bi)e^{rx} = re^{rx}.$

Performance, reliability, and the most power for your dollar—it's all yours with Maple V!

Maple V Release 3 Student Edition *for Macintosh or DOS/Windows* Just $99.00

Offering numeric computation, symbolic computation, graphics, and programming, **Maple V Release 3 Student Edition** gives you the power to explore and solve a tremendous range of problems with unsurpassed speed and accuracy. Featuring both 3-D and 2-D graphics and more than 2500 built-in functions, **Release 3** offers all the power and capability you'll need for the entire array of undergraduate courses in mathematics, science, and engineering.

Maple V's *vast* library of functions also provides sophisticated scientific visualization, programming, and document preparation capabilities, **including the ability to output standard mathematical notation.**

With Release 3, you can:

* create rectangular, polar, cylindrical, contour, implicit, and density plots in two or three dimensions
* apply lighting (or shading) models to 3-D plots and assign user-specified colors to each plotted 2-D function
* interactively manipulate 2-D and 3-D graphs
* use Release 3's animation capabilities to study time-variant data
* view and print documents with standard mathematical notation for Maple output (including properly placed superscripts, integral and summation signs of typeset quality, matrices, and more)
* save the state (both mathematical and visual) of a Maple session at any point—and later resume work right where you left off
* migrate Maple worksheets easily across platforms (This is especially valuable for students using Maple V on a workstation in a computer lab who then want to continue work on their own personal computers.)
* export to LaTeX and save entire worksheets for inclusion in a publication-quality document *(New!)*
* use keyword searches within help pages *(New!)*

ORDER FORM

Yes! Please send me Maple V Release 3 Student Edition

_____ For DOS/Windows Software only (ISBN: 0-534-25560-4) @ $99 _____

_____ For Macintosh Software only (ISBN: 0-534-25561-2) @ $99 _____

Subtotal _____

(Residents of AL, AZ, CA, CO, CT, FL, GA, IL, IN, KS, KY, LA, MA, MD, MI, MN, MO, NC, NJ, NY, OH, PA, RI, SC, TN, TX, UT, VA, WA, WI must add appropriate sales tax)

Tax _____

Payment Options

Handling _____

_____ Check or Money Order enclosed.

Total _____

_____ Charge my _____ VISA _____ MasterCard _____ American Express

Card Number _____ Expiration Date _____

Signature_____

Please ship to: (Billing and shipping address must be the same.)

Name_____

Department _____ School _____

Street Address _____

City _____ State_____ Zip+4_____

Office phone number (_____) _____

You can fax your order to us at 408-375-6414 or e-mail your order to: info@brookscole.com or detach, fold, secure, and mail with payment.

SECURE WITH TAPE

Now symbolic computation and mathematical typesetting are as accessible as your Windows™-based word processor

Scientific WorkPlace™ 2.0 Student Edition for Windows

Ideal for homework, projects, term papers, lab reports, or just writing home—choose from a variety of predesigned styles

"Scientific WorkPlace is a heavy-duty mathematical word processor and typesetting system that is able to expand, simplify, and evaluate conventional mathematical expressions and compose them as elegant printed mathematics." —Roger Horn, University of Utah

"The thing I like most about Scientific WorkPlace is its basic simplicity and ease of use." —Barbara Osofsky, Rutgers University

Easy access to a powerful computer algebra system
***inside* your technical word-processing documents!**

Scientific WorkPlace is a revolutionary program that gives you a "work place" environment—a single place to do all your work. It combines the ease of use of a technical word processor with the typesetting power of TeX and the numerical, symbolic, and graphic computational facilities of the **Maple**® **V** computer algebra system. All capabilities are included in the program—you don't need to *own* or *learn* TeX, LaTeX, or Maple to use ***Scientific WorkPlace***—**everything for super productivity is included in one powerful tool for just $162!**

With ***Scientific WorkPlace***, you can enter, solve, and graph mathematical problems right in your word-processing documents in seconds, with no clumsy cut-and-paste from equation editors or clipboards. ***Scientific WorkPlace*** calculates answers quickly and accurately, then prints your work in professional-quality documents using TeX's internationally accepted mathematical typesetting standard.

Install *Scientific WorkPlace* and watch your productivity soar! You'll be creating impressive documents in a fraction of the time you would spend using any other program!

ORDER FORM

Yes! Please send me Scientific Workplace 2.0 Student Edition for Windows

_____ copies (ISBN: 0-534-25597-3) @ $162 each   _____

(Residents of AL, AZ, CA, CO, CT, FL, GA, IL, IN, KS, KY, LA, MA, MD, MI, MN, MO, NC, NJ, NY, OH, PA, RI, SC, TN, TX, UT, VA, WA, WI must add appropriate sales tax.)   Tax   _____

Payment Options   Handling _____

_____ Check or Money Order enclosed.   Total   _____

_____ Charge my _____ VISA _____ MasterCard _____ American Express

Card Number _____ Expiration Date _____

Signature_____

Please ship to: (Billing and shipping address must be the same.)

Name_____

Department _____ School _____

Street Address _____

City _____ State_____ Zip+4_____

Office phone number (_____) _____

You can fax your order to us at 408-375-6414 or e-mail your order to: info@brookscole.com or detach, fold, secure, and mail with payment.